Clinical Brachytherapy Physics

Medical Physics Monograph No. 38

Clinical Brachytherapy Physics

Mark J. Rivard, Luc Beaulieu, and Bruce R. Thomadsen

Editors

American Association of Physicists in Medicine
2017 Summer School Proceedings

In conjunction with the American Brachytherapy Society

Lewis & Clark College
Portland, Oregon
June 10–14, 2017

Published for the
American Association of Physicists in Medicine
by Medical Physics Publishing, Inc.

To order American Association of Physicists in Medicine (AAPM) publications, contact:

Medical Physics Publishing, Inc.
4555 Helgesen Dr.
Madison, WI 53718
Phone: (800) 442-5778 or (608) 224-4508
Fax: (608) 224-5016
E-mail: mpp@medicalphysics.org
Web: www.medicalphysics.org

Published by:
Medical Physics Publishing, Inc.
Madison, Wisconsin 53718

Published for:
American Association of Physicists in Medicine (AAPM)
1631 Prince Street
Alexandria, VA 22314

Library of Congress Control Number: 2017940440

ISBN hardcover book: 978-1-936366-57-6
ISBN eBook: 978-1-936366-58-3

Printed in the United States of America

Contents

Preface

The first two American Association of Physicists in Medicine (AAPM) summer schools on brachytherapy physics occurred in 1994 and 2005. It can be argued that their focus was on historical aspects as well as future developments; the current summer school fills the gap by focusing mostly on the present and clinically relevant materials. This approach was thought to best serve the brachytherapy community due to the maturity of many brachytherapy-related technologies and an identified need for consistent and high-quality treatment delivery. In the dozen years since the last summer school on brachytherapy, training and education have been identified as key to continue the utilization of clinical brachytherapy. Therefore, the 2017 summer school included workshops to provide practical hands-on opportunities for attendees to gain experience on nine key aspects of clinical brachytherapy physics, with opportunities for feedback from the 20 faculty members. To our knowledge, these workshops encompassed the largest training opportunity in clinical brachytherapy physics outside of an accredited training program. Medical Physics Continuing Education Credits (MPCECs) and Self Assessment Modules (SAMs), accredited by the American Board of Radiology, were provided for the morning lectures as well as the nine afternoon workshops. Without patient treatments and radioactive sources, this summer school on clinical brachytherapy physics offered a close substitute to hand-on experience in a clinic.

This textbook captures the 2017 AAPM Summer School, which provided an intense experience to cover state-of-the-art clinical brachytherapy physics. It has become expected that low-dose-rate (LDR) prostate brachytherapy for early-stage disease can provide 10-year survival rates exceeding 95% with minimal morbidities. The convenience of this approach, yet with dwell-time modulation and no residual seed placement, is becoming widespread for high-dose-rate (HDR) prostate brachytherapy. Computerized treatment planning for brachytherapy has largely replaced nomograms and other methods that estimate treatment metrics without employing 3D patient-specific imaging. Several methods for dose optimization and treatment plan evaluation are now standard-of-care. For brachytherapy using HDR ^{192}Ir sources, advances in the past dozen years have included model-based dose calculation algorithms that evaluate dose while accounting for radiation scatter conditions during treatment and material properties of the patient and brachytherapy applicators. Breast brachytherapy has seen the widespread use of balloon-based brachytherapy applicators in North America and the popularization of catheter-based treatments for accelerated partial breast irradiation (APBI). Treatment of gynecological malignancies has progressed in the past several decades, and 3D image-guided planning is now the standard with greater utilization of MRI. Skin brachytherapy now has many choices for treatment delivery, such as interstitial implantation, surface molds, and collimated applicators. Electronic brachytherapy has been integrated into many of the aforementioned anatomic sites, with special applicators that draw on attributes of the high dose rate of treatment, yet the low energy of their photon emissions. Further efforts to move beyond the AAPM Task Group-43 dose calculation formalism are intrinsic to intensity-modulated brachytherapy, which contain sources or applicators with collimation or shielding. Brachytherapy now includes emerging technologies, such as focal therapy, 3D printing (additive manufacturing), new *in vivo* dosimetry techniques, live needle tracking, and robotics. This textbook discusses all those topics and also includes over 100 questions and answers to aid the reader in autodidactic learning based on the 10 chapters.

The editors acknowledge the yeoman's work of Todd Hanson of Medical Physics Publishing for crafting this summer school textbook. He created many proofs for several chapters toward assembling a quality product. From the AAPM, Karen MacFarland was instrumental in leading the school forward through her deep experience and devotion to our common goal. Her candor and positive spirit helped drive forward this largely volunteer effort. Jaime Hoza filled in to chase the faculty for necessary information to coordinate the large undertaking. Jackie Ogburn championed the opportunities for offering comprehensive MPCECs and SAMs for both the morning lectures and afternoon workshops. Laurie Allen established formal communications with the entire faculty for gathering their programmatic commitments via the AAPM AMOS system. Holly Lincoln, as Chair of the AAPM Summer School Subcommittee, assisted with our budget and built upon knowledge gained from prior schools. This textbook (and the entire summer school) could not have happened without the love, support, and understanding from Joanna Sganga, Chantal Landry, and Louise Goldstein.

Mark J. Rivard, Ph.D., FAAPM
Luc Beaulieu, Ph.D., FAAPM
Bruce R. Thomadsen, Ph.D., FAAPM

June 2017
Portland, Oregon

Contributors

Deidre Batchelar, Ph.D., FCCPM
Senior Medical Physicist
British Columbia Cancer Agency
399 Royal Avenue
Kelowna, British Columbia V1Y 5L3 Canada
dbatchelar@bccancer.bc.ca

Luc Beaulieu, Ph.D., FAAPM
Département de physique, de génie physique et d'optique
et Centre de recherche sur le cancer
Université Laval
and
Département de radio-oncologie et CRCHU de Québec
CHU de Québec–Université Laval
Québec, Canada
beaulieu@phy.ulaval.ca

Wayne M. Butler, Ph.D.
Schiffler Cancer Center
Wheeling Hospital
Wheeling, WV 26003
wbutler@wheelinghospital.org

Ivan Buzurovic, Ph.D
Dana-Farber/Brigham and Women's Cancer Center
Harvard Medical School
Boston, MA 02115
ibuzurovic@lroc.harvard.edu

Zhe (Jay) Chen, Ph.D., FAAPM
Professor and Smilow Chief Physicist
Department of Therapeutic Radiology
Yale University School of Medicine
P.O. Box 208040
New Haven, CT 06520
zhe.chen@yale.edu

J. Adam M. Cunha, Ph.D.
Department of Radiation Oncology
UC–San Francisco
San Francisco, CA 94115
JAdamMCunha@gmail.com

Antonio Damato, Ph.D.
Memorial Sloan Kettering Cancer Center
1275 York Avenue
New York, NY 10065
damatoa@mskcc.org

Regina K. Fulkerson, Ph.D.
Medical Physicist
RKF Consultants, LLC
Dundee, NY 14837
rmkenned@gmail.com

Christian Kirisits, Ph.D.
Associate Professor
Department of Radiotherapy
Medical University of Vienna
Wahringer Gurtel 18-20
Vienna 1090 Austria
christian.kirisits@akhwien.at

Bruce Libby, Ph.D., FAAPM
Department of Radiation Oncology
University of Virginia Health System
1335 Lee Street, Box 800375
Charlottesville, VA 22908
bl8b@virginia.edu

Firas Mourtada, Ph.D., FAAPM
Chief of Clinical Physics
Department of Radiation Oncology
Helen F. Graham Cancer Center
Christiana Care Hospital
4701 Olgetown-Stanton Road
Newark, DE 19713
fmourtada@christianacare.org

Zoubir Ouhib M.S., FACR
Lynn Regional Cancer Center
16313 S. Military Trail
Delray Beach, FL 33484
zouhib@brrh.com

Sujatha Pai, M.S., FAAPM
Regional Director of Medical Physics
Memorial Hermann Texas Medical Center, LMP
6400 Fannin Street
Houston, TX 77030
Sujathapai@yahoo.com

Susan L. Richardson, Ph.D., FAAPM
Swedish Medical Center
1221 Madison Street
Seattle, WA 98104
susan.richardson@swedish.org

Mark J. Rivard, Ph.D., FAAPM
Professor of Radiation Oncology
Department of Radiation Oncology
Tufts University School of Medicine
800 Washington Street
Boston, MA 02111
markjrivard@gmail.com

Jason J. Rownd, M.S.
Radiation Oncology Department
Medical College of Wisconsin
Milwaukee, WI 53226
jrownd@mcw.edu

Daniel J. Scanderbeg
Associate Professor
Radiation Medicine & Applied Sciences
University of California–San Diego
3855 Health Sciences Drive, Suite #0843
La Jolla, CA 92093
dscanderbeg@ucsd.edu

Frank-André Siebert, Ph.D., Prof.
Head of Medical Physics
Clinic of Radiotherapy
Department of Medical Physics
Universitätsklinikum Schleswig-Holstein
24105 Kiel, Germany
siebert@onco.uni-kiel.de

Ron S. Sloboda, Ph.D., FCCPM
Professor, Department of Oncology
University of Alberta
11560 University Avenue
Edmonton, AB, Canada T6G 1Z2
rsloboda@ualberta.ca

Amandeep Taggar, M.D., M.S.
Radiation Oncologist
Department of Radiation Oncology
Memorial Sloan Kettering Cancer Center
1275 York Avenue
New York, NY 10065
taggara@mskcc.org

Bruce R. Thomadsen, PhD, FAAPM
Professor of Medical Physics
University of Wisconsin
School of Medicine and Public Health
Madison, WI 53705
brthomad@wisc.edu

Dorin A. Todor, Ph.D.
Associate Professor
Department of Radiation Oncology
Virginia Commonwealth University Health System
401 College Street
Richmond, VA 23298
dorin.todor@vcuhealth.org

Chapter 1

General Planning

Bruce Libby[1], Zhe (Jay) Chen[2], Bruce R. Thomadsen[3], Susan L. Richardson[4], Mark J. Rivard[5], Frank-André Siebert[6], Wayne M. Butler[7], J. Adam M. Cunha[8], Dorin A. Todor[9], Jason J. Rownd[10], and Christian Kirisits[11]

[1]University of Virginia
Charlottesville, Virginia

[2]Yale University School of Medicine
New Haven, Connecticut

[3]University of Wisconsin
Madison, Wisconsin

[4]Swedish Medical Center
Seattle, Washington

[5]Tufts University School of Medicine
Boston, Massachusetts

[6]Universitätsklinikum Schleswig-Holstein
Kiel, Germany

[7]Wheeling Hospital
Wheeling, West Virginia

[8]University of California–San Francisco
San Francisco, California

[9]Virginia Commonwealth University
Richmond, Virginia

[10]Medical College of Wisconsin
Milwaukee, Wisconsin

[11]Medical University of Vienna
Vienna, Austria

1.1 Introduction

The treatment planning process for brachytherapy can be thought of as the entire list of steps required to perform the treatment that complies with the written directive. Published schema of this process is shown in Fig-

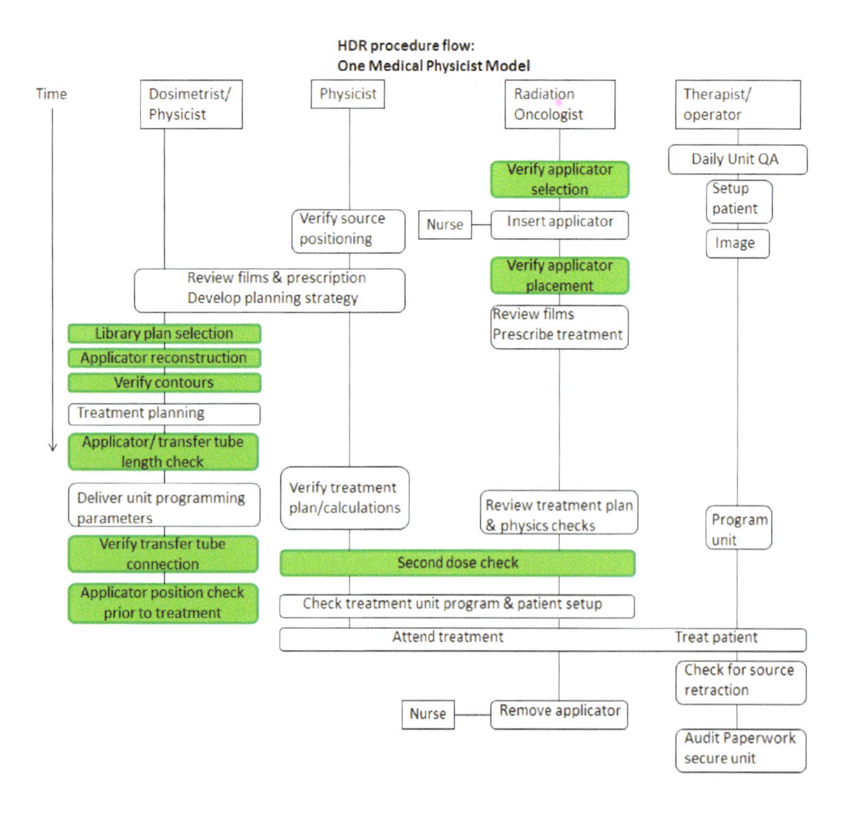

Figure 1–1 Published schema of the treatment planning process for brachytherapy (Kim et al. 2015; Kubo et al. 1998). Reprinted from *Brachytherapy* 14 (2015) pp. 834–839, Kim et al. "Parallelized patient-specific quality assurance for high-dose-rate image-guided brachytherapy in an integrated computed tomography-on-rails brachytherapy suite." © 2015, with permission from Elsevier.

ure 1–1 (Kim et al. 2015; Kubo et al. 1998) which shows the need of a large, well-trained staff in order to perform brachytherapy treatments. Even prior to performing the brachytherapy treatment, careful consideration of the radiobiology of the tumor and of brachytherapy treatments must be accounted for. Applicator selection, type of imaging, design of the implant, optimization of the treatment plan, evaluation of the treatment plan, and pretreatment quality assurance are just some of the steps that are considered prior to initiation of the treatment.

However, the treatment planning steps in Figure 1–1 cannot be performed without the treatment dose being prescribed by the radiation oncologist. The determination of the dose required to damage or kill the cancerous cells while sparing normal tissue requires detailed knowledge of the radiobiology of the brachytherapy treatment, which is discussed in the following section of this chapter.

1.2 Radiobiology

1.2.1 Introduction

Like external beam radiotherapy (EBRT), the goal of brachytherapy is to achieve a high level of tumor control with minimal normal tissue complications by delivering sufficient radiation dose to the tumor while minimizing radiation exposures to adjacent normal tissues. By positioning radiation source(s) directly in or at a short distance from the tumor, brachytherapy is theoretically more advantageous than EBRT at creating dose distributions with high conformality to tumor volume and rapid dose fall-off outside (Venselaar et al. 2016).

The resulting dose distributions, however, are typically more heterogeneous than those in EBRT. For techniques employing multiple sources (or source dwell positions) in a target volume, such as low-dose-rate (or high dose rate) interstitial brachytherapy for prostate cancer, the delivered dose within the target volume can vary from the level of a prescribed dose (usually the isodose surface that covers the target volume) to several times greater at points close to radiation source(s). The average dose given to the target volume is usually much greater than the prescribed dose. In addition, the temporal patterns of dose delivery vary widely between brachytherapy dose delivery techniques, ranging from temporary continuous low-dose-rate (LDR) irradiation at a nearly constant dose rate to permanent LDR irradiation with exponentially decreasing dose rate, and fractionated high-dose-rate (HDR) irradiations as a boost treatment to EBRT or as monotherapy.

Prospective assessment of the potential impact of such diverse spatial-temporal variations on tumor control or normal tissue complication is difficult using dose-based indices alone because the biologic effect of radiation is fundamentally determined by the interaction between a radiation field and the tissue being irradiated. This interaction can be influenced by factors related to the radiation field (e.g., total dose, spatial and temporal dose delivery pattern, and radiation quality) or to the tissue (e.g., intrinsic radiosensitivity, damage repair capacity, oxygenation status, and cell proliferation rate). The effectiveness of a brachytherapy technique for a given patient is, therefore, the result of a potentially complex interplay between the physical and biological factors involved.

Radiobiology, which specializes in the study of radiation action on biological systems, provides a useful scientific framework to study and model the interplay between brachytherapy dose delivery and underlying cellular processes. A growing number of studies have demonstrated that mathematical models capable of capturing the basic radiobiological interactions can offer a potentially useful tool for systematic elucidation of major radiobiological features of brachytherapy, for prospective evaluation of the relative effectiveness of different dose delivery techniques, and for treatment planning and optimization. This section provides a brief overview of commonly used radiobiological models and their use in brachytherapy.

1.2.2 Radiobiological Modeling for Brachytherapy

The process from the initiation of a radiation action to a manifested biologic damage in human cells typically progresses through three main phases. The initial physical phase ($\sim 10^{-18}$ s) involves energy transfer from radiation field to atomic electrons, leaving atoms in excited or ionized states. As the excited and ionized atoms interact in the subsequent chemical phase (~10^{-3} s), molecular damages in the form of chemical bond breakages emerge as a result of either direct action of radiation or indirect action mediated via free radicals formed by the interaction of the radiation field with other molecules in the cell (such as water). Among the molecular damages, double-strand breaks (DSBs) in DNA are considered to be the most significant molecular lesions that affect the fate of a living cell (Chadwick and Leenhouts 1981). Biological effects typically appear over a longer time scale (from seconds to years) as the cells react to the inflicted chemical damages. Some of the DNA damages inflicted by radiation could be successfully repaired, while others may be repaired incorrectly or cannot be repaired at all. Incorrectly repaired and irreparable lesions could lead to genetic alterations and cell death.

Since tumor control depends on how many cancer cells survive a given course of irradiation, radiation-produced cell kill (or survival) has often been used as a direct surrogate of the biologic effect of a given radiation (Barendsen 1982; Thames et al. 1982; Fowler 1989b). While radiation-induced cell death in normal tissue/organs is also detrimental, its effect on organ function may further depend on the specific organization of functional subunits (FSUs) in a given organ, which could have a nonlinear relationship with the number of cells killed. Nonetheless, it is generally recognized that the cells surviving a therapeutic dose of radiation play a central role on the subsequent biological response of an irradiated tissue/organ.

1.2.2.1 Cell Survival Modeling—Linear-quadratic (LQ) Model

Several mathematically more sophisticated cell survival models based on the kinetics of cellular damage production and repair—such as the repair–mis-repair (RMR) model (Tobias 1985), the lethal and potentially lethal (LPL) model (Curtis 1986), and the repair–mis-repair–fixation (RMF) model (Carlson et al. 2008)—have been proposed in the past for estimating the cell survival after a given radiation. Nonetheless, the LQ model, which describes the *in vitro* cell survival dose relationship well via a simple linear-quadratic function of dose (Chadwick and Leenhouts 1981), has remained as a most popular and most used model in clinical studies (Fowler 1989).

The LQ model can be derived from the kinetic models, such as LPL, in the limit of small doses and dose rates (Carlson 2013). In its simplest form for acute irradiation, the LQ model postulates that the average fraction (S) of cells surviving a clinical dose D can be described by

$$S = \exp[-\alpha D - \beta D^2], \qquad (1.1)$$

where α and β are two model parameters characterizing the cell's intrinsic radiosensitivity. The mechanistic basis for the LQ model has been extensively reviewed in literature (Chadwick and Leenhouts 1973; Sachs et al. 1997; Brenner 1998; Sachs and Brenner 1998; Zaider 1998b,a). It is generally accepted that the α term represents one-track lethal damage resulting from lethally mis-repaired DSB and lethal intra-track exchange-type chromosome aberrations, and the β term represents two-track lethal damage resulted from lethal inter-track exchange-type chromosome aberrations formed through binary mis-repair of two separate DSBs (two-track lethal damage) (Carlson et al. 2008). The incidences of these two modes of lethal damage are reflected, respectively, by the values of α and β. The ratio of α/β, which equals to the dose at which the lethal damages resulted from the two modes are equal, provides a measure of the relative importance of the two modes of lethal damages and the cell's radiosensitivity to dose fractionation.

For protracted dose delivery, the DNA DSBs produced by two separate tracks of radiation action (for the β term) could be repaired if the actions from the two tracks occurred at two separate times. The chance of a successful repair increases with increasing temporal separation between the two DSBs. To take into account DNA damage repair and cell proliferation during protracted irradiation, the LQ model can be generalized to

$$S = \exp[-\alpha D - \beta G(T) D^2 + \gamma \max\{0, (T - T_k)\}], \qquad (1.2)$$

where $\gamma \equiv \ln(2)/T_d$ is the cell proliferation rate, T_d is the effective cell doubling time (which also accounts for non-radiation-induced cell loss during therapy), T is total treatment duration, and T_k is the onset or lag-time of cell proliferation. $G(T)$ accounts for the effect of interaction between DNA damage repair and temporal dose delivery pattern, which has a numerical value between 0 (instant repair) and 1 (no repair).

When DNA damage repair follows a mono-phasic exponential kinetics, $G(T)$ becomes the Lea–Catcheside dose protraction factor (Lea and Catcheside 1942):

$$G(t) = \frac{2}{D^2}\int_{-\infty}^{\infty} dt\dot{D}(t)\int_{-\infty}^{t} dt'\dot{D}(t')\exp[-\mu(t-t')], \tag{1.3}$$

where μ is the time constant of damage repair (inversely proportional to the repair half-time, which ranges from minutes to hours for mammalian cells) and the temporal dose delivery pattern is characterized by the instantaneous dose rate $\dot{D}(t)$.

The mono-exponential repair kinetics and uniform cell proliferation rate used in equations (1.2) and (1.3) are simplistic depictions of the actual cellular processes. Nonetheless, they are useful for elucidating the key effects of temporal dose delivery pattern on cell survival and its potential impact on the relative effectiveness of different dose delivery techniques used in brachytherapy and EBRT.

1.2.2.2 Biologically Effective Dose (BED)

In a review article published in 1989, Fowler examined the role of the LQ model in stratifying radiotherapy fractionation schemes and introduced the concept of BED as an overall "one number" measure of the biological effectiveness of a radiotherapy scheme (Fowler 1989b). In this construct, BED relates to cell survival as follows,

$$S = \exp[-\alpha \cdot BED]. \tag{1.4}$$

Comparing the exponents in equations (1.4) and (1.2) yields

$$BED = D\left(1 + \frac{G(T)D}{(\alpha/\beta)}\right) - \frac{\gamma}{\alpha}\max\{0,(T-T_k)\} \tag{1.5}$$

The quantity within the angular brackets, $1 + G(T)D/(\alpha/\beta)$, is referred to as relative effectiveness (RE) of radiation dose D on a given tissue as the product of RE and D yields BED in absence of cell proliferation. In the absence of DNA damage repair and cell proliferation (or when the dose is delivered instantly), Equation (1.5) reduces to the more familiar form used in EBRT,

$$BED = D\left(1 + \frac{D}{(\alpha/\beta)}\right). \tag{1.6}$$

By definition then, different dose delivery schemes that give the same BED would produce the same level of cell killing in a given tissue. As such, BED can be used for comparing the relative effectiveness of different dose fractionation schemes. It can also be used to determine alternate fractionation schemes from a known fractionation scheme by matching the level of cell kills. BED has superseded the earlier concept of extrapolated response dose (ERD), nominal stand dose (NSD), and time–dose factor (TDF) tables that were traditionally used in comparing different fractionation schemes in EBRT. BED has been used widely in elucidating the effects of changing dose fractionation in EBRT (Fowler 2010). Its use in brachytherapy took off in late 1980s when analytical forms of $G(T)$ for several temporal patterns of brachytherapy dose delivery were worked out (see Section 1.2.3).

Nonetheless, the use of BED (Equation 1.5) on proliferating tissues could encounter a situation where BED would become negative (e.g., in the case of permanent interstitial brachytherapy for prostate cancer to be discussed later in Section 1.2.3.2). An iso-effective dose (IED) model proposed by Zaider and Minerbo and described in Section 4.9.2.2, based on the theory of cell birth-and-death stochastic processes (Zaider 2000) can also be used for brachytherapy with arbitrary temporal dose delivery patterns. When the LQ model is used for cell survival, the IED model is equivalent to the BED model for non-proliferating tumors, but is also mathematically well behaved in the limit of $T \rightarrow \infty$ for proliferating tumors.

It should also be noted that BED is not a deliverable physical quantity, although it shares the same unit with physical dose (Gy). Indeed, BED is often confused in the literature with "biologically *equivalent* dose" such as the equivalent dose in 2-Gy fractions (EQD2) discussed in the following subsection (Dale 2010). The confusion between "biologically effective" and "biologically equivalent" doses is easy to understand, given the similarity between the names and the fact that both are expressed in physical dose units (Gy). Fowler has recently proposed a change of the unit for BED to something other than the Gray to alleviate the confusion (Fowler 2010).

1.2.2.3 Equivalent Dose in 2-Gy Fraction (EQD2)

While the concept of BED arises from rearranging the exponent (i.e., dividing it by $-\alpha$) of the LQ model, one can take a step further in using "cell survival as a surrogate of radiation effect" to its full extent. In theory, one can always find a theoretically equivalent total dose when delivered in 2-Gy fractions that would produce the same level of cell survival as a given dose delivery technique. This equivalent total dose is often referred to as EQD2, where 2 signals the reference equivalent total dose is given in 2-Gy fractions. ICRU recommends using "equi-effective dose" in place of "equivalent total dose" (Bentzen et al. 2012).

By equating the exponent of the LQ cell survival expression for the treatment under study to that of a theoretically equi-effective treatment given in 2-Gy fractions, one obtains

$$EQD2 = BED / [1 + 2 / (\alpha / \beta)], \tag{1.7}$$

where BED is the biologically effective dose of the treatment technique under study, as given by Equation (1.5) or Equation (1.6). Strictly speaking, Equation (1.7) holds only in absence of cell proliferation or when the elapsed time of the whole treatment course is the same for the 2-Gy reference and the treatment technique under study.

Like BED, the concept of EQD2 is fundamentally the same as using cell survival as a surrogate of radiation effects. Treatment techniques producing the same EQD2 would yield the same level of cell kill. The model is, therefore, also useful for comparing the relative effectiveness of different dose fractionation schemes and for determining alternate fractionation schedules to match the cell killing of a known fractionation scheme. In addition to their dependence on dose delivery characteristics, both BED and EQD2 also depend on the radiobiological properties of the tissue under irradiation (e.g., α/β, DNA damage repair, cell proliferation, etc.).

Practical advantages of EQD2 include: (1) it relates directly to physical dose, (2) its value is more intuitive to interpret as it relates to the clinical experiences already gained in EBRT using 2-Gy fractions, and (3) it has been shown to be very practical and useful in the comparisons of different brachytherapy techniques for cervix cancer. For these reasons, the ICRU Committee on Bioeffect Modeling and Biologically Equivalent Dose Concepts in Radiation Therapy favors the use of the EQD2 concept for bioeffect modeling (Bentzen et al. 2012).

1.2.2.4 Other Models

Several other models are available for radiobiological analysis. For example, the concept of equivalent uniform dose (EUD) has been proposed and used for summarizing the biological effect of inhomogeneous dose distributions (Niemierko 1997). Using the LQ model, cell survival produced by an inhomogeneous dose distribution can be matched theoretically to that produced by a uniform dose distribution by adjusting the total dose of uniform distribution. Like EQD2, EUD is directly related to physical dose and is dependent on the dose delivery patterns and biological properties of the irradiated tissue.

In addition, high-level models—such as tumor control probability (TCP) and normal tissue complication probability (NTCP)—have also been used in radiation oncology. TCP and NTCP are theoretically more appealing if they can be reliably calculated. In principle, TCP can be calculated directly from cell survival if radiation-induced cell killing follows the Poisson statistics and the initial tumor burden is known. Calculation of NTCP is more complicated as, in addition to the information on cell killing, it often requires knowledge on the organization of functional subunits in the irradiated organ. TCP and NTCP models have not been used as often in brachytherapy compared to the LQ cell survival model (and its derived indices, such as BED and EQD2). Due to the limited scope of this section, these models are not reviewed further.

1.2.3 Radiobiology of Major Brachytherapy Dose Delivery Techniques

1.2.3.1 Protracted Dose Delivery with Constant Dose Rates

Brachytherapy dose can be delivered in single fraction or over multiple fractions. In the case of multi-fraction dose delivery, successive treatments are typically separated long enough (>12 h) to allow complete DNA damage repair before the next fraction (except in pulsed dose rate brachytherapy or in twice daily treatments, in which the treatments are separated by at least six hours). Under this condition, the biologic effects of individual fractions are generally considered as additive, provided that the separation between fractions is not unduly long. It is, therefore, interesting and instructive to first examine the radiobiology of single-fraction dose delivery.

For a single-fraction dose delivery in which the dose rate remains approximately constant (e.g., in intracavitary LDR brachytherapy using sources with long decay half-lives like ^{137}Cs or in HDR brachytherapy when the duration of dose delivery per fraction is short), the Lea–Catcheside dose protraction factor of Equation (1.3) can be expressed in a simple analytic form,

$$G(T) = \frac{2}{(\mu T)^2}(e^{-\mu T} + \mu T - 1). \tag{1.8}$$

The relative effectiveness factor in the BED equation of Equation (1.5) then becomes

$$RE = 1 + \frac{D}{(\alpha / \beta)} \frac{2}{(\mu T)^2}(e^{-\mu T} + \mu T - 1). \tag{1.9}$$

Several key radiobiological features of this type of dose delivery can be discerned using the equation above in the context of BED:

1. ***The biologic effect of a given dose is dependent on dose rate.*** Since the dose delivery time T for a given dose is inversely proportional to dose rate, changing dose rate will result in a different dose delivery time, which will affect the value of RE and, hence, the BED. This phenomenon is known as dose rate effect.

   Figure 1–2 plots the RE as a function of dose rates for early-reacting (high α/β ratio) and late-reacting (low α/β ratio) tissues. The ranges of typical dose rates used in permanent LDR brachytherapy (~0.07 Gy/h to ~0.4 Gy/h), temporary LDR brachytherapy (~0.2 Gy/h to ~2 Gy/h), and HDR brachytherapy (~10 Gy/h to ~50 Gy/h) are highlighted in the plot for reference. In the low-dose-rate limit ($T \rightarrow \infty$), RE approaches a limiting value of 1. At the high-dose-rate limit ($T \rightarrow 0$), RE reaches a high limiting value of $1 + D / (\alpha/\beta)$, which is dependent on the total dose and tissue type (α/β). In

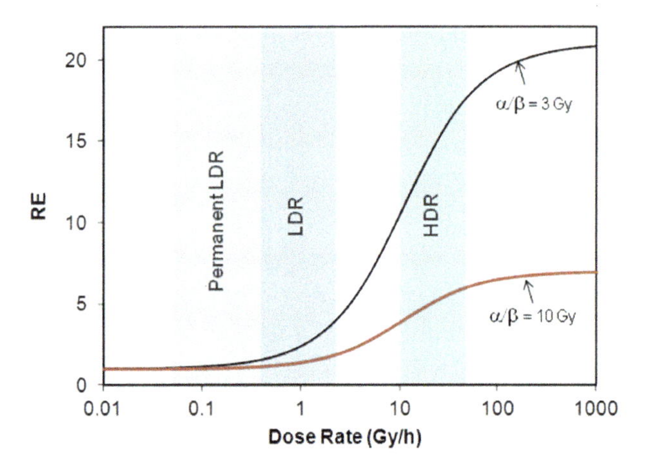

Figure 1–2 Relative effectiveness (Equation 1.9) as a function of dose rate for typical early-reacting (α/β = 10 Gy) and late-reacting (α/β = 3 Gy) tissues. Other parameters used in the calculation include $t_{1/2} = \ln 2/\mu$ = 1.5 h, γ = 0.0, and D = 60 Gy.

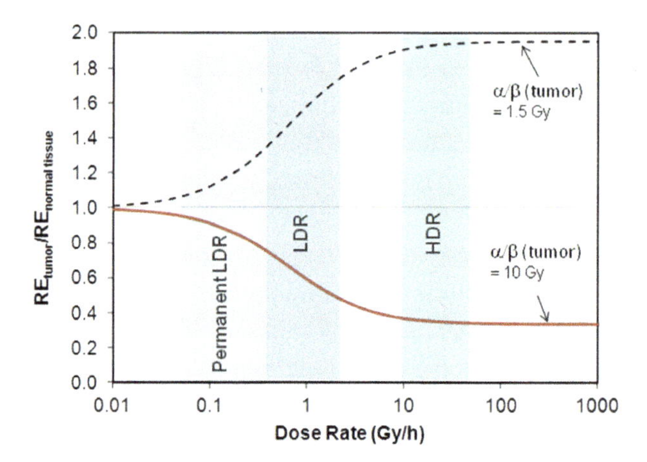

Figure 1–3 Plot of $RE_{tumor}/RE_{normal\ tissue}$, a surrogate of therapeutic ratio, as a function of dose rate. The solid curve represents a typical tumor with α/β = 10 Gy, while the dashed curve represents a hypothetical tumor with a low α/β of 1.5 Gy. Other parameters used in the calculation include $t_{1/2} = \ln 2/\mu$ = 1.5 h, γ = 0.0, D = 60 Gy, and normal tissue α/β = 3 Gy.

general, the relative effectiveness (hence, the biologic effect) of a given dose increases with increasing dose rate. Over the range of dose rates used in temporary LDR and HDR brachytherapy, the relative effectiveness can vary substantially.

The trend depicted in Figure 1–2 is consistent with the laboratory and clinical observations that reducing dose rate (hence, protracting dose delivery) generally leads to reduced biological damage (Hall and Bedford 1964, 1991; Mazeron et al. 1991; Lambin et al. 1993). For example, in the treatment of breast cancer using a combination of EBRT (45 Gy) and ^{192}Ir brachytherapy boost (37 Gy), the local recurrence rate increased from ~8% to ~31% when the dose rate of ^{192}Ir boost treatment was reduced from 0.75 Gy/h to 0.35 Gy/h, even though the total dose delivered was the same (Mazeron et al. 1991).

2. ***The biological impact of the dose rate effect is tissue dependent.*** As shown in Figure 1–2, the variation of RE is much greater for tissues with small α/β values than those with large α/β values. Since most late-reacting normal tissues have lower α/β and typical tumors have larger α/β, increasing dose rate will have a much greater effect on late-reacting normal tissues than on tumors for the same dose. In order to keep the normal tissue BED to an acceptable level when the dose rate is increased, the total dose needs to be reduced, which would lead to a lower tumor BED and a loss in therapeutic ratio. This observation is consistent with the long-established clinical observations that protracting a radiotherapy exposure (i.e., reducing the dose rate) can yield a therapeutic gain between tumor control and normal tissue complications (Coutard 1932).

   Using the ratio of RE of a tumor over a typical late-reacting normal tissue (α/β = 3 Gy) as a surrogate of potential therapeutic gain, Figure 1–3 illustrates its dose rate dependence for three hypothetical tumor α/β values. For a typical tumor with α/β of 10 Gy, the therapeutic gain reduces from a maximum value at low dose rate to a minimum value at the high dose rate limit. When a tumor's α/β value is similar to or lower than that of surrounding normal tissues, the advantage of dose protraction on therapeutic gain diminishes. In the case of prostate cancer with a reported α/β as low as 1.5 Gy (Brenner and Hall 1999) increasing dose rate would actually result in a relative sparing of normal tissue and potentially higher therapeutic ratio than low-dose-rate irradiation (the dashed curve in Figure 1–3). Prostate cancer is one of the very few diseases for which a radiobiological case for HDR or hypofractionated dose delivery can be made. This is also discussed in Section 1.2.3.3 and Chapter 5.9.

The discussions above assumed the normal tissues receive the same total dose as the tumor and there was no cell proliferation. In practice, if the dose to normal tissue can be made lower than that of the tumor, it could potentially lessen or compensate the loss of therapeutic ratio when delivering dose in high dose rate, even for tumors with α/β values greater than normal tissues (see additional discussion in Section 1.2.3.3).

3. ***Dose rate effect arises from the interplay between dose delivery and time-dependent cellular processes.*** Equations (1.5) and (1.9) capture two of the time-dependent cellular processes, i.e., DNA damage repair and cell proliferation. Other time-depended cellular processes—such as cell cycle redistribution and reoxygenation—could also cause dose rate effect if the characteristic time scale associated with these processes is comparable to that of dose delivery. While damage repair and cell repopulation generally reduce the cell killing, cell cycle redistribution and reoxygenation could be beneficial to increase cell killing. In general, DNA damage repair, which occurs on the time scale of minutes to hours in mammalian cells, is one of the main causes of dose rate effect in brachytherapy. The magnitude of the dose rate effect is determined by both the dose rate and the irradiation time for a given tissue. Because the dose rates used in brachytherapy vary significantly between different dose delivery techniques and within a target volume for a given technique, their impact should be carefully evaluated.

1.2.3.2 Protracted Dose Delivery with Exponentially Decreasing Dose Rates

One of the early and successful brachytherapy applications involves permanent implantation of radioactive sources directly into the tumor volume. Permanent interstitial brachytherapy uses low-energy, photon-emitting radioactive sources with short decay half-lives ($T_{1/2}$) ranging from ~2 days to 60 days. As a result, the target cells are subjected to continuous LDR irradiation with exponentially decreasing dose rates over time, which is very different from the near constant dose rate irradiation discussed in the previous section. For this type of dose delivery, the Lea–Catcheside dose protraction factor can also be expressed in analytic form (Dale 1985; Dale 1989). The relative effectiveness for dose delivered up to a total elapsed time T is given by

$$RE(T) = 1 + \left(\frac{\beta}{\alpha}\right)\frac{\dot{D}_0}{(\mu-\lambda)} \times \frac{1}{1-e^{-\lambda T}}\left\{1 - e^{-2\lambda T} - \frac{2\lambda}{\mu+\lambda}(1-e^{-(\mu+\lambda)T})\right\}, \quad (1.10)$$

where, λ denotes the radioactive decay constant and $\dot{D}_0$ the initial dose rate at a selected point of interest. A $T = \infty$, Equation (1.10) reduces to

$$RE = 1 + \frac{\lambda}{\mu+\lambda}\frac{D}{\alpha/\beta}. \quad (1.11)$$

RE and, hence, the biologic effect is dependent on the choice of radionuclide, the α/β ratio, the DNA damage repair half-time of the irradiated tissue, and the prescribed total dose.

For proliferating tumors, however, the cell repopulation term in Equation (1.5) would become increasingly large as treatment progresses and become infinite at $T \to \infty$. This creates a conceptual problem for BED because, due to source decay, the rate of cell killing decreases exponentially with time and will eventually become smaller than the rate of cell proliferation, leading eventually to a negative BED value as $T \to \infty$. To avoid this conceptual difficult, BED evaluated at a special time point, at which the instantaneous rate of cell kill from the irradiation equals to that of cell repopulation, has been used as the surrogate for permanent implants. This special time point is termed as "effective" treatment time, T_{eff}, beyond which the irradiation will cease to produce a net gain in cell reduction. While the definition of T_{eff} is physically intuitive, the need to use T_{eff} as *the* time point for BED calculation may introduce additional uncertainties in the application of the BED model to tumors that continuously repopulate in permanent implants.

Zaider and Hanin (2007) have shown that BED calculated at T_{eff} may underestimate the iso-effective dose. A recent systematic comparison between the BED and IED models for permanent brachytherapy with proliferating tumors indicated that the iso-effective prescription dose derived using the IED and BED models can be different (Chen 2012). Using ^{125}I implant with 145 Gy as a reference, the iso-effect prescription doses derived from BED model was 2.7% and 3.5% lower than those from the IED model for ^{103}Pd and ^{131}Cs, respectively, for slow-growing tumors (e.g., at T_d of 42 days). The difference increased to 8.4% and 13.4%, respectively, for ^{103}Pd and ^{131}Cs implants for fast-growing tumors (e.g., at T_d of 5 days) (Chen and Nath 2012). The clinical significance between the use of these two models remains to be tested.

The radiobiological features of permanent brachytherapy can be assessed using equations (1.5) and (1.10). Because permanent brachytherapy generally delivers dose over a relatively long duration, DNA damage repair, as well as cell proliferation, can significantly affect the biologic response. However, since the dose rate decreases exponentially with time, a significant portion of the total dose is typically delivered within the first few half-lives of radioactive decay. For example, implants using ^{131}Cs, ^{103}Pd, and ^{125}I sources would deliver the 90% of a prescribed dose in about one, two, and six months, respectively. The impact of damage repair and cell proliferation, therefore, can be quite different on permanent brachytherapy using sources of different decay half-lives.

Figure 1–4 plots the BED per unit dose as a function of half-life for three effective cell doubling times (T_d = 10, 42, and ∞ days) with a fixed DNA damage repair half-time of 0.27 h. In the absence of cell proliferation (T_d = ∞ days), BED/D increases gradually with decreasing half-life. The rate of increase is modest when $T_{1/2}$ is greater than 20 days, but it becomes rapid when $T_{1/2}$ is less than 20 days. The increase in BED/D is expected because the effective dose delivery time reduces as $T_{1/2}$ is reduced, which increases the chance of damage repair. The dose delivered by a radionuclide with a $T_{1/2}$ of 5 days is biologically more effective (by ~10%) than that delivered by a radionuclide with $T_{1/2}$ of 60 days. This qualitative trend remains for other values of α/β and μ.

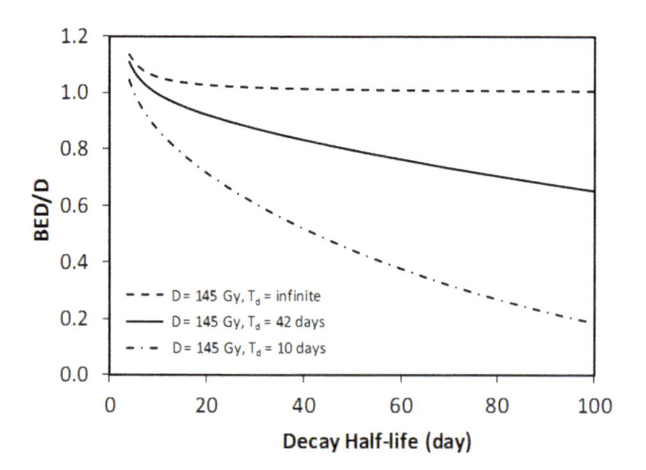

Figure 1–4 BED per unit physical dose as a function of radionuclide half-life. Parameters used in this plot are: D = 145 Gy, α = 0.15 Gy^{-1}, α/β = 3 Gy, DNA damage repair half-life = 0.27 h, and T_k = 0.

For proliferating cells, a strong interplay between radioactive decay and cell repopulation rate exists. BED/D decreases with increasing $T_{1/2}$, and the rate of decrease becomes more dramatic for tissues with fast cell proliferation rates. For cells with a doubling time of 42 days, a source with $T_{1/2}$ of 10 days (e.g., near that of ^{131}Cs) is approximately 30% more effective than a source with $T_{1/2}$ of 60 days (e.g., near that of ^{125}I) for the same total dose. This difference results from the initial dose rate being much higher for the shorter half-life radionuclide to deliver the same total dose. In this situation, the radionuclide with the shorter half-life delivers much more of the dose integrated over the time before reaching the T_{eff} than does that with the longer half-life and the lower initial dose rate. This difference increases to 130% for tissues with a faster doubling time of 10 days. Therefore, for tumor eradication, sources with shorter $T_{1/2}$ are more effective than those with longer $T_{1/2}$. However, the advantage of using shorter $T_{1/2}$ sources has to be balanced against its effect on normal tissues. Because late-reacting normal tissues usually have lower α/β ratio than that of early-responding normal tissues, the biological effect of a given total dose is generally larger on late-responding tissues.

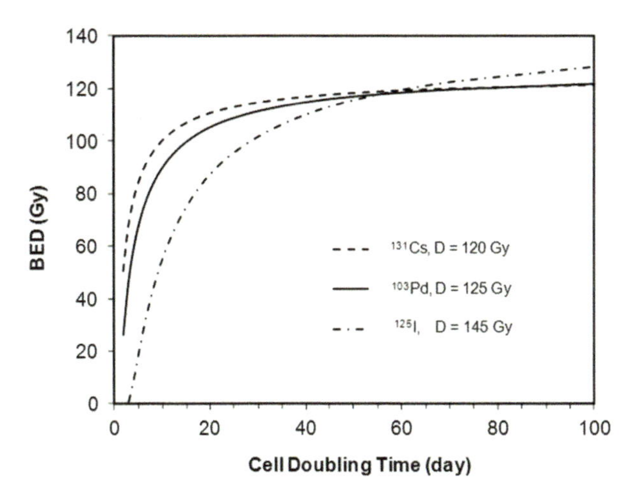

Figure 1–5 BED as a function of tumor cell doubling time for PIB using different sources.

Among the sources used in current clinical practice, ^{131}Cs would produce a greater biologic effect on both tumor and normal tissue than either ^{103}Pd or ^{125}I, assuming that all were prescribed to the same total dose. Using BED, one can perform detailed analysis of the effects of a prescribed dose on early- and late-responding normal tissues and assess potential complications associated with a particular treatment. In practice, the dose prescribed for a radionuclide with a shorter $T_{1/2}$ is generally smaller than that prescribed for a longer $T_{1/2}$ radionuclide. For example, in permanent implants for prostate cancer, the prescribed doses for monotherapy using ^{131}Cs, ^{103}Pd, and ^{125}I are approximately 120 Gy, 125 Gy, and 145 Gy, respectively, aimed to achieve similar biological effects on the tumor while maintaining acceptable normal tissue complications.

Because the cellular processes can vary between patients, one may appreciate that the nominal prescription dose currently used for each radionuclide does not provide a one-size-for-all solution to individual patients. Figure 1–5 illustrates the theoretical BED achieved by each radionuclide for tumors with different doubling times. For a tumor with T_d of 42 days, all three radionuclides produce similar BED, with the largest value achieved by ^{131}Cs. For slow-growing tumors ($T_d > 60$ days), ^{125}I is slightly more effective; however, for fast-growing tumors ($T_d < 30$ days), ^{125}I is least effective compared to ^{103}Pd and ^{131}Cs. A prescribed dose of 120 Gy with ^{131}Cs is more effective than ^{103}Pd and ^{125}I over a wide range of tumor doubling times. This advantage may be offset to some degree by edema-induced source variations, which is greater in permanent implants using ^{131}Cs than ^{125}I.

1.2.3.3 Fractionated Dose Delivery

As discussed in Section 1.2.3.1, single-fraction dose delivery is biologically more potent at high dose rates than low dose rates for both tumor and normal tissues. In addition, the increase in biological damage is much greater in late-responding normal tissues than in early-responding tumors when changing from low- to high-dose-rate irradiations. In general, to keep late-reacting normal tissue complications clinically acceptable, HDR brachytherapy has to be fractionated into multiple fractions. Even in LDR brachytherapy, some treatments are delivered in more than one fraction (e.g., two fractions in cervical brachytherapy using LDR ^{137}Cs sources).

When the separation between fractions is long enough (>12 h) to allow complete DNA damage repair, the BED for a fractionated dose delivery is given by

$$BED = D\left(1+\frac{g}{\alpha/\beta}d\right)-\frac{\gamma}{\alpha}\max\{0,(T-T_k)\}, \tag{1.12}$$

where D denotes the total dose, n the number of fractions, and d the dose per fraction ($D = nd$). The effect of intra-fraction DNA damage repair is captured by the dose protraction factor g. It should be noted that the dose protraction factor G(T) in equations (1.3) and (1.5) is for the entire course of dose delivery. In the fractionated case with not incomplete repair between fractions, g = nG(T), with g representing the dose protraction factor evaluated over the duration of dose delivery for one fraction only. The factor g is close to 1 in most EBRT and HDR cases because the fraction dose delivery is typically much shorter than the repair half-time of less than 10 minutes.

Key radiobiological features of fractionated dose delivery for HDR brachytherapy, discernible from Equation (1.12), are similar in principle to those observed in the single-fraction dose delivery, specifically

1. *The biologic effect of a given dose is dependent on how the dose is fractionated.* Based on Equation (1.12), fractionation schemes using larger dose per fraction (therefore, fewer fractions) would produce greater biological effect since BED increases with increasing dose per fraction, d. While this is beneficial for tumor effects, increasing the fraction size would also increase the biological damage to normal tissues, usually to a greater extent than for the tumor.
2. *A given dose and fractionation can produce different biologic effects on different types of tissue.* As shown in Equation (1.12), BED increases with decreasing α/β value. Hence, the biologic effect of a given dose and fractionation would be greater on tissues having smaller α/β values. Since late-reacting normal tissues usually have smaller α/β values, these normal tissues would suffer greater biologic damage than typical tumors if these tissues were to receive the same dose as the tumor.
3. *Altering dose fractionation has different biologic effects on early- and late-reacting tissues.* As alluded to in points 1 and 2 above, and apparent in Equation (1.12), the coefficient of fraction size is $[g / (\alpha/\beta)]$. Therefore, the rate of increase in BED would be greater in late-reacting normal tissues (smaller α/β values) than in early-reacting tissues with large α/β values, such as typical tumors, if they were given the same dose. This is similar to the effect of increasing dose rate in single-fraction dose delivery discussed in Section 1.2.3.1.
4. *Prolonged dose delivery at each fraction could reduce the biologic effect.* Prolonged dose delivery increases the chance of intra-fractional DNA damage repair, leading to smaller dose protraction factor g, which will lead to a reduced BED. This impact can be different on different types of tissue for the similar reasons discussed in points 2 and 3 above.
5. *Cell proliferation during a treatment course always reduces the biologic effect of radiation.* For tumor control, dose delivery schemes with shorter overall times would suffer less negative effects from cell proliferation.

The features illustrated above indicate that the clinical impact of a fractionated HDR brachytherapy has to be analyzed with full consideration of the dose to, and the radiobiological properties of, both the target and normal tissues involved. An example by Dale and Jones (1998) illustrates perfectly how such analysis could lead to different conclusions relevant to clinical practice:

> Assuming critical normal tissues near a tumor receive the same dose as tumor, tumor and normal tissue have α/β of ~10 Gy and ~3 Gy, respectively, and both tumor and normal tissue have repair half-life of 1.5 h, then 11 fractions of HDR would be needed to match both the early and late effects to a single-fraction continuous LDR dose delivery of 40 Gy in 48 hours. HDR treatments using 11 or more fractions are likely to improve the therapeutic ratio, but may not be practical. On the other hand, if the total dose to normal tissues is 20% less than tumor in HDR, the number of HDR fractions needed would be reduced from 11 to 5 fractions, even with no normal tissue dose sparing in the corresponding LDR regime. When this

level of dose sparing is combined with favorable repair kinetics in tumor (repair half-life of 0.5 h instead of 1.5 h), the number of HDR fractions needed would reduce from 11 to 1 fraction.

These analyses indicate that HDR treatment delivered in a small number of fractions may be more radiobiologically acceptable than previously thought (Dale 1990). An extensive analysis performed by Brenner and Hall (1991) comparing fractionated HDR and LDR regimens for intracavitary brachytherapy of the cervix also reached a similar conclusion.

Another factor that influences the choice of dose fractionation is the tumor α/β value. When a tumor's α/β value is similar to or lower than that of the surrounding normal tissue, the disadvantage associated with increasing fraction size (or dose rate in single-fraction dose delivery) would disappear. Indeed, when the tumor α/β is lower than that of the normal tissue, a gain in therapeutic ratio would be achieved by delivering dose in fewer fractions with large dose per fraction. This provides the radiobiological rationale for hypofractionated HDR brachytherapy (and EBRT) for prostate cancer, which has a reported α/β as low as 1.5 Gy (Brenner and Hall 1999). See additional discussions in Chapter 5.9.

In the absence of apparent biological rationale (such as low tumor α/β value), hypofractionation may still be used effectively in case of small target volumes, such as in the applications of stereotactic radiosurgery and stereotactic body radiosurgery. The use of interstitial and balloon brachytherapy for accelerated partial breast irradiation (APBI) follows mainly in this consideration. The α/β value estimated for breast cancer is approximately 10 Gy with a repair half-life of 1 h (Guerrero and Li 2003). On the higher level, the radiobiology of APBI is basically the same as fractionated HDR. The time separation between the treatment fractions is at least 6 h (which covers about six repair half-lives for typical breast cancer); incomplete repair between fractions may be neglected. A full radiobiological analysis of APBI, however, may require additional consideration of the heterogeneous distributions of both the dose and the cancer burden.

1.2.3.4 Pulsed Dose Delivery

To maintain the radiobiological advantage of continuous LDR dose delivery while taking advantage of the HDR afterloading technology, pulsed-dose-rate brachytherapy was developed to simulate the radiobiological characteristics of a continuous LDR irradiation using a large number of short HDR irradiations (on the order of minutes) over a time period similar to that used in LDR, thereby limiting the potential reduction in the therapeutic ratio (e.g., the ratio of TCP and NTCP at a specified dose level) associated with HDR dose delivery (Brenner 1997; Dale and Jones1998; Hall and Brenner 1996).

The dose rate within each radiation pulse is high, while the effective dose rate of the entire course of treatment is low due to the radiation-free periods between the pulses. For a pulsed dose delivery using n pulses, each of duration t and separated by radiation-free interval X, the relative effectiveness is given by (Dale and Jones 1998):

$$RE = 1 + \frac{2\dot{D}}{\mu(\alpha / \beta)}\left(1 - \frac{nY - SY^2}{n\mu t}\right) \tag{1.13}$$

where

$$S = \frac{nK - K - nK^2Z + K^{n+1}Z^n}{(1 - KZ)^2}, \tag{1.14}$$

in which $Z = e^{-\mu t}$, $Y = 1 - Z$, $K = e^{-\mu X}$, X is the radiation-free interval between pulses, and the total dose is $D = \dot{D}Nt$. The RE and, hence, BED for PDR dose delivery depends on the dose rate of each pulse, pulse duration, number of pulses, and inter-pulse interval, as well as their interplay with underlying cellular processes of the irradiate cells.

The PDR approach offers a unique opportunity to optimize the temporal pattern of dose delivery to achieve a similar or better therapeutic ratio as continuous LDR brachytherapy, while taking advantage of many of the logistical advantages of HDR brachytherapy (Brenner 1997). However, proper use of PDR brachytherapy, especially when designing pulse sequences to achieve a desired clinical objective, is more dependent on the understanding and use of radiobiological models. Similar radiobiological analyses discussed in sections 1.2.3.1 to 1.2.3.3 can be carried out for PDR dose delivery, which will not be presented here due to the limited scope of this section. PDR brachytherapy also carries with it the disadvantages of LDR brachytherapy, such as applicator movement on the order of centimeters through the treatment, patient discomfort due to the longer applicator placement, and the expenses of inpatient treatment.

1.2.4 Application of Radiobiological Models in Brachytherapy

The application of radiobiological modeling in brachytherapy was relatively sparse before the 1980s. Experimental radiobiological models (*in vitro* or *in vivo*) directed specifically to brachytherapy were also few. As a result, the radiobiological basis of brachytherapy was less well elucidated compared to EBRT, even though brachytherapy was used much more than EBRT in the early half of the twentieth century (Dale and Jones 1998).

In the late 1980s, Dale extended the BED model to protracted dose delivery patterns used in brachytherapy. Since then, the BED model has been used by Dale and others in the investigation of a variety of radiobiological questions regarding, for example, the relative effectiveness of LDR and HDR irradiations, the effect of source decay in permanent implants, the impact of using sources of different decay half-lives in the same implant, the effect of tumor shrinkage during permanent brachytherapy, the impact of prostate edema in permanent prostate brachytherapy, the effects of dose heterogeneity inherent in brachytherapy, and the biological effect of combining brachytherapy with EBRT. It is simply not possible to discuss all of these in this section due to the limited scope. Readers are referred to the original articles or related references for more detailed discussion and applications. In the following, we will discuss, in general, how radiobiological models may be used to address clinically relevant questions in brachytherapy treatment planning and evaluation.

1.2.4.1 General Utility

For reasons alluded to in the previous sections, the current LQ-based models cannot and should not be used to predict the treatment response of individual patients. This is because the modeling of biological processes and their radiobiological interactions is fairly simplistic, the model does not include all relevant biological processes, and there is currently no reliable way to determine patient-specific model parameters. The model predictions, however, may be used as a figure of merit for relative comparison of potential biological responses between different dose delivery techniques.

The relative comparisons may be focused on the relative effect of different treatment techniques by keeping the radiobiological parameters fixed to a set of representative values. For example, one can use a set of self-consistent parameters for prostate cancer to compare the relative effectiveness of different radioactive sources (e.g., between ^{125}I and ^{103}Pd or ^{131}Cs) used in permanent prostate implants (Nath et al. 2009). This approach is also useful in evaluating potential biological impact caused by technical variations (such as source or catheter movements) or in treatment technique optimization for better radiobiological response.

The relative comparisons may also be used to assess the relative impact of a given treatment technique on different biological targets. For this purpose, the spatial-temporal dose delivery pattern would be kept the same while the radiobiological model parameters are varied to simulate tumors or normal tissues of different characteristics. For example, the relative impact of a treatment technique on early- and late-reacting tissues can be assessed by changing the α/β value. This approach can also be useful for assessing the potential impact of uncertainties associated with the radiobiological parameters by comparing the relative response using model parameters sampled over a range of uncertainties.

In practice, these two types of relative comparisons are often performed within a given study. For example, one may study the potential impact of prostate edema on permanent prostate brachytherapy between ^{125}I and ^{131}Cs sources by comparing the BEDs of ^{125}I and ^{131}Cs implants calculated with the same level of prostate edema and the same set of nominal model parameters. After which, one can test if the conclusion drawn is applicable to tumors having different α/β ratios or if the conclusion is robust with respect to potential uncertainties associated with the model parameters by varying the value of the parameters under consideration. As such, radiobiological models provide a useful tool for performing pre-clinical evaluation of the relative effectiveness of different dose delivery patterns, for conducting meaningful comparisons of treatment outcomes from different brachytherapy techniques, and for optimizing the therapeutic efficacy of brachytherapy treatments.

1.2.4.2 Utilities in Treatment Planning and Evaluation

Planning for brachytherapy has been guided primarily by dosimetric considerations and clinical experiences. In general, key planning tasks may include selecting a treatment technique and, for the selected treatment technique, identifying treatment target volume, selecting appropriate radiation source, and optimizing source location and irradiation (dwell) time such that the target volume is adequately covered by a desired prescription dose with minimal dose to the adjacent critical organs and normal tissues.

Since physical dose alone is not adequate to determine the potential biologic effect and, for the reasons stated earlier, regarding the spatial-temporal variability in brachytherapy, radiobiological considerations should be ideally included in the design and evaluation of brachytherapy treatment plans. From a treatment planning and evaluation prospective, the radiobiological models could potentially be useful in (1) treatment technique selection, (2) treatment plan optimization and evaluation, (3) sensitivity analysis of a resulting plan with respect to potential variations in dose delivery and patient radiosensitivity, (4) quantifying the impact of dose heterogeneity, and (5) proper assessment of the combined effects with other treatments, such as EBRT.

While the general approaches outlined in Section 1.2.4.1 may be used for some of the analysis related to treatment planning, additional tools are needed for efficient use of radiobiological models in treatment planning. In particular, since the spatial and temporal dose delivery pattern is already known in the treatment planning system, it would be ideal to have the radiobiological models implemented within the treatment planning system based on voxel-by-voxel analysis. With such a system, 3D distributions of BED (or EQD2) and BED (or EQD2) volume histograms can be generated to allow quantitative evaluation of the effects of dose heterogeneity, which would be valuable when combining brachytherapy with EBRT or with another brachytherapy technique. The 3D BED (or EQD2) information is also useful for analyzing potential radiobiological regions of high or low radiobiological equivalent doses.

1.2.5 Best Practices in Using Radiobiological Models in Brachytherapy

Clinical application of radiobiological modeling in brachytherapy has been active since the mid-1980s. These models are becoming an increasingly useful tool for comparative analysis of potential biological and clinical responses from different dose delivery techniques. Nonetheless, the models presented here have their limitations and challenges. All modeling work and analysis, therefore, must be performed with the model limitations and their potential impact in mind.

For example, despite its plausible mechanistic underpinning, the description of the events that govern radiation response in the LQ model (such as DNA DSB repair and cell proliferation, etc.) is rather simplistic. There are data suggesting DNA DSBs are often repaired with bi-phasic kinetics rather than the mono-exponential process used in this section. Even in the cases where DSB repair kinetics is not a concern, the applicability of the LQ model has been questioned, especially at higher doses per fraction. When extrapolating to doses above 15 Gy, the LQ model can over-predict the level of cell killing by an order of magnitude compared with the LPL model. In addition, many biological factors—such as the presence of tumor hypoxia, cell cycle redistribution, cancer stem cells, low dose hyper-radiosensitivity, bystander effects, and the influence of tumor microenvironment on cell damage and damage repair—are not included in the LQ model described in this section. While some of these factors, e.g., hypoxia and cell cycle redistribution, may be incorporated into the existing model parameters, others may likely require new modeling efforts.

Another major challenge in using the current model or other models is the lack of reliable model parameters for all disease and tissue types. A set of "nominal" values is often used. For disease sites with available model parameters, these values are often derived from clinical outcomes observed among a population of patients. Inter-patient variability in the value of model parameters exists among patient populations. In addition, disease burden and radiobiological characteristics may also vary within tumors and, over time, for a given disease type. Treatment protocols derived from population-averaged model parameters may not be applicable to individual patients. It is, therefore, critical to assess the influence of model parameters on model predictions. While the sensitivity of a model prediction to model parameters may be estimated by examining the variations of predicted response over a range of model parameter values sampled over their observed or expected uncertainty range, methods for reliable and quick identification of disease- and patient-specific model parameters remain a big challenge and a research opportunity toward patient-specific treatment planning.

Although the predictions of current models are largely consistent with general expectations, careful and rigorous model validations for brachytherapy applications are still lacking. Clinical data from prospective randomized trials are the gold standard in medicine. The current models are most useful in assessing relative changes. It may be used to guide the selection of a new or alternate treatment regimen in a careful and methodical way for treatment sites that do not already have established reference treatment techniques. It is important to bear in mind that model limitations and uncertainties in radiosensitivity parameter estimates will likely have a greater impact on deriving alternate treatments that differ greatly from a reference treatment technique. It is, therefore, prudent to make small incremental changes away from established regimens instead of drastic changes to accepted clinical standards of care.

1.3 Targeting

1.3.1 Introduction

Historically, brachytherapy targeting evolved a century ago from manual palpation of intracavitary and interstitial targets and the manual verification of placement of devices and catheters. This was soon supplanted by planar x-rays and fluoroscopy and devices and implant rules for planar and volume implants to deliver consistent dosimetry, at least within a given system, such as Paris or Patterson–Parker. Two-dimensional imaging remained dominant until the 1990s, when 3D imaging gradually began to displace it. Various forms of 3D imaging are used in brachytherapy, including computed tomography (CT), magnetic resonance imaging (MRI), and ultrasound (US), along with image fusion among these various modalities.

1.3.2 Computed Tomography vs. Magnetic Resonance Imaging

CT has been the standard for radiation therapy treatment planning because of minimal geometric distortion and the relation between CT-estimated electron density and radiation therapy dose deposition. Modern MRI has superior soft-tissue contrast, which makes it preferred for target identification, but uncertainties in magnetic field gradients leads to geometric uncertainties, and the conversion of MR image voxels to electron density is not straightforward.

1.3.3 Radiographic Density vs. Soft Tissue Contrast

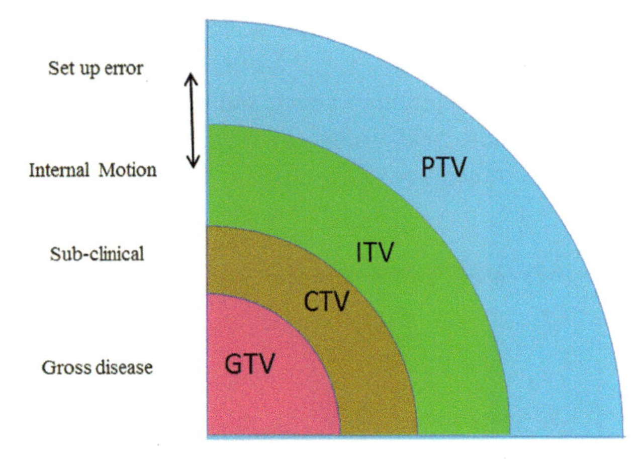

Figure 1–6 Idealized schematic of target volumes used in radiation therapy—GTV (gross tumor volume), CTV (clinical target volume), ITV (internal margin or internal target volume), and PTV (planning target volume).

Image fusion of MR onto CT via either rigid or deformable registration is one way to marry the best of both worlds—the tissue identification accuracy of MRI is warped onto the CT data set and the CT density used for dosimetric calculations. However, distortions in MR anatomy and errors in fusion may remain. Vendors of MRI and treatment planning systems have been refining software to correct for magnetic field gradient variations and to minimize susceptibility artifacts. If the MRI patient geometry differs only negligibly from the CT geometry, one may manually contour structures and assign them bulk densities for bone or tissue. Use of institution-specific electron density calibration curves or look-up tables provides additional accuracy (Maspero et al. 2017; Kim et al. 2015; Dowling et al. 2012). In 2016, the International Commission on Radiation Units and Measurements (ICRU) recommended MRI as the reference standard for gynecologic tumor assessment (ICRU Report 89). The recommendation was based on numerous studies verifying the accuracy of MRI in staging of disease against a surgical standard and the correlation of treatment response with MRI-defined tumor volume. Nevertheless, the recommendation to adopt expensive and sophisticated technology will not be implemented rapidly because of capital costs and the difficulties inherent in shifting a treatment paradigm.

1.3.4 Target Identification and Definitions—GTV, CTV, ITV, and PTV

The definitive statements for volume specification in brachytherapy are in ICRU reports 58 and 89 quoted liberally below (ICRU 1997; ICRU 2016). An idealized schematic of the various target volumes in common use is shown in Figure 1–6.

1.3.4.1 GTV—Gross Tumor Volume

The gross tumor volume corresponds to the gross palpable, visible or clinically demonstrable, location and extent of the malignant growth. Disease that is histologically proven, but non palpable and MRI invisible, does not have a defined GTV. There is no GTV when disease consists of only a few individual cells or microscopic involvement. Examples would be (1) prostate cancer in clinical stage T1c identified only by needle biopsy for suspiciously elevated PSA when disease is found in less than 50% of the length of a single core and (2) post-lumpectomy breast brachytherapy where all of the identifiable tumor has been removed.

When visible or palpable, the GTV should be contoured on the pretreatment images and, preferably, correlated with multiple modalities such as US, MRI, or CT-PET. After several fractions of brachytherapy or external-beam radiation therapy, the GTV, particularly for gynecological malignancies, can change in volume and topography. Significant volume regression during the first weeks of radiotherapy is generally more pronounced after combined chemo-radiotherapy. Because of this, tumor regression during treatment can impact treatment strategies. The tumor at diagnosis is referred to as the GTV_{init} and the residual gross tumor volume after a significant part of the treatment has been delivered as the GTV_{res}. In combined modality therapy, GTV_{res} is the volume after external beam therapy and prior to the brachytherapy boost.

After delivery of a dose of 45 to 50 Gy, the pathologic nature of the GTV may change, and clinical imaging of the GTV_{res} is less reliable than at the time of diagnosis. In a study assessing the accuracy of MRI after downsizing rectal tumors with about 50 Gy of preoperative radiation therapy, Dresen and colleagues (2009) found that the MRI-visible GTV_{res} could contain macroscopic, microscopic, or no disease at the time of surgery. An illustration of the changes in the GTV before and after a course of radiation therapy is in Figure 1–7.

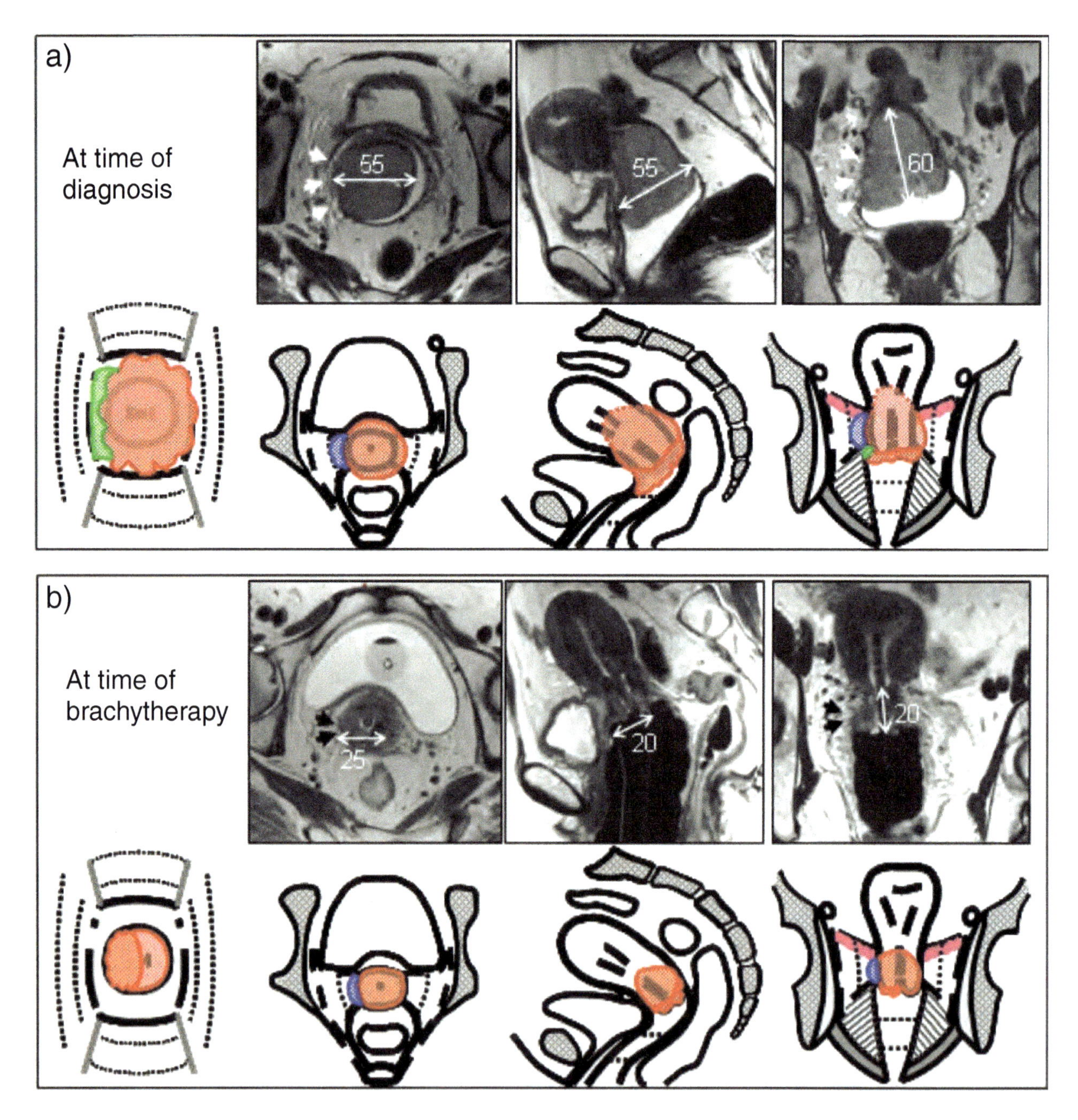

Figure 1–7 Changes observed in the MRI of a cervical cancer patient a) before and b) after 45 Gy of external beam therapy in 25 fractions. From left to right, each panel shows transverse, sagittal, and coronal T2-weighted MR images with schematic drawings of the GTV (gross tumor volume) below in red. The left-most drawing in each set is the speculum view of the tumor. The tumor dimensions shrink by more than a factor of 2 and the volume by more than a factor of 10. Reprinted with permission of the International Commission on Radiation Units and Measurements, http://ICRU.org.

1.3.4.2 CTV—Clinical Target Volume

The clinical target volume is the volume that contains the GTV and includes a volume of surrounding tissue with a high risk of microscopic, subclinical disease. This marginal region around the GTV should be treated with a dose sufficient to control microscopic malignant disease. Delineation of the CTV encompasses not only the probable extent of subclinical malignant cells outside the GTV, but also potential microscopic tumor spread into lymph nodes. The extent of the CTV beyond the GTV requires assumptions about the decrease in cancer cell density with distance from the GTV and knowledge about typical routes of microscopic spread into adjacent tissues.

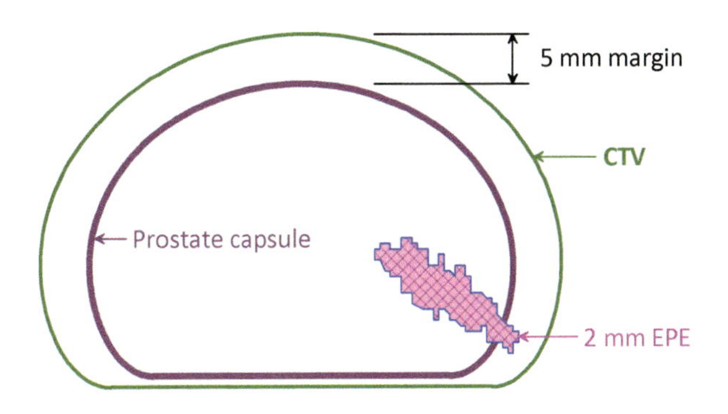

Figure 1–8 Enlargement of the prostate to create its clinical target volume (CTV) in order to encompass the likely extra-prostatic extension (EPE). The EPE is rarely visible on trans-rectal ultrasound imaging and is inferred statistically from a whole-mount radical prostatectomy series. The enlargement is three-dimensional and extends superiorly above the prostate base into the proximal seminal vesicles and inferiorly below the prostate apex.

Although some centers are investigating focal therapy for prostate cancer, in most cases of 12-core transrectal biopsy—and an even larger majority of cases of transperineal template-guided mapping biopsy—early stage prostate cancer presents as a multi-focal disease with cancer found in both lobes of the gland (Bittner et al. 2015). Given this behavior, at least the whole prostate gland must be included in the CTV. In multiple analyses of whole prostate specimens, the median extent of subclinical extraprostatic extension is generally less than 3 mm, but 5% of early-stage prostate cancer patients have extension beyond 5 mm, and more than 20% of high-risk patients exceed 5 mm (Chao et al. 2006). Therefore, the clinical target volume for prostate dosimetry should be the entire prostate expanded with a three-dimensional margin of 3–6 mm depending on patient risk group. This three-dimensional expansion is usually constrained posteriorly to avoid the anterior rectal wall and constrained superiorly to avoid the bladder neck. A schematic of a prostate CTV is shown in Figure 1–8. The posterior margin in prostate brachytherapy is usually set to 0–1 mm to avoid rectal toxicity.

1.3.4.3 ITV—Internal Target Volume

The PTV and ITV are sometimes defined separately but are often combined in one expansion. The ICRU defines the ITV as the CTV plus a margin which takes into account uncertainties and variations in size, shape, position, and movements of the CTV within the patient (ICRU Report 89). These uncertainties in the position of the CTV result essentially from normal physiological motions due to respiration or organ movement and variations in size, shape, and position of the GTV, the CTV, and organs in or adjacent to the CTV, such as movement resulting from bladder and rectum filling. This additional normal tissue volume to be irradiated constitutes the ITV or the internal margin.

1.3.4.4 PTV—Planning Target Volume

The PTV is determined by a geometrical expansion of the CTV to account for geometric and dosimetric uncertainties. In EBRT, the PTV is essential to ensure that the delivered dose to the CTV is within the limits defined in the dose prescription. The dosimetric uncertainties in brachytherapy differ from those in EBRT, and accounting for these uncertainties assures that the CTV receives the prescribed dose within a clinically acceptable probability. The delineation of a PTV takes into account two types of uncertainties: the movements of the CTV within the patient (the ITV) and uncertainty in radiation delivery, which depends on the brachytherapy technique. For interstitial or tandem and ring (T&R) treatments, the PTV may be equal to the

CTV in the anterior/posterior and lateral directions, but may require a small expansion in the cephalocaudal direction to account for the possible slippage of the applicators. With interstitial prostate seed implants, however, the prostate moves with each needle insertion and rebounds with variable elasticity as the needle is withdrawn and the seeds deposited. A further expansion of the prostate CTV to a PTV is necessary to account for needle placement error, source spacing differences, and seed splaying near the periphery (Roberson et al. 1997). Additionally, the PTV used to guide needle placement for interstitial treatments may be different from the dosimetric PTV. In order to properly cover the treatment dosimetric PTV, it may be necessary to place needles outside that region.

In intracavitary brachytherapy—such as in cervical cancer and post-lumpectomy breast cancer with commercial applicators such as MammoSite (Hologic, Inc, Bedford MA) and SAVI (Cianna Medical, Inc., Aliso Viejo, CA)—the tandem and catheters containing the HDR source are within the CTV. Application of any additional margin in the transverse direction (lateral and anterior–posterior) and normalizing the prescribed dose to the enlarged volume will result in an increased dose throughout the CTV and organs at risk. In such cases, defining any expanded volumes beyond the CTV to receive the prescribed dose must be done cautiously. Most of the uncertainties in such intracavitary applications are in reconstruction, which is dependent on the type of applicator, the imaging modality, and the image fusion procedures. Most reconstruction uncertainty is in the longitudinal direction parallel to the axis of the tandem or breast applicator, and expansion in this direction to a PTV will not adversely affect the dose to the CTV (Tanderup et al. 2010).

1.3.4.5 Use of Boolean Operators

The PTV is not always simply additive margins to the CTV, but is commonly reduced in order to avoid normal tissue at risk. For example, in partial breast brachytherapy after lumpectomy, the cavity (and the balloon applicator itself, if used) is of no dosimetric concern and should be omitted when optimizing the treatment plan. The surface of the cavity is contoured; it is then expanded by 1 cm to create a CTV. Avoidance volumes include the ribs and lung, as well as a 5 mm thickness of skin. Using Boolean operators, the latter volumes are subtracted from the CTV, a structure designated in the clinical trial NSABP B39/RTOG 0413 (see Chapter 8), as shown in Figure 1–9.

Sometimes Boolean avoidance is applied without user selection. A major software update to a popular prostate seed treatment planning system—VariSeed version 8.0 to VariSeed version 9.0 (Varian, Inc., Palo Alto, CA)—excluded the urethra by default as part of the prostate or CTV. This exclusion ignored 20 years of prior practice where the urethra was included in the primary target volume based on extensive surgical pathology evidence showing prostate cancer in contact with the urethra or involving the urethra (Huang et al. 2007). Although the urethral volume within the prostate is typically $<0.3\ cm^3$, negligible compared to the volume of the prostate, the vendor has indicated that the default will be to include the urethra as part of all defined target volumes in the next release of the software, VariSeed version 9.0.1.

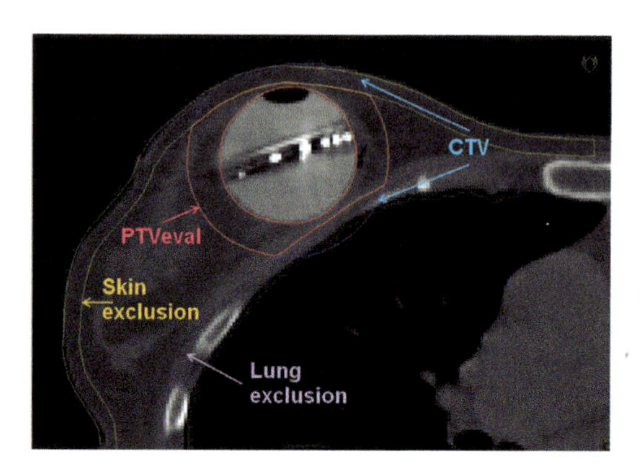

Figure 1–9 The breast lumpectomy cavity with a balloon catheter inserted (inner red contour) is enlarged by 1 cm to create a CTV pointed to by the blue arrows encompassing the likely extension of microscopic breast cancer. The final CTV, delimited by the inner and our red contours, is reduced by a Boolean subtraction of volumes, sparing the ribs and lung and a 5 mm thick skin volume.

1.4 Treatment Planning

1.4.1 Elements of Planning

1.4.1.1 General Type of Implant

Implants can be characterized as surface applications, intracavitary or intraluminal insertions, or interstitial implants, depending on the approach to the source distribution. Surface brachytherapy is used to treat skin cancer or other superficial lesions, either through the application of a custom mold or with specialized skin applicators, such as Leipzig or Valencia applicators. An example of a custom mold applicator used to treat the sole of a patient's foot is shown in Figure 1–10.

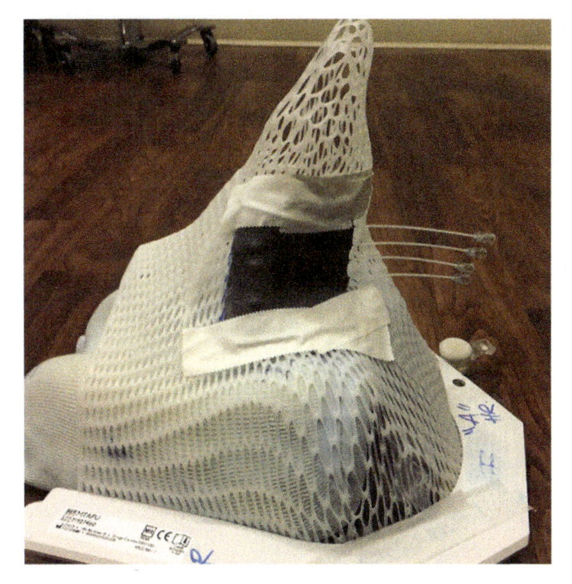

Figure 1–10 A custom mold applicator for surface brachytherapy to the sole of a patient's foot.

Examples of intracavitary brachytherapy include treatment to the vagina, vaginal cuff, and cervix, or balloon applicators for breast brachytherapy. In these treatments, the applicator is placed in an existing or created anatomical cavity. Intraluminal treatments are a special form of intracavitary brachytherapy where the cavity is a narrow tube, such as the esophagus or bronchus. These cases often use a long source train within a catheter. In interstitial implants, the sources are placed directly into the tissue, as in prostate implants, irradiation of the surgical bed after sarcoma removal, or many head and neck treatments. For large-volume disease or irregularly shaped targets, intracavitary insertions may require the introduction of supplementary needles or catheter implants throughout the treatment volume.

1.4.1.2 Directional Approach

While intracavitary brachytherapy relies on the use of anatomical structures for the placement of the treatment devices, interstitial implants require the placement of catheters or needles through the skin into the area to be treated, such as the base of the tongue or in the surgical bed after tumor removal. In designing an interstitial implant, care should be made as to the direction of the placement of the needles. The ABS has published guidelines for the treatment of sarcomas, which may also apply for other diseases treated in a similar method. (Holloway et al. 2013). In general, catheters should be placed parallel to the longest dimension of the treatment volume, reducing the number of needles required and possibly lessening tissue trauma from the punctures, as shown in Figure 1–11.

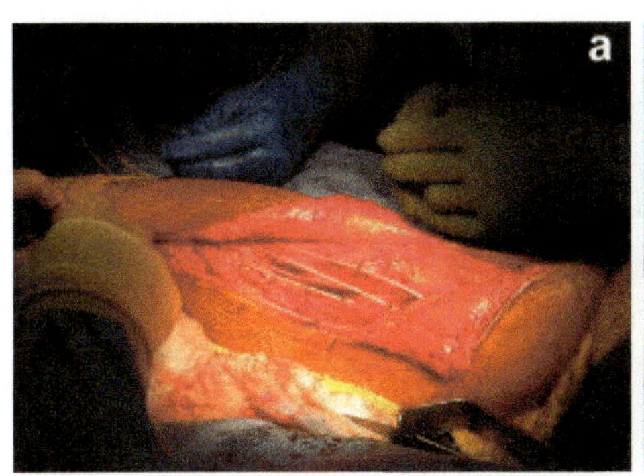

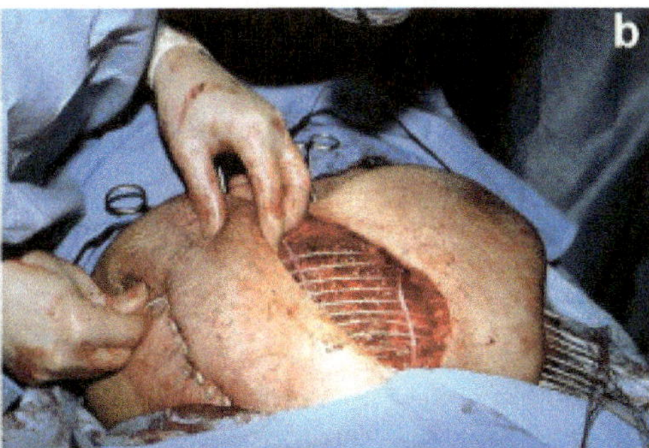

Figure 1–11 Placement of interstitial catheters a) parallel to and b) perpendicular to the longest dimension of the site to be treated. Reprinted from *Brachytherapy* 12(3), Holloway, et al., "American Brachytherapy Society (ABS) consensus statement for sarcoma brachytherapy," pp. 179–90, © 2013 with permission from Elsevier.

This often may not be applicable for various reasons, such as the location of normal tissue or neighboring structures. The location of critical structures should be evaluated prior to implant to ensure that catheters do not traverse them in order to treat the tumor. During treatment, the transfer guide tubes should not lie across critical structures and should be supported so that their weight does not affect the catheters.

1.4.2 Geometry of Implant

1.4.2.1 Format for Implant—Planar or Volume

The aim of catheter placement in brachytherapy is to allow the delivery of a homogeneous dose to the target volume, with the geometry of the implant designed to most effectively treat the tissue. Historically, ^{192}Ir implants were designed to treat sarcomas as well as breast tumors. See, for example, the work by Kwan and colleagues (1983) and the references within. If the volume of the tissue to be treated is small, a single plane of catheters can effectively treat the tissue. As the volume of treated tissue increases, a second plane of catheters can effectively treat in between the planes. For centrally located tumors, such as the prostate, the catheters or needles must be placed to treat a spheroid volume of tissue. This can most effectively be performed by placement of the needles throughout the tissue. There can be competition between the use of planar or volumetric approaches for the same disease site. For example, accelerated partial breast irradiation (APBI) can be delivered either via an interstitial implant using two planes of catheters or through an implanted balloon applicator that subsequently treats the volume of tissue around it (Shah et al. 2011). The approach that is used can depend on the experience of the group, the vagaries of the disease, or the wishes of the patient. For example, an interstitial implant may be more practical for a patient with breast augmentation as opposed to a balloon-based treatment (Kuske and Patel 2007).

1.4.2.2 Guidance for Needle Placement—Systems

The historical systems for guiding needle catheter placement for HDR brachytherapy can be a useful starting point in planning. While the historic systems may not be "used" in the literal sense in the modern, image-guided practice of brachytherapy, they can be used to guide source placement, to prevent a large deviation of the implant from what would be considered normal, and to aid in quality assurance (QA) of the treatment. Both the Manchester system (also known as the Patterson–Parker system) and the Quimby system were developed for planning planar implants with radium, and later with other high-energy radionuclide sources. The Manchester system used a peripheral loading technique, which led to a dose throughout most of the target volume within 10% of the prescribed dose. The Quimby system assumed that all sources were of equal activity, causing the central dose to be much higher than the peripheral dose. The Quimby system informed the guidelines that were later developed by Kwan et al. (1983) and Zwicker et al. (1995). The system that Zwicker developed for LDR implants was readily converted to a system for guiding HDR brachytherapy, with thickness of the volume to be implanted leading to guidance on the distance between the planes of catheters to treat the target effectively (Zwicker et al. 1999). Shown in Figures 1–12 are the isodose curves for a single plan implant with equal loading (left) and peripheral weighting (right). A single plan implant with equal dwell loading shows cool spots at the end of each catheter, whereas these low-dose regions can be remedied by increasing the dwell times at the ends of each catheter.

In a two-plane implant, shown in Figure 1–13, a uniform loading technique (left) shows higher dose in the center of the implant than a peripheral loading technique (center). Adding needles at the periphery of the implant (right) improves the dose coverage between the planes at the edge of the implant.

A modern example of these two systems would be a prostate seed implant using a peripheral loading technique versus that with an equal spacing technique. A disadvantage of the uniform loading technique is the creation of unacceptable hot spots within the implant. For computer-based treatment planning of HDR brachytherapy treatments, inverse planning techniques—in which the dose distribution is optimized to meet

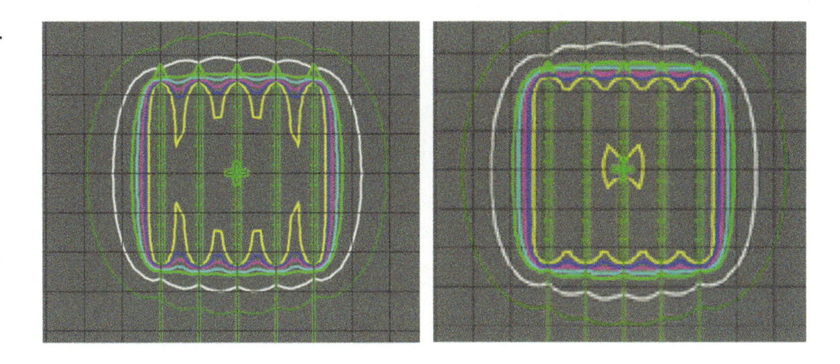

Figure 1–12 Isodose curves for a single-plane implant with equal (left) and peripheral (right) loading scenarios.

Figure 1–13 Isodose curves for a two-plane implant with equal (left) and peripheral-weighted (middle) loading scenarios. Two additional needles are located mid-plane (right) to decrease the region that is bowed in.

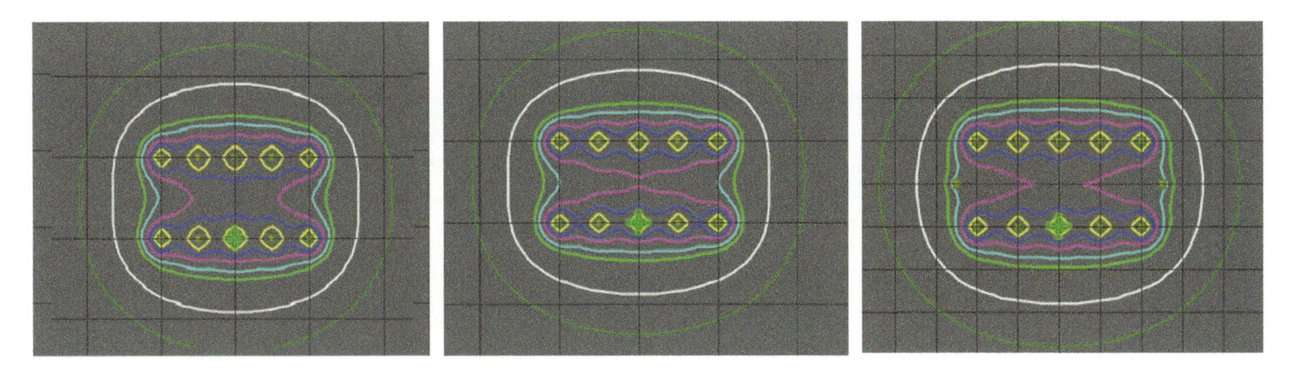

dosimetric goals in terms of tumor coverage and critical structure dose limits—will generally lead to a peripherally weighted dwell time distribution, analogous to the Manchester method.

1.4.3 Templates

Templates can be used to guide the placement of catheters or needles into the patient by providing a fixed geometry at the insertion point. Based on American Brachytherapy Society (ABS) consensus guidelines, a template should be used for the interstitial treatment of apical vaginal tumors, usually with a vaginal dilator in place (Beriwal et al 2012). Templates are also used for interstitial placement of needles or catheters for breast and prostate treatments. Templates have holes a fixed distance apart, usually 1 cm, to aid in guiding

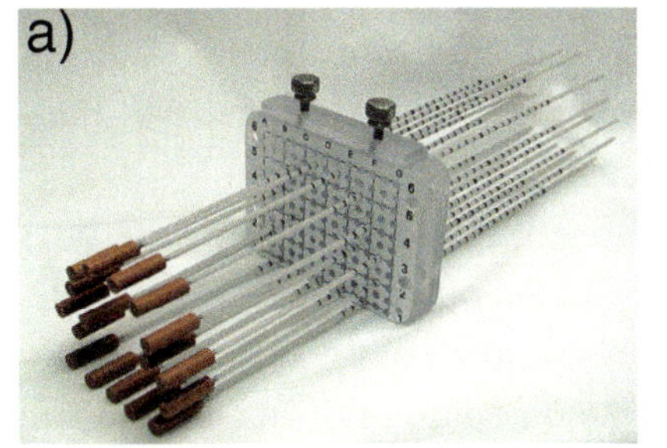

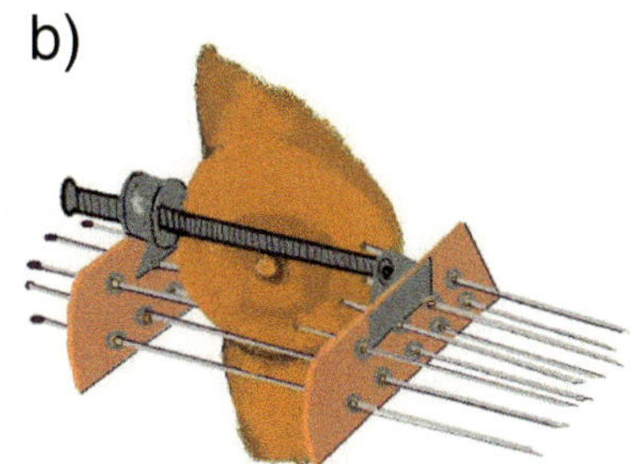

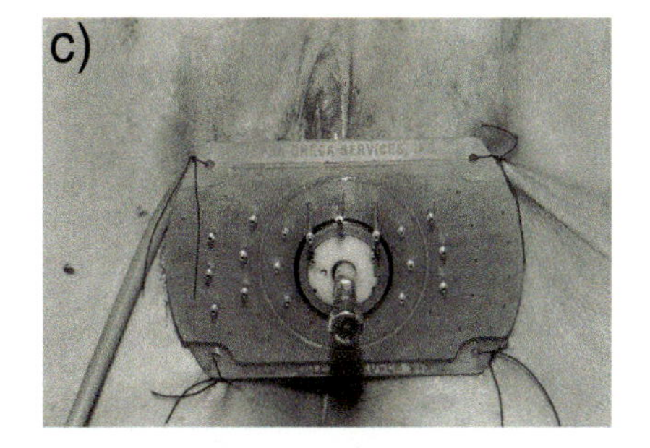

Fig 1–14 Template samples. a) A prostate template. b) Example of a breast brachytherapy template. c) A Syed template. [a) reprinted from *Brachytherapy* 9(3) Slessinger, "Practical considerations for prostate HDR brachytherapy," pp. 282–87, © 2010 with permission from Elsevier. c) reprinted from *Gynecol. Oncol.* 73(1) Paley et al., "A New Technique for Performing Syed Template Interstitial Implants for Anterior Vaginal Tumors Using an Open Retropubic Approach," pp. 121–25 © 1999 with permission from Elsevier.]

placement into the tumor. The regular hole spacing allows coverage of the treatment volume while avoiding high- or low-dose regions that could come from having the spacing too close together or too far apart.

Templates are typically used during prostate, gynecological, and breast interstitial brachytherapy. Figure 1–14 shows several different commercially available templates. The templates can either remain in place for the duration of the treatment, such as for prostate or gynecological treatments, or be removed after aiding the catheter placement, as with breast treatments.

The template needs to be compatible with the imaging technique that is used for simulation. Either CT- or MR-compatible templates should be used. With the advent of 3D printing technology, custom templates can be readily designed. These can have features like MR compatibility, which is not available in commercially available templates, although custom machined templates have been used for more than two decades (Lindegaard et al. 2016; Ritter et al. 1989).

For multi-fraction treatments, the use of a template can lead to other issues with the patient. Irritation can come from the suturing of the template in place or the pressure of the patient's skin against the template (for example, the thigh against a perineal template). Movement of the patient can cause the entire template or individual needles to shift, necessitating imaging prior to treatment to confirm needle positions. The use of templates is further discussed in Section 1.6.4.

1.4.4 Dose Coverage

Consensus guidelines for dose coverage of the treatment volume have been developed by both the ABS and Groupe Europeen de Curietherapie-European Society for Therapeutic Radiology and Oncology (GEC-ESTRO). For example, the ABS guidelines for cervical and interstitial vaginal brachytherapy state that a $D_{90\%}$ of greater than 100% of the prescribed dose should be the planning goal, keeping in mind the critical structure doses (Beriwal et al. 2012; Viswanathan et al. 2012). In pelvic irradiation, critical structures include the urethra, bladder, rectum, sigmoid, and bowel. The consensus guidelines state that the D_{2cm^3}, and $D_{0.1cm^3}$ of the critical structures should be tracked and used in determining trade-offs between CTV coverage and critical structure doses (Viswanathan et al. 2012). The dose tracking spreadsheets posted on the ABS website can be used to calculate the EQD2 of the CTV and critical structures to ensure that the tumor coverage is adequate, while keeping the critical structure doses within their constraints.

Tracking the doses for each treatment and updating the total EQD2 after each fraction is delivered allows knowledge of issues that may arise later in the treatment course. For example, for fractionated treatment with multiple applicator placements, this could mean that an individual treatment may have a higher-than-desired critical structure dose as long as the total dose for the treatment course is below the limit.

Various treatment types have customized dosimetric goals and constraints. For example, interstitial breast brachytherapy has different guidelines than balloon-based treatments. For interstitial breast brachytherapy, the CTV $D_{90\%}$ should be at least 90% of the prescribed dose, while the $V_{150\%}$ and $V_{200\%}$ of the CTV should be less than 70 cm^3 and 20 cm^3, respectively. For balloon-based treatments, these criteria are 50 cm^3 and 10 cm^3 (Shah et al. 2013). These parameters can be reviewed and the plan modified if the dosimetric constraints are not met. For interstitial breast implants, the maximum skin point dose should be $<100\%$ of the prescribed dose, which is less than that allowed for intracavitary, balloon-based implants, in which the maximum skin dose is usually required to be less than 145% of the prescribed dose. Additionally, new dosimetric parameters can be readily added to existing treatment reporting systems. For example, it has been suggested that the $D_{0.2\ cm^3}$ of an inner 2-mm rind of skin is more relevant and consistent than the maximum skin dose in breast brachytherapy (Hilts et al. 2015). This parameter can readily be calculated and added to any dose reporting schema, as shown in Table 1–1.

Table 1–1 Example of the dosimetric quality parameters used to evaluate a balloon-based breast brachytherapy treatment

Structure	Value	Result
CTV	$D_{99\%}$	316.67 cGy (99.23% of Rx)
CTV	$D_{95\%}$	331.10 cGy (97.38% of Rx)
CTV	$D_{90\%}$	343.95 cGy (101.16% of Rx)
CTV	$D_{10\%}$	608.66 cGy (179.02% of Rx)
CTV	$D_{100\%}$	125.57 cm^3 (91.65% of volume)
CTV	$D_{150\%}$	38.18 cm^3 (27.86% of volume)
CTV	$D_{200\%}$	7.15 cm^3 (5.22% of volume)
Inner skin dose	$D_{0.2\ cm^3}$	313.34 cGy (92.16% of Rx)

1.5 Image Guidance for Catheter Placement and Reconstruction

1.5.1 Introduction

The first priority for any patient procedure should be patient safety. Toward that end, a fully stocked brachytherapy procedure suite should have ready access to any and all equipment that would be needed for catheter placement, including any needed tools, anesthesia cart, sutures, and other medical equipment. While CT and US guidance have been common in brachytherapy for many years, only recently has MR become readily available within many radiation oncology departments. Many imaging suites are focused on the safe, efficient, and accurate acquisition of the images. In those cases, there may not be room to cross utilize these suites as surgical suites. Dedicated brachytherapy suites with MR scanners (Jaffray et al. 2014), CT-on-rails (Orcutt et al. 2014), CBCT (Al-Halabi et al. 2010)—which allow the placement of applicators in the same location as the imaging system and treatment—are becoming more common, but are still a rarity. The optimal work flow for applicator placement would allow imaging during the procedure, as discussed for MR-guided HDR ^{192}Ir prostate brachytherapy in Chapter 5, with a final treatment planning scan at the completion of the procedure. A more likely scenario, especially in a community environment, would be the placement of the applicator in a separate location, such as an operating room, with the patient then being moved for imaging or treatment. This could lead to a sub-optimal implant and sub-optimal treatment plan—most patients would not be moved back to the procedure room for implantation of additional needles should the imaging indicate such a need. General discussions regarding imaging methods as applied to brachytherapy treatments will be given; the reader is referred to specific chapters for site-specific information and further details on each site.

1.5.2 CT Guidance

The widespread prevalence of CT simulation suites in radiation oncology departments makes a logical argument for their cross-utilization as a brachytherapy procedure suite. However, due to the length of brachytherapy procedures, a dedicated CT or CBCT may be required. The goal of CT-guided catheter placement is the accurate visualization and placement of the treatment applicator within the patient. One way to achieve this is with frequent scans during the implant procedure, while avoiding excess scan volumes and unnecessary patient dose. Small fields of view and large slice thickness would minimize dose, but these may also not provide sufficient detail to judge the quality of the physical implant. Large fields of view and small slice thicknesses would give the best overall impression of the physical implant, but these are probably unneeded until the final planning scan. A reasonable compromise is to limit the number scans during placement and have these scans consist of a few narrow slices within the principle target volume and then subsequently obtain a more complete scan for planning.

Optimum patient positioning during catheter insertion is important. For gynecological procedures (interstitial needles or intracavitary applicators), the patient is usually placed in the dorsal lithotomy position for better visualization and access during catheter placement, but then returned to a more relaxed supine position for final imaging and subsequent treatments (most imaging equipment would not allow for the patient to pass through the bore in dorsal lithotomy). For interstitial or intracavitary breast procedures, the patient may be placed on a breast simulation board to allow better access to entrance and exit sites for each needle or applicator. Catheter placement for head and neck treatments is more often performed by physical inspection and adjustments in an operating room rather than in a CT simulator room. If done under CT guidance in the simulation suite, adjustments and access to the needle entrance and exit sites may be readily performed between scans. The planning scans must be acquired in the treatment position so that the final brachytherapy treatment plan can reflect the patient's geometry accurately at the time of treatment.

Since the routine implementation of CT simulators in radiotherapy clinics, they have been used for image guidance in brachytherapy for a multitude of treatment sites. In addition, the use of CT-based planning has allowed the transition from classical implant systems to patient-specific planning, optimized to deliver the prescribed dose to the target volume while sparing normal tissues. A variety of locations and uses will be described.

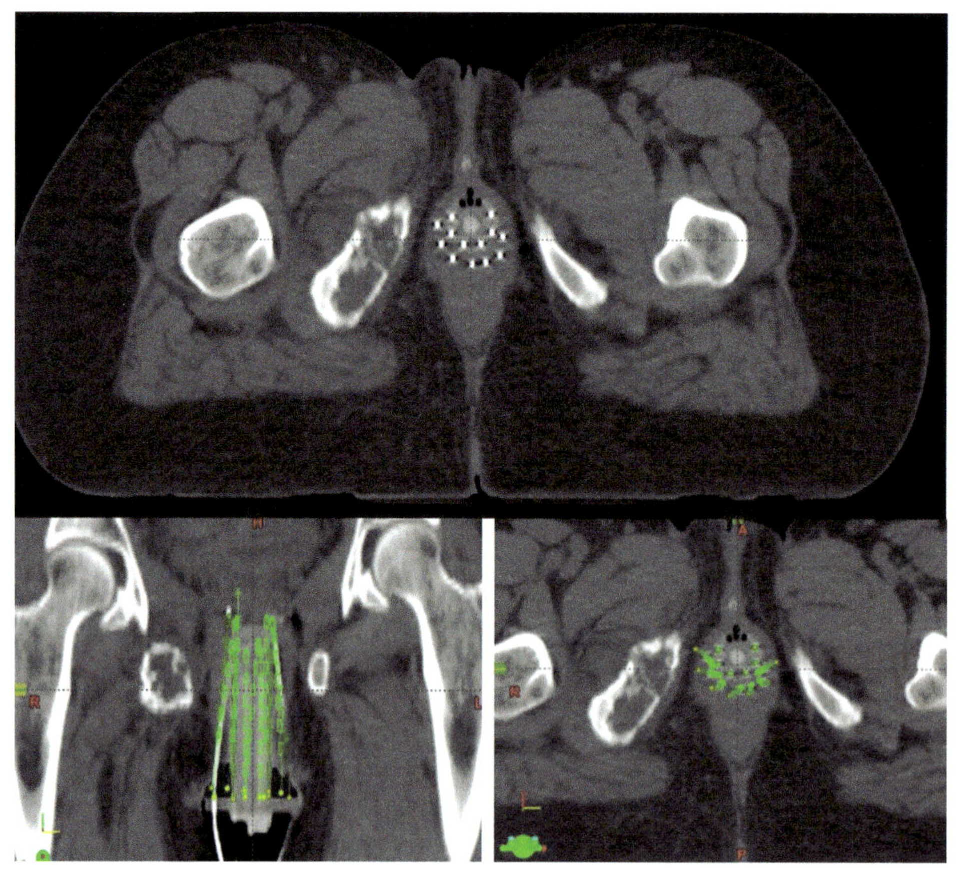

Figure 1–15 Interstitial gynecological implant using a cylinder with six needles placed around a vaginal cylinder and seven interstitial needles in a Syed-style template. The plastic catheters have metal stylets for visualization in CT.

1.5.2.1 Gynecological Sites

CT guidance in gynecological cases has a solid history of successful use, both with the use of standardized templates, such as the Syed template, and free-hand techniques (Dyk et al. 2015). Using CT guidance allows more accurate dosimetry to be performed, with the relationships between the dose and the patient's anatomy to be calculated explicitly (Erickson et al. 1996). Figure 1–15 shows an HDR interstitial implant performed under CT guidance.

1.5.2.2 Head and Neck Sites

For head and neck treatment, the proximity of bone to the target can lead to less-than-ideal catheter placement. Using some imaging equipment during the implantation allows for adjustments during the procedure before leaving the operating room (Figure 1–16).

1.5.2.3 Multi-catheter, Interstitial Breast Implant

A successful interstitial breast implant requires the ability to place the catheters and to be able to visualize them accurately within the acquired CT images. During catheter placement, it is only necessary to visualize that a catheter is placed appropriately with respect to the internal anatomy. This would include covering the CTV plus the desired margin, creating a PTV for implantation. Ultrasound can be used to visualize the cavity, or the placement can be performed under CT guidance. Needles should be angled away from the ribs and lungs. Partial visualization along the length of the catheters is preferable to across the small diameter of the catheters because one can see a longer length of the catheter and its placement relative to the regions of concern, such as the ribs and skin in Figure 1–17. This aids in judging the quality of the implant. The original

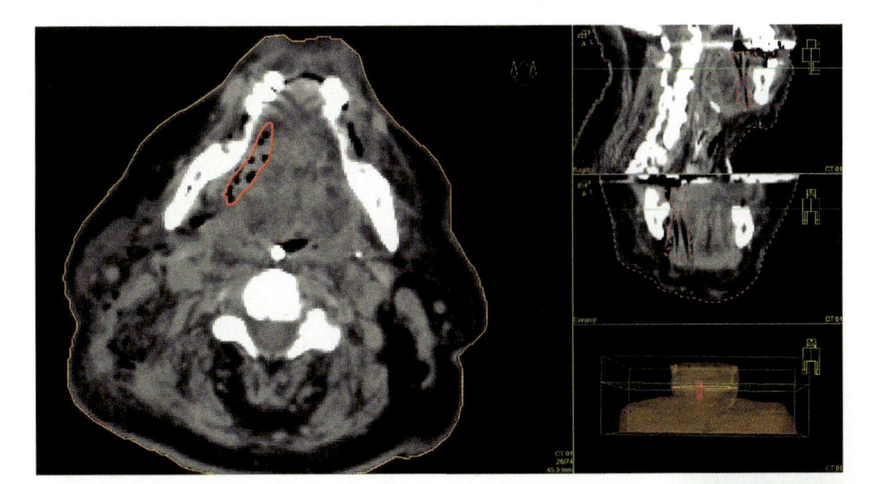

Figure 1–16 Transverse, sagittal, and coronal views of a head and neck implant.

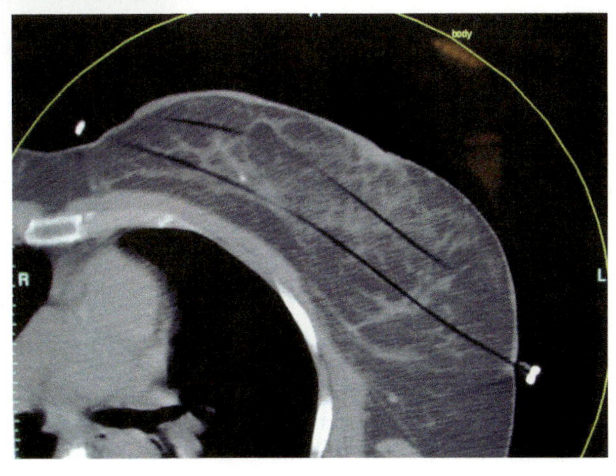

Figure 1–17 CT acquisition showing potential difficulty in catheter delineation.

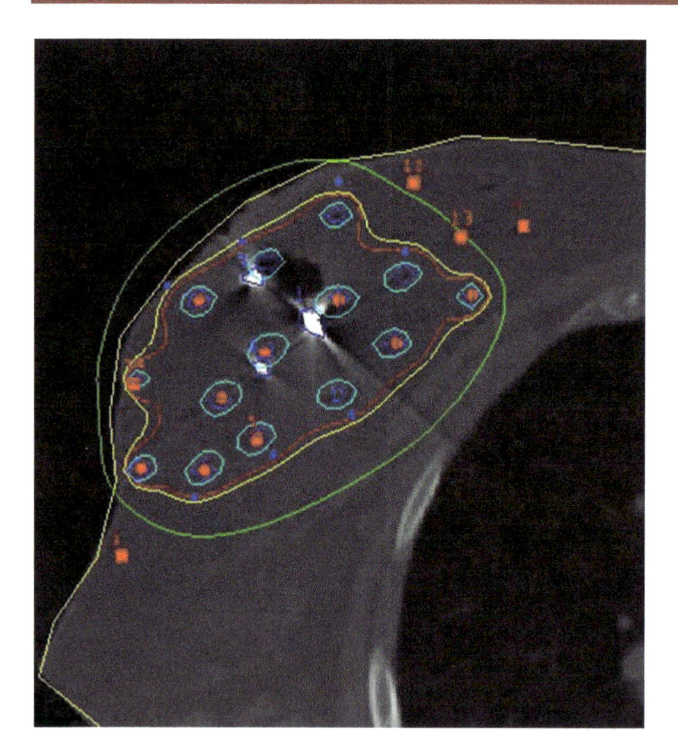

Figure 1–18 CT acquisition acquired in a plane transverse to the long axis of the catheters for ease in catheter reconstruction in treatment planning.

implantation of CT into brachytherapy required the scan direction to be transverse to the catheters, allowing visualization across the width of the catheters rather than the possibility of only partial visualization along their lengths, as shown in Figure 1–17. However, the current practice within commercial TPSs allows rotation of the imaging plans so that the catheters can be properly viewed regardless of their orientation with respect to the scanning plane. By changing the viewing plane, each catheter can be visualized and digitized into the TPS.

Figure 1–18 shows a different implant where the catheters were imaged in a different orientation, transverse to the catheters. Here, the individual catheters can be seen in cross section and are easier to reconstruct within the planning system.

An ideal implant from a planning reconstruction point of view has the catheters directly along the cephalad–caudal axis with transverse images. Often, an ideal implant from a surgical placement point of view would follow a purely medial–lateral needle placement to better judge target coverage and rib proximity. The optimal implant may be a compromise between these two approaches, as illustrated by the left image in Figure 1–19. The needles and catheters are placed at a slight angle to the medial–lateral access to allow for better imaging of the catheter cross sections during reconstruction and planning, as shown in the right panel of Figure 1–19.

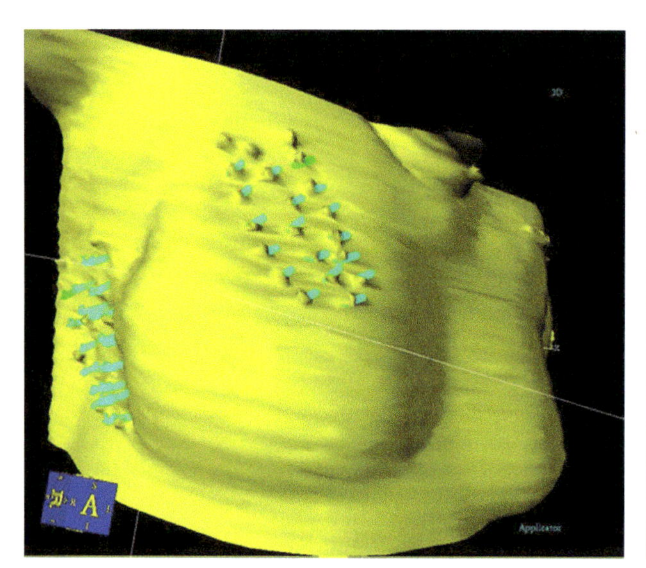

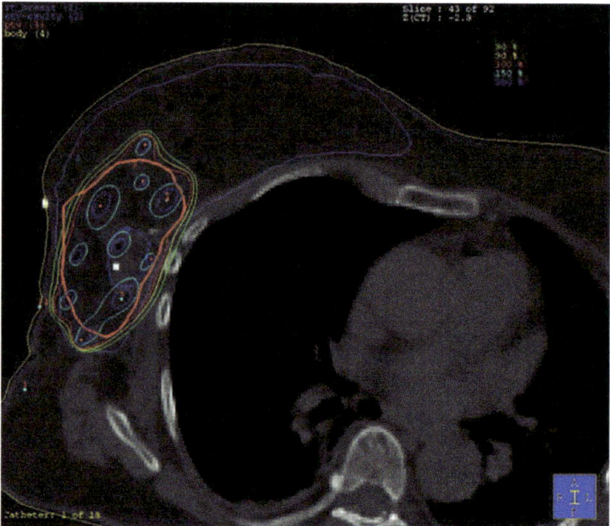

Figure 1–19 Reconstruction of the CT-based implant and subsequent dose distribution.

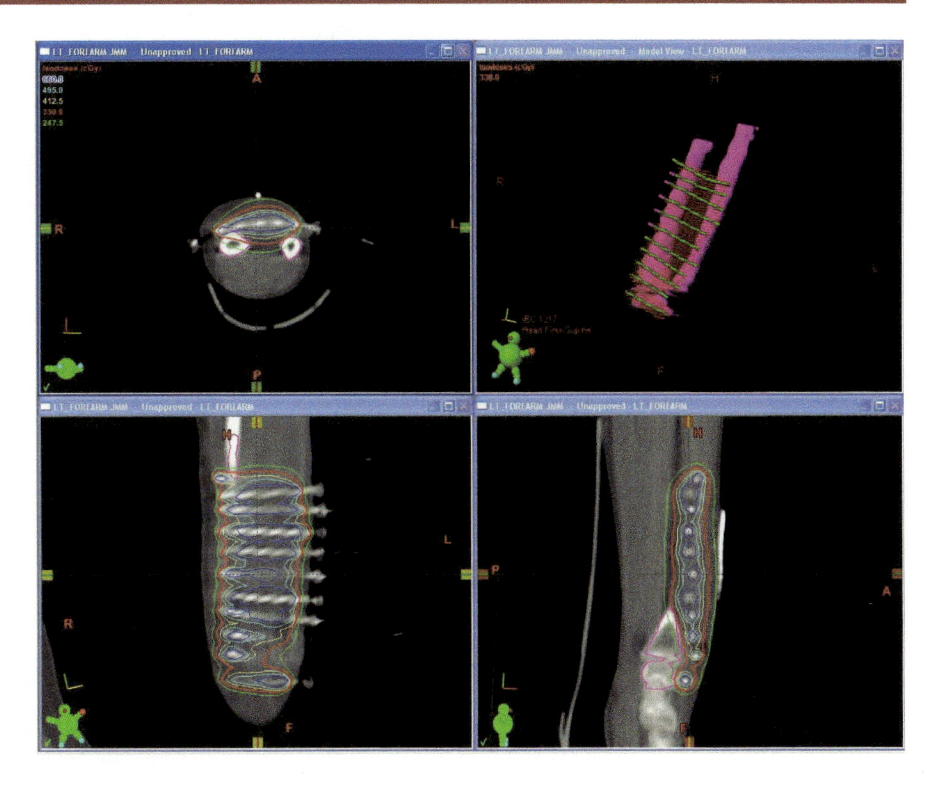

Figure 1.20 Images of a CT guide treatment for a forearm sarcoma.

1.5.2.4 Sarcoma

The registration of CT and MR image sets has been used for the planning the treatment of soft tissue sarcomas in a variety of body sites. As recommended by the ABS (Nag et al. 2001), MR provides excellent soft tissue contrast and allows for the determination of the clinical target volume prior to surgery. The catheters will then be placed in the OR after excision, and CT used for brachytherapy treatment planning and applicator localization (Figure 1–20). Particularly in the treatment of extremities, the catheters are placed perpendicular to the length of the disease. Usually a single, 1-cm spaced planar implant is sufficient to cover the CTV. With the advent of HDR rather than LDR, the use of cross-ends is no longer necessary, as manual or computerized optimization can ensure target coverage at the edges of the planes. Care must also be taken to minimize dose to the normal skin, such as not placing any dwell positions within 5 mm of the surface (Holloway 2013).

1.5.2.5 Prostate

CT-based planning for prostate brachytherapy is well described in Chapter 5.6 of this book. That section can be consulted for a detailed discussion. Briefly, the needle placement is usually fairly routine and based on the size of the prostate; it is the optimization of the dwell times that takes the most effort in creating an ideal plan. Obviously, good geometry of the needles helps the planner more easily meet the planning objectives; however, unlike in LDR brachytherapy, there are chances for partial compensation for a less-than-ideal implant. The target volume can be identified through a variety of imaging modalities, including US, CT, and MR. Usually inverse planning based on objectives placed into the planning system is used to create the dwell times for each needle.

1.5.2.6 Inoperable Endometrial Cancer

The primary treatment for endometrial cancer is a hysterectomy, with the possible use of vaginal cuff brachytherapy depending on the stage of the initial disease. For patients who present as medically inoperable due to other comorbidities, intracavitary brachytherapy can be performed by placement of a Y applicator into

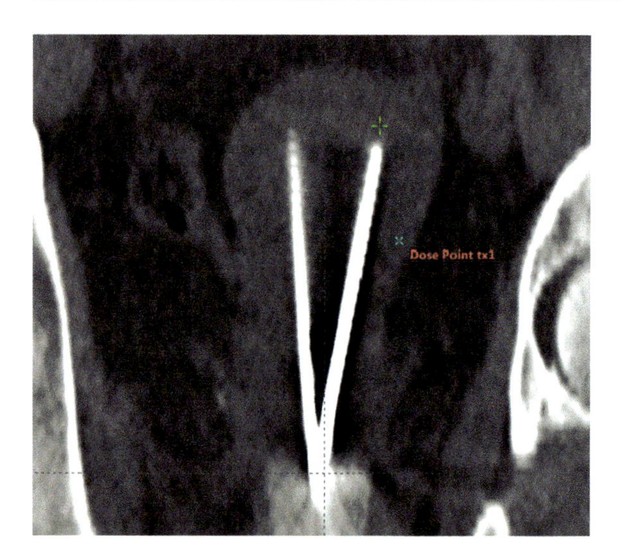

Figure 1–21 Coronal CT image of a Y applicator within the uterus.

the uterus (Figure 1–21). This can be accomplished either with or without a central uterine tandem. CT guidance will show whether the applicator has perforated the uterine wall. In these cases, there is a choice as to whether to remove the applicator for replacement or to set the dwell times at the tips of the applicator to zero to ensure that normal tissue is not unnecessarily dosed.

1.5.3 MR Guidance

1.5.3.1 Introduction

The use of MR for brachytherapy has expanded greatly in the last several years, and it is discussed at great length in chapters 5 (HDR prostate brachytherapy), and 6 (gynecological brachytherapy) of this book. What follows is a short discussion of the use of MR for various treatment sites.

1.5.3.3 Interstitial Sarcoma

The use of MR for guiding interstitial brachytherapy for soft tissue sarcomas is most common for the evaluation of the tumor prior to implant, or registered to CT images after implantation (Krempien et al. 2003). Similar techniques would be used as for prostate or cervical imaging.

1.5.3.3 Prostate

MR guidance for HDR prostate brachytherapy is extensively discussed is Chapter 5, Section 4. Briefly, the procedure is performed either in a brachytherapy suite in which the MR is present as an imaging device, or the catheters are placed under MR guidance and the patient is moved to the brachytherapy suite for treatment. The prostate is scanned using T2-weighted technique, although multiparametric scanning is also performed.

1.5.3.4 Cervix

A detailed discussion of the use of MR for the treatment of cervical cancer is presented in Chapter 6 of this book. The following is a brief overview of the topics presented in that chapter. MR-guided implantation of the cervix has reached critical mass in the past few years. While MR guidance for every fraction is still a work in progress for most institutions in the United States, many clinics have begun using MR for at least the first fraction of the patient's treatment, i.e., for target delineation. In Europe, MR has been routine for quite some time. The Vienna group has made numerous reports of their success with MR guidance for cervical cancer brachytherapy, including the use of the Vienna applicator, which is an interstitial/intracavitary hybrid applicator for treatment of bulky parametrial disease (Dimopoulos et al. 2006).

The development of CT/MR-compatible applicators has allowed for hybrid imaging to be used for patients undergoing tandem and ovoid or tandem and ring brachytherapy. A patient might receive an MR-based treatment plan for their first fraction, and then receive CT-based planning in their subsequent ones. Various techniques can be used, including the use of T1, T2, and proton density-weighted MR sequences for optimum imaging of both the applicators and the normal tissue (Zoberi et al. 2016; Hu et al. 2013).

The sagittal images below show the use of MR for tandem and ovoid brachytherapy. As can well be seen, the contrast between the tandem and normal tissue is sufficient to not only visualize the normal tissue and tumor regions, but to also make geometric assessments of the quality of the implant (Figure 1–22).

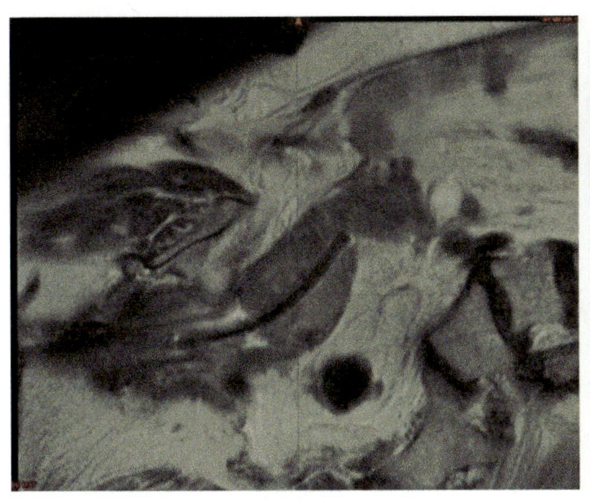
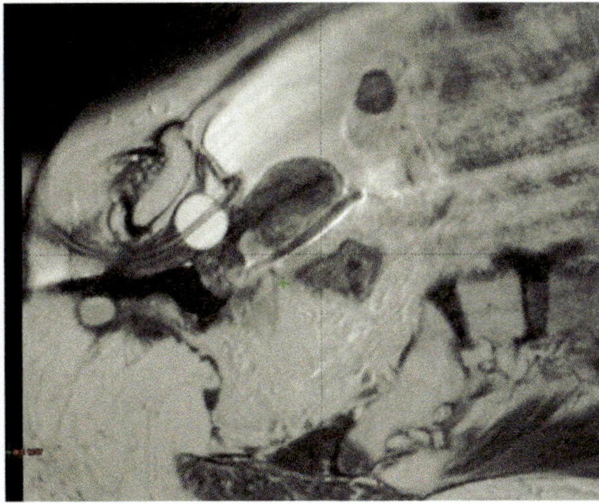

Figure 1–22 MR treatment planning images for a good implant (left) and an improper implant (right). The slight blooming of the tip of the tandem due to MR artifact is visible.

1.5.4 Ultrasound Guidance

While US guidance has been used to great effect for prostate brachytherapy, it has been limited as a treatment planning image acquisition method for cervical brachytherapy. However, the ease and relative availability of US units offer great help in the guiding of applicator and needle insertions. The use of US for this gynecological treatment is presented in more detail in Chapter 6.

1.5.4.1 Interstitial

For the guidance of interstitial needle placement, US is a safe and inexpensive method with proven results for a variety of treatment sites, including prostate and gynecological applications (Weitmann et al. 2006). It is particularly helpful with needles that are approaching the bladder and in determining the depth of needle insertion (Nag et al. 2004). HDR prostate needles may also be localized by the use of ultrasonic guidance (Lee et al. 2014). Batchelar and colleagues (2014) have described a retrospective study in which US was used to guide and plan the treatment and found the localization of the needle tips agreed between TRUS and CBCT on the order of a few millimeters. Transrectal US has been used for needle placement, with excellent visualization in both the transverse and longitudinal planes. The bladder and rectum can be directly imaged and subsequently avoided (Stock et al. 1997). LDR interstitial breast patients typically have US performed to determine eligibility, and investigations have been made into the practically of using it for treatment planning (Morton et al. 2016).

1.5.4.2 Intracavitary

Trans-abdominal US can be used to "sound" the uterus in the operating room and guide the placement of the tandem and ovoid applicators. Upon initiation of this technique, Davidson reported a reduction in the incidence of intrauterine perforation from 10% to 3% (Davidson et al. 2008). Not only were there no uterine perforations in 34 out of 35 insertions, the insertion time itself decreased from 34 minutes to 26 minutes, and the angle of the tandem was changed in almost 50% of cases. This can improve the quality of the implant and the quality of the patient experience since she will not be in the operating room as long. Ultrasound may also be used to help guide the percutaneous placement of breast applicators (Zannis et al. 2003).

1.6 Interstitial Equipment and Related Quality Assurance

Quality management often overlooks interstitial brachytherapy equipment, likely because it seems so simple and its operation so obvious. While the materials are simple, failures can be disastrous, and the quality checks are simple. A more detailed discussion can be found in Thomadsen (1999).

1.6.1 Needles, Catheters, and Accessories

Needles have several important properties, but they are usually taken for granted. To a great extent, the quality of implant needles is of such consistent quality that specific testing is not warranted, although users should be conscious of each of the properties while preparing for implantation.

1.6.1.1 Straightness

Executing the planned implant requires straight needles that are rigid enough to penetrate tissue without bending. Even with care, needles sometimes get bent during insertion, particularly during passage through dense tumors. Bent needles produce a force v in the direction perpendicular to the direction of insertion, increasing the probability of deviation from a straight path. Bent needles should be removed and discarded.

1.6.1.2 Integrity

For needles that carry sources trains, the needles must be closed to body fluids. While not an issue for steel needles, plastic needles sometime break or split during insertion. Often there is no way to detect ahead when a plastic needle will fail, so some quick checks before insertion may allow elimination of suspect needles:

- Check for cracks or nicks.
- Check the tip for sharpness and integrity.
- Check the gap between the obturator and the tip against a light.

1.6.1.3 Sharpness

Needles that will be replaced by a catheter should have very sharp tips to facilitate straight passage. Very sharp needles carry with them an increased probability of cutting through an artery, and where this is a potential hazard, duller needles prove a safer alternative. Very sharp needles produce intolerable pain in patients where the tips of the needles remain in the patient, such as gynecological template cases.

1.6.1.4 Bevel

Conical-pointed needles drive straighter than beveled needles (Meltsner 2007). For passing catheters through the needle, an open, flat-ended needle with a sharp conical obturator (stylet) works well. For beveled needles, larger degrees of bevel usually result in sharper cutting edges. The bevel of the needle tip results in a component of force perpendicular to the direction of insertion (see Figure 1–23) that pushes the needle away from the intended track, with the cross-needle force increasing with bevel angle. However, with large angles, the

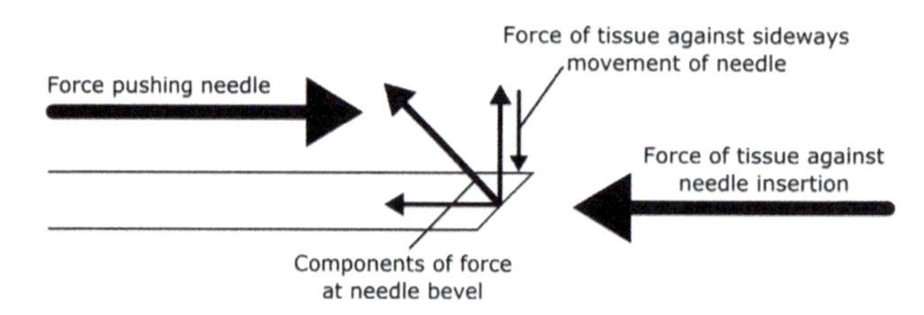

Figure 1–23 Forces on a needle tip passing through tissue. The bevel translates the force of the tissue opposing the passage of the needle into components along the needle (resistance) and perpendicular to the needle (deviation). The tissue also produces a force counter to the sideways movement of the needle. Figure used with permission (Thomadsen 1999).

increased cutting power of the tip reduces the effect of this force. While the conical tips drive straighter, some practitioners prefer the beveled needles, feeling that they provide the ability to steer the needle using the cross-needle force.

1.6.1.5 Diameter

Assuming that the inner diameter of the implant has been ordered to be compatible with either the part of a catheter that must pass or a ribbon containing sources, the main concern about diameter relates to the outer diameter and the holes in any template through which the needles must pass. This will be discussed below under template quality.

1.6.1.6 Length

Needle length only becomes important if the needle will be holding LDR sources or providing the path for an HDR source. The inner length controls the positioning of the source. Control of the implant during needle insertion most often uses some imaging to monitor the position of the tips of the needles. Thus, the difference between the needle tips and the deepest point in the needle to which the source can pass becomes critical for dosimetric quality.

1.6.2 Catheters

The quality of interstitial catheters has become very consistent. That being said, at the time of implant the integrity of catheters should be checked visually for any cracks, bends, or divots that may indicate a lack of integrity.

With most catheters, the length to the first possible source position is measured as part of the localization procedure. The exception is with the OncoSmart Catheter System™ (Elekta AB, Sweden) for use with HDR implants. This system uses double catheters: an outer catheter that functions as a normal implant catheter and an inner catheter insert that provides the source pathway. The button that fixes the outer catheter on the open end docks with clips on the inner catheter. The clip on the inner catheter is fixed at the appropriate length to place the tip at the end of the outer catheter. While the outer catheters are cut to arbitrary lengths to fit the patient, all of the inner catheters should have the same length. Quality control for the inner catheters entails ensuring that the outer lengths match the standard length, which requires only a normal ruler.

1.6.3 Buttons

Buttons serve to hold interstitial catheters in place. To do this, they must slide snuggly enough over the catheters to prevent that catheter from slipping along its length. The buttons may be plastic or metal. The plastic buttons use friction to hold the catheter in place, although a drop of quick-drying adhesive at the junction between the catheter and button helps the fixation. Metal buttons usually are lightly crimped on the catheter. For applications where both ends of the catheter extend outside the patient, a catheter with a button permanently attached on the closed end has become common. For LDR applications, the button on the open end of the catheter often also serves to clamp the sources in place so they do not slide along the inside of the catheter. HDR treatments require free access for the source to enter and retract from the catheter unimpeded.

Fixing both ends of the catheter tightly to the patient during the implant can cause discomfort due to edema from the procedure. Often, the catheter needs to be somewhat loose during the first day, with the buttons moved to tighten the following day. For implants lasting several days, there may be a second period of edema secondary to the radiation effects at about treatment day three, requiring a loosening of the buttons a second time. This may also require replanning.

Most metal buttons purchased with catheters work without concern. Production control for the plastic buttons seems to be a bit more difficult, and some batches may be too tight or too loose. When performing a procedure, it is good to have buttons on hand for a batch with proven efficacy as back-up.

1.6.4 Templates

Templates guide the needles to their planned destination and, in some cases, hold them in place during the treatments. QA for templates used for permanent source delivery is considered in Chapter 4. Verification for templates include checking the following:

1.6.4.1 Proper Hole Placement

The hole pattern must match that used by the computer in generating the treatment plan. More detail on checking this aspect of a template for US-guided implants can be found in Chapter 4. Verification of the hole pattern forms part of the acceptance testing for a template, but a part that seldom fails. The most important part of checking the hole pattern is verifying that the pattern is the same as in the treatment planning computer. Differences between the actual pattern and that in the computer can result in erroneous dose distributions. Failures of the hole pattern become less important if the catheters and source track are entered on images rather than from a library of template patterns. Improper hole spacing can result in degradation of dose homogeneity.

1.6.4.2 Proper Hole angulation

Templates with holes angled to cover hard-to-reach locations, such as behind bones or organs, can fail to achieve the desired geometry if the angles of the holes are wrong. Again, if the position of the source track is entered from images, the calculated dose distribution should match the delivered distribution, but the placement of the dose may end up in an organ at risk.

1.6.4.3 Sufficient Guidance for the Needles

Aligning the needle in the desired position requires sufficient material thickness in the template. As per Figure 1–24, the acceptance angle for the needle becomes

$$\theta = \cos^{-1}\left\{ \frac{\frac{d}{D} + \frac{L}{D}\sqrt{1+\left(\frac{L}{D}\right)^2 + \left(\frac{d}{D}\right)^2}}{1+\left(\frac{L}{D}\right)^2} \right\} \tag{1.15}$$

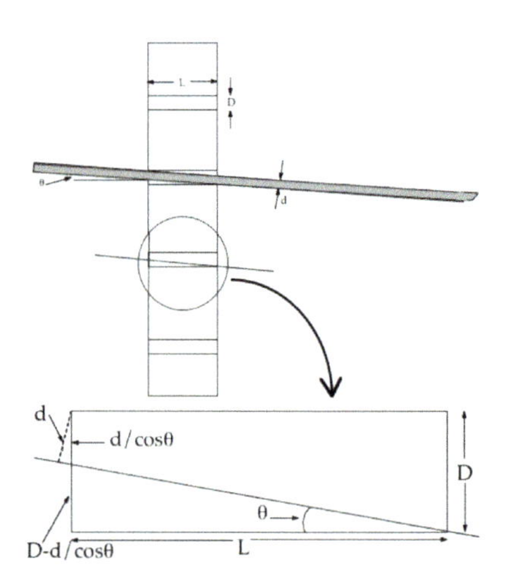

Figure 1–24 A schematic showing the acceptance angle for a needle passing through a template. Reprinted with permission from (Thomadsen 1999).

A sandwich template, with a plate on both sides of a target, serves very well to position the needles.

1.6.4.4 Sufficient Fixation for the Needles

Templates used to hold needles in place during treatment require some means to prevent slippage. Some templates use O-rings around each needle, while others fill the space between two plates with a gummy material, either of which increase the friction on the needles when the plates are squeezed, such as by screws between the two plates (Feder et al. 1978; Fleming et al. 1980; Ritter et al. 1989). Other templates use individual screws to hold the needles. Disposable templates consist entirely of soft plastic that grabs and hold needles when not lubricated. The condition of O-rings or other gripping mechanisms, as well as any screws or fasteners, should be part of checking the templates before sending them for sterilization. Even for templates that use a locking mechanism of some kind, the needles should be marked at the point they enter the template to provide visual confirmation that the needles have not moved. Additionally, a dab of quick-setting glue can be used with disposable templates to provide additional fixation to the needles, rather than just friction.

1.7 Localization and Reconstruction Issues

1.7.1 Localization and Reconstruction Issues—CT

While brachytherapy applicator placement has been considered to be the most readily visualized due to the fact that the imaging is performed at the time of placement, there have been few studies of setup and organ motion errors (Williamson et al. 2008). A main issue in the use of CT for treatment planning in brachytherapy is inter-observer differences in the determination of the volume to be treated.

With the advent of CT-compatible applicators, there is less of an issue with artifacts from high-density applicators leading the uncertainties in patient anatomy (Herron et al. 2008). Applicators can be reconstructed in the image either by direct reconstruction or by the overlay of a library applicators onto the treatment imaging scan (or by a combination of the two techniques) (Hellebust et al. 2010). It is still necessary to determine the path the source takes in the applicator during its commissioning (Kirisits et al. 2014) as well as the position of the first dwell position in relation to the end of the applicator. For proper visualization of both the applicator and the patient's anatomy, a CT slice thickness of ≤3 mm should be used (Kirisits et al. 2014). For CT-based applicator reconstruction, registration of the applicator is generally within 1 mm, which leads to small differences in the calculated dose (Kirisits et al. 2014). The use of titanium applicators with MR guidance leads to the introduction of susceptibility artifacts in the images. Because the artifacts depend on the image sequence, phantom scans should be performed with the clinical sequences. Registration be-

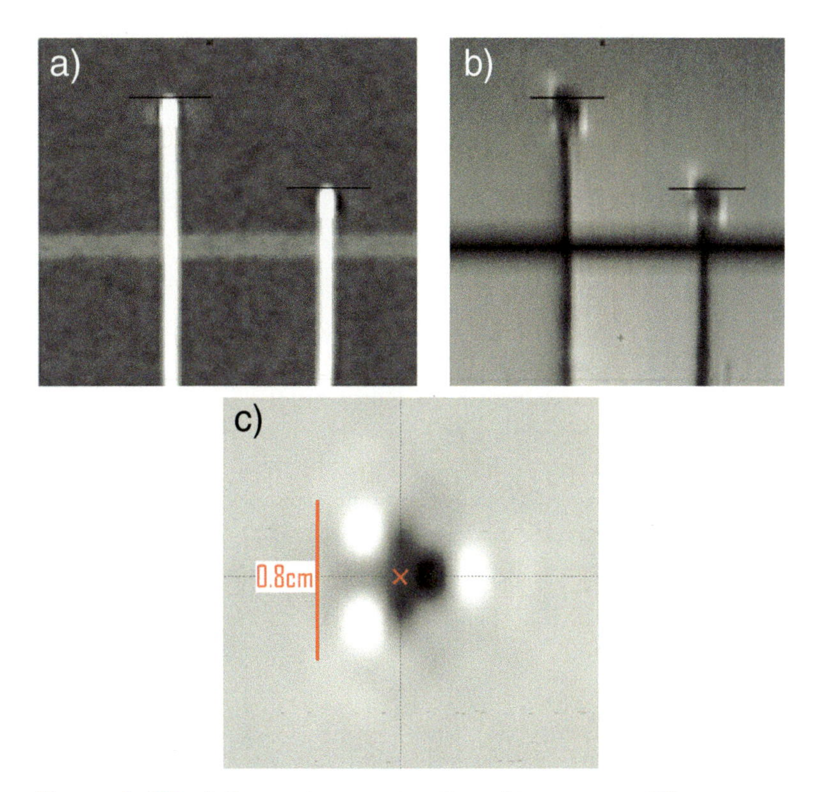

Figure 1–25 a) Coronal reconstruction of transverse CT scan and b) T2-weighted MR scan of a water phantom with titanium needles. MRI susceptibility artifacts are seen in relation to the needle tip. c) Para-transaxial alignment of the tandem applicator on MRI. Bright susceptibility artifacts are seen around the tandem. The red X marks the position of the center of the source channel. Reprinted from *Radiotherapy and Oncology* 96, T. Hellebust, C. Kirisits, D. Berger, J. Pérez-Calatayud, M. De Brabandere, A. De Leeuw, I. Dumas, R. Hudej, G. Lowe, R. Wills, K. Tanderup, "Recommendations from Gynaecological (GYN) GEC-ESTRO Working Group: Considerations and pitfalls in commissioning and applicator reconstruction in 3D image-based treatment planning of cervix cancer brachytherapy," 153–60, © 2010 with permission from Elsevier.

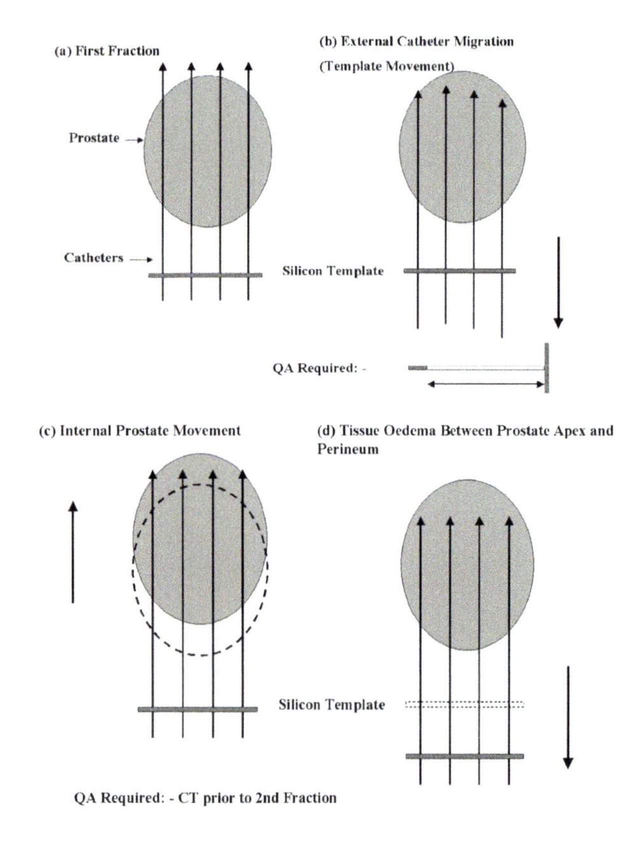

Figure 1–26 Possible sources of catheter migration. Reprinted from *Radiotherapy and Oncology* 68, P. J. Hoskin, P. J. Bownes, P. Ostler, K. Walker, L. Bryant. "High dose rate afterloading brachytherapy for prostate cancer: catheter and gland movement between fractions," pp. 285–88, © 2003 with permission from Elsevier.

tween MR and CT phantom scans can assist in the assessment of the MR image artifacts as compared to the applicator visualized on CT, as shown in Figure 1–25.

Of more concern is the migration of catheters during treatment. This migration can occur throughout a multi-fraction treatment or in the course of a single insertion during the time between imaging and treatment (Holly et al. 2011). In multi-fraction prostate brachytherapy, there have been many studies showing that needles can move up to one centimeter, even before issues with patient transport are dealt with. Shown in Figure 1–26 are various reasons for the migration of catheters during treatment (Hoskin et al. 2003). For migration of catheters more than a few mm, remediation of the plan would need to be performed or the delivered dose would not meet the planned dose (Holly et al. 2011). This remediation could consist of replanning the entire treatment or moving the catheters back to the planned position.

Another issue that can arise is the motion of the applicator during patient transfers, for example, to and from the CT couch for imaging after applicator placement. In this case, the imaged applicator geometry may not correspond to the treatment geometry (Herron, et al. 2008).

1.7.2 Localization and Reconstruction Issues—MR

MRI has become one of the standard imaging modalities for several brachytherapy indications. Treatment planning using MRI has been developed especially for gynecological indications, where it became a standard for cervical cancer brachytherapy but also prostate brachytherapy (see chapters 4, 5, and 6).

Anal canal and rectal cancer treatments present interesting challenges. The soft tissue contrast of MRI supports target definition better than CT. Tagliaferri and colleagues (2015) presented a feasibility study for anal canal brachytherapy. The increase distortions of MRI at the body periphery have to be taken into account. Unlike in gynecological or prostate brachytherapy, the target volume can be very close to the body surface, an area with often reduced geometric accuracy. This affects the correct dose reporting for the target structure, but also the organs at risk—in this case the anal canal and the portion of skin closest to the treated region. Of particular importance, the correct dwell positions have to be considered, preferably by carefully detecting the tips of the applied needles and the appropriate reconstruction of the straight (or maybe also bended) needles. The use of image registration should be considered with care, as the distortions can cause severe uncertainties. Figure 1–27 shows an example of anal cancer brachytherapy. The first issue is that applicator materials usually appear as a black region in T2-weighted images. Current templates on the body surface are not visualized directly, and it is not possible to distinguish them from the surrounding air. Thus,

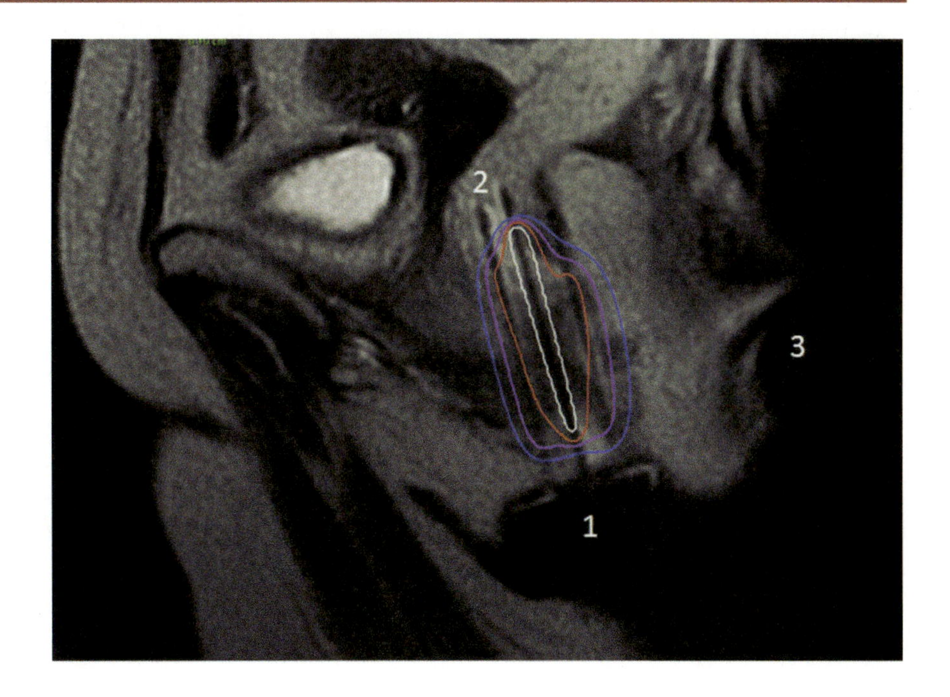

Figure 1.27 MRI-based treatment planning of anal cancer using a low magnet field strength. 1) A template outside the patient is used to guide needles. 2) The needle tips appearing as an artifact can be clearly visualized and used for reconstruction. 3) At one body surface, which is at the periphery of the scanning field, substantial distortions appear.

the reconstruction of the needles has to be based on the regions with contrast inside the patient. In this case, using a low field 0.35 T scanner, the titanium needles appear as an absence of signal. However, at the needle tip the black region is surrounded by a small white artifact, which helps to clearly visualize the needle tips. However, the appearance of this artifact needs to be studied during the commissioning of the needles with the MRI scanner in a phantom. In this case, it could be verified that the visible tip is identical to a real needle tip within ±1 mm. The example in Figure 1–27 also shows the limitations of MRI. While the accuracy could be verified with phantom checks in the center of the magnetic field, which is inside the patient, severe distortions appear at the field border, which is at the body surface. While this is a substantial problem for external beam planning, whether this effect poses a problem in brachytherapy needs assessment. In the figure, no applicator, target volume, or organ at risk is located within the distorted regions, which allows proper treatment planning.

Another new approach which benefits from MRI is brachytherapy in the management of metastatic malignant melanoma (Bretschneider et al. 2015). The use of plastic catheters and the location far from the body surface usually results in only small distortions or negative artifacts.

The reconstruction accuracy of brachytherapy applicator tips in MRI-guided liver tumor brachytherapy has been verified by Wybranski and colleagues (2015), and they reported reconstruction accuracy in agreement with the AAPM TG-56 standard. They also used Nitinol inserts to improve applicator visualization in MRI. However, this result has to be verified for higher field strengths, as their study used a 1.0 T MRI. Especially for 3T systems, most metal material that is used as MRI compatible (e.g., titanium) can cause severe artifacts.

1.8 Optimization

1.8.1 General Theory

Manual treatment planning is typically a forward-directed endeavor where choices are made to alter the dose distribution through refinements of controllable parameters. For brachytherapy, these parameters include the following:

1.8.1.1 Low Dose Rate Brachytherapy

In low-dose-rate brachytherapy, the source position is typically in 5 mm increments along the needle tracks. Uniform source strengths are most common in permanent implants. Variable source strength is sometimes used in temporary implants.

1.8.1.2 High-dose-rate Brachytherapy

In HDR brachytherapy, the source dwell positions are restricted to the catheters, needles, or applicators. The source dwell times can have a variable weight at 0.1 s time increments.

For pre-implant treatment plans, there is more freedom in selecting source positions than in post-implant treatment plans, where the goal is to reconstruct the implanted circumstances. This analysis assumes that aspects such as the choice of source model or applicator have already been decided.

Manual treatment planning includes an iterative process where adherence to planning goals is evaluated and changes are made to the source positions and weightings toward improving the plan quality according to these planning goals (Lessard and Pouliot 2001). These planning goals can be for uniform dose to a target (surface or volume), dose limits to several organs, dose limits to a defined volume, or other parameters. The problem is complex as there can be several planning goals, often directly competing with each other (Deasy 1997), and there are endless possibilities for selecting the source positions and weightings (van der Laarse and De Boer 1990) .

1.8.1.3 Optimization

An alternative to manual treatment planning is the use of optimization, first considered in the 1970s with simple programs for selecting the number and strength of radium or radium-equivalent sources for gynecological applicators and LDR interstitial ^{192}Ir implants. This was expanded in the 1990s for HDR and LDR sources (van der Laarse and De Boer 1990; Edmundson 1990; Roy et al. 1991; Sloboda 1992; Meertens et al. 1994; Kolkman-Deurloo et al. 1994; Niël et al. 1994; Thomadsen et al. 1994; Yu and Schell 1996; Anacak et al. 1997; Berns et al. 1997; Yang et al. 1998; Lahanas et al. 1999). Optimization does not seek to achieve a specific dose distribution; rather, it strives to achieve quantitative planning goals through changing the controllable parameters. In this way, planning reproducibility can be enhanced across different patients, and the plan quality can be improved for a less-experienced treatment planner.

A theoretical understanding of planning optimization can be gleaned through mathematically ascribing the quantitative planning goals. As an example, the difference in dose (D) between the prescription to the PTV (${}_{\mathrm{PTV}}D$) and points $i = \{0 \text{ to } N\}$ on the surface of the PTV can be minimized in a mathematical *objective function*, FN, toward achieving target dose conformity. The dose at points D_i will depend on the source positions and weightings, and the goal here is to alter these variables to minimize the sum of the differences for all chosen points.

$$FN = \sum_{i}^{N}(D_i - {}_{PTV}D)^2 \tag{1.16}$$

This example may be quickly complicated through consideration of the influence of organs at risk to produce a multiobjective problem (Milickovic et al. 2002). For the case of prostate brachytherapy (Lessard and Pouliot 2001), the physician may set dose limits (${}_{y}L_{x}$) to organs at risk (OARs) such as the rectum, bladder, urethra, penile bulb, etc., in addition to the desired ${}_{\mathrm{PTV}}D$. Further, the relative importance of each planning goal (w_y) will generally be different and may differ across different patients as the radiation oncologist devises a plan for a patient-specific treatment to account for individual variation in anatomy, disease extent, and personal goals.

$$FN = w_1\sum_{i}^{N}(D_i - {}_{PTV}D)^2 + w_2\sum_{j}^{N}({}_{rectum}D_j - {}_{rectum}L_j) + w_3\sum_{k}^{N}({}_{bladder}D_k - {}_{bladder}L_k) + \ldots \tag{1.17}$$

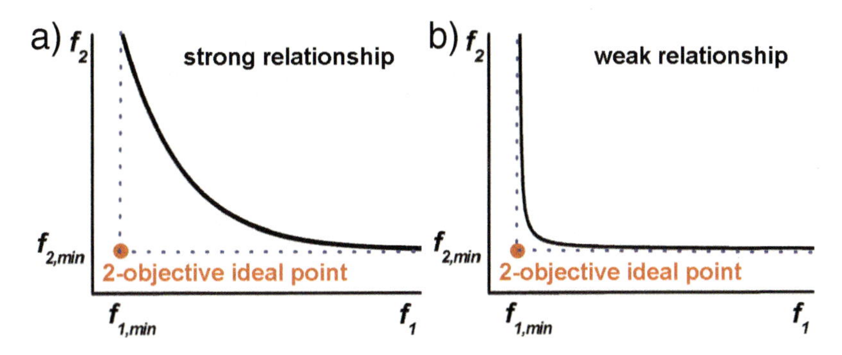

Figure 1–28 Graphical depiction of two objective functions showing a) strong and b) weak interrelationships between f_2 and f_1, where the subscript "min" indicates the minimum value for the variable *f*. Following the hyperbolic bold line shows tension between the two functions with minimization in one function resulting in an increase for the other function. The weak relationship in b) produces the result closest to the ideal point.

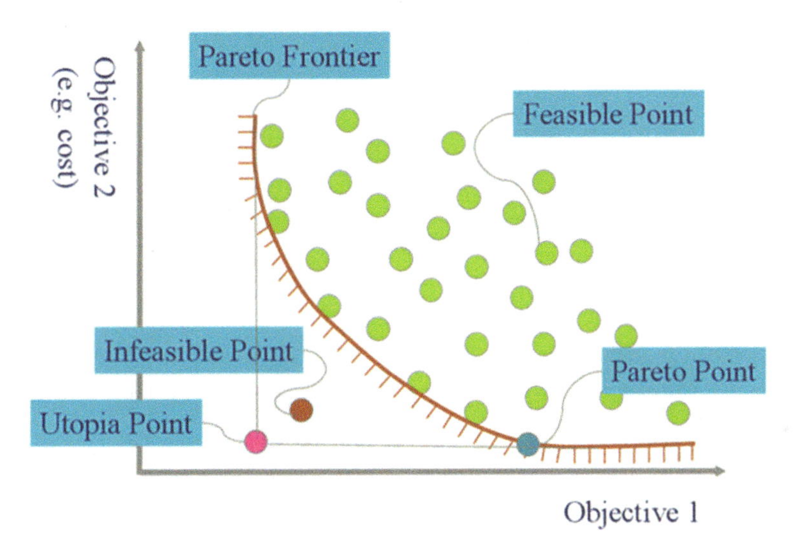

Figure 1–29 Graphical depiction of the interrelationships between possible objective 2 and objective 1. Following the hyperbolic bold line shows tension between the two functions with minimization in one function resulting in an increase for the other function. Image courtesy of www.ramweb.org.

A graphical depiction of the interrelationship of these mathematical objective functions is given in Figure 1–28. The tension between two functions is shown, where the left image depicts a strong relationship and the right image depicts a weak relationship. The circumstances with a strong relationship between the two functions can more closely approach the position of the ideal point.

A deeper analysis of this portrayal indicates that treatment planning optimization can result in numerous feasible points, with only a subset that are truly optimized. This bold line in Figure 1–29 indicates several choices of the planning parameters that achieve the objective function minimization, and is termed the Pareto front or Pareto frontier, which is named after Vilfredo Pareto, an Italian polymath who introduced the concepts of microeconomics and the power law probability distribution. This depiction is instructive as it can show that many possible solutions, i.e., feasible points, are not truly minimized objective functions. Also evident are that there can be infeasible points (such as a utopia point) which may appear desirable but is not possible given the planning constraints.

Some optimization tools are better than others for realizing a given planning goal for certain disease sites. In later sections in this chapter, specific optimization tools are explored and their applications to brachytherapy are discussed. It is important to remember that a bad implant (perhaps defined as source needle or catheter positions not adequately covering the target or not avoiding the OARs) cannot be corrected with planning optimization. Classic loading systems (e.g., Paris, Manchester, Quimby, Paterson–Parker) serve as a good foundation for setting source positions.

There is little guidance from professional societies on the quality management of computerized brachytherapy dose optimization algorithms (Nath et al. 1997; Kubo et al. 1998; Fraass et al. 1998). Brachytherapy dose optimization was covered in the 1995 and 2005 AAPM summer schools (Pouliot et al. 2005; Ezzell and Luthmann 1995) and in Baltas and Kolkman-Deurloo (Baltas and Kolkman-Deurloo 2013). There have since been substantial advancements in optimization in brachytherapy (Lapuz et al. 2013; Holm et al. 2013; Deufel et al. 2014; Dinkla et al. 2014; De Boeck et al. 2014; McGeachy et al. 2015; Cunha et al. 2016).

1.8.1.4 Work Flow

The work flow for the optimization generally follows the following steps:

1. Define an objective function.
2. Generate an initial solution.
3. Calculate the objective function value.
4. Make a change to the solution (a.k.a. an *iteration*).
5. Evaluate the effect of the change on the objective function.
6. Determine whether to keep the change or discard the change based on Step 4.
7. Repeat from 3 until the maximum number of iterations has been reached.

Step 1 is usually defined by the end user, while steps 2–7 occur behind the scenes. There are numerous approaches to implementing each of these seven steps. The details of how each are implemented define the type of optimization. Several of the more common optimization types for brachytherapy are described in the following sections: stochastic (1.8.2), binary heuristic (1.8.3), continuous analytic (1.8.4), continuous non-analytic (1.8.5), and deterministic (1.8.6).

A few items should be noted before delving into the particulars. (1) *A solution* to the optimization is, in general, not the *best* solution. The term *solution* is used very generally in optimization theory and can be thought of colloquially in the context of brachytherapy as a configuration of seed positions or dwell times that satisfies the specification of the problem within some limit. (2) Most algorithms in use in brachytherapy do not guarantee that they have found the absolute *best (a.k.a. optimal)* solution. Often this guarantee is academic, however, since real dosimetric difference between near optimal solutions and the optimal solution is often indiscernible in the dosimetric measures of the plan quality (e.g., $V_{100\%}$). This effect was demonstrated in the work of Alterovits and colleagues (2006). (3) An *iteration* is generally considered to be one calculation of the objective function for one source configuration. Each execution of the algorithm by the user (i.e., clicking the "optimize" button) generally executes hundreds of thousands of iterations.

1.8.2 Binary Stochastic and Continuous Stochastic Optimization

Optimization algorithms have three main components: the objective function, the variable, and the means for finding the optimal solution to the objective function by exploring changes in the variable. In a *binary* optimization, the variable is a set of binary options. In permanent seed implant brachytherapy, the set of positions where a seed can be placed is defined by the needle template and the seed spacing within each needle. This defines a set of positions where there can be either a seed or a spacer. Thus, the optimization sees a binary choice between placing a seed or not at any given position. This is in contrast to an afterloader-based optimization problem where the dwell time allows for a continuum of possibilities at any given dwell position. HDR brachytherapy can, therefore, be considered a *continuous* optimization.

A *stochastic optimization* is one that employs randomness when changing the variable to find the optimal solution. The IPSA (Inverse Planning Simulated Annealing) algorithm is an example of stochastic optimization (Pouliot et al. 1996; Lessard et al. 2001). In the binary version of this optimization algorithm, the initial solution (Step 2 in the work flow) is to place a seed in every other position in each needle, creating a seed-spacer-seed-spacer pattern. The stochastic nature of the optimization happens at two places, Step 4 and Step 6. During Step 4 a needle and position is chosen at random and a change is executed: either a seed is placed where there was none before (0 to 1 transition) or a seed is removed (1 to 0 transition). The new optimization function value is calculated given the new seed configuration.

In this case, it is desired to minimize the *FN*. Therefore, Step 6 evaluates the objective function from this iteration (i) and compares it the previous iteration (i–1). If FN(i) < FN (i–1), then the new configuration of seeds is kept; otherwise the seed distribution reverts back to the configuration of iteration i.

This progression is depicted in Figure 1–30 and allows the optimization to find the first local minimum.

Once the solution is at the first local minimum applications of the test in Equation (1.16), further calculations will be very likely to fail and further changes to the source configuration will not be accepted since they do not reduce the optimization function further. This can prevent the algorithm from finding an acceptable solution to the problem.

One stochastic process to overcome this limitation in the process is the simulated annealing method. This method enables the possibility of jumping out of local minima by allowing the algorithm to occasionally accept an iteration even if it fails the test in Equation (1.16). This is generally implemented after evaluating Equation (1.16) for a given iteration. If Equation (1.16) fails for iteration i (i.e., FN(i) > FN(i–1)), the algorithm generates a pseudo-random number. If this number is above a given threshold, the source configuration of i is accepted (even though the objective function is worse than the best configuration found so far). Iteration i+1 is then evaluated using Equation (1.16) against iteration i as the new best solution found so far. This is depicted in Figure 1–30 as the red dotted line jumps out of the local minimum.

If this process is allowed to continue for every iteration, an acceptable solution is unlikely to be found because the algorithm will never find a minimum. Thus, the simulated annealing process raises the threshold for the pseudo-random number as i increases. That is, the threshold is higher for i+1 than it is for i. In this way, the algorithm is able to probe a large part of the search space during the first fraction of iterations and hone in on the best solution during the later fraction of iterations.

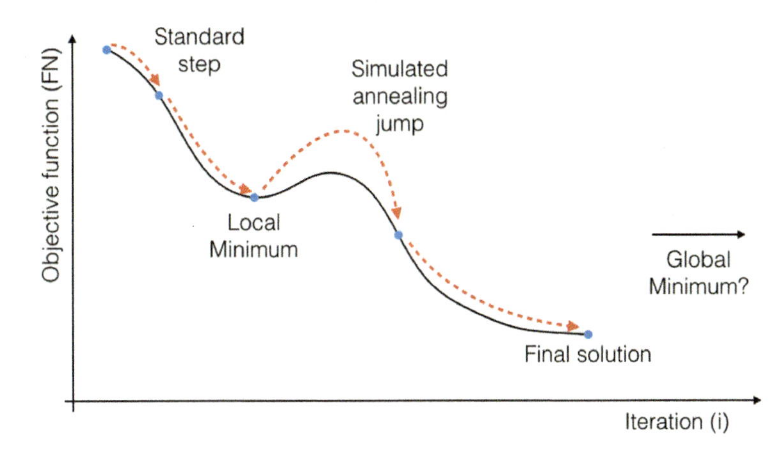

Figure 1–30 A qualitative depiction of the optimization process. Each iteration (blue circles) executes a change on the state of the system with the goal of minimizing the objective function. Without specific functionality to escape, simply choosing the better value for the objective function on each iteration can result in the final solution being "trapped" in local minima. This process does not guarantee that the final solution is the global minimum. See the discussion for the meaning of the red curve.

It is important to remember that for this type of optimization there is no guarantee that the solution returned at the end is at the global minimum (a.k.a. best solution). In theory, the global minimum exists and the returned solution *may* be the global minimum. But the best solution may not have been found for $i \le i_{max}$. Generally, however, the objective function plot will flatten as i increases. It is the responsibility of the user to choose an appropriate i_{max} so that the optimization has enough time to find an acceptable solution, but doesn't waste time on circling around the minimum with little improvement.

For HDR brachytherapy, in this example, the stochastic nature of the optimization algorithm isn't changed, but the seed-existence binary variable is replaced with dwell time in each dwell position. In the initialization phase (Step 2), the dwell times will be set, e.g., to one second for all dwell positions. In Step 4, the optimization choses a random dwell position and makes a change of 0.1 s. Since a Δt of 0.1 s in one dwell position makes a very small change in the dose distribution, the optimization is probing the search space in small enough steps that it can be considered continuous.

Stochastic optimization can incorporate other variables as well. For example, it can include the option to incorporate more than one seed strength (Cunha et al. 2010) or allow for ad hoc restrictions on the dwell time variance within a given needle (Cunha et al. 2016).

1.8.3 Binary Heuristic Optimization

A heuristic is an approach to solving a problem that returns a useful or adequate solution. It may not provide the exact solution or, in this case, *the* optimal solution, but produces a solution that satisfies the requirements. The object of using a heuristic is that process should be very much faster than solving for the true optimum. In this case, the solution is a pattern of source loading that provides the desired target coverage while holding OAR below their tolerance doses.

An example heuristic would be the Adjoint–Greedy Algorithm (Yoo et al. 2003; Chaswal et al. 2007; Yoo et al. 2007; Chaswal et al. 2012). The adjoint is the inverse of the dose distribution around a source. While the dose distribution begins with the source in space and calculates the dose that it produces at all points, the adjoint starts with a detector in space, which can simply be a point at which the dose can be determined in some manner, and then plots the signal in the detector from a source as the source occupies every place in space. For a prostate implant, useful detectors can be the CTV and the organs at risk, e.g., rectum, bladder, and urethra. The signal can be taken as the average dose to those volumes, admittedly a concept not usually considered. For each of the volumes, the average dose is calculated for a unit-strength source placed at each of the allowed positions. The allowed positions most often would be defined by the template and image planes, constrained to fall within the CTV. None of those conditions are necessary, and the source position can go forward without such constraints. The mapping of the average dose to one of the volumes for a source at the possible positions is the adjoint function, *AF,* for that volume.

After determining the adjoint function, *adjoint ratios, AR,* are calculated as:

$$AR_i = \frac{\sum_{j=1}^{n} w_j \, {}_j AF(i)}{{}_{CTV} AF(i)} \tag{1.18}$$

where i indicates a possible source position, j indicates each of the organs at risk, $AF(i)$ is the adjoint function for a unit-strength source at position i, and w_j is the importance weighting for a given OAR. The adjoint ratio tells how effectively a source delivers dose to the target while shielding the OAR.

The process begins by picking the source position with the minimum value of the adjoint ratio. Following the selection, the strength of the source is adjusted to satisfy the prescription, and then the doses to the OAR are determined. If the prescription conditions are satisfied, the process stops, which is unlikely after the first source selection. Part of the prescription conditions may be the homogeneity of the dose to the CTV.

Were the second step to pick the source position with the second best adjoint ratio, the second source likely would fall right next to the first position, resulting in almost the same dose distribution. The second source would better fall at a distance from the first to broaden the dose distribution. To accomplish this, the adjoint ratios at each point are multiplied by the dose distribution at the points. This product helps select a source position that would spread the dose across the target, so the selection goes to the minimum of the product. Again, as with each step, the source strength is set to deliver the dose to the CTV and the dose distribution criteria compared with the prescription. Assuming the process needs to continue, the original adjoint ratios are multiplied by the new dose distribution values and the process continues until the dose criteria are met.

The process is very fast since there is no iteration. The *greedy* moniker comes because once a source position has been chosen, it is never given up. Sometimes, instead of just assessing the dose distribution compared with the prescription, the process might use an objective function to determine whether to go on. The objective function can also incorporate other constraints, such as with the number of needles, where the function is multiplied by a factor such as (number of needles over some limit) which greatly penalizes selecting a source position that requires a new needle after a normal number of needles has been used.

1.8.4 Continuous Analytic

With a single source or single dwell position, the only dosimetric parameter is the distance to the desire isodose line, ignoring the anisotropy of the isodose distribution. This is a relatively simple dose calculation. Adding more sources or dwell positions to the brachytherapy implant increases the available parameters within the plan, especially if more complex isodose shapes are desired. The general isodose shape can be described by a set of optimization points placed about the implant. The points can be anywhere within the region of the brachytherapy applicator. Referring only to HDR brachytherapy, these points are some distance away from every defined dwell position within the applicator. One way to envision this complex relationship between dwell positions and optimization points is to use the following equation to calculate the dose at point a from dwell position i in place for duration t_i, where there are n dwell positions with a functional relationship defined by f:

$$D_a = \sum_{i=1}^{n} f_{i \to a} t_i \tag{1.19}$$

The problem may be defined by the constraint of doses desired at a set of m dose optimization points. Continuous analytic optimization refers to finding a solution to this set of simultaneous equations that meets the desired planning goals, within the limits of the math. In mathematical terms, this problem falls into one of three areas: if n equals m it is determined; if m is greater than n it is overdetermined; and if m is less than n it is undetermined.

With an overdetermined system (more dose specification points than dwell positions) there is usually not a solution. However, a possible practical solution could be found if one chooses the solution that minimizes the sum of the squares of the differences between calculated dose and desired dose for all dose optimization points.

With under-determined systems (fewer dose specification points than dwell positions), there are an infinite number of solutions, but the "best" answer, in this case, is found by having the minimum total dwell times, with no negative dwells times. This minimizes the integral dose to the patient.

With determined systems, the algebraic solution could still contain, unrealistically, negative dwell times. An additional factor can be applied to the above equation to account for dwell-to-dwell variations in time.

$$X^2 = \sum_{\substack{\text{all dose}\\ \text{points } a}} \left[\sum_{\substack{all\\ \text{dwells}\\ i}} \left| D_a - f_{i \to a} t_i \right| \right] + w_t \sum_{\substack{dwell\\ times\\ i<n\\ along\\ a\,catheter}} t_i - t_{i+1} \tag{1.20}$$

This equation would add the further constraint to minimize X^2 as part of the overall solution.

$$\frac{\delta X^2}{\delta a_j} = 0 \tag{1.21}$$

The w_t coefficient could be adjusted until a non-zero dwell time solution is found from every dwell position. The added result to this method is a relative smooth gradient of dwell times between adjacent dwell positions. This smoothly varying nature can then be approximated by a p-ordered polynomial function and is easily adapted to large implants that have many dwell positions n, and dwell times i, along with many optimization points within the treatment plan.

The calculation for time of dwell i at position x distance from first position is

$$t_i(x_i) = \sum_{j=1}^{p} a_j x_i^j \tag{1.22}$$

where

$$p = 2\sqrt{n} - 1 \tag{1.23}$$

and the previous constraint on X^2 from above is still valid.

The above polynomial optimization may still yield a clinically unacceptable implant depending on dwell position locations relative to optimization point locations. Too many optimization points, with the corresponding importance of their dose values, placed at the perimeter would yield a cold center, and the opposite if too many optimization points are placed in the central region of the implant.

In cases with many dwell positions, another type of optimization can also be applied. Geometric optimization refers to using the geometric relationships between the physical dwell positions to help determine the dwell times. The general principle, assuming all dwell times start equal, is that optimization points that are in a region crowded with dwell positions will have a higher dose than those in a region sparsely populated with dwell positions. Geometric optimization works by using a function that decreases dose if there are many dwells nearby and increasing the dose if there are few dwells nearby. One way to do this is by calculating the sum of the inverse square distances for all other dwell positions with strength s to dwell at position i, as shown below:

$$t_i = \sum_{j=1}^{p} \frac{s_j}{x_j^2} \tag{1.24}$$

The relative dwell time for a particular dwell position is then set inversely proportional to this sum. The result is a relative increase in the homogeneity of the dose distribution. The problem with geometric optimization is that the calculation of dwell times assumes the dwell times are all equal, but the result is a set of varying dwell times, which violates the initial assumption.

Analytic solutions fall into two classes: distance optimization and volume optimization. Distance optimization uses the relative dwell distances within individual catheters to determine the dwells times and creates acceptable dose distributions for implants with few catheters. Volume optimization places the importance of giving equal dose between catheters by using the relative dwell distances between different catheters, ignoring the catheter containing the dwell position of interest in the above equations, and this is appropriate for implants with large numbers of catheters.

1.8.5 Continuous Non-analytic Optimization

The heuristic discussed in Section 1.8.3 also has been applied to the continuous HDR brachytherapy model (Chaswal 2009). The implementation for HDR applications has been approached in two ways. The first method increments the time for a dwell position by 0.1 s each time that dwell position was selected, noting that a dwell position is selected many times. The other technique increments the dwell weight by a decreasing amount as the number of cycles increase.

1.8.6 Deterministic Optimization

Deterministic optimization is unique in the algorithm types described here because it provides theoretical guarantees that the solution returned is the global minimum within a predefined tolerance. Often the endeavored is to derive a rigorous method—an algorithm that returns the global optimum value in finite time. Deterministic optimization can be performed by solving integer programming, in which an effort is made to determine whether a solution exists that satisfies all dosimetric constraints (Siauw et al. 2012). For the purposes of brachytherapy treatment planning, however, being rigorous is not necessarily enough, since the time demands for optimization for an HDR treatment plan usually require a solution within minutes. However, even if obtaining the global optimal solution would take longer than the time available in the clinic, deterministic optimization algorithms will at least give a solution along with a distance to optimality.

Penalty-based optimization algorithms like those described above often use penalties as a surrogate for actual DVH-based constraints. Because of the ability to give distance-to-optimality information, deterministic algorithms are extremely useful if designed to perform dose optimization directly on the DVH metrics that are used for evaluation, such as $V_{75\%}$, $V_{100\%}$, $D_{90\%}$, or other parameters (Siauw et al. 2011). This is because the maximize-given-limits philosophy allows the user to see the absolute maximum target coverage, e.g., provided $V_{75\%}$ of the bladder must be less than 1 cm^3. A deterministic algorithm will tell the user whether the requested target coverage is achievable. One must be careful, however, when using a method designed in this way because the optimization algorithm has no incentive to not use the entire allowable dose to the OAR. If the maximum defined dose for an OAR dose is set to 75% of the prescribed dose to 1 cm^3 of the OAR, the maximize-given-limits optimization does not have an incentive to make the OAR $V_{75\%}$ anything less than 1 cm^3 and, therefore, will most certainly push the dose to the user-defined limit. This is not necessary a bad thing and, in fact, may be useful in the clinic if used properly.

Deterministic algorithms can also be used to perform needle planning (Siauw et al. 2012). In this case, the search space is comprised of a library of needle paths. The optimization algorithm is able to tie in the dose optimization with needle optimization. This can be very useful for robotic brachytherapy where the needle geometry is fully disassociated from the restrictions of a template as discussed in Chapter 10.6.

One major disadvantage of deterministic algorithms is that they usually require dedicated software to perform the data analysis in a reasonable amount of time. These software platforms are designed specifically to solve linear programs, mixed integer programs, or other types of deterministic optimization algorithms. These solvers are generally quite expensive and cannot be expected to be included in commercially available dose planning systems.

1.9 Plan Evaluation

Evaluation of a treatment plan is a necessary step after its creation—the assessment should lead to higher-quality treatment plans. In addition to verifying that the planning goals were achieved, the evaluation is performed in many clinics as four-eye principle—the verification of the treatment plan by another qualified individual.

The ultimate arbiter of optimal plan quality is patient outcome. Patient outcomes are determined by comparing pre- and post-treatment patient quality of life instruments: questionnaires, photographs for the evaluation of cosmesis, and the time to treatment failure. These data should be readily available for periodic retrospective analysis of weaknesses in the brachytherapy program.

1.9.1 Trade-offs between Spatial Information and Dosimetric Analysis

The isodose curves, reviewed slice by slice in the treatment planning system, provide detailed spatial dose information. However, the plethora of information makes it difficult to compare one plan to another, so the condensation of the 3D dose data into various 2D and 1D formats allows these comparisons to be performed.

1.9.1.1 3D Dose Displays

One of the first things to be checked is the dose coverage of the CTV and the PTV. Is the dose to sites of positive mapping biopsies, or multi-parametric MRI (mpMRI), adequate and perhaps boosted to levels appropriate to your institution? The dose levels to critical structures and OARs should be evaluated as well in this highly granular view. Are tolerance doses for OARs exceeded or could they be further minimized? As typical in radiotherapy, covering the targets and sparing the organs at risk is a balancing act.

A closer look at source positions should also be part of plan evaluation. For HDR, the dwell position spacing should be verified. While variable dwell position spacing is available for the Varisource afterloader, its use should be minimized to avoid unnecessary confusion during both the planning and delivery process. In general, the dwell times should not be located in a few positions only, but distributed over a longer distance. In particular, after the use of dose-shaping tools, discontinuous dwell-time patterns can appear. These discontinuities should be evaluated to determine whether they should be smoothed prior to final plan approval and the dose distribution reevaluated afterward. For permanent brachytherapy, the sources should follow standard spacing increments to simplify either the source deposition using a Mick applicator or the source loading into a needle. The use of single-source needles should be minimized due to uncertainties of source placement, while a plan with a high number of back-to-back seeds may lead to the violation of treatment planning parameters, such as OAR doses of $V_{200\%}$ of the PTV.

Notably in HDR interstitial implants, the catheter number or labeling in the treatment planning system (TPS) is of importance. A consistent system of labeling the catheters should be followed: for example, starting at 12 o'clock, counting counterclockwise, and ending with central catheters. This system may vary from one clinic to another, but if no system is used, the lack of consistency can lead to errors in transferring the catheter number from the plan to the patient (or vice versa). At the time of the connection of the transfer guide tubes to the patient, verification that the needle identification at the patient is identical to that in the TPS should be performed. If needles are inserted or removed at later stages of the implant, a further verification should be performed.

1.9.2 Dose Volume Histograms (DVHs)

Although the full isodose display provides complete information about dose deposition within the patient, there is a need for more focused evaluation pertaining only to specific anatomic structures or defined volumes. Creating an ordered list of all the dose voxels within a volume or structure according to the voxel's dose allows presentation of various dose-volume histograms (DVHs).

1.9.2.1 Cumulative or Integral DVHs

Cumulative, or integral, DVHs are the most common type of DVH. Usually, the ordered list of dose voxels within a structure is binned by some increment of dose, and the number of voxels (or volume) with that dose

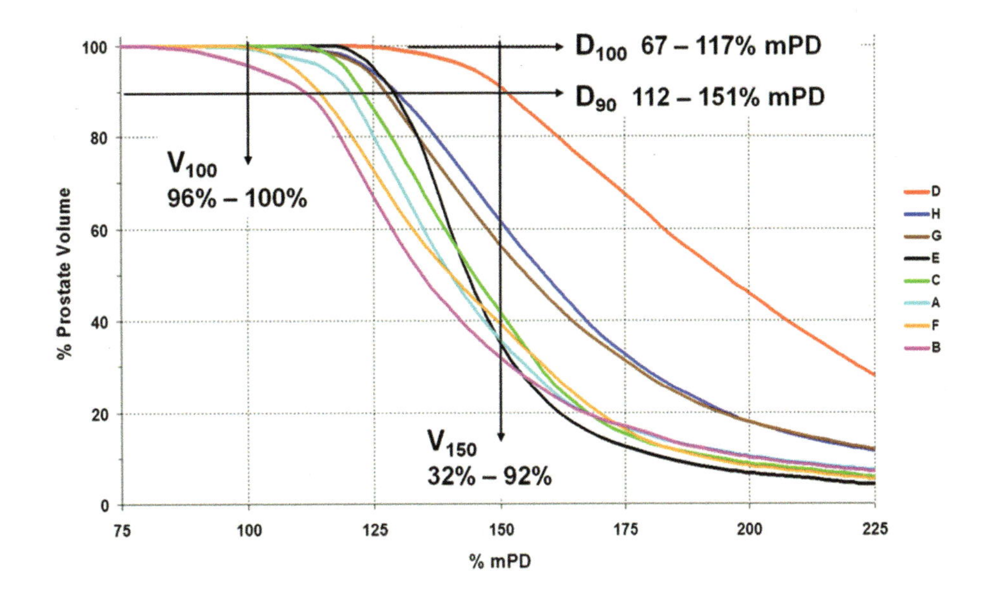

Figure 1–31 Cumulative or integral dose volume histograms for ^{125}I implants planned for the same prostate by eight experienced brachytherapists. Reprinted from *Brachytherapy,* 4(4), Merrick et al. "Variability of prostate brachytherapy preimplant dosimetry: A multiinstitutional analysis," pp. 241–51 © 2005 with permission from Elsevier.

or greater is tabulated. The graph of dose versus total volume at that dose is the cumulative DVH. This single two-dimensional curve discards all spatial information other than its origin within the structure of interest. Because of the loss of spatial information, DVH curves should be considered as determining the adequacy of a plan until the high- and low-dose regions within the 3D dose distribution have been deemed acceptable.

Figure 1–31 is a plot of DVHs for the same prostate planned by eight experienced brachytherapists. The brachytherapists apparently had different planning goals and different tolerance for variation from those goals. The histograms all start out covering 100% of the prostate volume at doses greater than 80% of the prescribed dose (designated as minimum peripheral dose, mPD, in the figure). The volume coverage then decreases at higher doses until leveling off at a small volume for doses greater than 200% of the prescribed dose. The slope of the DVH at the inflection point of the curve indicates the homogeneity of the implant, but that parameter is usually characterized by a measurement not requiring calculation of the slope.

The interplay between target coverage and dose to OARs is readily seen when they are all plotted together. If the sites of positive biopsies are separately targeted for a focal boost, the DVH curve for the focal volume should be shifted to the right of the rest of the target, i.e., to higher doses. The DVH curves for critical structures where complications may arise should lie to the left of the main target, but high-dose regions in an OAR are obvious.

Depending on the radiation tolerance and physical structure of the OAR, a tabulated contiguous volume analysis (CVA) may be called for. At user-defined increments, the CVA bins doses and calculates volumes by counting voxels just like a cumulative DVH; however, the CVA doesn't simply present the total volume of each dose bin but tabulates the size of sub-volumes containing contiguous voxels at the same dose range of the bin. Figure 1–32 illustrates the CVA for the prostate in a seed implant calculated by the VariSeed™ planning software version 8.0 (Varian, Inc., Palo Alto, CA). At lower doses, the entire prostate is covered, and there is no difference between the largest contiguous volume at a given dose level and the total volume covered. At higher doses, the volume at a given dose becomes subdivided into two or more contiguous sub-volumes, and these sub-volumes ultimately become localized to each individual source. For an organ at risk, at 150% of the prescribed dose, three separate contiguous volumes comprising 5 mm^3, 4 mm^3, and 2 mm^3 of a linear structure like the urethra would be of less concern than a single contiguous volume of 11 mm^3.

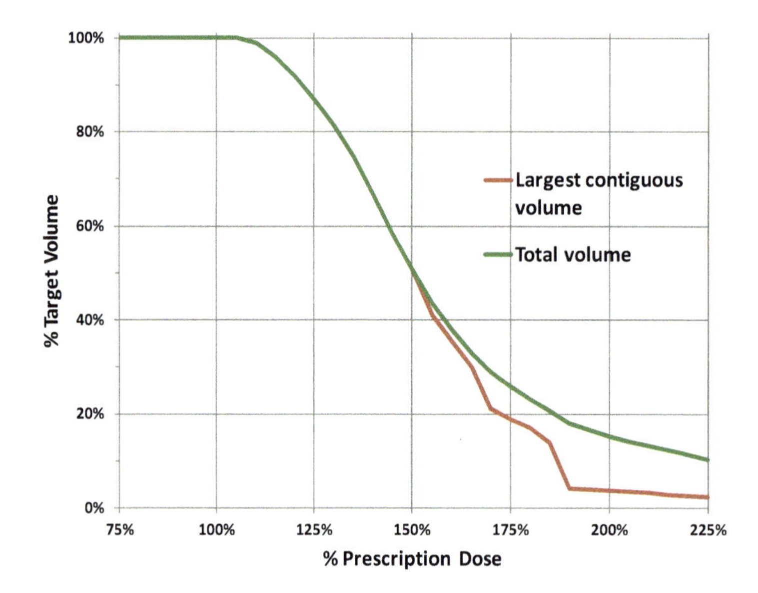

Figure 1–32 Contiguous volume analysis curve (red) for a prostate implanted with ^{103}Pd seeds. The 38 cm^3 prostate was within a 75 cm^3 PTV that required 105 seeds for complete coverage.

1.9.2.2 Differential Dose Volume Histograms

The derivative in calculus is essentially the slope of a function. The differential DVH (dDVH) calculates the slope of the DVH curve by taking the difference between the value of the cumulative DVH at one level and the next bin level. Within a target structure reasonably covered by the prescription dose at very low dose levels, all the voxels exceed the dose threshold, and the same is true for the next dose bin. For example, if the PTV is completely covered by the prescription dose, it is also covered by all doses less than the prescription dose. Because the entire volume of the PTV is covered by the 90% and 95% isodoses also, there is no volume difference between one dose bin and the next, so the dDVH is zero. The dDVH should be zero at low doses and increase to a maximum at the steepest part of the DVH curve, usually where the curve changes from concave down to concave up. An example of a dDVH is shown in Figure 1–33.

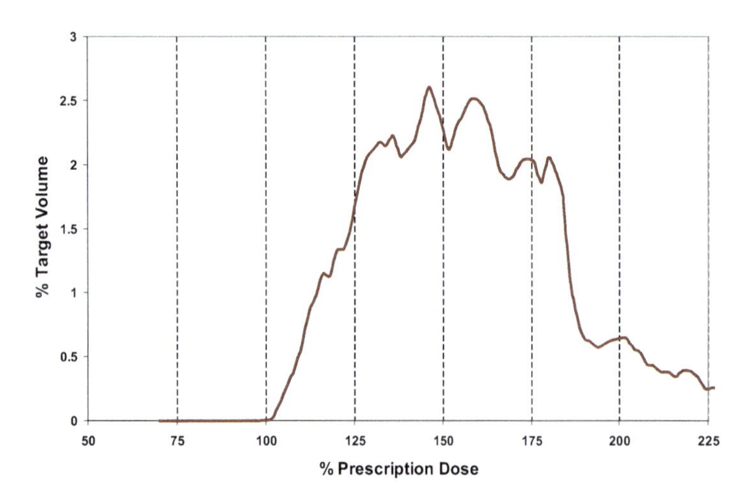

Figure 1–33 A differential dose volume histogram of a ^{103}Pd prostate implant derived from a cumulative DVH of the target volume.

Brachytherapy is inherently inhomogeneous because of the presence in the volume of sources whose dose deposition is dominated by inverse square effects. The implant homogeneity may be characterized by the peak height and width of the dDVH at a height of half the maximum volume.

The dDVH for the entire patient volume has a different appearance from the structure-specific dDVH. Low and very low doses occupy a large volume at considerable distance from the cluster of sources or dwell positions characterizing the target volume. Unlike the dDVH of the target volume, which is zero at low doses, the dDVH of the entire patient volume is large at low doses and decreases by a power law toward the prescribed dose. At higher doses, the dDVH approaches zero volume asymptotically.

1.9.2.3 Natural Dose Volume Histograms

To address the limitations of power law effects, Lowell Anderson derived an expression for the volume per unit dose, assuming a point source with a pure inverse square dose rate (Anderson 1986). Anderson defined a new unit, $u = -D^{-3/2}$, so the differential volume per u has the form of:

$$\frac{dV}{du} = \frac{dV/dD}{\frac{3}{2}D^{-\frac{5}{2}}} = \frac{2}{3}D^{\frac{5}{2}}\, dV/dD \qquad (1.25)$$

By defining what he called a Natural Dose Volume Histogram (nDVH) by plotting volume per –3/2 power of dose as a function of dose to the –3/2 power, Anderson demonstrated that the inverse square effects could be suppressed while retaining information about dose uniformity and over dosage (Anderson 1986). Moerland and colleagues (2000) extended this approach by using the nDVH to derive a natural prescription dose (NPD) and evaluate the quality of an implant in terms of the ratio of NPD to the given prescription dose called the natural dose ratio, NDR = NPD/PD, where PD is the prescribed dose. An ideal implant has NDR = 1, while NDR > 1 implies the target is overdosed and NDR < 1 is underdosed. The beauty of this approach is that the nDVH and NDR may be evaluated for the implant source distribution independent of the target volume definition. Figure 1–34 illustrates some of these quantities. For real clinical ^{125}I and ^{103}Pd implants with good homogeneity and >99% of the target covered by the prescribed dose, the NDR provides an excellent check on maintaining a consistent source placement philosophy.

1.9.2.4 Dose Surface Histograms and Linear Dose Traces

For some hollow organs such as the bladder or rectum, a DVH may be misleading because much of the volume evaluated is inert fill rather than tissue. Therefore, a dose surface histogram (DSH) or dose area histogram based on the voxels within a small incremental distance from the surface of a 3D structure of interest is of value. Some treatment planning systems can calculate a DSH directly, while others require creation of an annular volume by adding a small margin to the structure. The two-dimensional DSHs are similar to DVHs

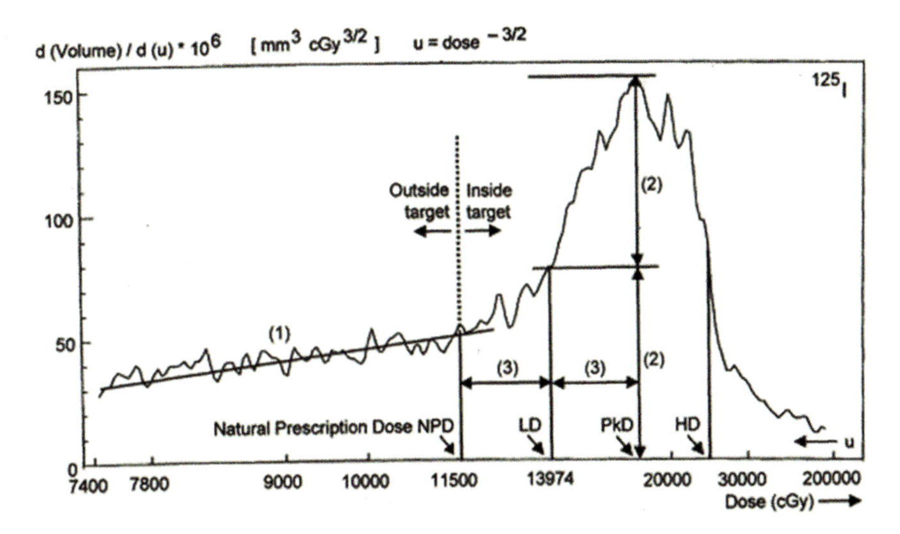

Figure 1–34 A natural dose volume histogram of the entire volume treated with ^{125}I. The peak dose is PkD, LD is the dose at the half peak height toward lower dose, and HD is the dose at half peak height toward higher dose. The natural prescription dose is an additional distance (3) from PkD on the Dose$^{-3/2}$ axis, relabeled as dose to make it more easily understood. Reprinted from *Radiother. Oncol.* 57(3), Moerland, et al. “The combined use of the natural and the cumulative dose–volume histograms in planning and evaluation of permanent prostatic seed implants,” pp. 279–84 © 2000) with permission from Elsevier.

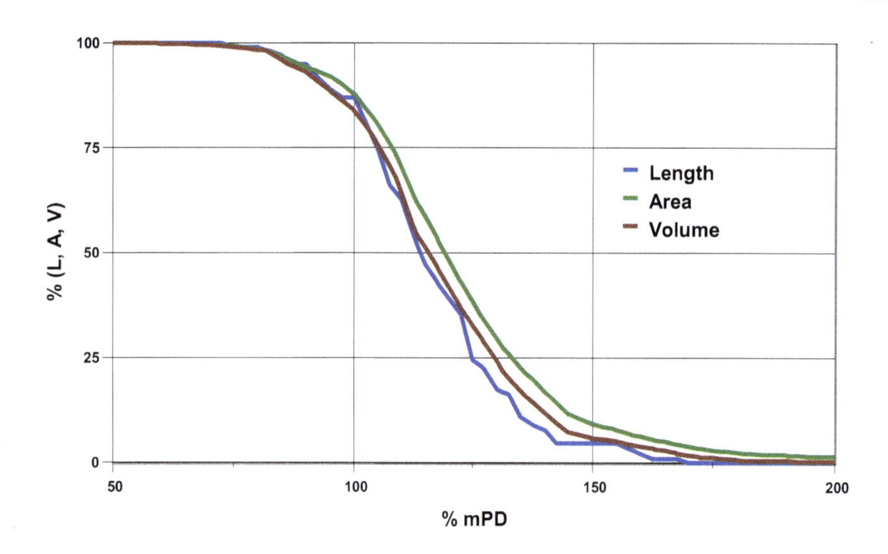

Figure 1–35 Dose volume, dose surface area, and dose length histograms of the urethra in a prostate with an ^{125}I implant. The dose length curve follows the center of the urethra, which has a lower dose than the urethral surface.

in that most spatial information is suppressed. However, by shrinking to one dimension as in a linear dose trace of a series of dose points marking the central lumen of the urethra, a linear dose histogram can display spatial information in terms of dose as a function of position along the line. For the urethra, with a diameter of about 5 mm, Butler and colleagues (2000) showed that there is no significant difference between the DVH, DSH, and linear dose trace (Figure 1–35).

1.9.3 Non-structure-based Volumes

Non-structure-based volumes are those volumes encompassed by a user-defined dose. The total tissue volume encompassed by the prescription dose, $V_{100\%}$, is dependent on the target volume, the radionuclide and source strength, the number of seeds, the loading pattern, and the prescribed dose. By generating a regression equation based on $V_{100\%}$ and target volume for a series of acceptable implants, $V_{100\%}$ can serve as a consistency check on the implant design. $V_{100\%}$ can stand on its own as a comparator, but it is most often used to calculate the conformation number (CN) or conformal index (COIN) (van't Riet et al. 1997; Baltas et al. 1998).

$$CN = \frac{(v_{100\%})^2}{CTVv_{100\%} \cdot CTVv} \tag{1.26}$$

where $CTVv_{100\%}$ is the volume of the CTV covered by 100% of the prescription dose and CTVv is the CTV volume.

1.9.4 Sector Analysis

An evaluation intermediate between the 3D dose distribution and summary dosimetric indices for the target is to subdivide the target into segments. Some treatment planning systems allow segmentation of any target volume into user-defined sectors. By judiciously choosing the number and placement of sectors appropriate to the target, useful information about the homogeneity of the target coverage may be gained. This information is usually gleaned retrospectively, particularly when comparing different techniques across a series of implants. However, it is most useful when applied prospectively to shorten the learning curve when initiating a new brachytherapy program and to maintain quality thereafter.

A study by Bice and colleagues (2001) of 118 prostate implants divided into 12 sectors found that seed delivery by a Mick applicator did not provide the same degree of coverage as they were able to achieve with loose seeds in needles. They divided the prostate first into thirds—superior (base), inferior (apex) and mid-

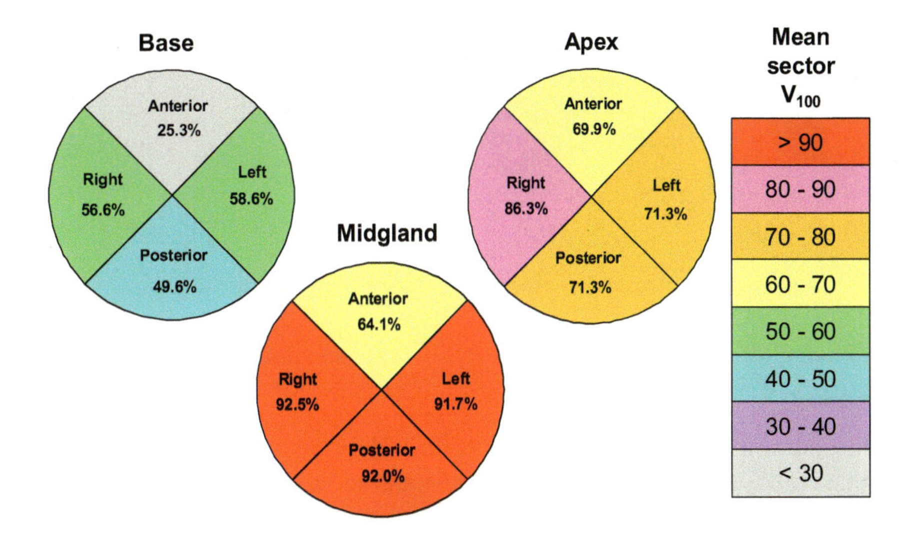

Figure 1–36 Analysis of 4,547 prostate implants by 129 community brachytherapists with the prostate segmented into 12 sectors and scored by mean $V_{100\%}$. Reprinted from *Brachytherapy* 13(2), Merrick, et al., "Multisector prostate dosimetric quality: Analysis of a large community database," pp. 146–51, © 2014 with permission from Elsevier

gland—and each third into fourths—anterior, posterior, left, and right about the center of gravity—creating 12 sectors, each with its own DVH.

A larger retrospective study of 4,547 implants performed by 129 community-based brachytherapists found no significant difference in quality between ^{125}I and ^{103}Pd implants or between monotherapy and brachytherapy boost treatments based on sectors and combinations of sectors. However, 59% of anterior base sectors and 30% of posterior base sectors were of substandard quality based on $V_{100\%}$ coverage <80%. The dosimetry of each sector and various combinations of sectors were compared as in Figure 1–36.

Stratifying sectors using dosimetric indices such as $V_{100\%}$ or $D_{90\%}$, particularly across a series of patients, allows quick visual confirmation of quality or highlights anatomic areas in need of further procedural refinement. This sector-based approach was used to analyze biochemical failures in a prostate brachytherapy series caused by inadequacies of extra-prostatic annular dosimetry.

1.9.5 Variation in DVHs

1.9.5.1 Target Volume Uncertainties

Whenever the edge of the target volume is in a high-dose-gradient region, there can be considerable variation in the DVH caused by small changes in the source distribution. Figure 1–37 illustrates the DVHs of a ^{103}Pd implant for the prostate, the prostate expanded by about 5 mm everywhere except posteriorly to a PTV, and the urethra. Movement of the four needles that are closest to the urethra an additional 2 mm away from it has a large effect on the urethra DVH, a modest effect on the PTV DVH, and a negligible effect on the prostate DVH.

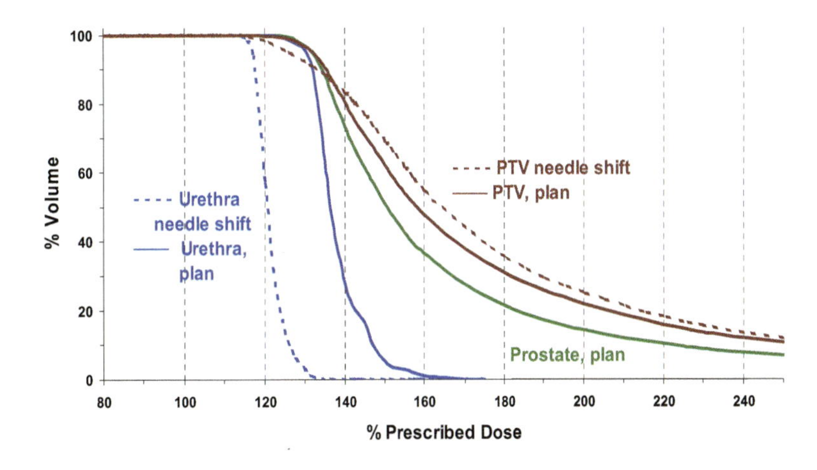

Figure 1–37 Shifts in the urethra, PTV, and prostate DVHs caused by moving 4 central needles near the urethra 2 mm farther away from the urethra.

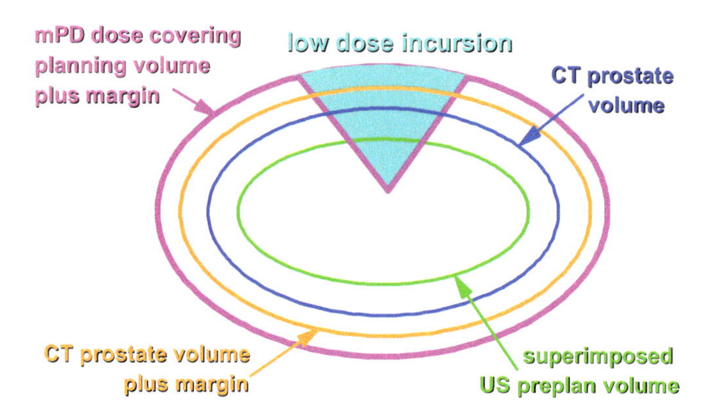

Figure 1–38 How a low dose incursion had proportionate effects on whichever volume was analyzed when the prescription isodose volume was greater than the superimposed ultrasound volume, the CT prostate volume, or even the CT prostate volume expanded to a PTV covering extra prostatic extension. Reprinted from *Int. J. Rad. Onc. Biol. Phys.* 44(7), Merrick et al., "The dependence of prostate postimplant dosimetric quality on CT volume determination," pp. 1111–17, © 1999 with permission from Elsevier

Although considerable DVH variation is expected due to contouring uncertainties, this too is highly subject to treatment planning approach. One study found that in prostate implants planned with seeds outside the PTV, there was almost no difference in dosimetric quantifiers between imposing the pre-implant US prostate volume on post-implant CT, drawing the actual edematous prostate on the CT, or expanding the edematous prostate to a PTV encompassing likely extra prostatic extension (Merrick et al. 1999). This paradoxical effect was explained by the fact that when the PTV itself is well covered by the prescription isodose, cold dose incursions, such as near the bladder neck, have a proportionate effect on whichever volume was analyzed, as in Figure 1–38.

1.9.5.2 Effect of Edema

Replicating a patient's treatment plan in the clinic is confounded by numerous factors, such as uncertainties in imaging and needle or source placement. The greatest difficulties in translating the plan to the patient occur in permanent seed brachytherapy because the dose is delivered over an extended period of time while the target undergoes edema from operative trauma and the higher dose rate at the start of the radionuclide decay curve. Edema is at its maximum about one day after the implant, but the average maximum seems to depend on the institutionally specific implant procedure and ranges from volume increase factors of 1.2 (Merrick et al. 1998) to 1.3 (Taussky et al. 2005) to 1.7 (Badiozamani et al. 1999). This topic is also discussed in Chapter 4 of this book.

Not only is the magnitude of edema patient specific, but so is the rate of edema resolution. An analytic solution balancing edema, resolution, and source decay to select an optimal time minimizing relative dosimetry error was published by Chen and colleagues (2000). These radionuclide-specific times were incorporated into the recommendations of AAPM TG-137 (Nath et al. 2009). For ^{125}I, the recommendation is that the post-implant dosimetry be perform one month ±1 week after the implant. For ^{103}Pd and ^{131}Cs, these times are 16 ±4 and 10 ±2 days, respectively. The task group also acknowledged that many institutions perform dosimetry evaluations on day 0—real-time in the operating room or within hours of the implant after the patient is released from the recovery room—or on the day after (i.e., day 1). The advantage of this approach is that if an implant meets the dosimetric requirements on day 0 or day 1, no further evaluation is required. At that time, curiosity about the quality of the implant is highest and procrastination of target segmentation and dosimetry calculations is unlikely. Learning is enhanced because detailed dosimetry can be correlated with what went right and what went wrong during the operative procedure. If doses to critical structures (such as urethra or rectum) are too high, the physician may apply prophylactic measures, such as alpha-blockers like tamsulosin to ease urination or stool softeners to ease bowel movements. The physician and nursing staff will also be ready to act proactively if follow-up quality-of-life measures show the patient showing signs of distress. If the dose distribution is less than desired, the patient may be invited to undergo additional imaging at the optimum time when edema has receded. Having two post-implant dosimetry studies separated in time is far more accurate than any single imaging procedure because the first imaging study helps identify the maximum amount of edema, and the second imaging study marks the rate of edema resolution. If the dose to the prostate is still unacceptably low after the second imaging study, the patient may be offered supplementary treatments.

1.9.6 Interstitial Indices

Plan evaluation is an important step after the creation and assessment of the treatment plan and indicates its quality. In addition to determining whether the planning goals are reached, in many clinics the indices serve as a check of the treatment plan.

Dose homogeneity, if we can say so in brachytherapy, is another aspect in plan evaluation. A measure should be used, like CN or the $V_{200\%} / V_{100\%}$ ratio, to assess the hot spots in the treatment plans and to minimize them when possible.

A closer look to the dwell positions and dwell times is also part of plan evaluation. Are the dwell positions equally spaced with the correct distance? The dwell times should not be located in a few positions only, but be distributed over a longer distance. In particular, after the usage of dose-shaping tools, discontinuous dwell time patterns can appear. These should be smoothed and the dose distribution reevaluated afterward.

Notably in interstitial implants, the catheter number or labeling in the TPS is of importance. The system of labeling the catheters should be followed; for example starting at 12 o'clock, counting counterclockwise, and ending with central catheters. This system can vary from one clinic to another. If no system is available, at least checking the spreadsheet is necessary. Prior to connecting the patient to the afterloader, the needle number in the TPS should be verified and adapted if necessary. Care should be taken if needles are inserted or removed in a later stage of the implant.

1.9.7 Intracavitary Cervical Brachytherapy Measures

The long history and evolution of intracavitary applications for gynecological cancers should be taken into account when describing evaluations of plans. At the time of publication of ICRU Report 38, treatment planning for cervical cancer was based on gynecologic examination at diagnosis and radiography. Plans were evaluated through their prescription point dose, the doses to Point A (with the definition evolving), Point B, and rectal and bladder point doses, which serve as surrogates for the dose to those organs. Reporting the maximum width, thickness, and height of the 60-Gy reference volume covering this target was also recommended.

Many of the developments in the field (engineering, computer technology, imaging, radionuclide, delivery methods, etc.) were incorporated in the more recent GEC-ESTRO recommendations. GTV, CVT, and PTV are the important volume definitions for intracavitary brachytherapy. Since these treatments are typically delivered in conjunction with EBRT, EQD2 is an important quantity that allows the addition (in a radiobiological effect sense) of doses delivered by EBRT and brachytherapy.

The GEC ESTRO report recommended the reporting of $D_{100\%}$ and $D_{90\%}$. These DVH parameters reflect the dose in the outer region of the target. $D_{90\%}$ is more robust with respect to target delineation uncertainties when compared with an absolute minimum target dose, $D_{100\%}$.

The rectum, bowel, and bladder are identified as OAR. Due to absorbed-dose heterogeneity within the organ walls, it is recommended to report at least two dose-volume values in the high-dose region. The dose values $D_{0.1cm^3}$ and D_{2cm^3} represent, respectively, the minimum doses to the 0.1-cm^3 and 2-cm^3 volumes of the OAR. These OAR parameters were recommended by the GEC ESTRO GYN group in 2006 and seem to be useful in clinical practice (Potter 2006). It is not clinical practice to report the contiguity of the volumes receiving these doses. The contiguity is also different for different OARs (D_{2cm^3}, for example, can likely represent dose in two different compartments of the sigmoid, but likely a contiguous volume in bladder or rectum).

Another OAR is the vaginal mucosa. When point doses are used for plan evaluation, the vagina points located at the lateral vaginal applicator surface and at 5-mm depth into the lateral walls of the vagina have traditionally been used for vaginal dose reporting. Dose-effect relationships based on these points have not been well established, and the points have served mainly as tools for prescribing dose constraints. Implementation and evaluation of valid and reliable dose-volume parameters for the high-dose region in the vagina ($D_{0.1cm^3}$ and D_{2cm^3}) are challenging due to the very high dose gradients near the vaginal sources and the difficulties of precisely delineating and reconstructing the thin organ walls on 3D images with the applicator in place using the currently available treatment planning systems. Recommendations of the relevant dosimetric parameters for the vaginal mucosa have recently been published (ICRU 2016). While not an indicator of plan quality, a useful quantity for reporting the overall "strength" of the treatment is TRAK, the integral of the reference air kerma rate at a distance of 1 m from the source from all sources over the treatment duration.

1.9.8 Quality Assurance

A part of the evaluation of the treatment plan is QA of the plan—ensuring that the treatment plan complies with the prescription, that the dose is calculated properly, and that the plan complies with the intent of the radiation oncologist. QA is discussed extensively in various publications, including AAPM TG-59 and TG-56 (Nath et al. 1997; Kubo et al. 1998). Prior to initiating a brachytherapy program, these reports, as well as the governing regulations, should be consulted for guidance in the establishment of a QA program. Review of various treatment errors, such as on the website of the Nuclear Regulatory Commission (www.nrc.gov), should also be performed so that the brachytherapy team is aware of common pitfalls that can lead to brachytherapy treatment errors.

For simple plans, such as a vaginal cylinder treatment, pre-calculated library plans can be used to simplify the treatment planning process. In this case, the plan QA could consist of a few steps, such as confirmation of the cylinder dimension and the constancy of the TRAK or Curie-seconds (Ci-s) of the treatment.

As the plans become more complex, the QA procedures also become more involved. Even prior to planning the treatment, verification that the applicator was properly placed is essential (Kim et al. 2015). For treatments with a large number of catheters, such as the interstitial sarcoma treatment discussed in Section 1.5.2.4, verification of the catheter reconstruction, numbering, and connection to the afterloader must be performed. Additionally, a check of the catheter length should be performed to ensure that the measured length corresponds to the catheter length in the treatment plan. While it is unlikely (if the source had been properly commissioned in the TPS) that the dose would be calculated incorrectly, a secondary dose calculation should be performed. Commercial software products, such as RadCalc (Lifeline Software, Inc. Austin, TX) or IMSure (Standard Imaging, Inc. Middleton, WI) are available for performing secondary calculations. It is also possible to design a simple spreadsheet in which the coordinates and dwell times of each dwell position can be used to calculate the dose to any dose point present in the treatment plan. The TRAK or Ci-s in a Contura treatment can be approximated by the use of a single source in the center of a spherical applicator, usually confirming the treatment plan within a few percent.

At the completion of any brachytherapy procedure, the patient must be surveyed to ensure that the source has returned to the safe within the afterloader, and the post-treatment record should be reviewed to ensure that the treatment was completed properly. All of these steps need to be performed to ensure that the brachytherapy treatment was performed properly. The use of pretreatment, treatment, and post-treatment checklists can ensure that all of the QA steps have been performed and the patient treatment was properly completed.

1.10 Conclusion

As presented in this chapter, the treatment planning process in brachytherapy can be quite complex with many aspects to consider. Prescription, implant design, optimal imaging and dosimetry, plan evaluation, and QA are all part of the process. With the advent of dedicated brachytherapy suites, the treatment planning steps can all be compressed into a shorter implantation-image-plan-treat work flow, necessitating the need for a large, well-trained staff to ensure that the patient is treated properly.

Since the last AAPM Brachytherapy Summer School in 2005, many advances have been made in the field, including switching away from radiographic to 3D imaging, implementing MR into regular clinical practice, and making advances in dose calculation and optimization systems. All of these have shown that brachytherapy will continue to be a part of the a full-service clinic. Brachytherapy will continue to evolve as the latest advances—discussed in Chapter 10 of this book—get implemented into regular clinical practice.

References

Al-Halabi, H., L. Portelance, M. Duclos, B. Reniers, B. Bahoric, and L. Souhami. (2010). "Cone Beam CT-based three-dimensional planning in high-dose-rate brachytherapy for cervical cancer." *Int. J. Radiat. Oncol. Biol. Phys.* 77(4):1092–97.

Alterovitz, R., E. Lessard, J. Pouliot, I. C. J. Hsu, J. F. O'Brien, and K. Goldberg. (2006). "Optimization of HDR brachytherapy dose distributions using linear programming with penalty costs." *Med. Phys.* 33:4012–19. doi:10.1118/1.2349685.

Anacak, Y., M. Esassolak, A. Aydin, A. Aras, I. Olacak, and A. Haydaro. (1997). "Effect of geometrical optimization on the treatment volumes and the dose homogeneity of biplane interstitial brachytherapy implants." *Radiother. Oncol.* 45 (71–76).

Anderson, L. L. (1986). "A 'natural' volume-dose histogram for brachytherapy." *Med. Phys.* 13(6):898–903.

Badiozamani, K. R., K. Wallner, S. Sutlief, W. Ellis, J. Blasko, and K. Russell. (1999). "Anticipating prostatic volume changes due to prostate brachytherapy." *Radiat. Oncol. Investig.* 7(6):360–64.

Baltas, D. and I-K. K. Kolkman-Deurloo. "Optimization and Evaluation." In *Comprehensive Brachytherapy: Physical and Clinical Aspects*, D. Baltas, J. L. M. Venselaar, A. S. Meigooni, and P. J. Hoskin, Eds. Boca Raton, FL: Taylor & Francis Group, 2013.

Barendsen, G. W. (1982). "Dose fractionation, dose rate and iso-effect relationships for normal tissue responses." *Int. J. Radiat. Oncol. Biol. Phys.* 8(11):1981–97.

Batchelar, D., M. Gaztañaga, M. Schmid, C. Araujo, F. Bachand, and J. Crook. (2014). "Validation study of ultrasound-based high-dose-rate prostate brachytherapy planning compared with CT-based planning." *Brachytherapy* 13(1):75–79.

Bentzen, S. M., W. Dorr, R. Gahbauer, R. W. Howell, M. C. Joiner, B. Jones, D. T. Jones, A. J. van der Kogel, A. Wambersie, and G. Whitmore. (2012). "Bioeffect modeling and equieffective dose concepts in radiation oncology—terminology, quantities and units." *Radiother. Oncol.* 15(2):266–68.

Beriwal, S., D. J. Demanes, B. Erickson, E. Jones, J. F. De Los Santos, R.A. Cormack, C. Yashar, J. J. Rownd, and A. N. Viswanathan. (2012). "American Brachytheapy Society consensus guidelines for interstitial brachytherapy for vaginal cancers." *Brachytherapy* 11:68–75.

Berns, C., P. Fritz, F. W. Hensley, and M. Wannenmacher. (1997). "Consequences of optimization in PDF brachytherapy—is a routine geometrical optimization recommendable?" *Int. J. Radiat. Oncol. Biol. Phys.* 37:1171–80.

Bice, W. S., Jr., B. R. Prestidge, and M. F. Sarosdy. (2001). "Sector analysis of prostate implants." *Med. Phys.* 28(12):2591–67.

Bittner, N., G. S. Merrick, A. Bennett, W. M. Butler, H. J. Andreini, W. Taubenslag, and E. Adamovich. (2015). "Diagnostic performance of initial transperineal template-guided mapping biopsy of the prostate gland." *Am. J. Clin. Oncol.* 38(3):300–3. doi: 10.1097/COC.0b013e31829a2954.

Brenner, D. J. (1997). "Radiation biology in brachytherapy." *J. Surg. Oncol.* 65(1):66–70.

Brenner, D. J. and E. J. Hall. (1991). "Fractionated high dose rate versus low dose rate regimens for intracavitary brachytherapy of the cervix. I. General considerations based on radiobiology." *Br. J. Radiol.* 64(758):133–41.

Brenner, D. J. and E. J. Hall. (1999). "Fractionation and protraction for radiotherapy of prostate carcinoma." *Int. J. Radiat. Oncol. Biol. Phys.* 43(5):1095–101.

Brenner, D. J., L. R. Hlatky, P. J. Hahnfeldt, Y. Huang, and R. K. Sachs. (1998). "The linear-quadratic model and most other common radiobiological models result in similar predictions of time-dose relationships." *Radiat. Res.* 150(1):83–91.

Bretschneider T, K. Mohnike, P. Hass, R. Seidensticker, D. Göppner, O. Dudeck, F. Streitparth, and J. Ricke. (2015). "Efficacy and safety of image-guided interstitial single fraction high-dose-rate brachytherapy in the management of metastatic malignant melanoma." *J. Contemp. Brachytherapy* 7(2):154–60.

Butler, W. M., G. S. Merrick, A. T. Dorsey, and B. M. Hagedorn. (2000). "Comparison of dose length, area, and volume histograms as quantifiers of urethral dose in prostate brachytherapy." *Int. J. Radiat. Oncol. Biol. Phys.* 48(5):1575–82.

Carlson, D. J., R. D. Stewart, V. A. Semenenko, and G. A. Sandison. (2008). "Combined use of Monte Carlo DNA damage simulations and deterministic repair models to examine putative mechanisms of cell killing." *Radiat. Res.* 169(4):447–59.

Carlson, D. J., Z. J. Chen, P. J. Hoskin, Z. Ouhib, and M. Zaider. "Radiobiology for Brachytherapy." In *Comprehensive Brachytherapy: Physical and Clinical Aspects*. D. Baltas, J. L. M. Venselaar, P. Hoskin, and A. S. Meigooni, Eds. Boca Roton, FL: CRC Press, 2013.

Chadwick, K. H. and H. P. Leenhouts. (1973). "A molecular theory of cell survival." *Phys. Med. Biol.* 18(1):78–87.

Chadwick, K. H. and H. P. Leenhouts. *The Molecular Theory of Radiation Biology.* Heidelberg: Springer-Verlag, 1981.

Chao, K. K., N. S. Goldstein, D. Yan, C. E. Vargas, M. I. Ghilezan, H. J. Korman, K. M. Kernen, J. B. Hollander, J. A. Gonzalez, A. A. Martinez, F. A. Vicini, and L. L. Kestin. (2006). "Clinicopathologic analysis of extracapsular extension in prostate cancer: should the clinical target volume be expanded posterolaterally to account for microscopic extension?" *Int. J. Radiat. Oncol. Biol. Phys.* 65(4):999–1007. doi: S0360-3016(06)00365-8 [pii] 10.1016/j.ijrobp.2006.02.039 [doi].

Chaswal, V., B. Thomadsen, and D. Henderson. (2009). "Use of the adjoint analysis based Greedy Heuristic algorithms in treatment planning for LDR brachytherapy of the prostate and HDR brachytherapy using multicatheter breast implant Technique." *Med. Phys.* 36(6):2423–24.

Chaswal, V., B. R. Thomadsen, and D. L. Henderson. (2012). "Development of an adjoint sensitivity field-based treatment-planning technique for the use of newly designed directional LDR sources in brachytherapy." *Phys. Med. Biol.* 57(4):963–82.

Chaswal, V., S. Yoo, B. R. Thomadsen, and D. L. Henderson. (2007). "Multi-species prostate implant treatment plans incorporating ^{192}Ir and ^{125}I using a Greedy Heuristic based 3D optimization algorithm." *Med. Phys.* 34(2):436–44.

Chen, Z. and R. Nath. (2012). "On the use of biologically effective dose (BED) and iso-effective dose (IED) in radiobiological evaluations of permanent brachytherapy with proliferating tumors." *Int. J. Radiat. Oncol. Biol. Phys.* 84:S755.

Chen, Z., N. Yue, X. Wang, K. B. Roberts, R. Peschel, and R. Nath. (2000). "Dosimetric effects of edema in permanent prostate seed implants: a rigorous solution." *Int. J. Radiat. Oncol. Biol. Phys.* 47(5):1405–19. doi: S0360-3016(00)00549-6 [pii].

Coutard, H. (1932). "Roentgentherapy of epitheliomas of the tonsillar region, hypopharynx, and larynx, from 1920 to 1926." *AJR* 28 (313–31):343–48.

Cunha, A., T. Siauw, I-C. Chow, and J. Pouliot. (2016). "A method for restricting intracatheter dwell time variance in high-dose-rate brachytherapy plan optimization." *Brachytherapy* 15:246–51.

Cunha, J. A., B. Pikett, and J. Pouliot. (2010). "Inverse planning optimization for hybrid prostate permanent-seed implant brachytherapy plans using two source strengths." *J. Appl. Clin. Med. Phys.* 11(3):64–77.

Curtis, S. B. (1986). "Lethal and potentially lethal lesions induced by radiation—a unified repair model." *Radiat. Res.* 106(2):252–70.

Dale, R. G. (1989). "Radiobiological assessment of permanent implants using tumour repopulation factors in the linear-quadratic model." *Br. J. Radiol.* 62(735):241–44.

Dale, R. G. (1990). "The use of small fraction numbers in high dose-rate gynaecological afterloading: some radiobiological considerations." *Br. J. Radiol.* 63(748):290–94.

Dale, R. G. (2010). "The BJR and progress in radiobiological modelling." *Br. J. Radiol.* 83(991):544–45. doi: 10.1259/bjr/52885245.

Dale, R. G. and B. Jones. (1998). "The clinical radiobiology of brachytherapy." *Br. J. Radiol.* 71(845):465–83.

Dale, R.G. (1985). "The application of the linear-quadratic dose-effect equation to fractionated and protracted radiotherapy." *Br. J. Radiol.* 58(690):515–28.

Davidson, M. T., J. Yuen, D. P. D'Souza, J. S. Radwan, J. A. Hammond, and D. L. Batchelar. (2008). "Optimization of high-dose-rate cervix brachytherapy applicator placement: the benefits of intraoperative ultrasound guidance." *Brachytherapy* 3(248–53).

De Boeck, L., J. Beliën, and W. Egyed. (2014). "Dose optimization in high-dose-rate brachytherapy: A literature review of quantitative models from 1990 to 2010." *Oper. Res. Health Care* 3(7):1157–61.

Deasy, J. O. (1997). "Multiple local minima in radiotherapy optimization problems with dose-volume constraints." *Med. Phys.* 24(7):1157–61.

Deufel, C. L. and K. M. Furutani. (2014). "Quality assurance for high dose rate brachytherapy treatment planning optimization: using a simple optimization to verify a complex optimization." *Phys. Med. Biol.* 59(3):525–40.

Dimopoulos, J.C., C. Kirisits, P. Petric, P. Georg, S. Lang, D. Berger, and R. Pötter. (2006). "The Vienna applicator for combined intracavitary and interstitial brachytherapy of cervical cancer: clinical feasibility and preliminary results." *Int. J. Radiat. Oncol. Biol. Phys.* 66(1):83–90.

Dinkla, A. M., R. van der Laarse, K. Koedooder, K. H. Petra, N. van Wieringen, B. R. Pieters, and A. Bel. (2014). "Novel tools for stepping source brachytherapy treatment planning: Enhanced geometrical optimization and interactive inverse planning." *Med. Phys.* 42(1):348–53.

Dowling, J. A., J. Lambert, J. Parker, O. Salvado, J. Fripp, A. Capp, C. Wratten, J. W. Denham, and P. B. Greer. (2012). "An atlas-based electron density mapping method for magnetic resonance imaging (MRI)-alone treatment planning and adaptive MRI-based prostate radiation therapy." *Int. J. Radiat. Oncol. Biol. Phys.* 83(1):e5–11. doi: 10.1016/j.ijrobp.2011.11.056.

Dresen, R. C., G. L. Beets, H. J. Rutten, S. M. Engelen, M. J. Lahaye, R. F. Vliegen, A. P. de Bruine, A. G. Kessels, G. Lammering, and R. G. Beets-Tan. (2009). "Locally advanced rectal cancer: MR imaging for restaging after neoadjuvant radiation therapy with concomitant chemotherapy. Part I. Are we able to predict tumor confined to the rectal wall?" *Radiology* 252(1):71–80. doi: 10.1148/radiol.2521081200.

Dyk, P. T., S. Richardson, S. N. Badiyan, J. K. Schwarz, J. Esthappan, J. L. Garcia-Ramirez, and P. Grigsby. (2015). "Outpatient-based high-dose-rate interstitial brachytherapy for gynecologic malignancies." *Brachytherapy* 14(2):231–37.

Edmundson, G. K. "Geometry Based Optimization for Stepping Source Implants." In *Brachytherapy HDR and LDR.* C. G. Orton, A. A. Martinez, and R. F. Mould, Eds. Columbia, MD: Nucletron Corporation, 1990.

Erickson, B., K. Albano, and M. Gillin. (1996). "CT-guided interstitial implantation of gynecologic malignancies." *Int. J. Radiat. Oncol. Biol. Phys.* 36(3):699–709.

Ezzell, G. "Optimization in Brachytherapy." In *Brachytherapy Physics, 2nd edition.* B. R. Thomadsen, M. J. Rivard, and W. M. Butler, Eds. Madison, WI: Medical Physics Publishing, 2005.

Ezzell, G. and R. W. Luthmann. "Clinical Implementation of Dwell Time Optimization Techniques for Single-Stepping Source Remote Applicators." In *Brachytherapy Physics*. J. F. Williamson, B. R. Thomadsen, and R. Nath, Eds. Madison, WI: Medical Physics Publishing, 1995.

Feder, B. H., A.M. Nisar Syed, and D. Neblett. (1978). "Treatment of extensive carcinoma of the cervix with the 'transperineal parametrial butterfly': a prelimary report on the revival of Waterman's approach." *Int. J. Radiat. Oncol. Biol. Phys.* 4(7–8):735–42.

Fleming, P., A. M. Nisar Syed, and D. Neblett. (1980). "Description of an afterloading ^{192}Ir interstitial-intracavitary technique in the treatment of carcinoma of the vagina." *Obstet. Gynecol.* 55(4):525–30.

Fowler, J. F. (1989). "The linear-quadratic formula and progress in fractionated radiotherapy." *Br. J. Radiol.* 62(740):679–94.

Fowler, J. F. (2010). "21 years of biologically effective dose." *Br. J. Radiol.* 83(991):554–68.

Fraass, B., K. Doppke, M. Hunt, G. Kutcher, G. Starkschall, R. Stern, and J. Van Dyke. (1998). "American Association of Physicists in Medicine Radiation Therapy Committee Task Group 53: Quality assurance for clinical radiotherapy treatment planning." *Med. Phys.* 25(10):1773–1829.

Guerrero, M. and X. A. Li. (2003). "Analysis of a large number of clinical studies for breast cancer radiotherapy: estimation of radiobiological parameters for treatment planning." *Phys. Med. Biol.* 48(20):3307–26.

Hall, E. J. (1991). "Weiss lecture. The dose-rate factor in radiation biology." *Int. J. Radiat. Oncol. Biol. Phys.* 59(3):595–610.

Hall, E. J. and J. S. Bedford. (1964). "Dose rate: Its effect on the survival of HeLa cells irradiated with gamma rays." *Radiat. Res.* 22:305–15.

Hall, E. J. and D. J. Brenner. (1996). "Pulsed dose rate brachytherapy: can we take advantage of new technology?" *Int. J. Radiat. Oncol. Biol. Phys.* 34(2):511–12.

Hellebust, T., C. Kirisits, D. Berger, J. Perez-Calatayud, M. De Brabandere, A. De Leeuw, I. Dumas, R. Hudej, G. Lowe, R. Wills, and K. Tanderup. (2010). "Recommendations from Gynaecological (GYN) GEC-ESTRO Working Group: Considerations and pitfalls in commissioning and applicator reconstruction in 3D image-based treatment planning of cervix cancer brachytherapy." *Radiother. Oncol.* 96:153–60.

Herron, B., D. Chin, and J. Pollock. (2008). "CT Imaging for gynecological HDR: Tools and Tricks." *Med. Dosim.* 33(4):310–14.

Hilts, M., H. Halperin, D. Morton, D. Batchelar, F. Bachand, R. Chowdhury, and J. Crook. (2015). "Skin dose in breast brachytherapy: Defining a robust metric." *Brachytherapy* 14:970–78.

Holloway, C. L., T. F. DeLaney, K. M. Alektiar, P. M. Devlin, D. A. O'Farrell, and D. J. Demanes. (2013). "American Brachytherapy Society (ABS) consensus statement for sarcoma brachytherapy." *Brachytherapy* 3:179–90.

Holly, R., G. C. Morton, R. Sankreacha, N. Law, T. Cisecki, D.A. Loblaw, and H. T. Chung. (2011). "Use of cone-beam imaging to correct for catheter displacement in high-dose-rate prostate brachytherapy." *Brachytherapy* 10:299–305.

Holm, Å, T. Larsson, and Å. Carlsson Tedgren. (2013). "A linear programming model for optimizing HDR brachytherapy dose distributions with respect to mean dose in the DVH-tail." *Med. Phys.* 40(8):081705–1.

Hoskin, P., P. Bownes, P. Ostler, K. Walker, and L. Bryant. (2003). "High dose rate afterloading brachytherapy for prostate cancer: catheter and gland movement between fractions." *Radiother. Oncol.* 68:285–88.

Hu, Y., J. Esthappan, S. Mutic, S. Richardson, H. A. Gay, J. K. Schwarz, and P. W. Grigsby. (2013). "Improve definition of titanium tandems in MR-guided high dose rate brachytherapy for cervical cancer using proton density weighted MRI." *Radiat. Oncol.* 8(1):16.

Huang, W. C., K. Kuroiwa, A. M. Serio, F. J. Bianco, Jr., S. W. Fine, B. Shayegan, P. T. Scardino, and J. A. Eastham. (2007). "The anatomical and pathological characteristics of irradiated prostate cancers may influence the oncological efficacy of salvage ablative therapies." *J. Urol.* 177(4):1324–29; quiz 1591. doi: S0022-5347(06)03110-7 [pii].

ICRU. "Prescribing, recording and reporting photon beam therapy, ICRU Report 50." Bethesda, MD: ICRU, 1993.

ICRU. "Dose and volume specification for reporting interstitial therapy, ICRU Report 58." Bethesda, MD: ICRU, 1997.

ICRU. "Prescribing, recording and reporting photon beam therapy (Supplement to ICRU Report 50), ICRU Report 62." Bethesda, MD: ICRU, 1999.

ICRU. 2016. "Prescribing, Recording, and Reporting Brachytherapy for Cancer of the Cervix, ICRU Report 89." *JICRU* 13(1–2):Np. doi: 10.1093/jicru/ndw027.

Jaffray, D. A., M. C. Carlone, M. F. Milosevic, S. L. Breen, T. Stanescu, A. Rink, H. Alasti, A. Simeonov, M.S. Sweitzer, and J. D. Winter. (2014). "A Facility for Magnetic Resonance-Guided Radiation Therapy." *Seminat. Rad. Oncol.* 24:193–95.

Kim, J., K. Garbarino, L. Schultz, K. Levin, B. Movsas, M. S. Siddiqui, I. J. Chetty, and C. Glide-Hurst. (2015). "Dosimetric evaluation of synthetic CT relative to bulk density assignment-based magnetic resonance-only approaches for prostate radiotherapy." *Radiat. Oncol.* 10:239. doi: 10.1186/s13014-015-0549-7.

Kim, T., T. N. Showalter, W. T. Watkins, D. M. Trifiletti, and B. Libby. (2015). "Parallelized patient-specific quality assurance for high-dose-rate image-guided brachytherapy in an integrated computed tomographyeon-rails brachytherapy suite." *Brachytherapy* 14:834–39.

Kirisits, C., M. Rivard, D. Baltas, F. Ballester, M. De Brabandere, R. van der Laarse, Y. Niatsetski, P. Papagiannis, T. P. Hellebust, J. Perez-Calatayud, K. Tanderup, J. L. M. Venselaar, and F. A. Siebert. (2014). "Review of clinical brachytherapy uncertainties: Analysis guidelines of GEC-ESTRO and the AAPMq." *Radiother. Oncol.* 110:199–212.

Kolkman-Deurloo, I-K. K., A. G. Visser, C. G. J. H. Niël, N. Driver, and P. C. Levendag. (1994). "Optimization of interstitial volume implants." *Radiother. Oncol.* 31:229–39.

Kubo, H., G. P. Glasgow, T. D. Pethel, B. R. Thomadsen, and J. F. Williamson. (1998). "High dose-rate brachytherapy treatment delivery: Report of the AAPM Radiation Therapy Committee Task Group No. 59." *Med. Phys.* 25(4):375–403.

Kuske, R. and R. Patel. (2007). "Breast brachytheapy and breast augmentation: Breast conservation with capsular contracutre." *Sem. Breast Disease* 10(1):42–9.

Kwan, D., R. Kagan, A. Olch, P. Chan, B. Hintz, and M. Wollin. (1983). "Single- and double-plan iridium-192 interstitial implants: Implantation guidelines and dosimetry." *Med. Phys.* 10(4):456–61.

Lahanas, M., D. Baltas, and N. Zamboglou. (1999). "Anatomy-based three-dimensional dose optimisation in brachytherapy using multiobjective genetic algorithms." *Med. Phys.* 26:1904–18.

Lambin, P., A. Gerbaulet, A. Kramar, P. Scalliet, C. Haie-Meder, E. P. Malaise, and D. Chassagne. (1993). "Phase III trial comparing two low dose rates in brachytherapy of cervix carcinoma: report at two years." *Int. J. Radiat. Oncol. Biol. Phys.* 25(3):405–12.

Lapuz, C., C. Dempsey, A. Capp, and P. C. O'Brien. (2013). "Dosimetric comparison of optimization methods for multichannel intracavitary brachytherapy for superficial vaginal tumors." *Brachytherapy* 12:637–44.

Lea, D. E. and D. G. Catcheside. (1942). "The mechanism of the induction by radiation of chromosome aberrations in Tradescantia." *J. Genetics* 44:216–45.

Lee, C. D. (2014). "Recent developments and best practice in brachytherapy treatment planning." *Br. J. Radiol.* 87(1041):20140146.

Lessard, É. and J. Pouliot. (2001). "Inverse planning anatomy-based dose optimization for HDR-brachytherapy of the prostate using fast simulated annealing algorithm and dedicated objective function." *Med. Phys.* 28(5):773–79.

Lindegaard, J. C., M. L. Madsen, A. Traberg, B. Meisner, S. K. Nielsen, K. Tanderup, H. Spejlborg, L. U. Fokdal, and O. Norrevang. (2016). "Individualised 3D printed vaginal template for MRI guided brachytherapy in locally advanced cervical cancer." *Radiother. Oncol.* 118:173–76.

Maspero, M., P. R. Seevinck, G. Schubert, M. A. Hoesl, B. van Asselen, M. A. Viergever, J. J. Lagendijk, G. J. Meijer, and C. A. van den Berg. (2017). "Quantification of confounding factors in MRI-based dose calculations as applied to prostate IMRT." *Phys. Med. Biol.* 62(3):948–65. doi: 10.1088/1361-6560/aa4fe7.

Mazeron, J. J., J. M. Simon, J. Crook, E. Calitchi, Y. Otmezguine, J. P. Le Bourgeois, and B. Pierquin. (1991). "Influence of dose rate on local control of breast carcinoma treated by external beam irradiation plus iridium 192 implant." *Int. J. Radiat. Oncol. Biol. Phys.* 21(5):1173–77.

McGeachy, P., J. Madamesila, A. Beauchamp, and R. Khan. (2015). "An open-source genetic algorithm for determining optimal seed distributions for low-dose-rate prostate brachytherapy." *Brachytherapy* 14:692–702.

Meertens, H., J. Borger, M. Steggerda, and A. Blom. "Evaluation and Optimization of Interstitial Brachytherapy Dose Distributions." In *Brachytherapy from Radium to Optimization*. J. J. Battermann, R. F. Mould, A. A. Martinez, and B. L. Speiser, Eds. Veenendaal, The Netherlands: Nucletron BV, 1994.

Meltsner, M. A., N. J. Ferrier, and B. R. Thomadsen. (2007). "Observations on rotating needle insertions using a brachytherapy robot." *Phys. Med. Biol.* 52(19):6027–37.

Merrick, G. S., W. M. Butler, A. T. Dorsey, and J. H. Lief. (1999). "The dependence of prostate postimplant dosimetric quality on CT volume determination." *Int. J. Radiat. Oncol. Biol. Phys.* 44(5):1111–17.

Merrick, G. S., W. M. Butler, A. T. Dorsey, and H. L. Walbert. (1998). "Influence of timing on the dosimetric analysis of transperineal ultrasound-guided, prostatic conformal brachytherapy." *Radiat. Oncol. Investig.* 6(4):182–90.

Merrick, G. S., W. M. Butler, P. Grimm, M. Morris, J. H. Lief, A. Bennett, and R. Fiano. (2014). "Multisector prostate dosimetric quality: Analysis of a large community database." *Brachytherapy* 13(2):146–51. doi: 10.1016/j.brachy.2013.08.003.

Merrick, G. S., W. M. Butler, K. E. Wallner, J. C. Blasko, J. Michalski, J. Aronowitz, P. Grimm, B. J. Moran, P. W. McLaughlin, J. Usher, J. H. Lief, and Z. A. Allen. (2005). "Variability of prostate brachytherapy pre-implant dosimetry: a multi-institutional analysis." *Brachytherapy* 4(4):241–51.

Moerland, M. A., R. van der Laarse, R. W. Luthmann, H. K. Wijrdeman, and J. J. Battermann. (2000). "The combined use of the natural and the cumulative dose-volume histograms in planning and evaluation of permanent prostatic seed implants." *Radiother. Oncol.* 57(3):279–84. doi: S0167814000002899 [pii].

Morton, D., D. Batchelar, M. Hilts, T. Berrang, and J. Crook. (2016). "Incorporating three-dimensional ultrasound into permanent breast seed implant brachytherapy treatment planning." *Brachytherapy* 16(1):167–73.

Nag, S., D. Shasha, N. Janjan, I. Petersen, and M. Zaider. (2001). "The American Brachytherapy Society recommendations for brachytherapy of soft tissue sarcomas." *Int. J. Radiat. Oncol. Biol. Phys.* 49(4):1033–43.

Nag, S., H. Cardenes, S. Chang, I. J. Das, B. Erickson, G. S. Ibbott, J. Lowenstein, J. Roll, B. Thomadsen, and M. Varia. (2004). "Proposed guidelines for image-based intracavitary brachytherapy for cervical carcinoma: report from Image-Guided Brachytherapy Working Group." *Int. J. Radiat. Oncol. Biol. Phys.* 60(4):1160–72.

Nath, R., L. L. Anderson, J. A. Meli, A. J. Olch, J. A. Stitt, and J. F. Williamson. (1997). "Code of practice for brachytherapy physics: Report of the AAPM Radiation Therapy Committee Task Group No. 56." *Med. Phys.* 24(10):1157–98.

Nath, R., W. S. Bice, W. M. Butler, Z. Chen, A. S. Meigooni, V. Narayana, M. J. Rivard, and Y. Yu. (2009). "AAPM recommendations on dose prescription and reporting methods for permanent interstitial brachytherapy for prostate cancer: report of Task Group 137." *Med. Phys.* 36(11):5310–22.

Niël, C. G., P. C. Koper, A. G. Viser, D. Simpkema, and P. C. Levendag. (1994). "Optimizing brachytherapy for locally advanced cervical cancer." *Int. J. Radiat. Oncol. Biol. Phys.* 29:873–77.

Niemierko, A. (1997). "Reporting and analyzing dose distributions: a concept of equivalent uniform dose." *Med. Phys.* 24(1):103–10.

Orcutt, K. P., B. Libby, L. L. Handsfield, G. Moyer, and T. N. Showalter. (2014). "CT-on-rails guided HDR brachytherapy: single-room, rapid-workflow treatment delivery with integrated image guidance." *Future Oncol.* 10(4):569–75.

Paley, P., W. J. Koh, K. J. Stelzer, B. A. Goff, H. K. Tamimi, and B. E. Greer. (1999). "A new technique for performing syed template interstitial implants fo anterior vaginal tumors using an open retropubic approach." *Gynecol. Oncol.* 73:121–25.

Potter, R., C. Haie-Meder, E. Van Limbergen, I. Barillot, M. de Bragandere, J. Dimopoulos, I. Dumas, B. Erickson, S. Lang, A. Nulens, P. Petrow, J. Rownd, and C. Kirisits. (2006). "Recommendations from gynaecological (GYN) GEC ESTRO working group (II): Concepts and terms in 3D image-based treatment plannning in cervix-ancer brachytherapy-3D dose volume parameters and aspects of 3D image-based anatomy, radiation physics, radiobiology." *Radiother. Oncol.* 78:67–77.

Pouliot, J., D. Trembley, J. Roy, and S. Filice. (1996). "Optimization of permanent 125-I prostate implants using fast simulated annealing." *Int. J. Radiat. Oncol. Biol. Phys.* 36(3):711–20.

Pouliot, J., É. Lessard, and I-C Hu. "Advanced 3-D Planning." In *Brachytherapy Physics,* 2nd edition. B. R. Thomadsen, M. J. Rivard, and W. M. Butler, Eds. Madison, WI: Medical Physics Publishing, 2005.

Ritter, M., S. Shahabi, M. Gehring, T. Shanahan, B. Thomadsen, and T. Kinsella. (1989). "Transperineal prostate implantation with three-dimensional, computed tomography-based preplanning and customized template design." *Endocurietherapy/Hyperthermia Oncol.* 5:254.

Roberson, P. L., V. Narayana, D. L. McShan, R. J. Winfield, and P. W. McLaughlin. (1997). "Source placement error for permanent implant of the prostate." *Med. Phys.* 24(2):251–7.

Roy, J. N., K. E. Wallner, S. T. Chiu-Tsao, L. L. Anderson, and C. C. Ling. (1991). "CT-based optimized planning for transperineal prostate implant with customized template." *Int. J. Radiat. Oncol. Biol. Phys.* 21:483–89.

Sachs, R. K. and D. J. Brenner. (1998). "The mechanistic basis of the linear-quadratic formalism." *Med. Phys.* 25(10):2071–73.

Sachs, R. K., P. Hahnfeld, and D. J. Brenner. (1997). "The link between low-LET dose-response relations and the underlying kinetics of damage production/repair/misrepair." *Int. J. Radiat. Oncol. Biol. Phys.* 72(4):351–74.

Shah, C., F. Vicini, D. E. Wazer, D. Arthur, and R. R. Patel. (2013). "The American Brachytherapy Society consensus statement for accelerated partial breast irradiation." *Brachytherapy* 12:267–77.

Shah, C., J. V. Antonucci, J. B. Wilkinson, M. Wallace, M. Ghilezan, P. Chen, K. Lewis, C. Mitchell, and F. Vicini. (2011). "Twelve-year clinical outcomes and patterns of failure with accelerated partial breast irradiation versus whole-breast irradiation: Results of a matched-pair analysis." *Radiother. Oncol.* 100:210–14.

Siauw, T., A. Cunha, D. Berenson, A. Atamturk, I-C Hsu, K. Goldberg, and J. Pouliot. (2012). "NPIP: A skew line needle configuration optimization system for HDR brachytherapy." *Med. Phys.* 39(7)4339–46.

Siauw, T., A. Cunha, A. Atamtürk, I-C Hsu, J. Pouliot, and K. Goldberg. (2011). "IPIP: A new approach to inverse planning for HDR brachytherapy by directly optimizing dosimetric indices." *Med. Phys.* 38(7):4045-51.

Slessinger, E. (2010). "Practical considerations for prostate HDR brachytherapy." *Brachytherapy* 9:282–87.

Sloboda, R. S. (1992). "Optimization of brachytherapy dose distribution by simulated annealing." *Med. Phys.* 19:234–44.

Stock, R. G., K. Chan, M. Terk, J. K. Dewyngaert, N. N. Stone, and P. Dottino. (1997). "A new technique for performing Syed-Neblett template interstitial implants for gynecologic malignancies using transrectal-ultrasound guidance." *Int. J. Radiat. Oncol. Biol. Phys.* 37(4):819–25.

Tagliaferri L., S. Manfrida, B. Barbaro, M. M. Colangione, V. Masiello, G.C. Mattiucci, E. Placidi, R. Autorino, M.A. Gambacorta, S. Chiesa, G. Mantini, G. Kovác, and V. Valentini. (2015). "MITHRA—multiparametric MR/CT image adapted brachytherapy (MR/CT-IABT) in anal canal cancer: a feasibility study." *J. Contemp. Brachytherapy* 7(5):336–45.

Tanderup, K., R. Potter, J. C. Lindegaard, D. Berger, A. Wambersie, and C. Kirisits. (2010). "PTV margins should not be used to compensate for uncertainties in 3D image guided intracavitary brachytherapy." *Radiother. Oncol.* 97(3):495–500. doi: 10.1016/j.radonc.2010.08.021.

Taussky, D., L. Austen, A. Toi, I. Yeung, T. Williams, S. Pearson, M. McLean, G. Pond, and J. Crook. (2005). "Sequential evaluation of prostate edema after permanent seed prostate brachytherapy using CT-MRI fusion." *Int. J. Radiat. Oncol. Biol. Phys.* 62(4):974–80. doi: S0360-3016(04)03040-8 [pii]. 10.1016/j.ijrobp.2004.12.012 [doi].

Thames, H. D., Jr., H. R. Withers, L. J. Peters, and G. H. Fletcher. (1982). "Changes in early and late radiation responses with altered dose fractionation: implications for dose-survival relationships." *Int. J. Radiat. Oncol. Biol. Phys.* 8(2):219–26.

Thomadsen, B. R. "Achieving quality in brachytherapy." In *Achieving Quality in Brachytherapy.* Philadelphia: Institute of Physics Publishing, 1999.

Thomadsen, B. R., P. V. Houdek, G. Edmundson, R. van der Laarse, I-K. K. Kolkman-Deurloo, and A. G. Visser. "Treatment Planning and Optimization." In *High Dose Rate (HDR) Brachytherapy: A Textbook.* S. Nag, Ed. Armonk, NY: Futura Publishing Company, 1994.

Tobias, C. A. (1985). "The repair-misrepair model in radiobiology: comparison to other models." *Radiat. Res. Supplement* 8:S77–95.

van der Laarse, R. and R. W. De Boer. (1990). "Computerized High Dose Rate Brachytherapy Treatment Planning." In *Brachytherapy HDR and LDR.* C. G. Orton, A. A. Martinez, and R. F. Mould, Eds. Columbia, MD: Nucletron Corporation, 1990

Venselaar, J. L. M., L. Beaulieu, Z. Chen, and R. L. Smith. "Advances in Brachytherapy Physics." In *Advances in Medical Physics.* J. Van Dyk, D. J. Godfrey, S. K. Das, B. H. Curran, and A. B. Wolbarst, Eds. Madison, Wisconsin: Medical Physics Publishing, 2016.

Viswanathan, A., S. Beriwal, J. De Los Santos, D. J. Demanes, D. Gaffney, J. Hansen, E. Jones, C. Kirisits, B. Thomadsen, and B. Erickson. (2012). "American Brachytherapy Society consensus guidelines for locally advanced carcinoma of the cervis. Part II: High-dose-rate brachytherapy." *Brachytherapy* 11:47–52.

Weitmann, H. D., T. H. Knocke, C. Waldhäusl, and R. Pötter. (2006). "Ultrasound-guided interstitial brachytherapy in the treatment of advanced vaginal recurrences from cervical and endometrial carcinoma." *Strahlenther. Onkol.* 182(2):86–95.

Williamson, J., P. Dunscombe, M. Sharpe, B. R. Thomadsen, J. A. Purdy, and J. A. Deye. (2008). "Quality Assurance Needs for Modern Image-Based Radiotherapy: Recommendations From 2007 Interorganizational Symposium on Quality Assurance of Radiation Therapy: Challenges of Advanced Technology." *Int. J. Radiat. Oncol. Biol. Phys.* 71(1, Supplement):S2–S12.

Wybranski C., B. Eberhardt, K. Fischbach, F. Fischbach, M. Walke, P. Hass, F. W. Röhl, O. Kosiek, M. Kaiser, M. Pech, L. Lüdemann, and J. Ricke. (2015). "Accuracy of applicator tip reconstruction in MRI-guided interstitial ^{192}Ir-high-dose-rate brachytherapy of liver tumors." *Radiother. Oncol.* 115(1):72–7.

Yang, G., L. E. Reinstein, S. Pai, Z. Xu, and D. L. Carroll. (1998). "A new genetic algorithm technique for in optimization of permanent ^{125}I prostate implants." *Med. Phys.* 25:2308–15.

Yoo, S., M. E. Kowalok, B. R. Thomadsen, and D. L. Henderson. (2003). "Treatment planning for prostate brachytherapy using region of interest adjoint functions and a greedy heuristic." *Phys. Med. Biol.* 48(24):4077–90.

Yoo, S., M. E. Kowalok, B. R. Thomadsen, and D. L. Henderson. 2007. "A greedy heuristic using adjoint functions for the optimization of seed and needle configurations in prostate seed implant." *Phys. Med. Biol.* 52(3):815–28.

Yu, Y. and M. C. Schell. (1996). "A genetic algorithm for the optimization of prostate implants." *Med. Phys.* 23:2085–91.

Zaider, M. (1998a). "Sequel to the discussion concerning the mechanistic basis of the linear quadratic formalism." *Med. Phys.* 25(10):2074–75.

Zaider, M. (1998b). "There is no mechanistic basis for the use of the linear-quadratic expression in cellular survival analysis." *Med. Phys.* 25(5):791–92.

Zaider, M. and G. N. Minerbo. (2000). "Tumour control probability: a formulation applicable to any temporal protocol of dose delivery." *Phys. Med. Biol.* 45(2):279–93.

Zaider, M. and L. Hanin. (2007). "Biologically-equivalent dose and long-term survival time in radiation treatments." *Phys. Med. Biol.* 52(20):6355–62.

Zannis, V. J., L. C. Walker, B. Barclay-White, and C. A. Quiet. (2003). "Postoperative ultrasound-guided percutaneous placement of a new breast brachytherapy balloon catheter." *Am. J. Surg.* 186(4):383–85.

Zoberi, J. E., J. Garcia-Ramirez, Y. Hu, B. Sun, C. G. Bertelsman, P. Dyk, J. K. Schwarz, and P. W. Grigsby. (2016). "Clinical implementation of multisequence MRI-based adaptive intracavitary brachytherapy for cervix cancer." *J. Appl. Clin. Med. Phys.* 17(1):121–31.

Zwicker, R. D. and R. Schmidt-Ulrich. (1995). "Dose uniformity in a planar interstitial implant system." *Int. J. Radiat. Oncol. Biol. Phys.* 31(1):149–55.

Zwicker, R. D., D. W. Arthur, B. D. Kavanagh, R. Mohan, and R. Schmidt-Ulrich. (1999). "Optimization of planar high-dose-rate implants." *Int. J. Radiat. Oncol. Biol. Phys.* 44(5):1171–77.

Example Problems

(Answers are found at the end of the book.)

1. The biologic effect of a given radiation dose can be affected by the rate of dose delivery.

   True

   False

2. The x-ray survival curve of a particular cell line can be described by the linear-quadratic model with $\alpha = 0.4\ \text{Gy}^{-1}$ and $\beta = 0.2\ \text{Gy}^{-2}$. What is the dose at which the amount of one-track lethal damage equals that of two-track lethal damage?

   a. 0.08 Gy

   b. 0.16 Gy

   c. 0.4 Gy

   d. 2.0 Gy

3. Brachytherapy treatment planning using MRI only to reconstruct applicators and patients' anatomy _____.

   a. is not possible due to the missing Hounsfield units

   b. needs mandatory fusion with CT

   c. can use artifacts to determine the position of applicators

   d. has usually the highest accuracy at the periphery of the scanning field

4. What does the Pareto front indicate in the context of radiotherapy treatment planning?

   a) a treatment plan that achieves all the optimization goals

   b) treatment plans that minimize the differences between the optimization goals and the metrics obtained for a specific treatment plan

   c) treatment plans where at least one optimization goal is achieved and the other metrics are close to the planning objectives

   d) a meteorological phenomenon where high pressure from the therapists to finish a plan is balanced by the low pressure of information provided in the physician's optimization goals.

5. In a routine HDR brachytherapy interstitial implant, dwell positions should not be activated closer than _____ mm to the skin to avoid potential skin reactions.
   a. 1 mm
   b. 2 mm
   c. 5 mm
   d. 10 mm
6. For CT-based planning for cervical brachytherapy, what is the slice thickness that should be used?
   a. 1 mm
   b. 3 mm
   c. 5 mm
   d. 7 mm
7. For an prostate seed implant performed with ^{103}Pd, what is the recommended period for the performance of the post-implant CT scan?
   a. 10 days after the implant
   b. 16 days after the implant
   c. 30 days after the implant
   d. 60 days after the implant
8. In 2016, the International Commission on Radiation Units and Measurements (ICRU) recommended which imaging modality as the reference standard for gynecologic tumor assessment?
   a. computed tomography (CT)
   b. fluoroscopy or planar x-rays (2D)
   c. magnetic resonance imaging (MRI)
   d. any of the above
9. If prostate implants for ^{131}Cs. ^{103}Pd, and ^{125}I were all prescribed to the same dose, which would produce the greatest biologic effect?
   a. ^{131}Cs
   b. ^{103}Pd
   c. ^{125}I
   d. All would produce the same effect because the prescription is the same.
10. For an interstitial breast brachytherapy treatment, the maximum skin dose should be _____.
   a. 145% of the prescribed dose
   b. 125% of the prescribed dose
   c. 100% of the prescribed dose
   d. 85% of the prescribed dose

Chapter 2

Model-based Dose Calculation Algorithms in Brachytherapy

Luc Beaulieu[1], Firas Mourtada[2], Susan L. Richardson[3], Mark J. Rivard[4], and Ron S. Sloboda[5]

[1]Université Laval
Québec, Canada

[2]Christiana Care Hospital
Newark, Delaware

[3]Swedish Medical Center–Tumor Institute
Seattle, Washington

[4]Tufts University School of Medicine
Boston, Massachusetts

[5]University of Alberta
Edmonton, Canada

2.1 Introduction

For over two decades now, brachytherapy dose calculation has been relying on the AAPM Task Group No. 43 report (Nath et al. 1995), its 2004 update, TG-43U1 (Rivard et al. 2004), as well as its extension to high-energy brachytherapy sources (Perez-Calatayud et al. 2012). The widespread adoption of the TG-43 formalism brought uniformity and consistency to brachytherapy dose calculation worldwide. It further provided us with numerous advances over previous protocols:

- a primary standard and traceable source strength measurements (DeWerd et al. 2011);

- detailed, source-specific dose distribution in a homogeneous, infinite water medium, including attenuation and filtration from the source's internal components and encapsulation; and
- tabulated values of the various brachytherapy dosimetry parameters to make computation extremely fast and thus allowing dose optimization with thousands of iterations to be performed in one second.

Except very close to the source or for extended sources (active length >1 cm), the recommended TG-43 line-source approximation (or point-source approximation at large distances) provides an excellent description of the dose distribution at all points of the calculation geometry (Rivard et al. 2004).

A study by Perez-Catalayud et al. (2004) demonstrated that the homogeneous, infinite water medium condition is reached as long as there is at least 5 cm of water-like medium past the last point of interest for low-energy ^{125}I seeds and 20 cm of water or tissue for high-energy sources like ^{192}Ir. However, in many clinical situations these conditions are not fulfilled, and full scatter conditions are not established. Furthermore, the physical attenuation properties of the clinical medium can differ from that of water. This is especially true when shielding—either from one source by another (inter-source attenuation) or from metallic components or bones—is present in the region of interest. Departure from the TG-43 formalism in terms of scatter conditions, attenuation, and energy depositions were reviewed by Rivard et al. (2009b) for various treatment sites and separated in terms of low-energy (^{103}Pd, ^{125}I, ^{131}Cs, and electronic sources) vs high-energy (^{192}Ir, ^{137}Cs, and ^{60}Co) brachytherapy. From that review, it becomes quite clear that the condition of 20 cm of backscatter past the ROI boundary is a major constraint for high-energy brachytherapy. It thus affects strongly many treatment sites such as breast or superficial therapies. On the other hand, most soft tissues are water equivalent (within 1%) in that energy range. The situation is very different for low-energy brachytherapy as depicted in terms of energy deposition in Figure 2–1 (Beaulieu et al. 2012). The difference between various tissues exhibits a maximum around 25 keV, where the spread reaches an order of magnitude.

Radiation therapy is based on the fact that dose-outcome relationships guide our treatment decisions, in particular prescription doses to targets and limiting dose levels to organs at risk (OARs). As a consequence, it is likely that for numerous brachytherapy treatment sites, the dose parts of those relationships are currently wrong, and that the solution to this problem does not consist of a simple dose offset correction. Rather, accurate and patient-specific dose calculations will be required (Rivard et al. 2009b; Beaulieu et al. 2012).

2.2 Model-based Dose Calculation

2.2.1 A Radiation Physics Perspective

Photon sources used in modern brachytherapy treatments generally have relatively low photon energy (<400 keV). Hence, it is fairly accurate to assume photon attenuation to be negligible within the range of the most energetic secondary electrons liberated in soft tissue or water. In other words, photon mean free paths are much longer than secondary electron ranges, and those secondary electron ranges are, in turn, much smaller than the calculation voxel sizes used for clinical purposes (1 mm^3 or larger). Under these conditions, one can posit electronic equilibrium and, thus, collisional kerma is equal to dose. It also turns out that radiative energy loss of secondary electrons is negligible in water (and soft tissues) in the energy range of brachytherapy and is 1% at most for ^{60}Co. Thus, it can be further posited that energy transfer equals energy absorption, or $\mu_{tr} = \mu_{en}$, so that the dose is equal to kerma, and thus dose can be related to photon fluence.

For any given point (or voxel) at a distance r from the source in a specific geometry, the dose at that distance can be approximated by the kerma, and the latter can be calculated exactly if the energy fluence, Φ_E, is known at all points of the geometry:

$$D(r) = K(r) = \int_E E\Phi_E(r)\frac{\mu_{en}}{\rho}dE \tag{2.1}$$

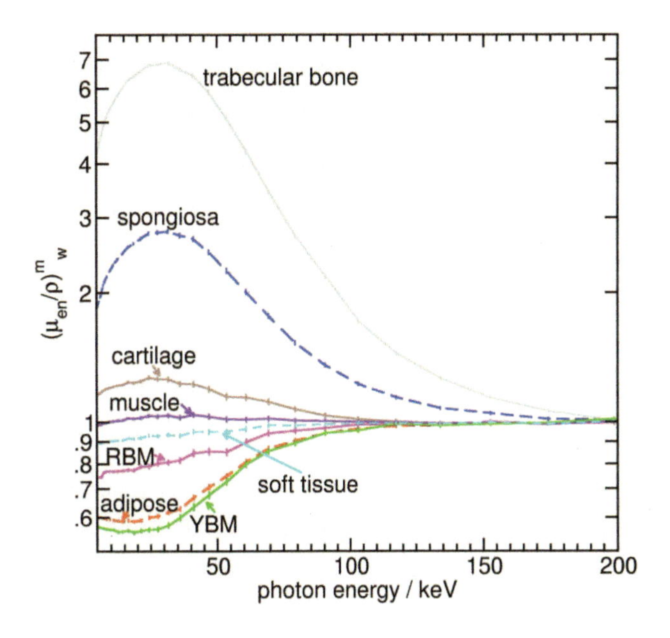

Figure 2–1 Mass-energy absorption coefficients for the materials indicated relative to those for water for energies from 5 to 200 keV, calculated with the EGSnrc user-code *g*. Details on atomic compositions and densities used for the calculation can be found in Beaulieu et al., *Med. Phys.* 39 (2012) 6208–36. (Reproduced with permission.)

For a perfect point source, calculating the primary energy fluence in any voxels in space is trivial, and one can further take into account heterogeneities by using scaling rules. However, the total energy fluence in the voxel of interest is the sum of the primary fluence and the scatter energy fluence coming from all other voxels, in which primary photons will undergo numerous possible types of interaction that will change their direction and energy. The latter process can be repeated multiple times within the geometry of interest. As such, from a theoretical perspective, the only way to have full knowledge of the energy fluence in a given voxel is to know the results of all interactions in all the other voxels of the geometry. This kind of recursive calculation makes it impossible to accomplish the task analytically, so one must resort to numerical methods, and sometimes further assumptions (Papagiannis et al. 2014). The numerical methods used in radiation therapy rely strongly on interaction physics and are called model-based dose calculation algorithms, or MBDCAs.

2.2.2 Overview of Leading Algorithms

Over the last decade, several MBDCAs have been implemented for clinical use for high-dose-rate (HDR) photon brachytherapy. In particular, the three most common MBDCAs are Monte Carlo, Scatter-Separation, and explicit deterministic methods. In this section, the basic concepts and clinical implementation of these MBDCAs for brachytherapy photon sources are described. All MBDCAs presented have the same assumption—that dose is equal to kerma under conditions of charged particle equilibrium as explained above.

2.2.2.1 Monte Carlo

Monte Carlo (MC) methods solve the linear Boltzmann transport equation (LBTE) "inexplicitly" by simulating many particles' behavior to stochastically infer the average behavior of all particles in the transport medium (using the central limit theorem). MC simulations are stochastic in nature, with most common codes used in medical physics applications being PTRAN (Williamson 1987), EGS (Kawrakow and Rogers 2000), and MCNP (Briesmeister 1999). These codes have been well published for brachytherapy dosimetry. More recently, GEANT (Agostinelli et al. 2003) and PENELOPE (Sempau et al. 1997) have also been shown to meet the accuracy requirements. These are general-purpose MC codes that differ in several aspects. While MC is considered the gold standard MBDCA, Rivard and colleagues stressed that extensive validation and comparison with prior results is necessary preceding the use of any MC code for clinical dosimetry (Rivard et al. 2004).

For routine clinical practice, the implementation of MC methods into clinical brachytherapy platforms has been demonstrated, but none has reached commercialization (Chibani and Williamson 2005; Taylor et al. 2007; Pratx and Xing 2011; Hissoiny et al. 2011; Afsharpour et al. 2012). The slow MC calculation speed is the main reason for the lack of integration into clinical TPS. This is inherent to the need for a large number of independent histories (or particles) N required to converge with a small statistical uncertainty. The efficiency ε of a MC calculation is given by Sheikh-Bagheri et al. (2006):

$$\varepsilon = \frac{1}{\sigma^2 T} \tag{2.2}$$

where T is the CPU time and σ^2 is the estimated variance of the quantity of interest. T is proportional to N, and σ^2 decreases with $1/N$. The efficiency of MC can be improved by simulating more independent particles for a given CPU time; these are called variance reduction techniques.

AAPM Monograph 32 has described variance reduction techniques (Sheikh-Bagheri et al. 2006). In brachytherapy, the most obvious and most-used speed-up techniques are the approximation of dose by using kerma and track-length estimators (Williamson 1987; 2005). The kerma approximation assumes charged-particle equilibrium. This condition is satisfied for clinically used voxelized geometries (≥ 1 mm^3 voxels) at ^{192}Ir photon energy or lower. As a result, electron transport can be omitted and their energies can be considered to be deposited locally at the interaction point, enabling a direct increase of N per unit of CPU time. In the track estimators, all photons whose trajectories traverse scoring voxels are used to calculate the kerma, whether or not collisions occur within the voxels. This approach greatly improves the scoring statistics and, thus, the variance. It leads to an increase in efficiency by one to two orders of magnitude over the analog scoring method (Williamson 1987; Hedtjärn et al. 2002). These approaches have been implemented in brachytherapy-specific MC codes such as PTRAN (Williamson 1987), MCPT (Hedtjärn et al. 2002), MCPI (Chibani and Williamson 2005), BrachyDose (Taylor et al. 2007), Geant4-based ALGEBRA (Afsharpour et al. 2012), and, more recently, the EGSnrc-based egs_brachy (Chamberland et al. 2016). Other approaches—such as particle recycling, in which photons emitted by one seed are used for every seed in a multi-seed implant—have also been shown to be highly effective (Thomson et al. 2008). MC methods relying on pre-calculated data that are application-specific (Rivard et al. 2009a; Poon and Verhaegen 2009) have also been proposed.

2.2.2.2 Advanced Scatter-separation Techniques

A popular MBDCA method separates the absorbed dose distribution in terms of the contributions of the primary photon dose and that coming from the scatter particles (Russell and Ahnesjo 1996). Here, primary dose relates to all photons that leave the source capsule surface and contribute to energy deposition from the first interaction. The scatter dose is the dose that results from subsequent interactions. The total dose can be written as

$$\begin{aligned} D &= D_{prim} + D_{scat} \\ &= D_{prim} + (D_{1sc} + D_{msc}) \end{aligned} \tag{2.3}$$

Generally, D_{prim} and D_{scat} are obtained from MC calculations that include detailed simulations of the 3D geometry. It follows that D_{prim} can be calculated without any approximation through the use of ray-tracing techniques that propagate the photon energy fluence from the source through the geometry. As such, the calculation of D_{prim} will be equivalent for any MBDCA. D_{scat} can be modeled using exponential functions or using pre-calculated MC kernels in water (Russell and Ahnesjo 1996; Carlsson and Ahnesjo 2000a). Russell et al. (2005) have shown that the primary-scatter separation approach can be used to derive TG-43 parameters and vice-versa. Therefore, the primary-scatter separation formalism contains all the necessary information for extensive comparisons of various dose calculation methods around a given source in water.

One highly efficient way to deal with the scatter component is the collapsed-cone convolution technique (Carlsson and Ahnesjo 2000b; Carlsson Tedgren and Ahnesjo 2003; Tedgren and Carlsson 2009; Tedgren et al. 2015) which uses a successive-scattering superposition method. MC methods are used to generate the once-scatter and multi-scatter dose point kernels that are fit to analytical mono- or bi-exponential functions (Carlsson and Ahnesjo 2000a). This method, called Advanced Collapsed-cone Engine or ACE, is included in the OncentraBrachy^{TM} TPS from Elekta (Nucletron BV, The Netherlands) and further details on that specific implementation can be found in the literature (Veelen et al. 2014; Ma et al. 2015).

2.2.2.3 Explicit Deterministic Dose Calculation Methods

Unlike MC, deterministic methods solve the LBTE "explicitly" where the average particle behavior converges using the LBTE differential form in the limit of very fine phase-space mesh spacing. Because these methods are based on all phase-space discretization, such methods are referred to as the grid-based Boltzmann equation solvers (GBBS). In general, the method of characteristics, spherical harmonics, and discrete ordinates are all classified as deterministic, with the later being most commonly applied to medical physics applications (Shapiro et al. 1976; Nigg et al. 1991; Borgers 1998; Daskalov et al. 2000 and 2002; Gifford et al. 2006 and 2008; Vassiliev et al. 2008; Gifford et al. 2010; Mikell and Mourtada 2010; Han et al. 2011). These methods solve the LBTE by discretizing spatial (via finite difference or element meshes), angular (via discrete ordinates, spherical harmonics, etc.), and energy variables (via the multi-group method), which results in a system of linear equations that are iteratively solved. Further details can be found in the TG-186 report (Beaulieu et al. 2012).

The general GBBS equations used for neutral particle transport are briefly described with the focus on brachytherapy photon emitters where secondary charged particle equilibrium is a valid approximation. GBBS solutions for coupled photon-electron radiation transport using the Boltzmann–Fokker–Plank equation (Lewis and Miller 1984) that have been used for external photon beam dose calculations are not discussed.

GBBS solves the 3D linear Boltzmann transport equation, a six-variable integro-differential governing equation for radiation transport (Lewis and Miller 1984). Briefly, for volume *V* with surface *δV*, the linear Boltzmann transport equation (along with vacuum boundary conditions) is given by:

$$\hat{\Omega}\cdot\vec{\nabla}\Psi(\vec{r},E,\hat{\Omega})+\sigma_t(\vec{r},E)\Psi(\vec{r},E,\hat{\Omega})=Q^{scat}(\vec{r},E,\hat{\Omega})+Q^{ex}(\vec{r},E,\hat{\Omega}),\vec{r}\in V \tag{2.4}$$

$$\Psi(\vec{r},E,\hat{\Omega})=0,\ \vec{r}\in\delta V,\ \hat{\Omega}\cdot\vec{n}<0 \tag{2.5}$$

Here $\Psi(\vec{r},E,\Omega)$ is the angular flux at position $\vec{r}=(x,y,z)$, energy E, direction $\hat{\Omega}=(\mu,\eta,\xi)$, and $\vec{n}$ is the normal vector to surface δV. The first term on the left-hand side of Equation (2.4) is termed the *streaming operator*. The second term on the left-hand side of Equation (2.5) is termed the *collision operator*, while $\sigma_t(\vec{r},E)$ is the macroscopic total cross section. The right-hand side of Equation (2.5) includes the *source* terms, where $Q^{scat}(\vec{r},E,\Omega)$ is the *scattering source* and $Q^{ex}(\vec{r},E,\Omega)$ is the *extraneous source*. The scattering source is explicitly given as:

$$Q^{scat}(\vec{r},E,\vec{\Omega}\ddot{x})=\int_0^{\infty}dE'\int_{4\pi}\sigma_s(\vec{r},E'\rightarrow E,\Omega\cdot\Omega')\Psi(\vec{r},E',\Omega')d\hat{\Omega}' \tag{2.6}$$

where $\sigma_s(\vec{r},E'\rightarrow E,\Omega\cdot\Omega')$ is the macroscopic differential scattering cross section. For most deterministic transport methods, it is customary to expand the macroscopic differential scattering cross section in Legendre polynomials and to expand the angular flux appearing in the scattering source in spherical harmonics. Further details on the expansion operators are found in the literature (Lewis and Miller 1984; Wareing et al. 1998 and 2001). Once the solution to the transport equation has been obtained, any reaction rate, such as kerma rate or absorbed dose rates to medium, can be obtained from:

$$\dot{R}(\vec{r}) = \int_{0}^{\infty} \left(\frac{\sigma(\vec{r})}{\rho} \right) \phi_{0,0}(\vec{r}) dE. \quad (2.7)$$

Typically $\phi_{0,0}(\vec{r})$ is referred to as the "scalar flux" and often times is just given by $\phi(\vec{r})$.

As discussed, when deterministically solving the LBTE (Equation (2.4)), one must discretize all variables: energy (E), angle ($\hat{\Omega}$), and space ($\vec{r}$).

For brachytherapy sources, several authors studied the impact of the number of energy groups on both accuracy and speed of convergence (Daskalov et al. 2000 and 2002; Gifford et al. 2006 and 2008). Cross sections produced by CEPXS4 are typically used and suitable for brachytherapy source energy (Lorence et al. 1989). CEPXS includes all photon interactions with the exception of Rayleigh scatter, the effect of which is insignificant for dose distributions at energies produced by brachytherapy sources such as ^{192}Ir.

For clinical use of GBBS, Mourtada et al. in 2004 evaluated the 3D Attila® radiation transport code (LANL, Los Alamos, New Mexico) for the dosimetry of a pulsed dose rate (PDR) ^{192}Ir source in water (Mourtada et al. 2004; Gifford et al. 2006 and 2008). Based on the Attila work, a clinical GBBS platform called Acuros was developed (Transpire Inc, Gig Harbor, WA) and licensed for use by Varian BrachyVision TPS.

Major differences between deterministic and Monte Carlo solvers include:

1. Deterministic methods are non-stochastic, so solution errors arise from systematic sources rather than statistical sources.
2. Deterministic methods provide full solution for the entire space rather than for specific regions (or tally location) done in Monte Carlo.
3. Deterministic solutions can be more efficient than Monte Carlo once derived for similar problems solved previously, i.e., similar brachytherapy sources and patient volumes.

2.3 Current Clinical Implementations and Limitations

There are currently only two commercially available general-purpose advanced dose calculation algorithms used for brachytherapy dose calculations: Elekta ACE and Varian AcurovBV™. These two algorithms were released for high-energy HDR brachytherapy, in particular ^{192}Ir, with ACE also able to perform calculations for ^{60}Co. From a theoretical perspective, both of these approaches could also efficiently tackle low-energy sources. For example, Sloboda and colleagues (2017) have explored, as a proof-of-principal, the use of ACE for ^{125}I eye brachytherapy plaque dose calculations.

The latest versions of these two algorithms follow the basic recommendations from the TG-186 report, namely having predefined materials for various organs to be assigned based on the physician's delineated structures, use of CT number for density assignment (if no artifacts are present), and reporting the dose as $D_{M,M}$, i.e., radiation transported in the medium and energy deposition calculated, also in the medium of interest (Beaulieu et al. 2012). They also share common calculation steps, in particular the calculation of the primary dose without any approximation using ray-tracing with scaling. As such, both of these algorithms are expected to match the ground truth at that stage; they essentially differ in the way the scatter dose is calculated, as described in Section 2.2.

2.3.1 ACE

ACE implementation follows Equation (2.3) and will, therefore, carry two more calculation steps in addition to the primary dose. All calculation parameters are dictated by a single user-selectable option: standard or high resolution. These options, in turn, control the angular discretization for both the *1sc* and *msc* doses and the adaptive voxel sizes. In standard mode, 320/180 directions are used for *1sc/msc* between 1 and 50 dwell positions, numbers that will go down to 180/72 when over 300 dwell positions are involved (Veelen et al. 2014). In the high-resolution mode, the number of directions will be 1620/240 for a single dwell position, 720/240 for 2–50 dwell positions, and decreasing to 240/128 for 300 or more dwell positions. Similarly, ACE subdivides a calculation geometry in up to four regions, the first one encompassing all of the dwell positions and having 1 mm^3 voxel size. However, as the distance from this first box increases, the voxel size also increases to $2 \times 2 \times 2$ mm^3 at 8 (20) cm, $5 \times 5 \times 5$ mm^3 at 20 (35) cm, and $10 \times 10 \times 10$ mm^3 at 50 (50) cm for standard (high) resolution mode (Veelen et al. 2014). This implementation, similar to the octree method (Hubert-Tremblay et al. 2006), reduces the calculation time in regions where the dose is a fraction of the prescription dose, but for which the calculation volume is large.

Figure 2–2 demonstrates that ACE accurately renders the primary dose, as expected, as well as the *1sc* dose over a wide range of distances from the source. Small differences appear beyond 8 cm. Large differences are seen for the *msc* component already at 6 cm.

It turns out that for most clinical situations, the standard resolution gives excellent agreement with MC. However, the most difficult case for the algorithm is a calculation involving a single dwell where ray-effects due to angular discretization can be clearly seen even in the high-resolution mode at large distance (low-dose region) (Ma et al. 2015). Both studies by Papagiannis and colleagues (2014) and Ma and colleagues (2015) demonstrated the good performance of the ACE algorithm compared to Monte Carlo for a variety of clinical cases. They also pointed out the limitation related to the approximations made in generation of the *msc* kernels, which are seen in areas far from the sources, in low-dose regions, and in proximity to high-Z structures such as bone.

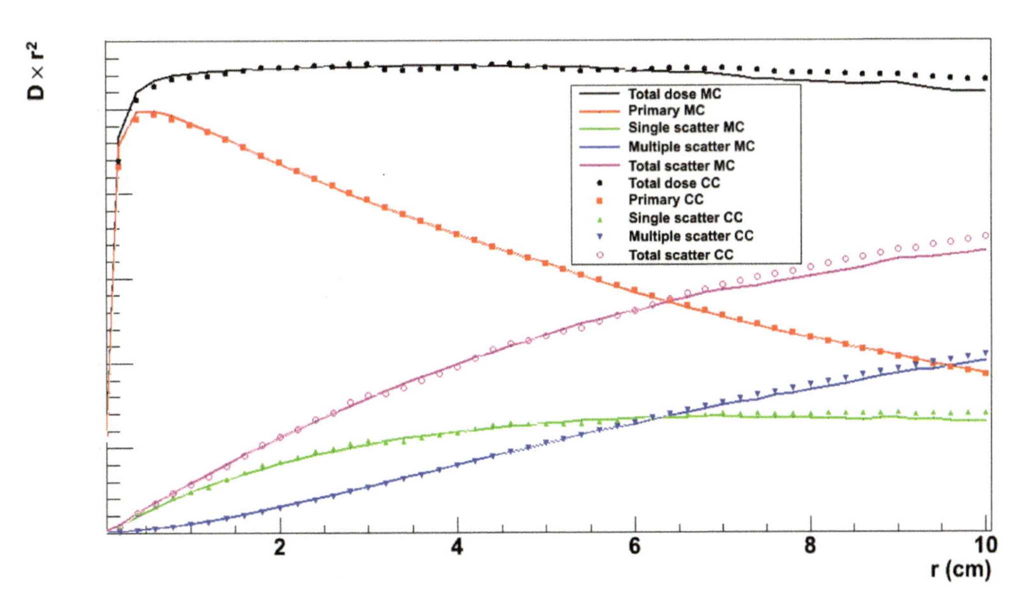

Figure 2–2 Comparison of the various components calculated by ACE (Equation 2.3) to Monte Carlo for a single dwell position in a homogeneous water geometry with full scatter conditions.

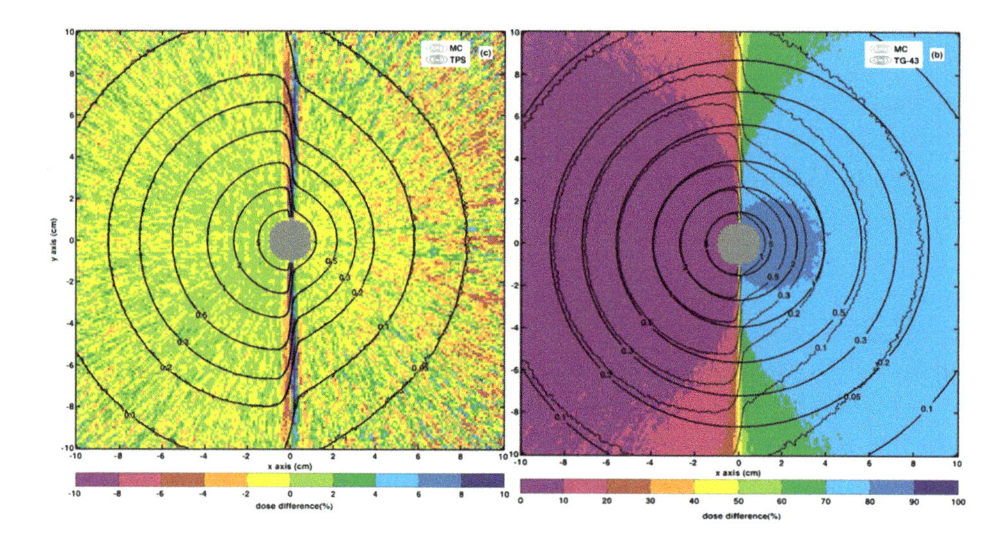

Figure 2–3 Dose distribution in the axial central plane of a 180-degree shielded vaginal cylinder comparing MC and AcurosBV (left) and MC and TG-43 (right). The color map gives the percentage differences between the two algorithms in each plot. (From Petrokokkinos et al. 2010 with permission.)

2.3.2 AcurosBV

Similarly to ACE, end users do not have access to AcurosBV algorithm parameters. As such, energy groups (related to interaction cross sections) are fixed to 37 and angular discretization with an adaptive scheme using between 24 and 960 directions (with the higher energy associated with a large number of directions). Using a low number of angular directions can lead to significant discritization artifacts or ray effects (Papagiannis et al. 2012).

Acuros was first made commercially available for brachytherapy in 2009 and has been extensively validated against Monte Carlo and experimental measurements in well-controlled geometry (Zourari et al. 2010; Petrokokkinos et al. 2011; Zourari et al. 2013) as well as clinical geometries (Mikell et al. 2012; Siebert et al. 2013; Papagiannis et al. 2014; Zourari et al. 2015). The advantages of an MBDCA such as AcurosBV is clearly demonstrated in Figure 2–3 where a shielded applicator is used. While the percentage difference between MC and AcurosBV is within 2% for almost all voxels, it is easy to see the large overdosage predicted by TG-43 behind the shield, with differences of 70% or more. There is also an effect of 10% to 20% on the unshielded side.

Papagiannis and colleagues (2014) further demonstrated some differences between AcurosBV and MC in the penumbra region around high-Z materials, as well as some differences in scatter-dominated regions (e.g., behind bones). Overall, however, AcurosBV performs on par with MC.

2.4 Impact of MBDCA for Selected Clinical Tumor Sites

Rivard and colleagues considered the field of brachytherapy physics and created an important table where the anatomical treatment sites are categorized into which were potentially influenced by heterogeneous dose calculations (Rivard et al. 2009b). The mechanism of dose calculation variability changes depending on the clinical site and whether the treatment energy is high or low (low energy typically referring to <50 keV). The mechanisms of different calculation variables are shown in Table 2.1 for the main clinical treatment sites.

Much work has been done by a variety of authors and institutions to quantify and predict for each clinical site the changes in anticipated dose distribution and absorbed dose. The following sections give a brief overview of the results of this work. Most authors used a combination of MC simulation, commercial treatment planning utilizing MBDCAs (e.g., Acuros), and measurements to evaluate the dosimetry of various clinical situations.

Table 2–1 Sensitivity of commonly treated anatomic sites to dosimetric limitations of the current brachytherapy dose calculation formalism

Clinical Site	Dominant Calculation Variable
Prostate (LDR)	Interseed attenuation Tissue composition
Breast (HDR)	Backscatter conditions Air/contrast in applicator
Breast (LDR)	Tissue composition
Gyn	High-Z materials, including contrast agents

2.4.1 HDR and LDR Prostate Implants

LDR prostate dosimetry, while seemingly simple from a heterogeneity standpoint, is influenced by tissue heterogeneity and inter-seed attenuation (Beaulieu et al. 2012). Since the typical photon energy for these procedures is low compared to other types of brachytherapy, the change in deposited dose may be significant, particularly in areas of large inhomogeneity, like in prostatic calcifications (Collins Fekete et al. 2015; Miksys et al. 2017).

Inter-seed attenuation and the effects of seed anisotropy have been studied by Chibani and colleagues, who found that the absolute total dose differences between a Monte Carlo simulation and a point-source dose-kernel superposition were in the range of 6% to 7% for ^{103}Pd implants and 4% to 5% for ^{125}I (Chibani et al. 2005). Tissue inhomogeneities were studied by Carrier and colleagues, who found them to affect the dose by approximately 2.5% (Carrier et al. 2007). Similarly, Oliveria and colleagues studied tissue inhomogeneity and the resulting change in DVH parameters. They found the mean absorbed dose was overestimated by 3% to 4% depending on prostate composition. D_{90} was overestimated by 2.8% to 3.9%, and maximum doses in the rectum were changed in the range of 6% to 8% (Landry et al. 2010; Oliveira et al. 2014; Miksys et al. 2017).

2.4.2 Breast Brachytherapy

Breast brachytherapy has been identified as one of the treatment areas where heterogeneous dose calculation algorithms show a significant dose difference from their homogeneous counterparts. One of the main factors contributing to this phenomenon is the physical construction of the applicators—whether they are filled with air and seroma like in a SAVI™ applicator (Cianna Medical, Aliso Viejo, CA) or filled with a contrast/water solution like in a balloon-based applicator. The other factor is the finite dimensions of the patient, with the dose being delivered near the skin surface and near the ribs and the patient's lungs. Consequently, doses to the PTV or critical structures may be either increased or decreased depending on the patient presentation and applicator chosen. Surface and maximum skin doses are, in general, reduced due to the lack of backscatter.

The original heterogeneity in APBI brachytherapy was the influence of contrast (high-Z iodine) in the MammoSite™ balloon (Hologic, Inc., Bedford, MA). Depending on the balloon diameter and concentration used, the influence could create a dose reduction through the entire target from 1% to 6% (Zhang et al. 2007). Recommendations were incorporated in the use of the device to use as little contrast as necessary to visualize the balloon/PTV interface.

The dose modulation factor for the Contura™ applicator (Hologic, Inc., Bedford, MA) was studied explicitly for the case of insufficient backscatter under asymmetric loading conditions (Pearson and Williams 2013). The MCNP5 radiation transport code was used to calculate the effect and was found to underdose the PTV from 7% to 12% relative to a TG-43 calculation depending on the lack of scatter conditions. Dose perturbations caused by small air pockets and the influence of contrast in the solution were investigated and found to cause approximately 2.25%/cm^3 and dose reductions from 1% to 4%, respectively (Slessinger et al. 2011).

The SAVI applicator differs from other breast APBI applicators in that it is not a balloon-based applicator, and it is not filled with anything on purpose by the physician. The cavity typically consists of some combination of air and seroma. It also comes in different sizes and can be expanded to different diameters; hence, the anticipated dose distribution may be extremely patient-specific and complex. In fact, an entire thesis was devoted to studying the effects (Koontz 2013). Other studies investigating heterogeneous dose calculation using the Acuros TPS with the SAVI applicator included MC simulations and measurements by film, ion chamber, and TLD (Richardson et al. 2013). They found that dose to the prescription point was increased between 2% to 5% depending on the size of the device modeled. The maximum skin dose calculated with the MBDCA-based Acuros TPS was approximately 4% to 5% lower than that calculated with the TG-43 formalism. Additionally, the maximum OAR doses were also lower with heterogeneous calculation: 9.8% lower in the lung and 5.3% lower in the rib. Similarly, Lamberto and colleagues used the Oncentra TPS to retrospectively recalculate SAVI treatment plans and found that patients with applicators close to the skin (<1 cm) had double the skin dose reduction (6% vs 3%) when compared with deep-seated applicators, showing the impact of radiation scatter considerations (Lamberto et al. 2016).

Interstitial breast brachytherapy was studied by Sinnatamby and colleagues. They found that the Acuros-calculated dose was around 4% less to the CTV due to attenuation through the metal catheters and due to tissue inhomogeneity and boundary conditions (Sinnatamby et al. 2015). OAR doses were also reduced compared to TG-43 calculations. While some differences were large in terms of a percentage, the absolute dose difference was not clinically significant. Their findings were similar to interstitial brachytherapy for head and neck, which found D_{90} and V_{100} for the target were about 3% lower when calculated with Acuros (Siebert et al. 2013). In a retrospective study comparing the Oncentra treatment planning system (OncentraBrachy v4.4) to a TG-43 dose calculation, Zourari and colleagues found only small discrepancies in the PTV region (1% to 2%) but a change of 4% for D_{max} on the rib, 5% for V_{10Gy} of the lung, and a 6% change for D_{10cm^3} on the skin (Zourari et al. 2015). In comparing Acuros to a GBBS calculation, Hofbauer and colleagues found similar results, with a decrease of 8% to the skin D_{10cm^3} and 5% for D_{2cm^3} to the rib (Hofbauer et al. 2016).

When using low-energy sources such as electronic brachytherapy, MBDCAs can produce substantial differences to a TG-43 dose calculation. White and colleagues (White et al. 2014), modeled breast tissues from seven patients who underwent APBI treatment and examined the effects of heterogeneous dose calculation. They found the average D_{90} to the PTV was reduced anywhere between 4% and 40% depending on the scenario. The maximum skin dose was also reduced by 10% to 15% due to lack of backscatter (White et al. 2014). While higher-energy APBI dosimetry is less sensitive to the patient-specific tissue composition, the variation of glandular and adipose tissue within the normal breast can substantially impact the dose when delivered with lower-energy radionuclides or electronic brachytherapy sources. In permanent breast brachytherapy with low-energy ^{103}Pd sources, the PTV dose was found to be reduced from 4% to 35% depending on breast composition (Afsharpour et al. 2010; Landry et al. 2010). The authors concluded an accurate patient-specific model is needed to be used in the dosimetry for this kind of low-energy implant (Afsharpour et al. 2011). Shi and colleagues described a method of tissue heterogeneity corrections and biologically effective dose-volume histograms to assess the doses for APBI using an electronic brachytherapy source (Shi et al. 2010). On average, they found the target dose coverage was reduced from 95.0% in water phantoms (planned) to only 66.7% in virtual patient phantoms (actual), exemplifying the differences between a TG-43 dose calculation and full MC simulation. The calculated maximum dose to the ribs was 3.3 times higher than the planned dose; the calculated mean dose to the ipsilateral breast and maximum dose to the skin were reduced by 22% and 17%, respectively. Segala and colleagues also showed that the barium impregnation of the Xoft Axxent^{TM} balloon applicator (Xoft, Inc., Sunnyvale, CA) caused a 6% decrease at the calculation point 1 cm past the balloon (Segala et al. 2011).

2.4.3 Gynecological Applications

In gynecological brachytherapy, material heterogeneities are typically found in the applicators used to deliver the radiation. For cervix cancer brachytherapy, the applicators are typically stainless steel or titanium tandems and either a ring or ovoids, which can also be shielded (see the discussion in Chapter 9). A stainless steel or metal tandem may be preferable to some practitioners due to the rigidity of the applicator, though plastic applicators can be used in MR imaging if the department lacks expensive CT-MR compatible titanium applicators. When using a stainless steel tandem as opposed to a plastic one, the dose reduction to a classic point A using the Manchester system is reduced by at least 3% (Parsai et al. 2009). Similarly, another study found smaller effects for cervix cases with a plastic applicator, with up to –2% (–0.2 Gy EQD2) per fraction for OARs and –0.5% (–0.3 Gy EQD2) per fraction for CTV_{HR}. The shielding effect of the titanium applicator resulted in a decrease of 2% for D_{2cm^3} to the OAR versus 0.7% for plastic (Hyer et al. 2012). Other studies have found a 1% to 2% reduction in point A and ICRU bladder and rectum points when calculated with an MDCBA-based TPS (Hofbauer et al. 2016). Another group investigated cervix cancer brachytherapy where patients had contrast in the Foley balloon of the bladder and contrast deposited via a rectal tube. Radiopaque gauze was also used for packing of the ovoids. Mikell and colleagues investigated the three different factors contributing to dose differences and found dose differences ranging from –0.5% to 2.5% for source and boundary conditions, –1% to 2% for applicator attenuation, and the influence of material heterogeneity ranging from –3.5% to 1.5% (Mikell et al. 2012). They concluded that the MBDCA had minimal clinical impact.

2.4.4 Surface and Skin Brachytherapy

Jeong and colleagues measured the dose profiles from the Varian Leipzig-style HDR ^{192}Ir skin applicators using radiochromic EBT3 film (Ashland, Inc., Bridgewater, NJ), a photon diode, and optically stimulated luminescence dosimeters for three different GammaMedplus HDR ^{192}Ir afterloaders. They found that measured depth-doses agreed well with Acuros dose calculations. Additionally, they found that with a very short source-to-surface distance, the small source sag inside the applicator has a significant dosimetric impact that is not accounted for when treatment planning with the applicator positioned upside-down (Jeong et al. 2016).

2.5 Commissioning Process

The application of MBDCAs in brachytherapy treatment planning is at an early stage. Hence, it is recommended these algorithms be used in parallel with the TG-43 dose calculation formalism (Nath et al. 1995, Rivard et al. 2004) only for dose comparison purposes at the present time. Accordingly, their clinical commissioning should only be done after regular commissioning of a TPS has been completed.

It is important that MBDCA commissioning follow the orderly process outlined in the AAPM TG-186 report (Beaulieu et al. 2012) and described in a recent presentation made by the AAPM Working Group on Dose Calculation Algorithms in Brachytherapy (Beaulieu et al. 2016). The aim of such an approach is to establish uniform commissioning procedures for various MBDCAs and across institutions. This process ideally begins with familiarization with the dose calculation algorithm and its specific implementation in a particular treatment planning system, proceeds to Level 1 commissioning, and concludes with Level 2 commissioning. Each of these distinct phases is addressed below.

2.5.1 Pre-commissioning Tasks

Regular commissioning of the brachytherapy TPS must be done first. This includes performing general tasks described in the AAPM TG-40 (Kutcher et al. 1994), TG-56 (Nath et al. 1997a), and TG-64 (Yu et al. 1999) reports, and more specific tasks for HDR ^{192}Ir (Kubo et al. 1998) and permanent prostate brachytherapy sys-

Table 2–2 Dosimetric uncertainty components for MBDCAs. From (Beaulieu et al. 2012), with permission.

#	Dosimetric Uncertainty Components	Responsible Party
1	Source (seed) modeling such as geometry, material specifications, etc. For HDR and PDR ^{192}Ir sources, the length of cable modeled with the source also leads to uncertainties (Ref. 71). Also for HDR ^{192}Ir sources, often times virtual sources are used where such geometric approximations can lead to dosimetric uncertainties.	Vendor
2	Applicator modeling (if applicable) such as geometries, material specifications (especially denser materials such as lead or tungsten alloys).	Vendor
3	Source and applicator characterization grid size resolution if not explicitly modeled.	Physicist
4	Source and/or applicator positioning in the patient anatomy, which is extremely important for shadowing effects behind dense materials, such as additional seeds (if applicable) dense shielding materials, or light material streaming paths such as through catheters.	Physicist
5	Modeling of brachytherapy source radiation emissions.	Vendor
6	Patient geometry and limitations to simulate the realistic geometry.	Physicist
7	Material voxels grid resolution size (based on CT scanning).	Physicist
8	Determination of patient composition through correlation of CT HU to material and density mapping.	Vendor and Physicist
9	Fundamental materials interaction data used (cross sections).	Vendor
10	MBDCA model and potential simplifications away from realistic physics models.	Vendor
11	Use of collision kerma versus absorbed dose, especially in tissue areas where CPE assumption is not true (near interfaces of applicator and gas/tissue interface).	Vendor
12	Output voxel grid resolution size.	Physicist

tems (Yu et al. 1999). The majority of recommendations made in these reports pertain to treatment-specific quality assurance, and all of the reports presume that the TG-43 formalism is being used for dose calculation.

Taking time to familiarize oneself with the characteristics of an MBDCA embedded in a TPS is not only valuable preparation for commissioning the algorithm, but will continue to pay dividends later when the algorithm is applied and clinical dose distributions require evaluation. User manuals, technical documents, training courses, and the peer-reviewed literature are typical sources of necessary and useful information. To help focus familiarization activities, Table IV from the TG-186 report (Beaulieu et al. 2012) itemizing dosimetric uncertainty components for MBDCAs is reproduced here as Table 2–2.

Seeking to understand the potential limitations on dose calculation accuracy associated with each item in the table is definitely a worthwhile exercise. Figure 2–4 provides an illustration of such a potential limitation for the geometry aspect of item #2, applicator modeling. Detailed guidance in the task of validating a TPS-based applicator is available (Mikell et al. 2013).

Several of the items in Table 2–2 are identified as the responsibility of the TPS vendor. MBDCA users are encouraged to request information about these items directly from the vendor if they cannot be found in product or peer-reviewed literature.

2.5.2 Level 1 Commissioning

This phase of commissioning involves comparing MBDCA dosimetry for a single source at the center a large water phantom providing essentially full scatter conditions with TG-43 dosimetry obtained using consensus parameters, as indicated in Table 2–3. The spatial dose distribution in water primarily depends on the physi-

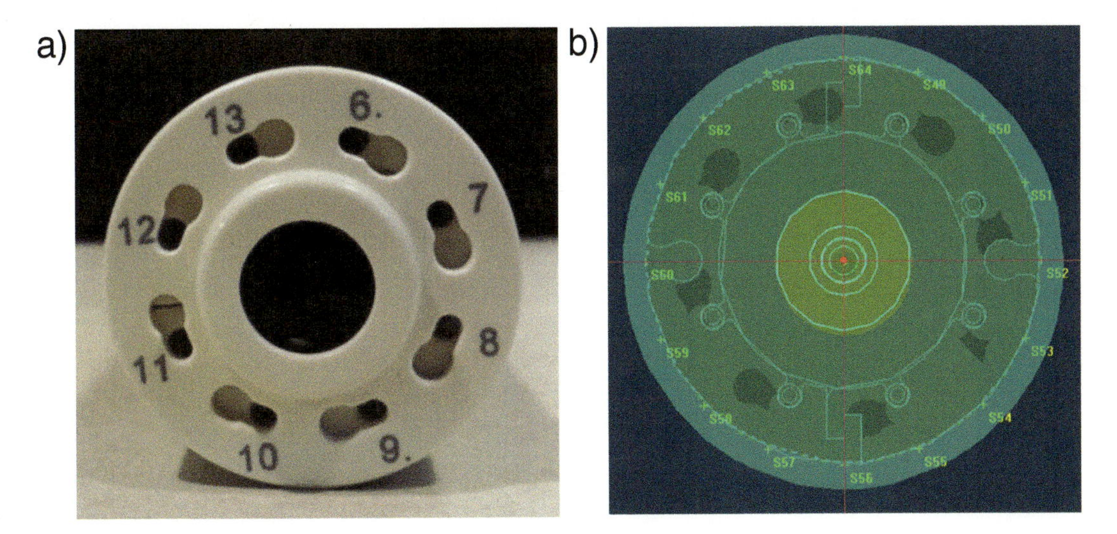

Figure 2–4 Multi-channel vaginal cylinder (a) connector-end photo and (b) cross-sectional graphic from the vendor's applicator modeling library. It is apparent from inspecting the graphic that discrete geometric elements are used to form an approximate representation of the applicator.

cal source model without detailed consideration of the surrounding environment and, as such, is useful to verify that the dose distribution obtained using an MBDCA is consistent with TG-43 for this simple geometry.

Attention should be given to the following important points when doing the comparison:

1. AAPM consensus TG-43 dosimetry parameters for the brachytherapy source model of interest should be used in the TPS for the TG-43 calculation.
2. A virtual water phantom approximating full scatter conditions—equivalent to a sphere having a 40 cm or 15 cm radius for high- or low-energy photon emitters, respectively—should be used for the MBDCA calculation.
3. MBDCA and TG-43 dose distributions should be compared on the same size (spatial resolution and extent) dose grid spanning clinically relevant distances from the source (i.e., potentially up to 10 cm away).
4. Additionally, TG-43 hand calculations should be performed for several points of interest covering the range of radius r and polar angle θ to be used in clinical practice.
5. As recommended in the AAPM TG-43U1 report (Rivard et al. 2004), agreement of the TG-43 and MBDCA dose distributions should ideally be within 2%. Larger differences should be carefully examined, and any potential clinical consequences understood and documented before clinical use, as illustrated in Example 1 on p. 6225 of the TG-186 report (Beaulieu et al. 2012).

Table 2–3 MBDCA commissioning schema described in the TG-186 report

Level	Source Position	Phantom(s)	Reference Dose Distribution
1	Single	H_2O approximating full scatter	TG-43 (consensus parameters)
2	Single, multiple	Virtual geometries mimicking clinical scenarios	Monte Carlo derived from same geometry

2.5.3 Level 2 Commissioning

Level 2 commissioning is intended to test the capability of an MBDCA to adequately account for radiation scatter conditions and the effects of material heterogeneities in and around a given treatment site, as these affect the dose distribution calculated for an individual patient. Accordingly, the TG-186 report advises that a number of well-defined virtual phantoms mimicking clinical scenarios be created, and each one used to compare the 3D dose distribution calculated with an MBDCA-based TPS against a reference dose distribution obtained for the identical phantom geometry. The reference dose distribution should only be generated using a Monte Carlo code that has been well benchmarked for brachytherapy application.

The following subsection discusses the generic work flow for MBDCA commissioning proposed by TG-186 to perform the comparison. As with any TPS commissioning, each comparison must be done locally by the physicist responsible for commissioning the MBDCA-based TPS. Any deviations between the MBDCA-based TPS results and the reference dose distribution should be documented, and their clinical significance understood.

It is acknowledged that a virtual phantom and associated reference dose distribution for a given clinical site might not be available. Hence, early adopters of MBDCA-based TPSs should be prepared to perform their own verifications using independent theoretical or experimental methods as necessary. For any independent method, dosimetric uncertainties must be small enough to permit dose comparisons to be made at clinically meaningful levels of accuracy. Examples 2–4 on p. 6226 of the TG-186 report provide illustrations (Beaulieu et al. 2012).

Based on peer-reviewed literature and observations made by the authors of the TG-43U1 report (Rivard et al. 2004), the TG-186 report indicates that at the present time a gamma-index criteria of 2.0% and 2.0 mm with a ≥99% pass rate for points within a specified volume may generally be used to evaluate agreement of MBDCA-based and reference dose distributions. The TG-186 report goes on to say that more refined or alternate acceptability criteria ought to be established through clinical experience and research. A practical issue with the proposed gamma-index criteria is that TPSs incorporating MBDCAs do not currently support gamma-index analysis comparison.

Work Flow

Figure 2–5 reproduces the generic work flow appearing as Figure 4 in the TG-186 report, which summarizes the main process components involved in Level 2 MBDCA commissioning. With reference to the TG-186 figure, these components are as follows:

a. An independent, widely accessible, web-based data repository or registry serving as a source of treatment planning data and associated reference dose distributions in DICOM format for virtual phantom geometries. Each distinct phantom geometry is referred to as a test case.

b. The download and importation of treatment planning data for a test case, including virtual phantom geometry, precontoured ROIs, points of interest, material assignments, source strength(s), dwell positions, dwell times, the plan report, and other information contained in the DICOM-RT structure. Also, the download and importation of the associated reference dose distribution.

c. The performance of TG-43 dose calculation locally for the test case and comparison with the plan report downloaded from the registry to verify treatment plan parameters.

d. The performance of MBDCA dose calculation locally for the test case and comparison with the 3D reference dose distribution downloaded from the registry using TPS evaluation tools.

e. Particularly for low-energy sources, the download and input of the reference CT scanner calibration curve into the TPS. This curve is to be used only for commissioning purposes. Also, checking that materials available in the TPS are consistent with those specified in the test case.

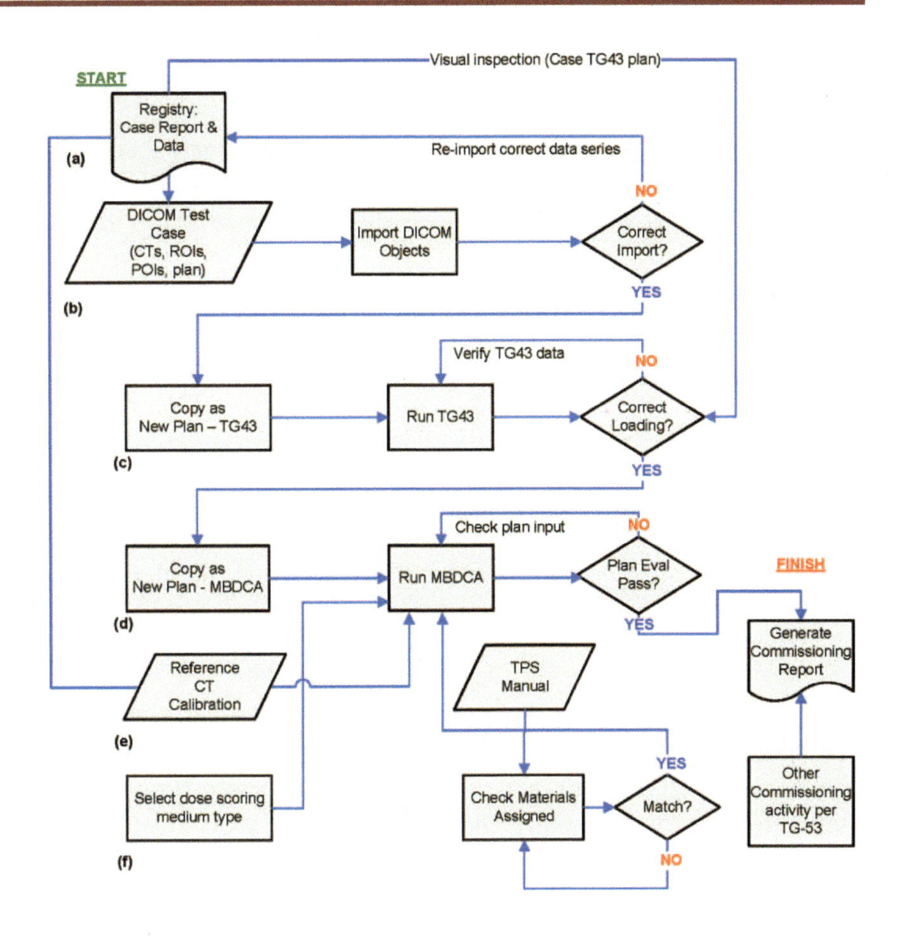

Figure 2–5 MBDCA Level 2 commissioning work flow. The main process components (a–f) are briefly described in the text. From (Beaulieu et al. 2012) with permission.

f. The specification of dose scoring to water or medium for MBDCA calculation (if the option is available).

The joint AAPM/ESTRO/ABG WG-DCAB (WG-DCAB 2011) was established primarily to implement the work flow depicted in Figure 2–5 and populate the registry with an initial set of test cases. Some key aspects of their completed and ongoing work to facilitate MBDCA commissioning by clinical physicists are highlighted in the next section.

2.5.4 Implementation of Level 2 Commissioning

At present, MBDCA-based TPSs only offer support for HDR ^{192}Ir brachytherapy sources. As these high-energy sources can be collectively characterized by their similar photon emission properties, a generic virtual HDR ^{192}Ir source model was created by the WG-DCAB expressly for the purpose of MBDCA commissioning (Ballester et al. 2015). The generic source has been implemented by two TPS vendors (Varian and Elekta) and incorporated in the WG-DCAB implementation of the TG-186 commissioning process. A registry for MBDCA commissioning data was also created and populated with four virtual phantom geometries, designated test cases 1–4. The registry, initial test cases, registry-based commissioning process, and some areas of ongoing research are described below.

2.5.4.1 MBDCA Commissioning Registry

An MBDCA commissioning registry was created by the WG-DCAB in collaboration with the AAPM-ESTRO Brachytherapy Source Registry Work Group (WG-BSR 2005) and made publicly available in June 2016. This registry can be accessed from the IROC-Houston Quality Assurance Center home page at http://

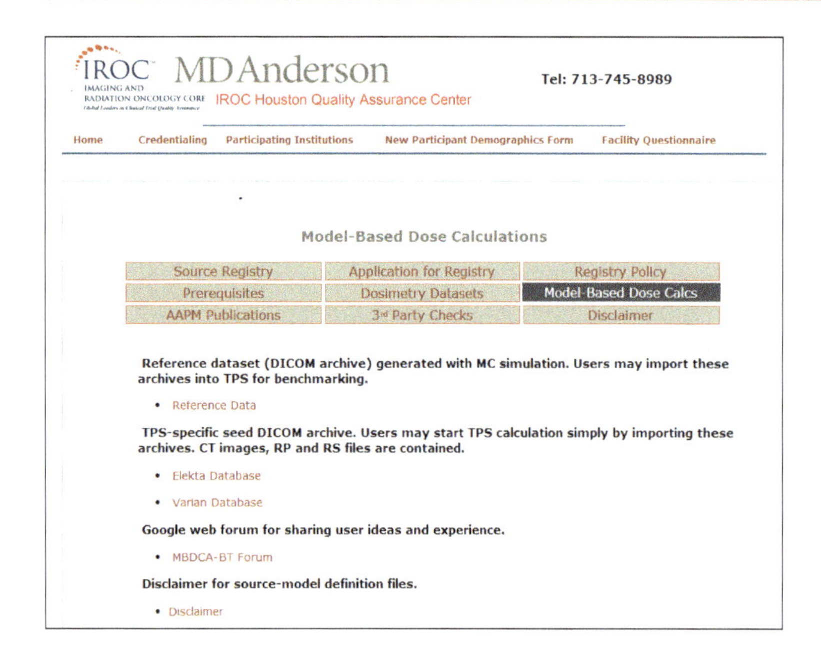

Figure 2–6 MBDCA commissioning registry home page (accessed March 8, 2017).

rpc.mdanderson.org/RPC/home.htm by selecting the "Brachy Sources" tab on the main page, followed by the "Source Registry" link, and finally the "Model-Based Dose Calcs" tab. The registry home page is shown in Figure 2–6. The "Reference Data" link provides access to reference dose distributions (dose to medium in medium, $D_{m,m}$) for each of the test cases in DICOM-RT format, generated using the MCNP6 radiation transport code (Ballester et al. 2015). There are separate datasets for the Elekta OncentraBrachy and Varian BrachyVision TPSs only because the current DICOM-RT standard is not sufficiently well-defined to enable both systems to import the same dataset. The "Elekta Database" and "Varian Database" links direct the user to TPS-specific treatment plan data for the test cases in DICOM-RT format (zipped), any auxiliary data that may be required for commissioning (zipped), and detailed user guides. The latter should be consulted for a step-by-step, TPS-specific description of the generic work flow shown in Figure 2–5. There is also a link to a web forum for sharing user ideas and experience.

2.5.4.2 Initial Test Cases

Test cases 1–3 (Ballester et al. 2015) are designed to evaluate the basic performance of an MBDCA. Each is based on a voxelized computational model of a homogeneous water cube (20.1 cm sides) set inside either a water or an air cube (51.1 cm sides). Both cubes have a common center located at (x, y, z) = (0, 0, 0) cm and their sides are parallel. The dimensions, in-plane resolution, and number of images were chosen so that 511 × 511 × 511 cubic voxels $(1\ \text{mm})^3$ fill the space. An odd number of voxels was chosen so that the geometrical center of the phantom coincides with a voxel center. The geometries for test cases 1–3 are summarized in Table 2–4. Test case 4 (Ma et al. 2017), illustrated in Figure 2–7, is a generic virtual shielded cylinder incorporating elements of a clinical applicator. All four test cases are represented as CT DICOM image series with $(1\ \text{mm})^3$ resolution.

Table 2–4 WG-DCAB test case 1–3 geometries

Test Case	Inner Cube Side, Material	Outer Cube Side, Material	Source Center Location	Applicator
1	20.1 cm, H_2O	51.1 cm, H_2O	(0, 0,0) cm	None
2	20.1 cm, H_2O	51.1 cm, air	(0, 0,0) cm	None
3	20.1 cm, H_2O	51.1 cm, air	(7, 0,0) cm	None

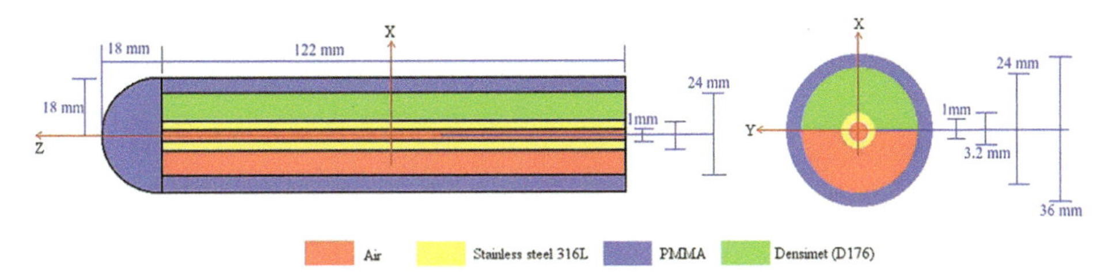

Figure 2–7 Generic virtual cylindrical applicator incorporating a 180° shield (not to scale) incorporated in test case 4. From (Ma et al. 2017) with permission.

2.5.4.3 Registry-based Commissioning Process

As mentioned above, the implemented registry-based MBDCA commissioning process closely follows the work flow proposed in the TG-186 report. TPS-specific descriptions of the steps involved are contained in the user guides found in the registry, to which the reader is referred for full details. Here we present the fundamental basis upon which the comparison between MBDCA-based and reference dosimetry datasets rests, and we also look briefly at selected results for test case 3 (scatter conditions differ from those for TG-43) and test case 4 (applicator material heterogeneities are present).

Level 2 commissioning involves comparing a TPS-calculated dose distribution to a reference one, with both distributions ideally having been obtained for the same value of total reference air kerma (TRAK) emitted by the radiation source. TRAK can be expressed as a product of the source air kerma rate constant [$\mu Gy \cdot m^2 \cdot h^{-1} \cdot MBq^{-1}$], source activity [MBq], source total dwell time [s], and a units conversion factor $(3.6 \times 10^8)^{-1}$ [$Gy \cdot \mu Gy^{-1} \cdot h \cdot s^{-1}$]. The air-kerma rate constant for the generic ^{192}Ir WG source is 0.098 [$\mu Gy \cdot m^2 \cdot h^{-1} \cdot MBq^{-1}$] (Ballester et al. 2015). For purposes of MBDCA commissioning, the source activity is specified as 3.7×10^5 MBq (approximately 44 kU). Accordingly, preparing for local dose calculation requires manually setting a single dwell time for a single dwell position for each of the four initial test cases. In this manner, the TRAK is fixed to the same value used to normalize the corresponding reference dose distribution, enabling direct comparison of the locally calculated and reference dose distributions. The normalized reference dose distribution for test case 3 (source offset in water cube) in the plane bisecting the active volume of the generic ^{192}Ir source is shown in Figure 2–8.

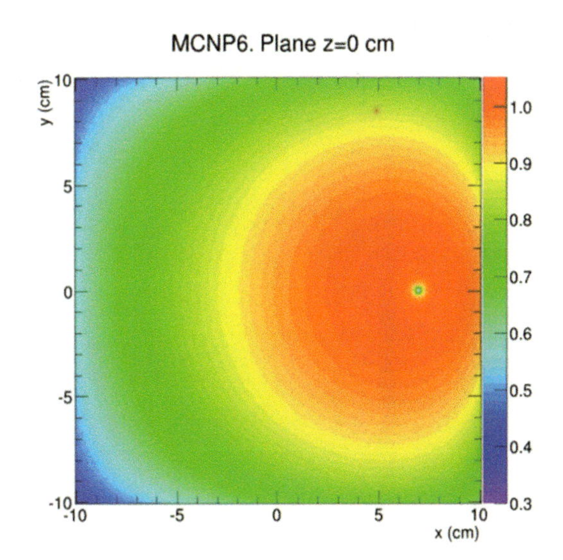

Figure 2–8 Normalized reference dose distribution for test case 3 in the plane z = 0 containing the source, calculated as $[(D(r) \times (r - r_s)^2] / [D(r_{ref}) \times (r_{ref} - r_s)^2)]$ for r_{ref} = (6,0,0) cm and r_s = (7,0,0) cm. From (Ballester et al. 2015) with permission.

The influence of the water-air interface located at x = 10.05 cm results in a clearly visible left-right asymmetry in this 2D dose distribution within the inner water cube (Table 2–2 gives the phantom geometry). To see how well the two commercial MBDCAs predict this change, relative local dose differences ΔD_{local} =

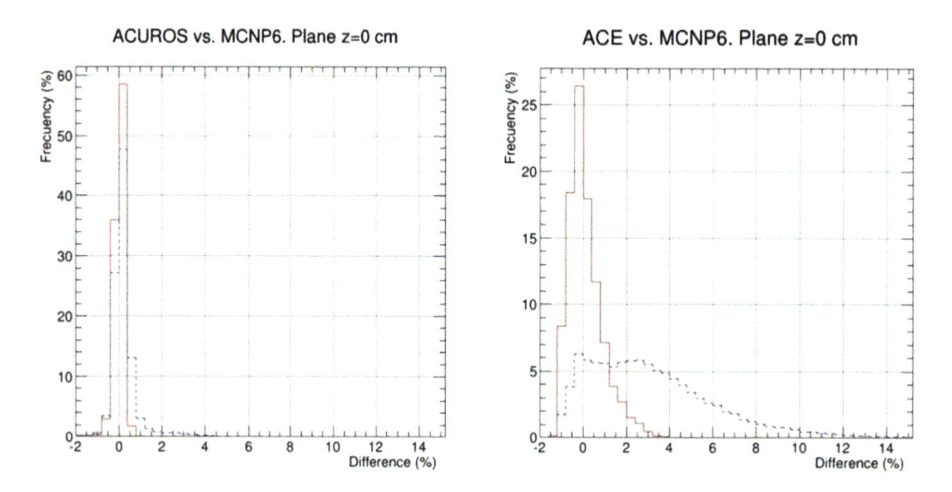

Figure 2–9 Relative local dose difference histograms for Acuros and ACE dose calculations relative to MCNP6 reference dosimetry for test case 3 in a 0.1 cm thick region of interest of size 10 × 10 cm^2 (blue dotted line) and 5 × 5 cm^2 (red solid line) centered on the source. The relative dose difference is defined as $\Delta D_{\text{LOCAL}} = (D(r) - D_{\text{ref}}(r)) / D_{\text{ref}}(r)$. From (Ballester et al. 2015) with permission.

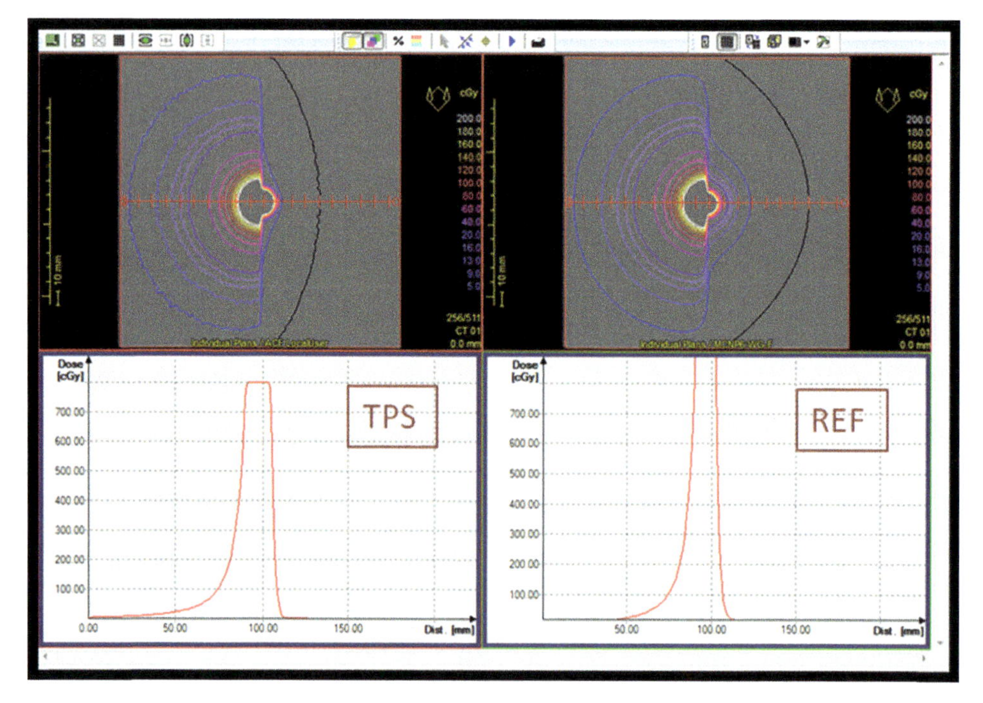

Figure 2–10 Side-by-side isodose displays (upper panels) and dose profiles (lower panels, corresponding to the red lines in the upper panels) in the z = 0 plane containing the source and transecting the shielded applicator for test case 4, calculated using ACE (TPS) and imported from the MBDCA commissioning registry reference data repository (Ma et al. 2017).

$[D(r) - D_{ref}(r)] / D_{ref}(r)$ between TPS-based and reference dose distributions for two differently sized ROIs in the z = 0 plane are shown in Figure 2–9. It is apparent that both MBDCAs yield results within 2% (and mostly within 1%) inside a 5 × 5 cm^2 ROI centered on the source.

The absolute dose distribution for test case 4 (source in shielded applicator) calculated using the ACE algorithm—displayed in the plane bisecting the active volume of the source and alongside the associated reference dose distribution—is shown in Figure 2–10. Several differences between the distributions are readily apparent, including the locations and shapes of isodose lines behind the shield and in the shield penumbra, the shapes of the dose profile tails in the unshielded region, and the "flat top" of TPS dose profile.

The latter feature stems from a limit imposed by the TPS on the maximum value of dose being a factor of eight times more than the entered prescribed dose which, in this case, was 100 cGy. As Figure 2–9 for test case 3 illustrates, the relative local dose difference metric $\Delta D_{LOCAL} = [D(r) - D_{ref}(r)] / D_{ref}(r)$ has diminished clinical utility in regions farther away from the source where dose values are low. This limitation is even more evident in test case 4, as dose values in the shielded region are quite low. To address this shortcoming, the WG-DCAB defined a complementary dose comparison metric, the relative global dose difference $\Delta D_{GLOBAL} = [D(r) - D_{ref}(r)] / D_{ref}(r_{ref})$, where $D_{ref}(r_{ref})$ is the dose value at a clinically relevant reference point. Its intended purpose is to indicate relative clinical relevance. Using either of the two dose-difference metrics in combination with point-to-source distance and region (unshielded or shielded) indicator, 2D histograms showing the region-specific correlation of relative dose with distance can be calculated. Figure 2–11 shows such histograms for test case 4 for both ΔD_{LOCAL} and ΔD_{GLOBAL} calculated using Acuros and ACE.

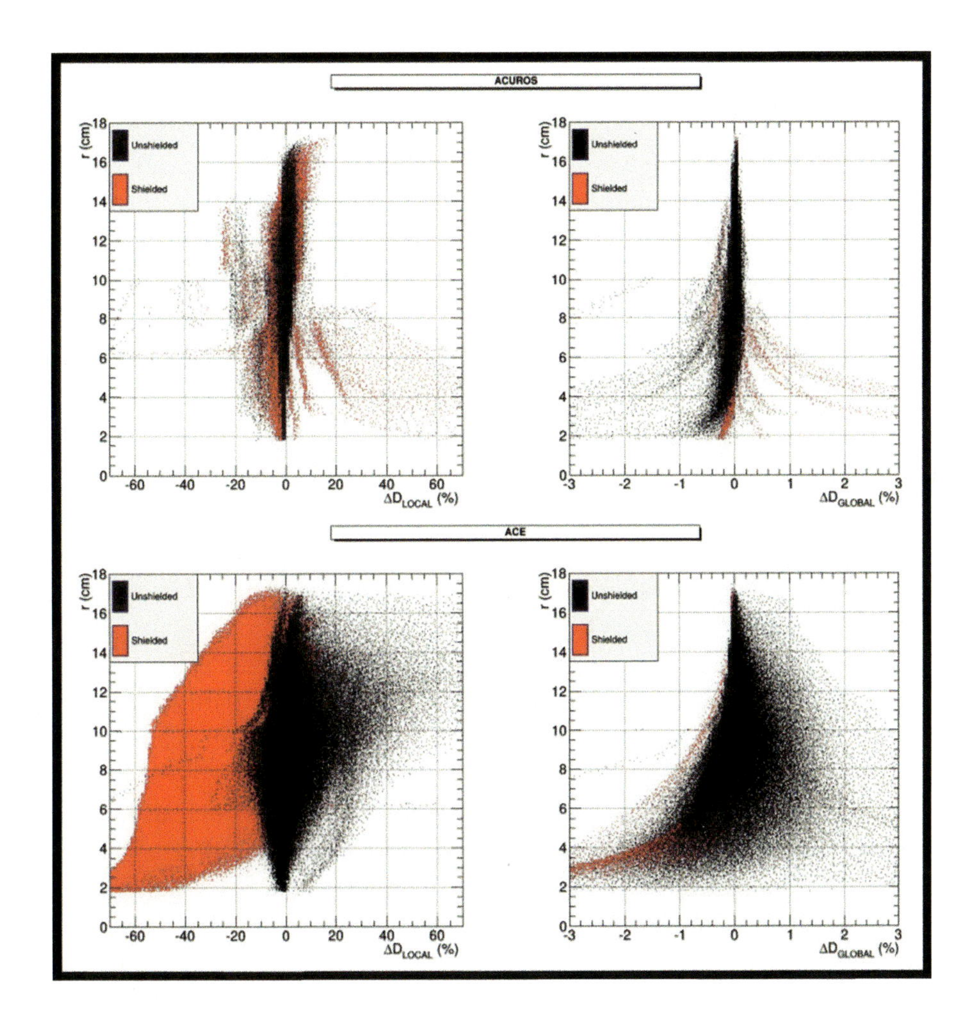

Figure 2–11 2D histogram of distance to the source center, r(cm), vs. ΔD_{LOCAL} (left) and ΔD_{GLOBAL} (right) for test case 4. The reference point for ΔD_{GLOBAL} is located 0.5 cm away from the applicator side at r_{ref} = (–2.3 cm, 0, 0). Comparisons for ACUROS (top) and ACE (bottom) with respect to MCNP6 are shown; red dots correspond to voxels located in the shielded region, black dots to voxels in the unshielded region. From (Ma et al. 2017) with permission.

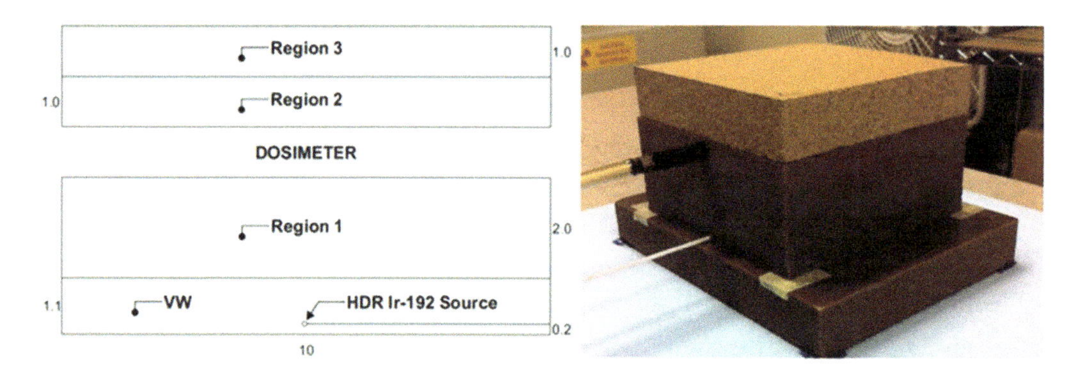

Figure 2–12 Phantom designed for validation of a TPS-based HDR MBDCA that can accommodate different materials (Regions 1–3) and dosimeters. From (Moura et al. 2015) with permission.

The reference point for ΔD_{GLOBAL} is located 0.5 cm away from the applicator side along the –x direction, i.e., r_{ref} = (–2.3 cm, 0, 0). It can be seen that ΔD_{GLOBAL} values for both Acuros and ACE in the unshielded region are, for the most part, <2% different from the reference dose distribution.

2.5.5 Areas of Future Research

Considerable work remains to augment, refine and extend the TG-186 MBDCA commissioning process implemented for HDR ^{192}Ir brachytherapy. Additional clinically oriented test cases are required on the registry to enable clinical physicists to perform commissioning that is specifically geared to the treatments for which an MBDCA will be used. On the dose distribution comparison and evaluation front, there is a need to clearly establish gamma-index acceptance criteria on a treatment site-specific basis (and possibly technique-specific basis) and to identify and evaluate other potentially clinically useful metrics, such as ΔD_{GLOBAL}. There is also a need to design and build physical phantoms having precise geometries and known materials that can be used to obtain reference dose distributions experimentally, such as that shown in Figure 2–12. Finally, it should be noted that MBDCAs are being extended to the low-energy brachytherapy domain (Sloboda et al. 2017), where they are expected to impact clinical practice in a major way for some treatments (Rivard et al. 2009b).

2.6 Clinical Use

2.6.1 Review and Interpretation of TG-186 Recommendations

The TG-186 report included several recommendations for users of MBDCAs for brachytherapy treatment planning (Beaulieu et al. 2012). Accounting for material heterogeneities, clinical conditions for radiation scatter, and intersource attenuation will cause differences between dose distributions obtained using the TG-43 formalism. These differences will not be scalar in nature, and they will depend on the brachytherapy modality and the anatomic site being treated (Rivard et al. 2009b; Beaulieu et al. 2012). Furthermore, the absorbed dose will depend on both the radiation transport medium as well as the target medium via the mass-absorption coefficient and the mass-energy absorption coefficient, respectively. The TG-186 report recommended that there be a minimal number of materials associated with a given MBDCA treatment plan and that the mass density derived from CT imaging should be used with special measures to be taken if metal streaking and other CT reconstruction artifacts are present (Beaulieu et al. 2012). Example material definitions are given with careful concern for the composition of breast and prostatic tissues due to adipose/gland ratios and the presence of calcifications, respectively. Where material compositions were not specified, the physicist is referred to the ICRU 46 report (ICRU 1992). If still not available, the physicist should contact the manufacturer of brachytherapy applicators and devices to account for the dosimetric perturbations.

The TG-186 report (Beaulieu et al. 2012) makes six key recommendations in its Section IV.B that are summarized below and apply to the physicist, MBDCA vendors, and manufacturers of brachytherapy sources and applicators:

1. The physicist is responsible to confirm that MBDCA dose predictions are based upon accurate and spatially resolved applicator and source models (including correct material assignments) to avoid dose-delivery errors prior to implementing the MBDCA in the clinic.
2. Patient CT grids ~1 mm^3 are spatially inadequate for accurate modeling on brachytherapy sources and applicators. Dose distributions from low-energy photon-emitting sources are particularly sensitive to simplistic geometric approximations. MBDCA vendors should use analytic modeling schemes or recursively specify meshes with ≥10 μm spatial resolution to accurately depict low-energy source geometries.
3. Manufacturers of sources and applicators should disclose the geometries, material assignments, and manufacturing tolerances to physicists and MBDCA vendors to permit accurate dose modeling. MBDCA vendors should incorporate sources and applicators into their TPS or permit the physicist to accomplish this by a simple method.
4. Sources and applicators incorporated into an MBDCA TPS should be independently verified.
5. Vendors of TPS applicator libraries should provide a means for the physicist to verify the source or applicator characterization.
6. The physicist should compare MBDCA-generated single-source dose distributions in water to directly calculated TG-43 benchmarks.

The report provides examples and details on how to accomplish the above tasks.

As an additional resources to the physicist, the joint Working Group on Model-Based Dose Calculation Algorithms in Brachytherapy (by the AAPM, GEC-ESTRO, and the ABG) has established an Internet web forum to promote dialogue among MBDCA users, vendors, and manufacturers. As of this writing, the forum discussions have helped to clarify means of generating tests to independently verify MBDCA dose calculations.

2.6.2 Independent Check Methods

Commissioning MBDCAs for brachytherapy dosimetry was a new endeavor upon release of the TG-186 report (Rivard et al. 2010; Beaulieu et al. 2012). The report recommended two levels of MBDCA commissioning:

1. Compare MBDCA results to TG-43 results for common conditions (i.e., a single brachytherapy source centered in a water phantom of the same size).
2. Compare MBDCA results to a benchmark case for conditions. Here, the brachytherapy source may be offset in an irregular phantom or placed in an applicator with an assortment of material compositions.

While the TG-43 dose calculation method is simple enough to use on a hand calculator or user-generated spreadsheet, no sophisticated benchmark cases existed at the time of publishing the TG-186 report (Rivard et al. 2010; Beaulieu et al. 2012). Since then, several benchmark cases have appeared in the literature; see, for example, Ballester et al. (2015) and Ma et al. (2017). These have been made available on the Imaging and Radiation Oncology Core Houston Quality Assurance Center (IROC Houston) Brachytherapy Source Registry for users of the Acuros (BrachyVision™) and ACE (OncentraBrachy™).

While these resources aim to facilitate TPS commissioning, to our knowledge there are no commercially available tools for independent checks of MBDCA-based brachytherapy treatment plans for everyday clinical use. Such software would presumably function in a manner similar to current radiotherapy independent-check software, such as software based on TG-43 for brachytherapy or with a sophisticated algorithm for linac-based external beam radiation therapy. Independent checks of brachytherapy treatment planning calculations are recommended in the AAPM TG-53 report (Fraass et al. 1998) and by the ACR-AAPM performance standards (ACR 2015a, 2015b).

2.6.3 Tolerances

Like the AAPM TG-56 report (Nath et al. 1997b)—and subsequently the TG-43U1 report (Rivard et al. 2004) and the AAPM/ESTRO Report #229 (Perez-Calatayud et al. 2012)—the TG-186 report recommends a tolerance of 2.0% for MBDCA agreement with societal-issued TG-43 dosimetry parameters (Beaulieu et al. 2012). This tolerance is for Level 1 commissioning dose calculations anywhere in the vicinity of the brachytherapy source and is used to define the region (near and far boundaries) for which the brachytherapy source is commissioned. Comparing MBDCA-generated, single-source dose distributions in water to directly calculated TG-43 benchmarks, differences exceeding 2% were observed with 95% of the points in the considered volume being within 1% and 3% for the Acuros BV and ACE MBDCA TPSs, respectively (Ballester et al. 2015). Evaluating a Level 2 benchmark, Ma and colleagues observed 95% of the points in a water phantom surrounding a shielded vaginal applicator were within 4.2% and 8% to 15% for the Acuros BV and ACE MBDCA TPSs, respectively (Ma et al. 2017). Therefore, application of a simple 2% tolerance to all circumstances involving MBDCA results may not be achievable, especially in regions of high dose gradients. To ameliorate this situation, the use of a combination of percentage tolerance and distance to agreement for complex brachytherapy dose distributions has been considered (Yang et al. 2011). The physicist will need to evaluate the balance of improved accuracy of MBDCA results over the TG-43 formalism with higher tolerances in comparison to the reference data.

References

ACR Resolution 50. (2015a). *ACR–AAPM Technical Standard for the Performance of High- Dose-Rate Brachytherapy Physics.* Reston, VA: American College of Radiology, 2015.

ACR Resolution 51. (2015b). *ACR–AAPM Technical Standard for the Performance of Low- Dose-Rate Brachytherapy Physics.* Reston, VA: American College of Radiology, 2015.

Afsharpour, H. H., G. G. Landry, M. M. D'Amours, S. S. Enger, B. B. Reniers, E. E. Poon, J-F. J. Carrier, F. F. Verhaegen, and L. L. Beaulieu. (2012). "ALGEBRA: ALgorithm for the heterogeneous dosimetry based on GEANT4 for BRAchytherapy." *Phys. Med. Biol.* 57:3273–80.

Afsharpour, H., G. Landry, B. Reniers, J.-P. Pignol, L. Beaulieu, and F. Verhaegen. (2011). "Tissue modeling schemes in low energy breast brachytherapy." *Phys. Med. Biol.* 56:7045–60.

Afsharpour, H., J.-P. Pignol, B. Keller, J.-F. Carrier, B.Reniers, F. Verhaegen, and L. Beaulieu. (2010). "Influence of breast composition and interseed attenuation in dose calculations for post-implant assessment of permanent breast ^{103}Pd seed implant." *Phys. Med. Biol.* 55:4547–61.

Agostinelli, S., J. Allison, K. Amako, J. Apostolakis, H. Araujo, P. Arce, M. Asai, D. Axen, S. Banerjee, G. Barrand, F. Behner, L. Bellagamba, J. Boudreau, L. Broglia, A. Brunengo, H. Burkhardt, S. Chauvie, J. Chuma, R. Chytracek, G. Cooperman, G. Cosmo, P. Degtyarenko, A. Dell'Acqua, G. Depaola, D. Dietrich, R. Enami, A. Feliciello, C. Ferguson, H. Fesefeldt, G. Folger, F. Foppiano, A. Forti, S. Garelli, S. Giani, R. Giannitrapani, D. Gibin, J. J. Gómez Cadenas, I. González, A. G. Gracia, G. Greeniaus, W. Greiner, V. Grichine, A. Grossheim, S. Guatelli, P. Gumplinger, R. Hamatsu, K. Hashimoto, H. Hasui, A. Heikkinen, A. Howard, V. Ivanchenko, A. Johnson, F. W. Jones J. Kallenbach, N. Kanaya, M. Kawabata, Y. Kawabata, M. Kawaguti, S. Kelner, P. Kent, A. Kimura, T. Kodama, R. Kokoulin, M. Kossov, H. Kurashige, E. Lamanna, T. Lampén, V. Lara, V. Lefebure, F. Lei, M. Liendl, W. Lockman, F. Longo, S. Magni, M. Maire, E. Medernach, K. Minamimoto, P. Mora de Freitas, Y. Morita, K. Murakami, M. Nagamatu, R. Nartallo, P. Nieminen, T. Nishimura, K. Ohtsubo, M. Okamura, S. O'Neale, Y. Oohata, K. Paech, J. Perl, A. Pfeiffer, M. G. Pia, F. Ranjard, A. Rybin, S. Sadilov, E. Di Salvo, G. Santin, T. Sasaki, et al. (2003). "GEANT4: a simulation toolkit." *Nucl. Instrum. Methods Phys. Res. A.* 506:250–303.

Ballester, F., A. C. Tedgren, D. Granero, A. Haworth, F. Mourtada, G. Paiva Fonseca, K. Zourari, P. Papagiannis, M. J. Rivard, F-A. Siebert, R. S. Sloboda, R. Smith, R. M. Thomson, F. Verhaegen, J. Vijande, Y. Ma, and L. Beaulieu. (2015). "A generic high-dose rate ^{192}Ir brachytherapy source for evaluation of model-based dose calculations beyond the TG-43 formalism." *Med. Phys.* 42:3048–62.

Beaulieu L., F. Ballester, A. Carlson-Tedgren, G. Fonseca, A. Haworth, J. Lowenstein, Y. Ma, F. Mourtada, P. Papagiannis, M. J. Rivard, F.-A. Siebert, R. Sloboda, R. Smith, R. Thomson, J. Vijande, and F. Verhaegen. (2016). "Implementation and Validation of an End-to-End Commissioning Process for Model-Based Dose Calculation Algorithms in Brachytherapy." *Brachytherapy* 15:S172 (abstract).

Beaulieu L., T. A. Carlsson, J.-F. Carrier, S. D. Davis, F. Mourtada, M. J. Rivard, R. M. Thomson, F. Verhaegen, T. A. Wareing, and J. F. Williamson. (2012). "Report of the Task Group 186 on model-based dose calculation methods in brachytherapy beyond the TG-43 formalism: Current status and recommendations for clinical implementation." *Med. Phys.* 39:6208–36.

Borgers, C. (1998). "Complexity of Monte Carlo and deterministic dose-calculation methods." *Phys. Med. Biol.* 43:517–528. http://www.iop.org/EJ/abstract/0031–9155/43/3/004/.

Briesmeister, J. F. *MNCP—A General Monte Carlo N–Particle Transport Code, Version 4C*. Los Alamos National Laboratory, 1999.

Carlsson, A. K. and A. Ahnesjo. (2000a). "Point kernels and superposition methods for scatter dose calculations in brachytherapy." *Phys. Med. Biol.* 45:357–82.

Carlsson, A. K. and A. Ahnesjo. (2000b). "The collapsed cone superposition algorithm applied to scatter dose calculations in brachytherapy." *Med. Phys.* 27:2320–32.

Carlsson, T. A. and A. Ahnesjo. (2003). "Accounting for high Z shields in brachytherapy using collapsed cone superposition for scatter dose calculation." *Med. Phys.* 30:2206.

Carrier, J., M. D'Amours, F. Verhaegen, B. Reniers, A. Martin, E. Vigneault, and L. Beaulieu. (2007). "Postimplant Dosimetry Using a Monte Carlo Dose Calculation Engine: A New Clinical Standard." *Int. J. Radiat. Oncol. Biol. Phy.* 68:1190–8.

Chamberland, M.J., R. E. Taylor, D. W. Rogers, and R. M. Thomson. (2016). "Egs_Brachy: a Versatile and Fast Monte Carlo Code for Brachytherapy." *Phys. Med. Biol.* 61(23): 8214–31. doi:10.1088/0031-9155/61/23/8214.

Chibani, O. and J. F. Williamson. (2005). "MCPI: a sub-minute Monte Carlo dose calculation engine for prostate implants." *Med. Phys.* 32:3688–98.

Chibani, O., J. F. Williamson, and D. Todor. (2005). "Dosimetric effects of seed anisotropy and interseed attenuation for ^{103}Pd and ^{125}I prostate implants." *Med. Phys.* 32:2557–66.

Collins Fekete, C.-A., M. Plamondon, A.-G. Martin, E. Vigneault, F. Verhaegen, and L. Beaulieu. (2015). "Calcifications in low-dose rate prostate seed brachytherapy treatment: Post-planning dosimetry and predictive factors." *Radiother. Oncol.* 114:339–44.

Curran, B. H., J. M. Balter, and I. J. Chetty. *Integrating New Technologies into the Clinic: Monte Carlo and Image-Guided Radiation Therapy.* Madison, WI: Medical Physics Publishing, 2005.

Daskalov, G.M., R. S. Baker, D. W. Rogers, and J. F. Williamson. (2002) "Multigroup discrete ordinates modeling of ^{125}I 6702 seed dose distributions using a broad energy-group cross section representation." *Med. Phys.* 29:113–24.

Daskalov, G., R. Baker, and D. Rogers. (2000). "Dosimetric modeling of the microSelectron high-dose rate ^{192}Ir source by the multigroup discrete ordinates method." *Med. Phys.* 27:2307–19.

DeWerd, L. A., G. S. Ibbott, A. S. Meigooni, M. G. Mitch, M. J. Rivard, K. E. Stump, B. R. Thomadsen, and J. L. M. Venselaar. (2011). "A dosimetric uncertainty analysis for photon-emitting brachytherapy sources: Report of AAPM Task Group No. 138 and GEC-ESTRO." *Med. Phys.* 38:782–801.

Fraass, B., K. Doppke, M. Hunt, G. Kutcher, G. Starkschall, R. Stern, and J. Van Dyke. (1998). "American Association of Physicists in Medicine Radiation Therapy Committee Task Group 53: quality assurance for clinical radiotherapy treatment planning." *Med. Phys.* 25:1773–829.

Gifford, K. A., J. L. Horton, T. A. Wareing, G. Failla, and F. Mourtada. (2006). "Comparison of a finite-element multigroup discrete-ordinates code with Monte Carlo for radiotherapy calculations." *Phys. Med. Biol.* 51:2253–65.

Gifford, K. A., M. J. Price, J. L. Horton, T. A. Wareing, and F. Mourtada. (2008). "Optimization of deterministic transport parameters for the calculation of the dose distribution around a high dose-rate ^{192}Ir brachytherapy source." *Med. Phys.* 35:2279–85.

Gifford, K. A., T. A. Wareing, G. Failla, J. L. Horton, P. J. Eifel, and F. Mourtada. (2010). "Comparison of a 3-D multi-group SN particle transport code with Monte Carlo for intracavitary brachytherapy of the cervix uteri." *J. Appl. Clin. Med. Phys.* 11:3103.

Han, T., J. K. Mikell, M. Salehpour, and F. Mourtada. (2011). "Dosimetric comparison of Acuros XB deterministic radiation transport method with Monte Carlo and model-based convolution methods in heterogeneous media." *Med. Phys.* 38:2651–64.

Hedtjärn, H., G. A. Carlsson, and J. F. Williamson. (2002). "Accelerated Monte Carlo based dose calculations for brachytherapy planning using correlated sampling." *Phys. Med. Biol.* 47:351–76.

Hissoiny, S., B. Ozell, P. Després, and J.-F. Carrier. (2011). "Validation of GPUMCD for low-energy brachytherapy seed dosimetry." *Med. Phys.* 38:4101.

Hofbauer, J., C. Kirisits, A. Resch, Y. Xu, A. Sturdza, R. Pötter, and N. Nesvacil. (2016). "Impact of heterogeneity-corrected dose calculation using a grid-based Boltzmann solver on breast and cervix cancer brachytherapy." *J. Contemp. Brachytherapy* 8:143–9.

Hubert-Tremblay, V., L. Archambault, D. Tubic, R. Roy, and L. Beaulieu. (2006). "Octree indexing of DICOM images for voxel number reduction and improvement of Monte Carlo simulation computing efficiency." *Med. Phys.* 33:2819–31.

Hyer, D. E., A. Sheybani, G. M. Jacobson, and Y. Kim. (2012). "The dosimetric impact of heterogeneity corrections in high-dose rate ^{192}Ir brachytherapy for cervical cancer: Investigation of both conventional Point-A and volume-optimized plans." *Brachytherapy* (6):515–20.

ICRU (International Commission on Radiation Units and Measurements). *Photon, electron, proton and neutron interaction data for body tissues, ICRU Report No. 46.* Bethesda, MD: 1992.

Jeong, J., C. A. Barker, M. Zaider, and G. N. Cohen. (2016). "Impact of source position on high-dose-rate skin surface applicator dosimetry." *Brachytherapy* 15:650–60.

Kawrakow, I. and D. Rogers. *The EGSnrc code system: Monte Carlo simulation of electron and photon transport.* Ottawa: National Research Council of Canada, 2000.

Kubo, H. D., G. P. Glasgow, T. D. Pethel, B. R. Thomadsen, and J. F. Williamson. (1998). "High dose-rate brachytherapy treatment delivery: report of the AAPM Radiation Therapy Committee Task Group No. 59." *Med. Phys.* 25:375–403.

Kutcher, G., L. Coia, M. Gillin, W. Hanson, S. Leibel, R. Morton, J. Palta, J. Purdy, L. Reinstein, G. Svensson, M. Weller, and L. Wingfield. (1994). "Comprehensive QA for Radiation Oncology—Report of AAPM Radiation Therapy Committee Task Group-40." *Med. Phys.* 21:581–618.

Lamberto, M., D. Jacob, J. Strasser, C. Koprowski, and F. Mourtada. (2016). "Dosimetric Impact of SAVITM Breast Applicator Position Near Tissue Heterogeneities." *Brachytherapy* 15:S164.

Landry, G., B. Reniers, L. Murrer, L. Lutgens, E. Bloemen-Van Gurp, J. P. Pignol, B. Keller, L. Beaulieu, and F. Verhaegen. (2010). "Sensitivity of low energy brachytherapy Monte Carlo dose calculations to uncertainties in human tissue composition." *Med. Phys.* 37:5188–98.

Lewis, E. E. and W. F. Miller. *Computational methods of neutron transport.* New York: John Wiley and Sons, Inc., 1984.

Lorence, L. J., J. E. Morel, and G. D. Valdez. *Physics guide to CEPXS: A multigroup coupled electron-photon cross-section generating code.* Albuquerque, NM: Sandia National Labs, 1989.

Ma, Y., F. Lacroix, M.-C. Lavallée, and L. Beaulieu. (2015). "Validation of the Oncentra Brachy Advanced Collapsed cone Engine for a commercial ^{192}Ir source using heterogeneous geometries." *Brachytherapy* 14:939–52.

Ma Y., J. Vijande, F. Ballester, A. Carlson-Tedgren, D. Granero, A. Haworth, G. Fonseca, K. Zourari, P. Papagiannis, M. J. Rivard, F.-A. Siebert, R. S. Sloboda, R. Smith, M. J. P. Chamberland, R. M. Thomson, F. Verhaegen, and L. Beaulieu. (2017). "A generic TG-186 shielded applicator for the commissioning of model-based dose calculation algorithms for high-dose-rate ^{192}Ir brachytherapy" *Med. Phys. In press.*

Mikell, J. K. and F. Mourtada. (2010). "Dosimetric impact of an ^{192}Ir brachytherapy source cable length modeled using a grid-based Boltzmann transport equation solver." *Med. Phys.* 37:4733–43.

Mikell, J. K., A. H. Klopp, G. M. N. Gonzalez, K. D. Kisling, M. J. Price, P. A. Berner, P. J. Eifel, and F. Mourtada. (2012). "Impact of heterogeneity-based dose calculation using a deterministic grid-based Boltzmann equation solver for intracavitary brachytherapy." *Int. J. Radiat. Oncol. Biol. Phys.* 83:e417–22.

Mikell, J. K., A. H. Klopp, M. Price, and F. Mourtada. (2013). "Commissioning of a grid-based Boltzmann solver for cervical cancer brachytherapy treatment planning with shielded colpostats." *Brachytherapy* 12:645–53.

Miksys, N., E. Vigneault, A.-G. Martin, L. Beaulieu, and R. M. Thomson. (2017). "Large-scale Retrospective Monte Carlo Dosimetric Study for Permanent Implant Prostate Brachytherapy." *Int. J. Radiat. Oncol. Biol. Phys.* 97:606–15.

Moura, E. S., J. A. Micka, C. G. Hammer, W. S. Culberson, L. A. DeWerd, M. E. C. M. Rostelato, and C. A. Zeituni. (2015). "Development of a phantom to validate high-dose-rate brachytherapy treatment planning systems with heterogeneous algorithms." *Med. Phys.* 42:1566–74.

Mourtada, F., T. Wareing, J. Horton, J. McGhee, D. Barnett, G. Failla, and R. Mohan. (2004). "A deterministic dose calculation method with analytic ray tracing for brachytherapy dose calculations." *Med. Phys.* 31:1807.

Nath, R., L. L. Anderson, J. A. Meli, A. J. Olch, J. A. Stitt, and J. F. Williamson. (1997a). "Code of practice for brachytherapy physics: report of the AAPM Radiation Therapy Committee Task Group No. 56. American Association of Physicists in Medicine." *Med. Phys.* 24:1557–98.

Nath, R., L. Anderson, G. Luxton, K. Weaver, J. Williamson, and A. Meigooni. (1995). "Dosimetry of Interstitial Brachytherapy Sources—Recommendations of the AAPM Radiation-Therapy Committee Task Group No 43." *Med. Phys.* 22:209–34.

Nath, R., L. Anderson, J. Meli, A. Olch, J. Stitt, and J. Williamson. (1997b). "Code of practice for brachytherapy physics: Report of the AAPM Radiation Therapy Committee Task Group No. 56." *Med. Phys.* 24:1557–98.

Nigg, D. W., P. D. Randolph, and F. J. Wheeler. (1991). "Demonstration of three-dimensional deterministic radiation transport theory dose distribution analysis for boron neutron capture therapy." *Med. Phys.* 18:43–53.

Oliveira, S. M., N. J. Teixeira, L. Fernandes, P. Teles, and P. Vaz. (2014). "Dosimetric effect of tissue heterogeneity for ^{125}I prostate implants." *Rep. Pract. Oncol. Radiother.* 19:392–8.

Papagiannis, P., L. Beaulieu, and F. Mourtada. "Computational Methods for Dosimetric Characterization of Brachytherapy Sources." In *Comprehensive Brachytherapy: Physical and Clinical Aspects.* J. Venselaar, D. Baltas, A. S. Meigooni, and P. J. Hoskin, Eds. Boca Raton, FL: CRC Press, 2013.

Papagiannis, P., E. Pantelis, and P. Karaiskos. (2014). "Current state of the art brachytherapy treatment planning dosimetry algorithms." *Br. J. Radiol.* 87:20140163.

Parsai, E. I., Z. Zhang, and J. J. Feldmeier. (2009). "A quantitative three-dimensional dose attenuation analysis around Fletcher-Suit-Delclos due to stainless steel tube for high-dose-rate brachytherapy by Monte Carlo calculations." *Brachytherapy* 8:318–23.

Pearson, D. and E. A. Williams. (2013). "Dose modification factor analysis of multilumen balloon brachytherapy applicator with Monte Carlo simulation." *J. Appl. Clin. Med. Phys.* 15:54–62.

Perez-Calatayud, J., F. Ballester, R. K. Das, L. A. Dewerd, G. S. Ibbott, A. S. Meigooni, Z. Ouhib, M. J. Rivard, R. S. Sloboda, and J. F. Williamson. (2012). "Dose calculation for photon-emitting brachytherapy sources with average energy higher than 50 keV: Report of the AAPM and ESTRO." *Med. Phys.* 39:2904–29.

Perez-Calatayud, J., D. Granero, and F. Ballester. (2004) "Phantom size in brachytherapy source dosimetric studies." *Med. Phys.* 31:2075–81.

Petrokokkinos, L., K. Zourari, E. Pantelis, A. Moutsatsos, P. Karaiskos, L. Sakelliou, I. Seimenis, E. Georgiou and P. Papagiannis. (2011). "Dosimetric accuracy of a deterministic radiation transport based ^{192}Ir brachytherapy treatment planning system. Part II: Monte Carlo and experimental verification of a multiple source dwell position plan employing a shielded applicator." *Med. Phys.* 38:1981–92.

Poon, E. and F. Verhaegen. (2009). "A CT-based analytical dose calculation method for HDR ^{192}Ir brachytherapy." *Med. Phys.* 36:3982–94.

Pratx, G. and L. Xing. (2011). "GPU computing in medical physics: A review." *Med. Phys.* 38:2685.

Richardson, S., K. L. Chen, R. Pino, C. Bloch, and P. Parikh. (2013). "Dosimetric Comparison of TG-43 Formalism with BrachyVision Acuros and Monte Carlo Method for Patients Treated with the SAVI Partial Breast Applicator." *Brachytherapy* 12:S22–3.

Rivard, M. J., L. Beaulieu, and F. Mourtada. (2010). "Enhancements to commissioning techniques and quality assurance of brachytherapy treatment planning systems that use model-based dose calculation algorithms." *Med. Phys.* 37:2645–58.

Rivard, M. J., B. M. Coursey, L. A. DeWerd, W. F. Hanson, M. S. Huq, G. S. Ibbott, M. G. Mitch, R. Nath, and J. F. Williamson. (2004). "Update of AAPM Task Group No. 43 Report: A revised AAPM protocol for brachytherapy dose calculations." *Med. Phys.* 31:633–74. Online: http://link.aip.org/link/MPHYA6/v31/i3/p633/s1&Agg=doi.

Rivard, M. J., C. S. Melhus, D. Granero, J. Perez-Calatayud, and F. Ballester. (2009a). "An approach to using conventional brachytherapy software for clinical treatment planning of complex, Monte Carlo-based brachytherapy dose distributions." *Med. Phys.* 36:1968–75.

Rivard, M. J., J. L. M. Venselaar, and L. Beaulieu. (2009b). "The evolution of brachytherapy treatment planning." *Med. Phys.* 36:2136–53.

Russell, K. R. and A. Ahnesjo. (1996). "Dose calculation in brachytherapy for a ^{192} Ir source using a primary and scatter dose separation technique." *Phys. Med. Biol.* 41:1007–24.

Russell, K. R., A. K. C. Tedgren, and A. Ahnesjo. (2005). "Brachytherapy source characterization for improved dose calculations using primary and scatter dose separation." *Med. Phys.* 32:2739–52.

Segala, J. J., G. A. Cardarelli, J. R. Hiatt, and B. H. Curran. (2011). "Interface dosimetry for electronic brachytherapy intracavitary breast balloon applicators." *J. Appl. Clin. Med. Phys.* 12:293–300.

Sempau, J., E. Acosta, and J. Baro. (1997). "An algorithm for Monte Carlo simulation of coupled electron-photon transport." *Nucl. Instrum. Methods* 132:377–90.

Shapiro, A., B. Schwartz, J. P. Windham, and J. G. Kereiakes. (1976). "Calculated neutron dose rates and flux densities from implantable californium-252 point and line sources." *Med. Phys.* 3:241–7.

Sheikh-Bagheri, D., I. Kawrakow, B. Walters, and D. W. O. Rogers. (2006). "Monte Carlo Simulations: Efficiency Improvement Techniques and Statistical Considerations." In *Integrating New Technologies into the Clinic: Monte Carlo and Image-Guided Radiation Therapy.* Madison, WI: Medical Physics Publishing, 2005.

Shi, C., B. Guo, C.-Y. Cheng, T. Eng, and N. Papanikolaou. (2010). "Applications of tissue heterogeneity corrections and biologically effective dose volume histograms in assessing the doses for accelerated partial breast irradiation using an electronic brachytherapy source." *Phys. Med. Biol.* 55:5283–97.

Siebert, F.-A., S. Wolf, and G. Kóvacs. (2013). "Head and neck ^{192}Ir HDR-brachytherapy dosimetry using a grid-based Boltzmann solver." *J. Contemp. Brachytherapy* 5:232–5.

Sinnatamby, M., V. Nagarajan, and K. S. Reddy. (2015). "Dosimetric comparison of Acuros^{TM} BV with AAPM TG43 dose calculation formalism in breast interstitial high-dose-rate brachytherapy with the use of metal catheters." *J. Contemp. Brachytherapy* 7:273–9.

Slessinger, E. D., R. Fletcher, and I. J. Das. (2011). "Dose perturbation study in a multichannel breast brachytherapy device." *J. Contemp. Brachytherapy* 3 220–3.

Sloboda, R. S., H. Morrison, G. Cawston-Grant, and G. V. Menon. (2017). "A brief look at model-based dose calculation principles, practicalities, and promise." *J. Contemp. Brachytherapy* 9:79–88.

Taylor, R. E. P., G. Yegin, and D. W. O. Rogers. (2007). "Benchmarking BrachyDose: Voxel based EGSnrc Monte Carlo calculations of TG-43 dosimetry parameters." *Med. Phys.* 34:445.

Tedgren, A. C. and G. A. Carlsson. (2009). "Influence of phantom material and dimensions on experimental Ir dosimetry." *Med. Phys.* 36: 2228.

Tedgren, Å. C., M. Plamondon, and L. Beaulieu. (2015). "The collapsed cone algorithm for ^{192}Ir dosimetry using phantom-size adaptive multiple-scatter point kernels." *Phys. Med. Biol.* 60:5313–23.

Thomson, R. M., R. E. P. Taylor, and D. W. O. Rogers. (2008). "Monte Carlo dosimetry for ^{125}I and ^{103}Pd eye plaque brachytherapy." *Med. Phys.* 35:5530–43.

Vassiliev, O. N., T. A. Wareing, I. M. Davis, J. McGhee, D. Barnett, J. L. Horton, K. Gifford, G. Failla, U. Titt, and F. Mourtada. (2008). "Feasibility of a Multigroup Deterministic Solution Method for Three-Dimensional Radiotherapy Dose Calculations." *Int. J. Radiat. Oncol. Biol. Phy.* 72:220–227. http://www.sciencedirect.com/science/article/B6T7X-4T85W5C-P/2/a8ad26cdcffb00dd65bd63e92ede7989.

Veelen, B. V., Y. Ma, and L. Beaulieu. *ACE—Advanced Collapsed Cone Engine.* Elekta White Paper. Elekta AB, Stockholm, Sweden: 2014.

Wareing, T. A., J. M. McGhee, J. E. Morel, and S. D. Pautz. (2001). "Discontinuous Finite Element Sn Methods on Three-Dimensional Unstructured Grids." *Nucl. Sci. Engr.* 138:(2).

Wareing, T. A., J. E. Morel, and D. K. Parsons DK. *A first collision source method for ATTILA, an unstructured tetrahedral mesh discrete ordinates code.* Los Alamos, NM: Los Alamos National Lab, 1998.

WG-BSR. (2005). Brachytherapy Source Registry work group. *https://aapm.org/org/structure/default.asp?committee_code=WGBSR* Online: https://aapm.org/org/structure/default.asp?committee_code=WGBSR.

WG-DCAB. (2011). Working Group on Model-Based Dose Calculation Algorithms in Brachytherapy *http://www.aapm.org/org/structure/default.asp?committee_code=WGDCAB* Online: http://www.aapm.org/org/structure/default.asp?committee_code=WGDCAB.

White, S. A., G. Landry, G. P. Fonseca, R. Holt, T. Rusch, L. Beaulieu, F. Verhaegen, and B. Reniers. (2014). "Comparison of TG-43 and TG-186 in breast irradiation using a low energy electronic brachytherapy source." *Med. Phys.* 41:061701.

Williamson, J. F. (1987). "Monte Carlo evaluation of kerma at a point for photon transport problems. *Med. Phys.* 14:567–76.

Williamson, J. F. (2005). "Semi-empirical dose-calculation models in brachytherapy." In *Brachytherapy Physics*, 2nd ed. B. R. Thomadsen, M. Rivard, and W. M. Butler, Eds. Madison, WI: Medical Physics Publishing, 2005.

Yang Y., C. S. Melhus, S. Sioshansi, and M. J. Rivard. (2011). "Treatment planning of a skin-sparing conical breast brachytherapy applicator using conventional brachytherapy software." *Med. Phys.* 38:1519–25.

Yu Y., L. L. Anderson, Z. Li, D. E. Mellenberg, R. Nath, M. C. Schell, F. M. Waterman, A. Wu, and J. C. Blasko. (1999). "Permanent prostate seed implant brachytherapy: report of the American Association of Physicists in Medicine Task Group No. 64." *Med. Phys.* 26:2054–76.

Zhang, Z., E. I. Parsai, and J. J. Feldmeier. (2007). "A 3-D quantitative dose reduction analysis in MammoSite balloon by Monte Carlo calculations." *J. Appl. Clin. Med. Phys.* 8:139–51.

Zourari, K., T. Major, A. Herein, V. Peppa, C. Polgár, and P. Papagiannis. (2015). "A retrospective dosimetric comparison of TG43 and a commercially available MBDCA for an APBI brachytherapy patient cohort." *Eu. J. Med. Phys.* 31:669–76.

Zourari, K., E. Pantelis, A. Moutsatsos, L. Petrokokkinos, P. Karaiskos, L. Sakelliou, E. Georgiou, and P. Papagiannis. (2010). "Dosimetric accuracy of a deterministic radiation transport based ^{192}Ir brachytherapy treatment planning system. Part I: single sources and bounded homogeneous geometries." *Med. Phys.* 37:649–61.

Zourari, K., E. Pantelis, A. Moutsatsos, L. Sakelliou, E. Georgiou, P. Karaiskos, and P. Papagiannis. (2013). "Dosimetric accuracy of a deterministic radiation transport based ^{192}Ir brachytherapy treatment planning system. Part III. Comparison to Monte Carlo simulation in voxelized anatomical computational models." *Med. Phys.* 40:011712.

Example Problems

(Answers are found at the end of the book.)

1. Which radiological physics interaction did the TG-186 report not indicate as being of concern for advanced brachytherapy dosimetry calculations?

    a. intersource attenuation

    b. material heterogeneities

    c. clinical conditions for radiation scatter

    d. deformable image registration

2. Which is not a recommendation from the TG-186 report to apply to physicists, MBDCA vendors, and manufacturers of BT sources and applicators?

    a. The physicist is responsible to confirm that MBDCA dose predictions are based upon accurate and spatially resolved applicator and source models (including correct material assignments) to avoid dose-delivery errors prior to implementing the MBDCA in the clinic.

    b. Patient CT grids ~1 mm^3 are spatially inadequate for accurate modeling on BT sources and applicators. Dose distributions from low-energy, photon-emitting sources are particularly sensitive to simplistic geometric approximations. MBDCA vendors should use analytic modeling schemes or recursively specify meshes with ≥10 μm spatial resolution to accurately depict low-energy source geometries.

    c. Sources and applicators incorporated into a MBDCA TPS need not be independently verified by the physicist as the Registry will indicate which sources and applicators have met the AAPM dosimetric prerequisites.

    d. The physicist should compare MBDCA-generated single-source dose distributions in water to directly calculate TG-43 benchmarks.

3. There are currently adequate tools available to perform independent checks for brachytherapy model-based dose calculations.

    a. True

    b. False

4. Which is not a reason for why a 2% tolerance between expected and planned dose calculations is infeasible for MBDCAs?
   a. single-source dose calculations with ACE can have angular artifacts based on the number of tessellations used
   b. differences between dose in medium to water or to medium can exceed this tolerance for ^{192}Ir photons
   c. dose calculations with a truncated image set will not utilize the clinical radiation scatter conditions
   d. regions of high-dose gradients (such as from high-Z shields) may not be depicted with adequate spatial resolution
5. The dominant calculation variable when considering heterogeneous dose calculation for LDR breast brachytherapy is the composition of the breast tissue itself.
   a. True
   b. False
6. In HDR gynecological brachytherapy, one may expect the most significant dosimetric difference from the TG-43 calculation when using a MBDCA when:
   a. The patient is very thin.
   b. There is a variety of different tissues present in the calculation region.
   c. There is air in the rectum.
   d. The applicator includes the use of shielded ovoids.
7. The maximum separation of various tissue relative to water for energy deposition occurs around 25–30 keV.
   a. True
   b. False
8. ACE and AcurosBV both rely on this approximation to calculate dose:
   a. assume a homogeneous infinite water geometry
   b. use the Kerma approximation to relate primary fluence to dose and extract first scatter dose (Scerma).
   c. Track all secondary electrons throughout all generations to extremely small 1 KeV cut-off.
   d. Simulate the internal structure of the source for each calculation.
9. Deterministic methods are stochastic, so solution errors arise from statistical sources rather than systematic sources.
   a. True
   b. False
10. Deterministic methods provide full solution for the entire space rather than for specific regions (or tally location) as done in Monte Carlo approaches.
    a. True
    b. False

Chapter 3

Electronic Brachytherapy Sources: Features, Dosimetry, and Clinical Use

Mark J. Rivard[1], Regina K. Fulkerson[2], and Sujatha Pai[3]

[1]Tufts University School of Medicine
Boston, Massachusetts

[2]RKF Consultants, LLC
Dundee, New York

[3]Memorial Hermann Texas Medical Center, LMP
Houston, Texas

3.1 Introduction

3.1.1 Description of Electronic Brachytherapy

Electronic brachytherapy is a means of delivering radiation therapy with miniature x-ray sources. Like conventional x-ray tubes used in radiography or radiotherapy, the x-rays are generated via bremsstrahlung when electrons are directed to a high-Z target by an electric field. The transmission target is on the order of 1 μm thick. The electric potential is on the order of tens of kilovolts, and the target is typically an electrically conductive metal with a high melting point. However, unlike conventional x-ray tubes, the entire design is miniaturized for temporary insertion within an applicator or directly inside the patient. As such, the dose rate needs to be sufficiently high to provide a therapeutic dose within a short period of time (i.e., minutes).

Due to the nature of bremsstrahlung, the resulting photon spectrum has a maximum energy equivalent to the electric potential. However, the mean photon energy is lower due to two processes. First, there is a low probability that the electron will exchange all of its kinetic energy as it decelerates in the vicinity of the target nucleus. In fact, the probability of a given bremsstrahlung energy increases with decreasing photon energy, where the mean photon energy is approximately one-third of the electric potential. The second reason why the mean photon energy is lower than the accelerating potential is due to resultant interactions of the bremsstrahlung photons within the electronic brachytherapy source and the surrounding materials, such as a treatment applicator and adjacent tissues. In the vicinity of 30 keV, interactions in most materials will predominantly be the photoelectric effect with lower likelihoods for incoherent (Compton) or coherent (Rayleigh) scattering. Photons with energies less than approximately 5 keV are filtered out by the electronic brachytherapy source.

3.1.2 Comparison to Conventional Radionuclide-based Sources

The radiological interactions in tissue for photons from electronic brachytherapy sources are the same as from radionuclide-based, low-energy, photon-emitting brachytherapy sources, such as ^{103}Pd, ^{125}I, and ^{131}Cs with principal emission energies spanning 20 keV to 35 keV. However, there are substantial differences between the designs of electronic brachytherapy sources in comparison to radionuclide-based brachytherapy sources. A key difference is the ability to turn off the electronic brachytherapy source due to the electrical nature of radiation generation, which differs from radionuclide sources that are always emitting. The dose-rate output of electronic brachytherapy sources tends to be fairly constant, while those from radionuclides will only diminish over time following exponential decay. With electrically generated radiation comes a new concern for electrical safety of hospital staff and especially the patient. While radiation therapy centers are highly reliant on electricity for the function of numerous pieces of medical and support equipment, they have little experience having high-voltage devices placed within the patient—this is more the domain of surgery.

Another consideration is the difference in size between electronic brachytherapy and radionuclide-based sources; electronic brachytherapy sources are larger (on the order of ½ cm in diameter instead of ~ 0.1 cm for high-dose-rate (HDR) ^{192}Ir sources). This is due to the need for a vacuum across the voltage potential so electrons can accelerate and to have coolant in the vicinity of the anode target so that the electronic brachytherapy source does not fail. Radionuclide-based sources typically consist of a radioactive pellet surrounded by encapsulation to prevent radiation contamination (i.e., a sealed source).

Unlike radionuclide-based sources, electrically generated radiation can be altered by varying the current and voltage. In turn, the radiation dose distribution resulting from the bremsstrahlung x-rays may be customized for a given patient, although neither of the available systems makes use of this possibility. Source output tends to increase as the square of the electrical current. To minimize treatment times, a balance is made on maximum output and source longevity. There is a linear relationship of the photon energy with voltage, but this may be complicated by production of characteristic x-rays that have thresholds for excitation. These characteristic x-rays are dependent on the composition of the electronic brachytherapy source. However, the source output still tends to increase as the square of the voltage potential.

Having electrically generated radiation avoids special licensing required for radionuclide-based sources containing nuclear byproduct materials. There are also fewer regulator concerns, although many states have put into place regulations specifically for electronic brachytherapy units (Thomadsen et al. 2009). Further, the useful combination of HDR delivery with low-energy photons allows the choice of a variety of treatment rooms without requiring a massive treatment room bunker, as is required for high-energy HDR brachytherapy sources such as ^{192}Ir. Another advantage is that shorter irradiation times will minimize the likelihood for implant spatial variation from a pre-treatment plan. However, this will not be a linear effect since there many other tasks between pre-treatment imaging and completion of treatment delivery (such as treatment planning, plan check, possibly patient transport, etc.).

3.2 Descriptions of Available Devices

3.2.1 Axxent

3.2.1.1 Unit Description

Electronic brachytherapy was designed to provide source strength similar to that of an HDR ^{192}Ir source, but the shielding requirements are more similar to an LDR source. Originally Xoft Axxent® electronic brachytherapy source (iCAD Inc., Nashua, NH) was developed for intravascular brachytherapy and treating coronary artery disease (Rusch and Rivard 2004). However, the initial clinical use of the Axxent source was in the treatment of superficial lesions and balloon-based APBI (Rusch 2007; Dickler et al. 2009). The dosimetric equivalency of the target volume coverage was also shown for endometrial cases, with increased sparing of the rectum and bladder (Dickler et al. 2008)

The model S700 Axxent x-ray source is a disposable miniaturized x-ray tube that measures about 15 mm in length and 2.2 mm in diameter (Park et al. 2008). The source is typically operated at 50 kVp with 300 µA of beam current striking a tungsten thin film on the inner surface of a ceramic x-ray-transparent anode. The vacuum tube is integrated into a water-cooled, flexible probe assembly shown in Figure 3–1a, measuring 250 mm in length and 5.4 mm in diameter. Details of the x-ray tube located at the tip of the source assembly are shown in Figure 3–1b and Figure 3–1c. The source assembly is connected to a high-voltage cable that is directed into the lumen of the applicator and enables the controller to step the source to preprogrammed dwell positions within the applicator.

When the source is active, the radiation output is 0.6 Gy/min at 3 cm from the source axis, as measured in water. During 50 kVp operation, the x-ray spectrum in air has a broad bremsstrahlung distribution with an average energy of 28 keV, which does not include the characteristic lines from the tungsten thin film and yttrium in the anode body (Liu et al. 2008; Hiatt et al. 2016). These energies are similar to those of ^{125}I (mean 28 keV). Power to the source reaches a maximum of 15 watts. The dose distribution from the bare Axxent

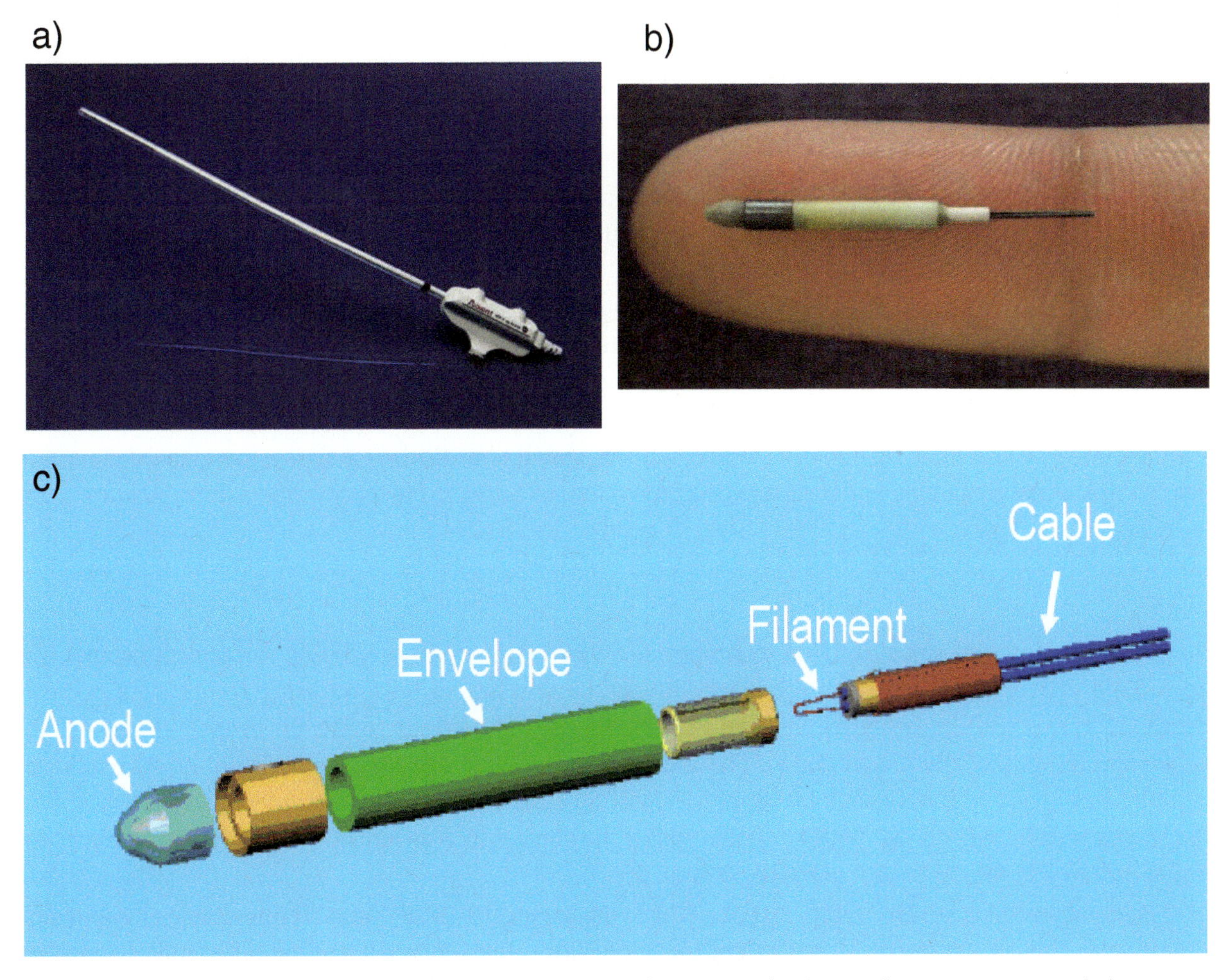

Figure 3–1 a) Xoft Axxent x-ray tube source assembly, b) source tube internal components, and c) source schematic. (Images courtesy of Rusch 2007.)

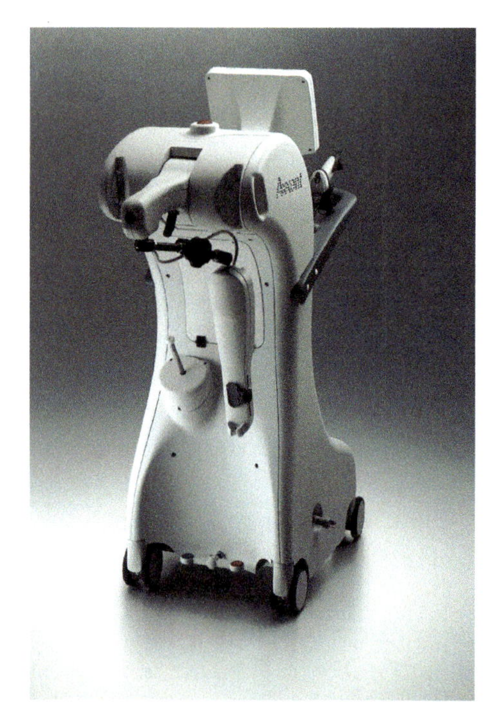

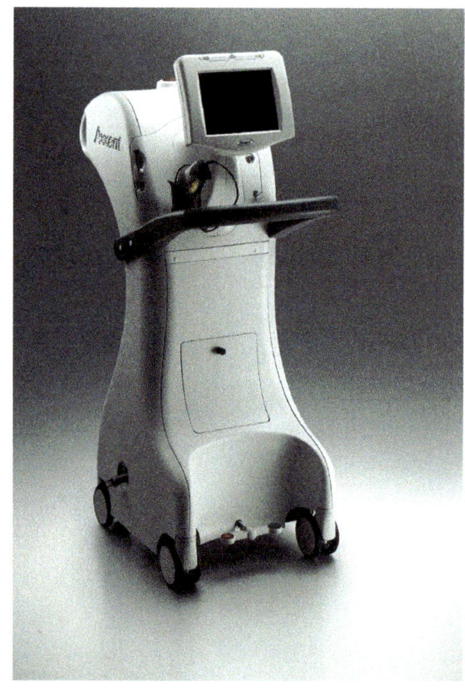

Figure 3–2 Xoft Axxent afterloader system. (Image courtesy of Rusch 2007.)

source is hemispherical in the distal direction and reduces in intensity in the proximal direction due to x-ray absorption within the source body. The nominal dose rate in water is 780, 125, 40, and 18 Gy/h at 10, 20, 30, and 40 mm, respectively (Rusch et al. 2007).

The Xoft Axxent system (Figure 3–2) consists of an afterloader system, an electronic controller, and the miniature electronic x-ray source contained inside a flexible catheter or a balloon applicator, depending on the type of application (Eaton 2015). Similar to an HDR ^{192}Ir afterloader, the x-ray tube can move within the catheter for dwell times determined at millimeter intervals. The source is encased in the cooling sheath, through which water is pumped continuously during treatment to provide cooling. Any malfunction in either the high-voltage circuit (including the x-ray tube) or the cooling system results in immediate treatment termination, and parameters of the treatment delivered thus far are recorded. The treatment time for each fraction is almost always less than 10 minutes. The controller contains a display showing the elapsed time, total planned time, time remaining at the current dwell position, and a visual display of the source position.

3.2.1.2 Treatment Sites and Applications

a. Skin

The Xoft source is well-suited for skin treatments due to the high dose-rate. Using surface applicators (Figure 3–3) to deliver dose, the electronic brachytherapy system has shown promising results with no recurrences, good to excellent cosmesis, and acceptable toxicities at ≥1-year follow-up in many forms of non-melanoma skin cancer (Bhatnagar 2013). The standard surface applicator set for use with the Axxent source includes four cones with diameters of 10, 20, 35, and 50 mm. Each cone has an integrated flattening filter. The source is centered over the flattening filter inside the cone, which ensures the delivery of a uniform dose profile of ±10% at the depth of 2 mm (Rusch et al. 2007).

The nominal SSDs are 20.7, 20.6, 20.6, and 30.3 mm for the 10, 20, 35, and 50 mm diameter applicators, respectively. The cone connects to one end of the applicator source channel, and the other end connects to a Tuohy–Borst adapter that is designed to lock the source in place for treatment. The source is oriented as with the V-type radionuclide-based applicators (see Chapter 7 for a discussion of skin applicator types). There is a thin polycarbonate window attached to the distal side of each insert which helps to ensure flat contact during treatment.

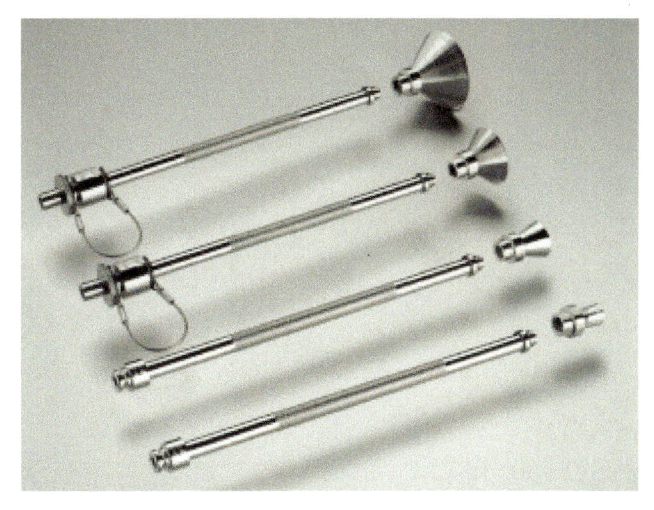

Figure 3–3 Surface applicators used with the Xoft Axxent source. (Image courtesy of Rusch 2007.)

b. Breast

Balloon catheters (Figure 3–4) have been used to treat early-stage breast cancer, initially with multiple fractions and then with single-fraction IORT. Dose distributions are similar to the MammoSite balloon catheter (Cytec Industries Inc., Mountain View, CA) used with HDR ^{192}Ir sources, but with higher doses close to the applicator surface (Dickler et al. 2008), which may increase the risk of fat necrosis. Conversely, doses to surrounding organs at risk are lower because of the relatively rapid dose falloff, as is the dose to breast tissue beyond the prescription distance. The placement of the breast balloon applicator can be performed at the time of surgery or under local anesthesia in an out-patient suite, depending on the procedure, i.e., APBI or IORT. The applicator is positioned within the breast cavity and inflated with sterile saline. Ultrasound, radiography, or CT is used to verify the position of the applicator, ensure that the cavity is filled, and check that the surgical margin conforms to the applicator. The applicator shaft is taped to the external skin of the breast for repeated access to the cavity.

Treatment planning is done in conjunction with CT images using a conventional brachytherapy treatment planning system (TPS), incorporating parameters describing the electronic source data. Library or the template plans are generally used for IORT treatments. The BrachyVision™ (Varian Medical Systems, Palo Alto, CA) TPS has been validated and can be used for fractionated APBI treatments. Plan details, in the form of source dwell positions and dwell times, are downloaded directly to the controller. Based on the preset treatment plan, the controller manages source movement through the programmed dwell positions by stepping the source back along the shaft in millimeter increments. The cable can negotiate up to a 15° curve, but requires a fairly straight pathway within the treatment region (Turian et al. 2006).

Prior to each treatment, the probe containing the electronically activated source is advanced into the central lumen of the applicator shaft. Once treatment is initiated, the controller moves the source to the farthest distal point inside the shaft of the saline-filled balloon and stops when the source comes in contact with the back wall of the lumen. A typical treatment plan requires the source to then be stepped through 5 to 10 dwell positions.

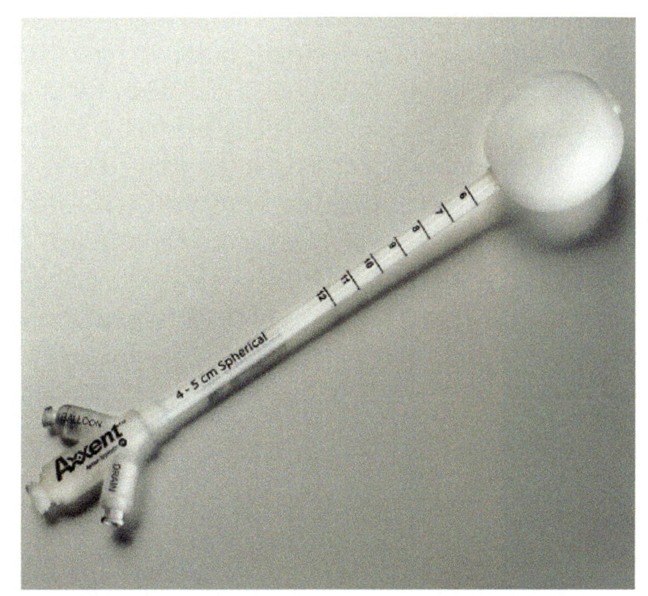

Figure 3–4 Xoft Axxent balloon applicator. (Image courtesy of Rusch 2007.)

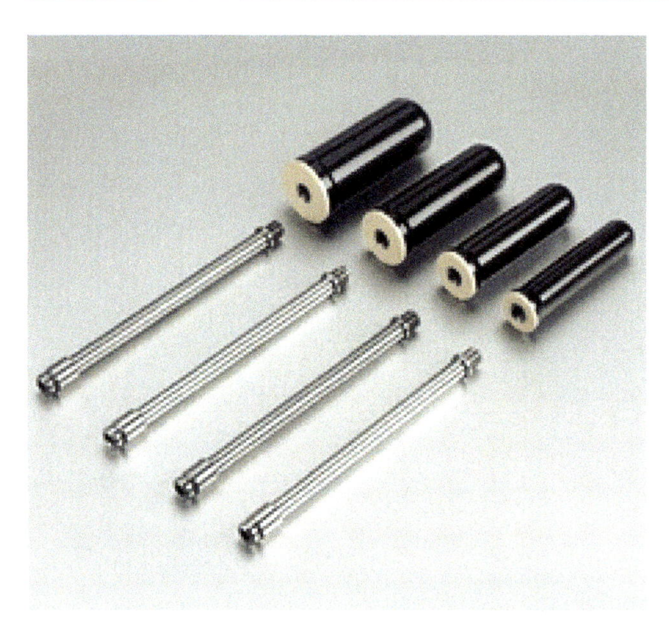

Figure 3–5 Xoft Axxent Vaginal applicators. (Image courtesy of Rusch 2007.)

c. Gynecological

Similar to radionuclide-based HDR ^{192}Ir vaginal cylinders for vaginal-cuff treatment following hysterectomy for endometrial cancer, the Axxent cylinders are available in various diameters. The available sizes are 20, 25, 30, and 35 mm in diameter. The vaginal applicators (Figure 3–5) are made of medical-grade polymers and provide transmission characteristics specifically for the low-energy x-rays emitted by the electronic brachytherapy source. The applicator gets inserted just prior to treatment and removed following treatment on each treatment visit. The base plate and clamp provide stabilization of the applicator during radiation treatment. The prescription dose and brachytherapy treatment plans get prepared individually for each patient, typically based on CT scans. BrachyVision software has been validated (incorporating parameters describing the Axxent electronic brachytherapy source data) for treatment planning.

d. Future Sites

Future applications of Xoft devices include head and neck, sarcoma, colorectal, gastric, pancreas, genitourinary, thoracic, and palliative applications such as kyphoplasty (Debenham et al. 2013).

3.2.2 INTRABEAM

3.2.2.1 Unit Description

The first reports on an electronic brachytherapy device were in 1996 (Beatty et al. 1996; Dinsmore et al. 1996; Douglas et al. 1996; Yanch and Harte 1996) as the INTRABEAM™ radiosurgery device. This was previously available from the Photoelectron Corporation (Waltham, MA) and now currently available by Carl Zeiss Meditec, Inc. (Oberkochen, Germany). Due to its internal use and interstitial application, such as for treating spinal metastases (Dinsmore et al. 1996), the source was later considered as a brachytherapy source. The INTRABEAM system is a portable soft x-ray source (XRS4) producing 50 kVp photons (Figure 3–6a,b). To produce the low-energy bremsstrahlung photon spectrum, electrons are produced and accelerated through the probe to strike a gold target (Figure 3–6c). The XRS4 source generates a spherical dose distribution at the tip of a thin probe (diameter = 3.2 mm) due to lateral dithering of the electron beam onto the gold target as monitored with five internal diodes (Armoogum et al. 2007).

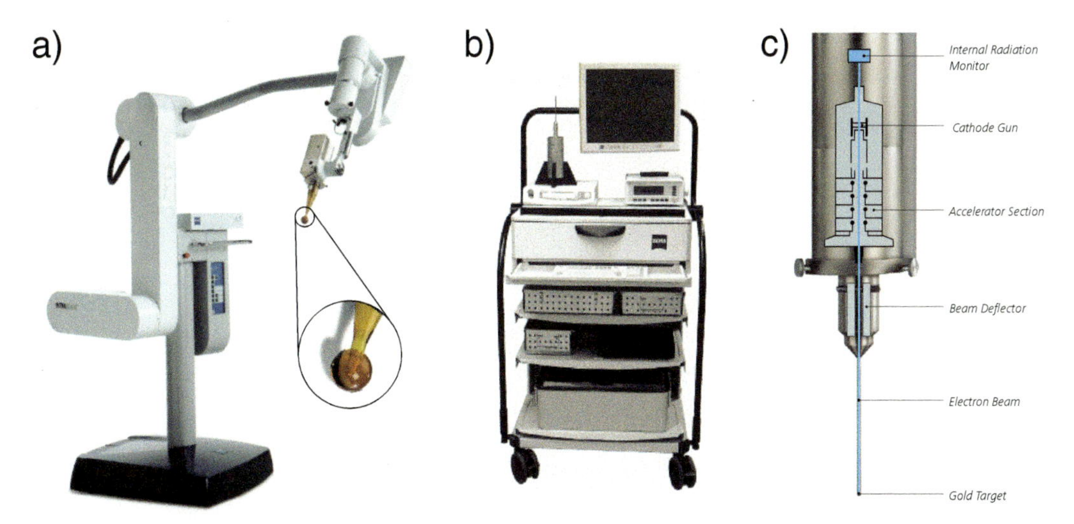

Figure 3–6 a) INTRABEAM system, b) treatment controller computer, and c) model XRS4 source rendering. (Images courtesy of Zeiss 2011.)

3.2.2.2 Treatment Sites and Applications

a. Skin

The INTRABEAM system has two types of applicators available for the treatment of skin lesions: "flat" and "surface." The surface applicator was designed to create a flat treatment field at the exit surface of the applicator, and the flat applicator was designed to create a flat treatment field at a depth of 5 mm from the applicator exit surface. The flat applicator has a flattening filter within it to create the flattened field at depth. The surface applicator has cone sizes ranging from 10–40 mm in diameter, and the flat applicator has cone sizes 10–60 mm in diameter. The SSDs vary with the applicator diameter, and these applicators have also been used for intraoperative radiotherapy to treat tumor beds and the body surface (Schneider et al. 2014; Goubert et al. 2015).

b. Breast

The INTRABEAM system was originally developed for intracranial stereotaxy (Eaton et al. 2011; Eaton 2015). Following this, solid spherical applicators were developed for the treatment of breast cancer. These applicators range in diameter from 1.5 to 5 cm (Wenz et al. 2010b). Immediately following a lumpectomy, the patient remains in the operating room, and the source (inside an applicator) is positioned in the tumor cavity for radiation delivery.

c. Gynecological

The INTRABEAM system has also been used in gynecological applications. For gynecological applications, a cylindrical applicator was developed for post-operative intravaginal treatments. Like radionuclide-based gynecological treatments, the INTRABEAM source is stepped through the cylindrical applicator for a predetermined time at each position (Schneider et al. 2009).

d. Other Intraoperative Sites

Spine metastases have also been treated intraoperatively with the INTRABEAM system (Wenz et al. 2010a; Schmidt et al. 2011). Unlike the treatment of a lumpectomy cavity, where the source tip is covered with a spherical applicator, spinal metastases are treated by first placing custom metallic sleeves into the treatment area, and the source tip is then guided into position before radiation is delivered. The INTRABEAM system has also been used intraoperatively for surgically resected, solitary brain metastases (Weil et al. 2015), and for rectal cancers immediately following radical resection using the spherical tip applicators of various diameters (Potemin et al. 2015; Guo et al. 2012).

3.3 Dosimetry

3.3.1 Methods for System Commissioning and Source Calibration

3.3.1.1 Axxent

The necessary tasks for system commissioning and source calibration are outlined in the paper by Hiatt and colleagues (Hiatt et al. 2008). They performed the following as a model set of tests:

1. Well chamber constancy was demonstrated through inter-comparison of the manufacturer-provided instrumentation (HDR 1000 Plus well chamber and MAX 4000 electrometer, both from Standard Imaging (Middleton, WI) with instrumentation existing already in the clinic. While an LDR ^{137}Cs source was used, other brachytherapy sources, such as a high-strength LDR ^{125}I source, would better match the energy range of the model S700 source. Note, however, that well-chamber calibration coefficients are specific to the paired well-chamber insert for a given source model, which may result in inter-comparison differences larger than the 0.14% observed change.
2. Stability of the voltage and current applied by the controller to the x-ray tube was evaluated through examining the Axxent controller log file from a test run. The manufacturer stipulates voltage constancy with a mean difference of 0.3 kV from the nominal 50 kV, and current constancy with a mean difference of 10 μA from the nominal 300 μA.
3. Source position accuracy was determined through measurements with radiochromic film in the manufacturer-provided QA test fixture. A specific value for the source guide tube length should be determined. At full current (0.3 mA), irradiation times on the order of 10 s per dwell position are needed for a clear result on the film. The source was translated through several dwell positions for coincidence with the shadowing on the film caused by Pb graticules in the QA test fixture. Alignment within 0.5 mm should be expected.
4. Unlike the voltage and current measurements in part 2 above, output stability for a given x-ray tube was measured in a QA test fixture and a small air ionization chamber located 1 cm from the source. Any detector that has high sensitivity and the ability to provide time-resolved readings would serve, for example, the well chamber. For their results at the dwell position having the highest current reading, tube output varied by 5% across five trial runs. While measuring output constancy for a single source under periodic testing will not demonstrate such constancy for all future sources, it will give the user some sense of the expected variations over time for individual sources.
5. Timer linearity of the system was determined by measuring output from a tube for several beam-on times. The times used covered the range expected for patient treatments and were measured in a non-monotonic order so that any variations of the system during the tests did not systematically influence the results. The times can be entered with 0.1 s precision. A timer error of 2.8 s and a correlation coefficient >0.9999 for timer linearity were determined.
6. Coincidence of the dummy marker positions with source dwell positions was demonstrated using Ag-based radiographic and radiochromic films, where the former demonstrated dummy marker positions following irradiation with a conventional simulator and the latter demonstrated source dwell positions following irradiation with the model S700 source. These films were chosen for their differing sensitivities. This work was performed 10 years ago, and most radiotherapy centers now do not

have access to radiographic film or conventional simulators. However, the principles of their test can be applied solely with radiochromic film. Positioning coincidence within 1 mm was demonstrated.

7. Several tests of controller functionality and safety interlocks were reported. These included validating a planned dwell file, verifying functionality of the status indicator light to reveal when radiation is being emitted, testing of the emergency off button, testing the treatment recovery procedure, and verifying functionality of the controller force pullback to see if an error resulted if a catheter is bent or obstructed.

8. Treatment planning was performed on a brachytherapy TPS not provided by the manufacturer. The PLATO TPS™ (Nucletron BV, now an Elekta company, Veenendal, the Netherlands) was compared to an in-house program for independent calculation checks. The maximum dose difference between the two was 1.2%. As the PLATO TPS is no longer available and Elekta's recent TPS offering (OncentraBrachy™) does not permit users to enter TG-43 parameters for custom sources, clinical users may perform treatment planning with another brachytherapy TPS, such as BrachyVision.

Additional tests for system commissioning recommended by the manufacturer include tests of all system functionality and buttons, validating source guide tube connections, source coolant flow obstruction, and tests of the well chamber and electrometer interlocks. As there have been changes to the Axxent system and the field of brachytherapy in general since the publication by Hiatt et al. (2008), clinical users should develop a modern commissioning procedure to capture their specific equipment and the environment in which it will be used, if possible using the techniques described in the AAPM TG-100 report (Huq et al. 2016).

Source calibration is performed by the clinical user at the time of source exchange and for every treatment. The initial measured result should match that reported by the manufacturer within 5%. The manufacturer-supplied well chamber, source holder insert, and electrometer are used for the National Institute for Standards and Technology (NIST)-traceable clinical measurements. Results are entered into the Axxent controller to combine with the treatment plan to determine the necessary dwell times for each patient treatment.

3.3.1.2 INTRABEAM

The INTRABEAM system is delivered to the customer with a set of instruments from the manufacturer. These instruments include a photodiode array comprised of five photodiodes positioned orthogonally from one another, which is used for an isotropy check. In addition, there is a probe adjuster and ionization chamber holder (PAICH) that is used to adjust the straightness of the probe manually, and an inbuilt temperature and pressure module and ionization chamber mount. There is also a water phantom (Figure 3–7) that is available on loan for commissioning purposes. This water tank is highly specialized for the INTRABEAM system, with radiation shielded lead glass and specialized geometry. The preferred ionization chamber used for commissioning and annual calibration is the PTW model 23342 (PTW Freiburg GmbH, Freiburg, Germany). This chamber is not inherently waterproof, so a specialized waterproof holder made of solid water is available for water phantom measurements. The PTW model 34013 ionization chamber is suggested for performing source isotropy measurements.

Upon commissioning, the manufacturer recommends depth dose curve measurements, source isotropy measurements, measurements of the applicator transfer function(s) (*ATF* defined as the ratio between the dose rates in the presence and in the absence of a specific applicator as a function of the distance from the target), applicator isotropy measurements for the spherical applicators, and source output measurements (using the PAICH). The system comes with a factory-calibrated PAICH-based dose rate D_{Original} determined as:

$$\dot{D}_{Original}\left[\frac{\text{Gy}}{\text{min}}\right] = I_{T,P}(PAICH)[\text{A}]\cdot N_K\left[\frac{\text{Gy}}{\text{C}}\right]\cdot k_Q \cdot 60\left[\frac{\text{s}}{\text{min}}\right] \qquad (3.1)$$

Figure 3–7 Water phantom and source mount, allowing for the x-ray source to be moved inside the water tank. (Image courtesy of Zeiss 2011.)

where $I_{T,P}$ is the corrected ionization chamber current (based on the sensor inside the PAICH). N_K and k_Q are taken from the air kerma calibration certificate of the factory ionization chamber, which is calibrated at PTW (a secondary-standards dosimetry lab) for air kerma at the T30 and T50 beam qualities and traceable to the German national metrology institute (PTB in Braunschweig, Germany). It should be noted that this dose rate is only valid in air and does not provide information about the dose output in water. A PAICH output measurement is performed during commissioning on site and before each treatment. For U.S. users, NIST traceability is obtained by having the PTW chamber used in the PAICH calibrated for air kerma at an ADCL and by utilizing the dosimetry protocol from the AAPM TG-61 report for determination of the reference dose rate in water. Respective dose rates in water at depth or with an applicator in place are determined based on measured depth dose curves with and without applicators in place. The *ATF* values for each applicator are stored in the treatment computer and applied to determine the treatment time. Initial commissioning measurements are completed by a qualified medical physicist on site and compared with any manufacturer-provided values. In agreement with accepted practices, independent verification of any manufacturer-provided dosimetry values should be performed.

3.3.2 Methods for Validating Source Output

3.3.2.1 Axxent

A NIST traceable air kerma standard has been established for the Axxent model S700 source (Seltzer et al. 2014). Since electronic brachytherapy is a low-energy (50 kVp) source, use of the Lamperti free-air chamber is appropriate. This calibration standard is based on realization of the air kerma produced by the x-ray beam at a reference distance in air of 50 cm (in air, not in vacuo). A free-air chamber measures the ionization in air without the secondary electrons hitting a wall of the chamber (DeWerd et al. 2015). The measured air kerma rate at 50 cm in air at NIST was independently validated through Monte Carlo simulations of radiation transport with mean results in agreement within 3.3% (Hiatt et al. 2015; Hiatt et al. 2016).

A well chamber provided by Axxent—a Standard Imaging model HDR 1000 Plus with the Axxent source holder—has been evaluated for use as a transfer instrument. The NIST calibration coefficients of the well chamber have units of Gy/(A s) normalized to reference temperature and pressure conditions. It has been demonstrated that the well chamber is an appropriate transfer standard for the electronic brachytherapy sources, as well as an efficient means of determining the stability of the sources (DeWerd et al. 2015).

A slight modification of the TG-43 dose calculation formalism (Rivard et al. 2004) is needed to accommodate the new source strength metric of air kerma rate at 50 cm in air (Simiele et al. 2016) as NIST does not provide a traceable calibration in terms of air kerma strength, like for low-energy, photon-emitting, radionuclide-based sources. New studies are underway to streamline implementation of the new NIST standard by proposing a new dose conversion coefficient, χ, and also investigate applicator effect on electronic brachytherapy dosimetry (not described in the TG-43 report series).

Each Axxent afterloader system comes with an HDR 1000 Plus well chamber (Standard Imaging, Middleton, WI) for verification of air kerma strength (Nath et al. 1997; Kubo et al. 1998; Thomadsen et al. 2009; DeWerd et al. 2015; Irwin 2015). The air kerma strength of the source is first determined by the manufacturer and then verified by the physicist upon delivery using the NIST calibration coefficient explained above.

Additional output verification considerations must be taken when using the electronic brachytherapy source in an applicator. This is particularly true in the case of a surface applicator. For Axxent surface applicators, the dose output at the surface is greatly altered due to the integrated flattening filter and, therefore, the qualified medical physicist (QMP) shall determine the applicator-specific dose rate at the treatment surface.

According to the resource availability, there are three options:

1. One option is to measure air kerma rate at the exit window of the surface applicator using an ADCL-calibrated small-volume parallel plate ionization chamber. To obtain the absolute dose rate in water, it is necessary to apply the methodology from the TG-61 report (Ma et al. 2001), applying the backscatter correction and mass attenuation coefficient ratio correction, in addition to the inverse square to take account for the effective point. The corrected absorbed dose rate at the phantom surface is given by:

$$\dot{D} = \frac{M_Q N_K B_w P_{stem,air}}{t}\left[\left(\frac{\mu_{en}}{\rho}\right)_{air}^{w}\right]\left(\frac{SSD + d_c}{SSD}\right)^2 \tag{3.2}$$

   where B_w is the backscatter factor, which depends on SSD, field size (collimator diameter), and HVL, and can be determined, for each available collimator, using Table V from the TG-61 report. The term $P_{stem,air}$ is the stem correction factor, which accounts for the change in photon scatter from the chamber stem between the calibration and measurement (mainly due to the change in field size). The estimation of $P_{stem,air}$ requires a comparison between the chamber used and a reference chamber for which $P_{stem,air}$ is known.

2. The second option is to measure the output in solid water using a chamber calibrated in absorbed dose to water (Candela et al. 2015). Special caution is required with solid water (designed to be water-equivalent for a specific energy range) as it has been reported that deviations larger than 5% are present for some materials (Hill et al. 2010). The absorbed dose in water divided by the radiation time t is given by the following modified expression:

$$\dot{D} = \frac{M_Q N_{D,w,Q_0} k_{Q,Q_0}}{t}\left(\frac{SSD + d_c}{SSD}\right)^2 \tag{3.3}$$

   where M_Q is the reading of the detector at the depth of interest, corrected for pressure, temperature, and electrometer calibration. Given that the polarity used in the dose measurements is the same as in the calibration, and given that the dose rate is lower than a few Gy/s, the polarity and ion recombination effects can be neglected.

3. The third option is to perform an in-air measurement using a calibrated chamber and then applying the correction factors to liquid water calculated using MC methods (Fulkerson et al. 2014a; Fulkerson et al. 2014b).

3.3.2.2 INTRABEAM

The XRS4 source comes with a factory-calibrated PAICH output value. It is the responsibility of the qualified user to verify this value independently, either with the PAICH system or some other measurement apparatus. Before each treatment session, the output is verified using the PAICH and should be within 5% of the value determined at commissioning as stored in the treatment computer. Deviations between 5% to 10% are reported to the user through a warning message and must be actively acknowledged. If the source output deviates >10% from the commissioned value, the software does not permit treatment.

3.3.3 Dose Calculation Methods

3.3.3.1 Axxent

Like most radionuclide-based brachytherapy sources, dose calculations for the model S700 source are performed using the TG-43 dose calculation formalism. The 2D calculation formalism is used with angular-dependent dose anisotropy as the source orientation within the patient is known. In 2006, Rivard and colleagues published TG-43 dosimetry parameters based on dose measurements and radiation transport simulations with Monte Carlo (MC) methods (Rivard et al. 2006). The MC results were updated in 2015 by Hiatt and colleagues based on manufacturing changes and a better understanding of the source design (Hiatt et al. 2015). Differences between the two papers in the radial dose functions were less than 2% over a range of $0.4 \leq r \leq 15$ cm. Differences between the two papers in the 2D anisotropy functions exceed a factor of two in contact with the source at $\theta = 150°$ and by more than 5% for one-third of the solid angle surrounding the source. The model S700 TG-43 radial dose function diminishes with increasing distance at a rate between that for ^{125}I and ^{103}Pd seeds. The model S700 TG-43 2D anisotropy function generally takes a shape similar to an inverted tomato (rounded on top toward $\theta = 0°$ and indented at the bottom toward $\theta = 180°$). For sources positioned within clinical brachytherapy applicators, the manufacturer provides TG-43 dosimetry parameters to account for material heterogeneities that influence the resultant dose distribution. A dosimetry formalism modifying the TG-43 approach was proposed by DeWerd and colleagues to account for the differing source calibration metrics between radionuclide-based brachytherapy sources and electronic brachytherapy sources, and to allow applicator-specific TG-43 parameters (DeWerd et al. 2015).

3.3.3.2 INTRABEAM

For the INTRABEAM electronic brachytherapy system, it is common practice to use the published atlas of the fractional depth dose (FDD) and profiles for treatment planning purpose (Zeiss, INTRABEAM Dosimetry, User Manual). The use of CT data for planning is not common practice and is not available at the present time. Intracavitary implant treatment planning is performed using CT imaging.

In the case of a typical breast IORT implant a single source position is used. The dose distribution is therefore mostly spherical. The radiation has the typical inverse square law behavior ($1/r^2$). Attenuation in the tissue introduces an additional attenuation factor governed by an approximate inverse linear law ($1/r$). Therefore, the radial dose attenuation decreases as the inverse cubic law ($1/r^3$).

The treatment time, run time (minutes), is calculated by taking the ratio of the prescribed dose D_{Rx}(Gy) with the dose rate in water determined during system daily QA D_{SQA}(Gy per minute) and the appropriate ATF value.

$$\text{Run time[min]} = D_{Rx}[\text{Gy}] / (\dot{D}_{SQA} \cdot ATF) \quad (3.4)$$

Typical run times vary from 16 to 33 minutes for 2.5 cm and 5.0 cm applicators, respectively.

3.3.4 Methods for Commissioning Dose Calculations

3.3.4.1 Axxent

Like most radionuclide-based brachytherapy sources, dose calculations for the model S700 source rely on the TG-43 dose calculation formalism (Rivard et al. 2004). Recommendations are given in the AAPM TG-53 report (Fraass et al. 1998) for commissioning a brachytherapy TPS, the AAPM TG-56 report (Nath et al. 1997) on brachytherapy code of practice, the AAPM TG-59 report (Kubo et al. 1998) for HDR ^{192}Ir brachytherapy safety, and, more recently, by Kim and colleagues with additional details specific for cervical cancer brachytherapy (Kim et al. 2016). A summary follows of these tasks specific to the model S700 electronic brachytherapy source.

Unlike for radionuclide-based HDR sources, the Axxent manufacturer does not provide brachytherapy TPS, so data entry of TG-43 dosimetry parameters will need to be performed by the clinical user. Records should be kept of which data were entered and the source/s of these data. These records should be maintained for the entire lifetime of the TPS to facilitate future QA (or if there are software or hardware changes) to demonstrate constancy. A qualified medical physicist should prepare a report documenting the suitability of the TPS for the electronic brachytherapy application and clarify any necessary workarounds to circumvent the TPS for characterizing the electronic brachytherapy source dose distribution. For example, the model S700 electronic brachytherapy source is not radionuclide-based, and the physicist should be concerned about entering some value for the half-life (to differ from ^{192}Ir), which will influence calculations of dose and dose-rate.

Commissioning should start with checks of dose calculations for a single dwell position, then proceed to multiple dwell positions to check dose summation. As with any radiotherapy dose calculations, there should be a means of performing independent calculations to check the results of the primary TPS-based method. From the onset, it should be known what specific errors can and cannot be identified by the independent calculation method. In addition to tests of dose calculations, tests of volume calculations and cumulative dose volume histograms should also be performed. Comparisons can be made of calculated and reference doses on all points of a plane (Yang et al. 2011).

Conventional HDR TPSs use S_K for source strength, which differs from that used to specify source strength for the model S700 electronic brachytherapy source (Seltzer et al. 2014; DeWerd et al. 2015; Hiatt et al. 2016). DeWerd and colleagues have proposed a means of rectifying this (DeWerd et al. 2015), but currently no TPS permits such nomenclature. Therefore, a workaround is necessary to cope with the differing units. The medical physicist should establish a standard method of entering source strength into the TPS so that errors are prevented. This could include simply by replacing the product of air kerma rate at 50 cm and χ with the air kerma strength and a unity value for the dose-rate constant. Also, specific to the model S700 source, care must be given to ensure that the radial dose function data are used in conjunction with the point-wise geometry function (i.e., $1/r^2$) to replicate the reference dose distribution (Hiatt et al. 2015). This will also require characterization of the source dimensions, permissible source dwell spacing, and the source guide tube length.

HDR TPSs use linear or bilinear interpolation of TG-43 dosimetry parameters to calculate dose at points not falling on the input values. It is crucial that data be entered in the TPS with high resolution as the model S700 electronic brachytherapy source exhibits dose falloff faster than an ^{125}I seed and with larger dose anisotropy, especially when present within certain applicators. Otherwise, interpolation errors will cause differences between the calculated and expected doses. The range of calculated dose should cover all clinical possibilities, from close to or far from the source. There should be no extrapolations (near or far) performed during clinical treatment planning. Similarly, the TPS should be commissioned for the range of dwell times to be used for patient treatments. Specific to the model S700 source, it is possible with HDR TPS optimization tools to have short dwell times where the few-second time error is not negligible. The choice of optimizers should be assessed, and the range of optimizer settings should be examined during TPS commissioning. Limits to the optimizer settings should be set specific to applicator types and disease sites.

If the influence of material heterogeneities are to be accounted for with dosimetry parameters that differ from those obtained for liquid water (Rivard et al. 2009), care must be taken to ensure that all the correct data are used for a given source model without accidentally combining dosimetry parameters from differing materials. Source modeling should be clear, such that clinical treatment planning will not mistakenly use the incorrect source for a given circumstance. Chapter 2 on model-based dose calculation algorithms describes these aspects in greater detail.

3.3.4.2 INTRABEAM

The determination of treatment time (and subsequently the dose delivered) can be based on a TG-61 reference dose rate and then corrected by measured depth dose curves and applicator transfer functions. Outside of an annual check of these values, no additional dose optimization is currently employed. The treatment planning computer is able to visualize anticipated dose distributions based on the measured source output, depth dose values, and isotropy values.

With ever-increasing computing capabilities available for clinical treatment planning, Zeiss recently introduced a hybrid MC algorithm for treatment planning with the INTRABEAM system. This algorithm is integrated into the *radiance* treatment planning software (GMV Innovating Solutions, Rockville, MD) and utilizes experimental data determined by the user, e.g., PDD curves and dose distributions, to optimize/fit the energy spectra of the source via a genetic algorithm. The hybrid MC algorithm accounts for the photoelectric effect, first- and second-order Compton scattering, and has been validated against full MC simulations in homogeneous and heterogeneous environments (Ibáñez et al. 2014). Monoenergetic phase-space files are generated for each type of applicator and then optimized to the data measured by the user (Figure 3–8). Dose computation is then completed based on the optimized phase-space files within minutes (Udías et al. 2014).

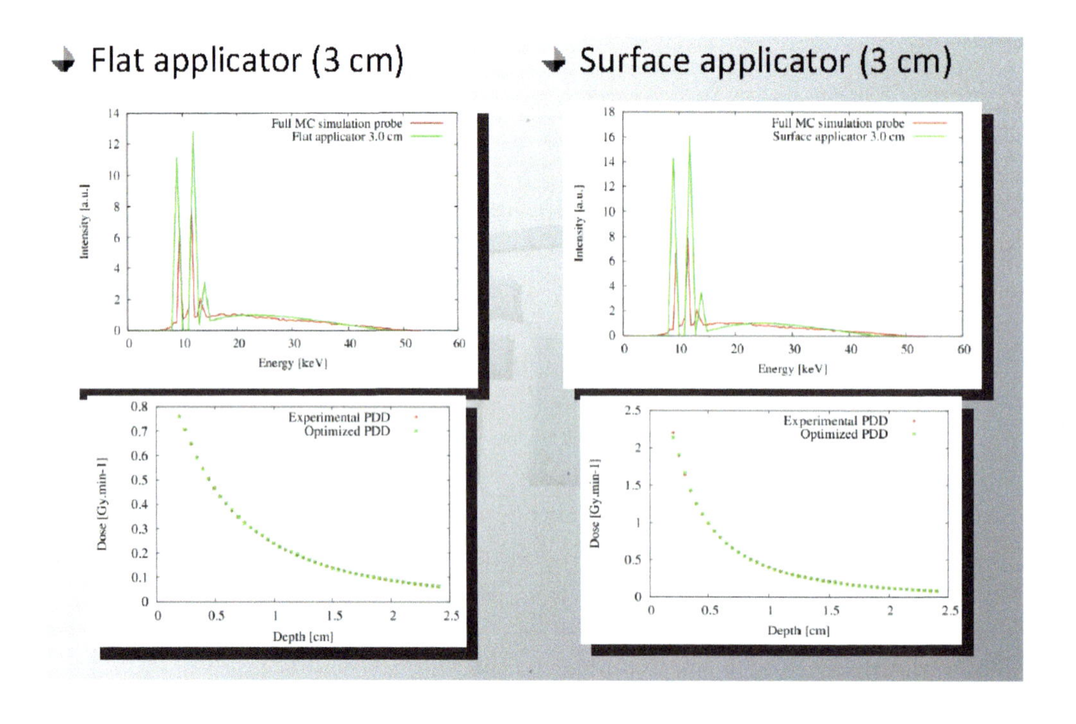

Figure 3–8 Energy spectrum optimization example for the flat and surface applicators. (Images courtesy of Zeiss.)

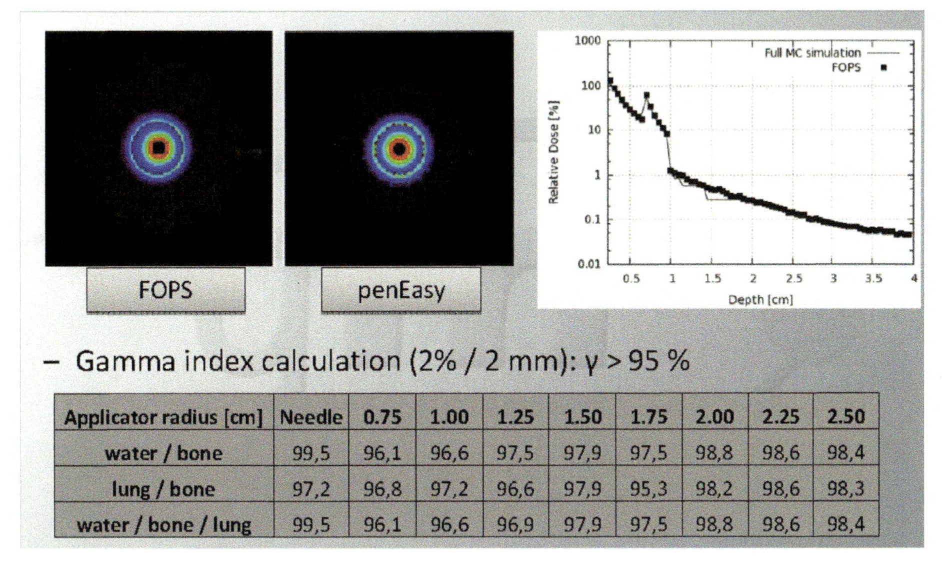

Applicator radius [cm]	Needle	0.75	1.00	1.25	1.50	1.75	2.00	2.25	2.50
water / bone	99,5	96,1	96,6	97,5	97,9	97,5	98,8	98,6	98,4
lung / bone	97,2	96,8	97,2	96,6	97,9	95,3	98,2	98,6	98,3
water / bone / lung	99,5	96,1	96,6	96,9	97,9	97,5	98,8	98,6	98,4

Figure 3–9 Dose calculation validation data for the hybrid Monte Carlo (FOPS) and full Monte Carlo (penEasy) for a needle applicator in a lung/bone/lung phantom. (Images courtesy of Zeiss 2011.)

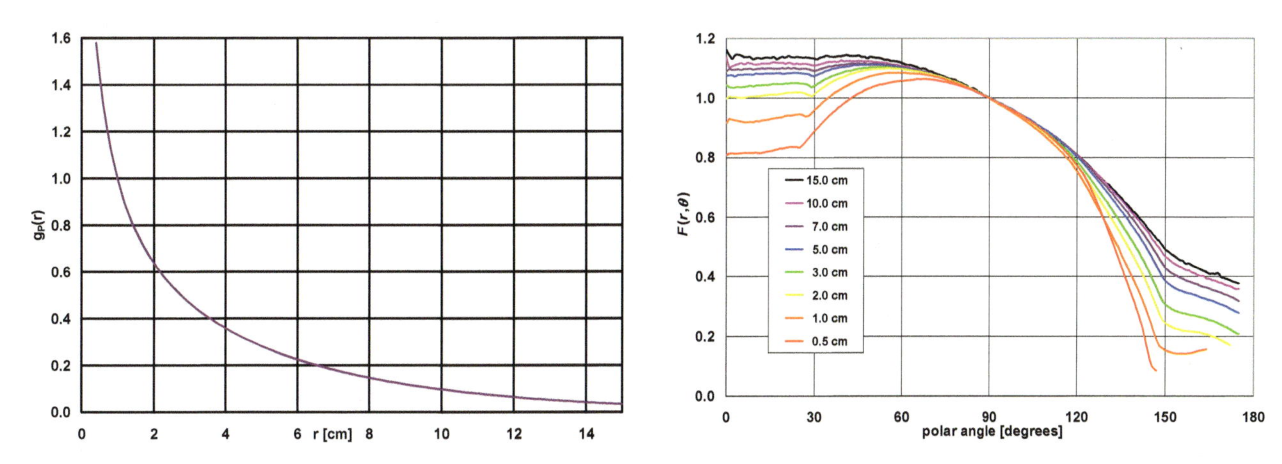

Figure 3–10 Radial dose function in liquid water for the model S700 source (Hiatt et al. 2015).

Figure 3–11 2D anisotropy function for the model S700 source (Hiatt et al. 2015).

A few groups have investigated the accuracy of this new treatment planning capability for INTRABEAM and found that the dose distributions determined with the fast hybrid MC tool are within 1% / 1 mm (standard 95% gamma-index assessment) compared with penEasy full MC simulations in a homogeneous environment (Vidal et al. 2014) and within 2% / 2 mm for heterogeneous environments (Figure 3–9). At the time of writing, there are no available scientific papers describing or validating the hybrid MC TPS. Also, it is unknown what the manufacturer-recommended procedures are for commissioning this algorithm; however, one can predict that a phantom study using film or other 2D dosimeter would be good practice.

3.3.5 Available Reference Data for Patient Dose Calculations for Each Source

3.3.5.1 Axxent

Because patient dose calculations for the model S700 source are based on the TG-43 dose calculation formalism, TG-43 dosimetry parameters are required for use within a brachytherapy TPS. The figures below of the radial dose function (Figure 3–10) and 2D anisotropy function (Figure 3–11) are taken from the supplementary materials from the 2015 paper by Hiatt and colleagues, which contains high-resolution tabulations for facilitating TPS data entry (Hiatt et al. 2015). These data are not copied here due to brevity, but are available in their reference 26 (Hiatt et al. 2015). Given the absence of societal-issued consensus datasets, these data may be taken as the current reference for clinical brachytherapy treatment planning.

3.3.5.2 INTRABEAM

To the best of our knowledge, each center maintains its own reference data for the INTRABEAM system. There is no manufacturer-provided dose distribution information, although the system does come with a factory-calibrated value of output (based on air kerma) for comparison purposes.

3.4 Clinical Use

3.4.1 Pertinent Tips/Tricks

3.4.1.1 Axxent Skin Applicators

While treatments of skin lesions are generally less complicated than other clinical sites, there are several safety and efficiency items to consider:

1. Use the plastic end cap to outline the area of treatment on the patient prior to treatment. The end cap can be detached from the selected cone applicator and centered on the lesion. A thin-tipped permanent marker can then be used to trace or dot an outline of the circumference on the patient's skin, leaving a template on the skin for the clinician to refer to. Ensure that the end cap is reattached to the cone before treatment.
2. Place the source in the cone applicator *before* placing the applicator on the patient. This allows the clinician to position the applicator on the ink template on the skin and fine tune the setup in one step, preventing the possibility of the applicator moving if the source is inserted after placement.
3. Make sure the wheels on the controller are locked *before* placing the applicator on the patient. The action of locking the wheels may cause movement of the controller and, hence, the setup.
4. Although full contact of the applicator to the treatment surface is desired, there should be no compression of the treatment area. A slight compression during the clamping process is acceptable since the clamp relaxes after releasing it.
5. Thin plastic sheets can be used to cover the lesion before treatment to prevent bodily fluid contamination. The presence of this sheet has a negligible effect on dose delivery.
6. Cutout factors can be efficiently performed with an Exradin A20 chamber. A jig for this chamber can be provided by Xoft, and the Xoft "Flexishield" material is commonly used for cutouts. This material is durable and stays in place during immobilization. It is suggested to position the cutout on the lesion, then center the end cap on the cutout and trace the circumference of the cone.

3.4.1.2 Axxent Breast IORT Treatments

For IORT using the Xoft system, it is recommended that the physicist move the unit into the operating room (OR) as soon as it is cleaned and before the patient is wheeled in. The physicist should perform daily QA on the unit. If there are any issues that are discovered during the QA that cannot not be resolved, the physicist can relay this information to the radiation oncologist/breast surgeon so a decision can be made as to whether to proceed with the case.

A 15-inch flexible drape (of 0.4 mm lead equivalent) placed over the breast helps to lower exposure levels. In addition, movable lead shields are used as personnel shielding for the anesthesiologist, radiation oncologist, and the physicists during the IORT treatment.

Physicists should ascertain in advance which breast is receiving the IORT. They should perform QA and place the controller on the same side of the room as the breast being treated in order to make the approach of the controller and connection between the balloon applicator and unit easier. There may be instances where the above suggestion may not work. That is, although the controller is placed on the same side of the room as the breast being treated, the surgeon may find that it is easier to implant and orient the balloon applicator shaft from the medial side of the breast instead of from the lateral side or from ant/post orientation. In that case, the controller may have to be moved to the opposite side of the patient for the applicator connection. Because of this, the physicist should survey the room for the best location to plug the controller in before the QA. This way, they can move the controller from one side or the other without having to unplug the controller and possibly performing another QA. If there are no suitable outlets in the OR for this, then one can use any outlet in conjunction with a suitable length of extension cord/cable. The physicist should request this from the OR staff or clinical engineering.

Due to sharp needles and objects around the seroma cavity, special precautions need to be taken to make sure that the implanted balloon's integrity does not get compromised. Since the balloon applicator is filled with sterile saline for the treatment, the physicist should record the volume of saline the surgeon used to inflate the applicator. At the end of the IORT treatment, the physicist should record the amount of saline withdrawn. If there is any difference in the saline amount, it will indicate a leak in the balloon, which implies increased delivered dose. The physicist will have to calculate the approximate delivered dose after determining the new volume of the balloon.

Since the procedure takes place in an OR environment, the physicist(s) should take care as to not contaminate any sterile areas with the Controller or accessories while in the room. It is recommended that the physicist leave their equipment (with the exception of the Controller) outside of the OR until it is needed. They may also ask the OR staff for a small cart for their laptop or equipment.

3.4.1.3 INTRABEAM

As discussed in Section 3.2.2.1, INTRABEAM provides a mobile radiotherapy solution for multiple indications, allowing radiation oncologists to cater to various surgical specialties with minimum interference to established surgical work flows. It is used in treating patients with neurological tumors (glioblastoma, brain, and spine metastases); ear, nose, head and neck, gastric and retroperitoneal areas; as well as breast cancer, soft tissue sarcoma, and skin cancer. Figure 3–12 indicates a generic work flow for INTRABEAM treatment delivery, which varies slightly depending on the type of treatment. However, the

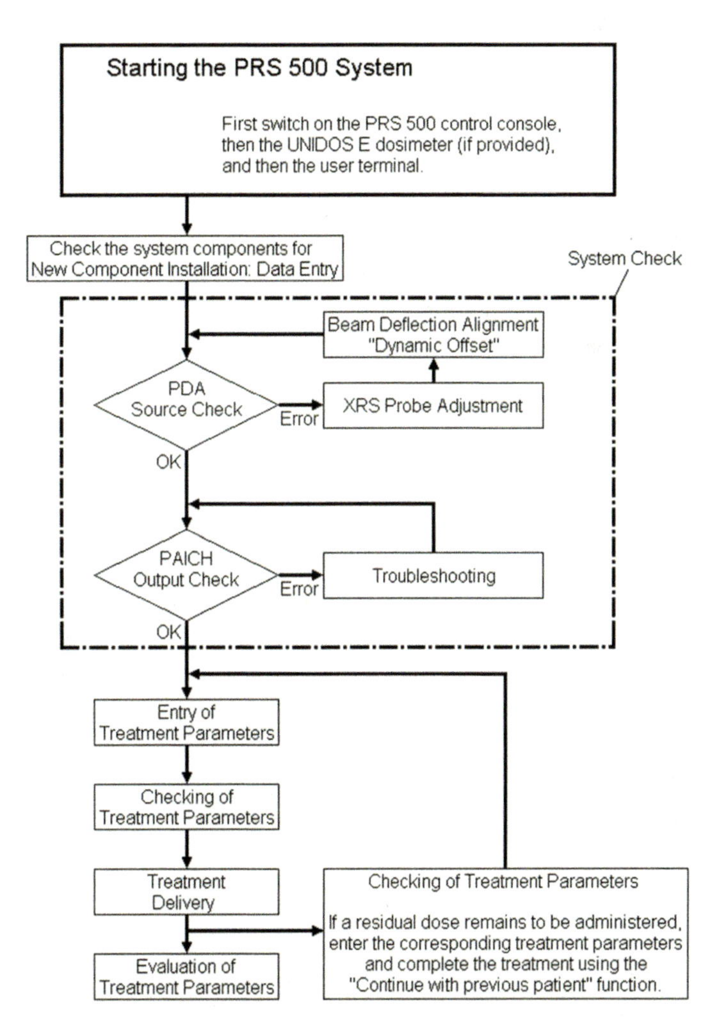

Figure 3–12 Schematic diagram of the INTRABEAM clinical work flow. (Image courtesy of Zeiss.)

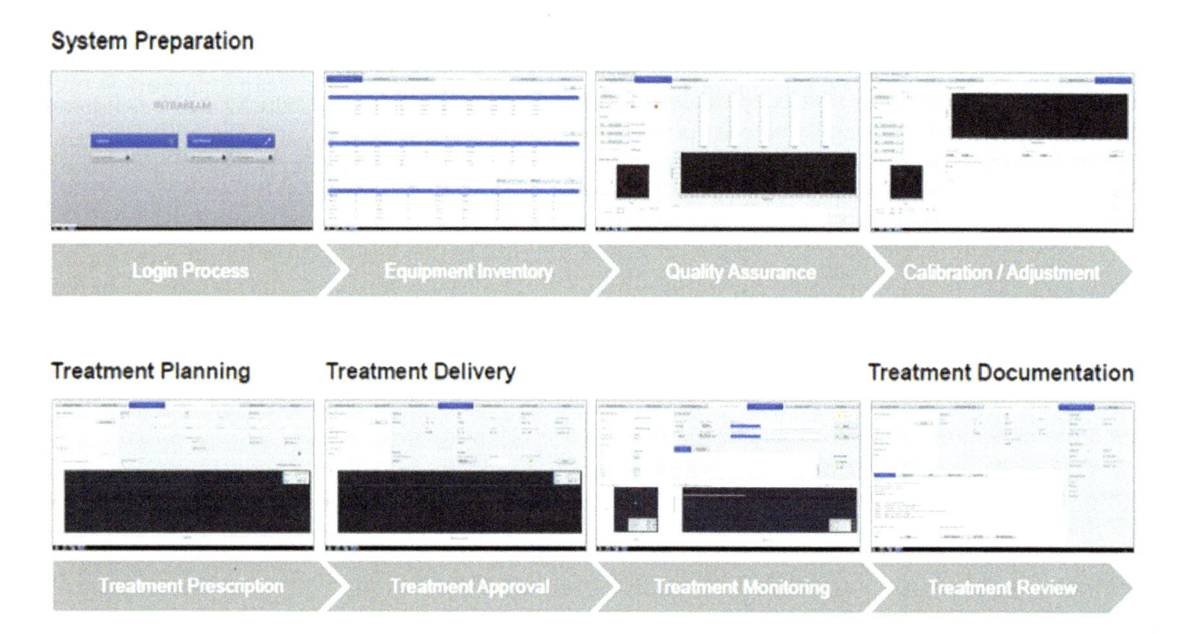

Figure 3–13 Schematic diagram of the work flow for the INTRABEAM system software. (Image courtesy of Zeiss 2011.)

physics work flow (Figure 3–13) evaluating the system quality assurance (prior to patient treatment) remains the same across all INTRABEAM treatments.

3.4.2 Electronic Brachytherapy QA Tests

3.4.2.1 Xoft Axxent QA Tests for IORT

The following is a summary of Xoft Axxent QA tests:

1. Source Change QA (done when a source is changed)
    a. Verify the length of the new source.
    b. Recalibrate the source to determine the new calibration factor.
    c. Perform film QA with single dwell or multiple dwell positions.
2. Daily QA (prior to surgery, prepare to do daily QA on the unit)
    a. Inspection of cooling system (check for leaks
    b. Applicator connection interlock
    c. Source connection interlock check
    d. Wheel lock interlock
    e. Obstruction test / Pull back test:
    f. Emergency stop
    g. Timer test
3. Periodic QA (monthly/quarterly)
    a. Daily QA
    b. Position verification (film)
    c. Emergency stop button

   d. Flow sensor test
   e. Well chamber interlocks
   f. Electrometer interlocks
   g. Timer test
   h. Drain water from pump
4. Annual QA
   a. Periodic QA
   b. Check flexi-shield for integrity
   c. Check all applicators
   d. Check calibration certificates

3.4.2.2 INTRABEAM QA Tests

INTRABEAM system QA is made efficient by building it into the software work flow. The daily and periodic QA consists of following tests:

1. *Probe adjuster.* The test checks the straightness of the probe. As everything from cathode to target is under vacuum and very thin, the probe must be very straight to achieve near 100% hit rate of electrons onto the target. During this test, the PAICH has to be rotated completely (≥360°) around the x-ray source (XRS4) probe with measurements at the cardinal points to verify straightness of the probe.
2. *Dynamic offsets.* This test should be always performed after the mechanical re-centering of the XRS4 probe. It is important that the photo diode array (PDA) device and the XRS4 are aligned to each other. This is why both devices have imprinted the X-Y coordinates and these +X and +Y markings have to be aligned on both the XRS4 and the PDA while performing the test.
3. *Isotropy check.* The PDA source check function is used to verify and, if necessary, adjust the XRS4 probe x-ray emission pattern to make it as isotropic as possible. As explained earlier, the +X and +Y markings have to be aligned on both the XRS4 and the PDA before performing the test.
4. *Output check.* This is the test where the internal radiation monitor will be checked and the source will be calibrated according to an independent ionization chamber which is placed on top of the probe of the x-ray source.

The above four tests have to be performed before every treatment is administered to ensure that the system runs within specifications.

3.5 Conclusions

Electronic brachytherapy sources are low energy yet with high dose-rate, have several attributes over radionuclide-based sources, and have been commercially available for over 20 years. The systems by Xoft and Zeiss are mature, having several applicator families to treat cancers over a variety of anatomic sites. The dosimetric characterizations and possibilities for treatment planning differ between the two systems. Increasingly there is literature to document clinical experience and physics practice for quality management. Reports from the AAPM are forthcoming to provide guidance to physicists on clinical practice and quality standards.

References

Armoogum, K. S., J. M. Parry, S. K. Souliman, D. G. Sutton, and C. D. Mackay. (2007). "Functional intercomparison of intraoperative radiotherapy equipment—Photon Radiosurgery System." *Radiat. Oncol.* 2:11. doi: 10.1186/1748-717X-2-11.

Bhatnagar, A. (2013). "Nonmelanoma skin cancer treated with electronic brachytherapy: Results at 1 year." *Brachytherapy* 12(2):134–40.

Beatty, J., P. J. Biggs, K. Gall, P. Okenieff, F. S. Pardo, K. J. Harte, M. J. Dalterio, and A. P. Sliski. (1996). "A new miniature x-ray source for interstitial radiosurgery: Dosimetry." *Med. Phys.* 23(1):53–62. doi: 10.1118/1.597791.

Candela-Juan, C., Y. Niatsetski, Z. Ouhib, F. Ballester, J. Vijande, and J. Perez-Calatayud. (2015). "Commissioning and periodic tests of the Esteya® electronic brachytherapy system." *J. Contemp. Brachytherapy* 7(2):189–95.

Debenham. B. J., K. S. Hu, and L. B. Harrison. (2013). "Present status and future directions of intraoperative radiotherapy." *Lancet Oncol.* 14:e457–e464.

DeWerd, L. A., W. S. Culberson, J. A. Micka, and S. J. Simiele. (2015). "A modified dose calculation formalism for electronic brachytherapy sources." *Brachytherapy* 14(3):405-8. doi: 10.1016/j.brachy.2015.01.003.

Dickler, A., M. C. Kirk, A. Coon, D. Bernard, T. Zusag, J. Rotmensch, and D. E. Wazer. (2008). "A dosimetric comparison of Xoft Axxent Electronic Brachytherapy and iridium-192 high-dose-rate brachytherapy in the treatment of endometrial cancer." *Brachytherapy* 7(4):351–54.

Dickler, A., O. Ivanov, and D. Francescatti. (2009). "Intraoperative radiation therapy in the treatment of early-stage breast cancer utilizing Xoft Axxent electronic brachytherapy." *World J. Surg. Oncol.* 7(24). doi: 10.1186/1477-7819-7-24.

Dinsmore, M., K. J. Harte, A. P. Sliski, D. O. Smith, P. M. Nomikos, M. J. Dalterio, A. J. Boom, W. F. Leonard, P. E. Oettinger, and J. C. Yanch. (1996). "A new miniature x-ray source for interstitial radiosurgery: Device description." *Med. Phys.* 23(1):45–52. doi: 10.1118/1.597790.

Douglas, R. M., J. Beatty, K. Gall, R. F. Valenzuela, P. Biggs, P. Okunieff, and F. S. Pardo. (1996). "Dosimetric results from a feasibility study of a novel radiosurgical source for irradiation of intracranial metastases." *Int. J. Radiat. Oncol. Biol. Phys.* 36(2): 443–50. doi: 10.1016/S0360-3016(96)00293-3.

Eaton, D. J., R. Gonzalez, S. Duck, and M. Keshtgar. (2011). "Radiation protection for an intra-operative x-ray device." *Br. J. Radiol.* 84(1007): 1034-1039. doi: 10.1259/bjr/29466902.

Eaton, D. J. 2015 "Electronic brachytherapy—current status and future directions." *Br. J. Radiol.* 88:20150002. doi: 10.1259/bjr.20150002.

Fraass, B., K. Doppke, M. Hunt, G. Kutcher, G. Starkschall, R. Stern, and J. Van Dyke. (1998). "American Association of Physicists in Medicine Radiation Therapy Committee Task Group 53: Quality assurance for clinical radiotherapy treatment planning." *Med. Phys.* 25(10):1773–1829. doi: 10.1118/1.598373.

Fulkerson, R. K., J. A. Micka, and L. A. DeWerd. (2014a). "Dosimetric characterization and output verification for conical brachytherapy surface applicators. Part I. Electronic brachytherapy source." *Med. Phys.* 41(2):022103.

Fulkerson, R. K., J. A. Micka, and L. A. DeWerd. (2014b). "Dosimetric characterization and output verification for conical brachytherapy surface applicators. Part II. High dose rate ^{192}Ir sources." *Med. Phys.* 41(2):022104.

Goubert, M. and L. Parent. (2015). "Dosimetric characterization of INTRABEAM® miniature accelerator flat and surface applicators for dermatologic applications." Phys. Medica. 31(3):224. doi:10.1016/j.emp.2015.01.009.

Guo, S., C. Reddy, M. Kolar, N. Woody, A. Mahadevan, F. C. Deibel, D. Dietz, F. Remzi, and J. Suh. (2012). "Intraoperative radiation therapy with the photon radiosurgery system in locally advanced and recurrent rectal cancer: retrospective review of the Cleveland clinic experience." Radiat. Oncol. 7:110 http://www.ro-journal.com/content/7/1/110. DOI: 10.1186/1748-717X-7-110.

Hiatt, J., G. Cardarelli, J. Hepel, D. Wazer, and E. Sternick. (2008). "A commissioning procedure for breast intracavitary electronic brachytherapy systems." *J. Appl. Clin. Med. Phys.* 9(3):58–68. doi: 10.1120/jacmp.v9i3.2775.

Hiatt, J. R., S. D. Davis, and M. J. Rivard. (2015). "A revised dosimetric characterization of the model S700 electronic brachytherapy source containing an anode-centering plastic insert and other components not included in the 2006 model." *Med. Phys.* 42(6):2764–76. doi: 10.1118/1.4919280.

Hiatt, J. R., M. J. Rivard, and H. G. Hughes. (2016). "Simulation evaluation of NIST air-kerma rate calibration standard for electronic brachytherapy." *Med. Phys.* 43(3):1119–29. doi: 10.1118/1.4940791.

Hill, R., Z. Kuncic, and C. Baldock. (2010). "The water equivalence of solid phantoms for low energy photon beams." *Med. Phys.* 37(8):4355–63.

Huq, M. S., B. A. Fraass, P. B. Dunscombe, J. P. Gibbons, Jr., G. S. Ibbott, A. J. Mundt, S. Mutic, J. R. Palta, F. Rath, B. R. Thomadsen, J. F. Williamson, and E. D. Yorke. (2016). "The report of Task Group 100 of the AAPM: Application of risk analysis methods to radiation therapy quality management." *Med. Phys.* 43(7):4209–62. doi: 10.1118/1.4947547.

Ibáñez, P., M. Vidal, R. García-Marcos, E. Herranz, P. Guerra, J. A. Calama, M. A. Infante, M. E. Lavado, and J. M. Udías. (2014). "Validation of a phase space determination algorithm for intra-operative radiation therapy." *Radiother. Oncol.* 111(S1):S71 (abstract).

Irwin, J. "Characterization of the New Xoft Axxent Electronic Brachytherapy Source Using PRESAGE ™." *Texas Medical Center Library, Digital Commons.* July, 2015.

Kim, Y., J. M. Modrick, E. C. Pennington, and Y. Kim. (2016). "Commissioning of a 3D image-based treatment planning system for high-dose-rate brachytherapy of cervical cancer." *J. Appl. Clin. Med. Phys.* 17(2):405–46. doi: 10.1120/jacmp.v17i2.5818.

Kubo, H., G. P. Glasgow, T. D. Pethel, B. R. Thomadsen, and J. F. Williamson. (1998). "High dose-rate brachytherapy treatment delivery: Report of the AAPM Radiation Therapy Committee Task Group No. 59." *Med. Phys.* 25(4):375–403. doi: 10.1118/1.598232.

Liu, D., E. Poon, M. Bazalova, B. Reniers, M. Evans, T. Rusch, and F. Verhaegen. (2008). "Spectroscopic characterization of a novel electronic brachytherapy system." *Phys. Med. Biol.* 53(1):61–75.

Ma, C. M., C. W. Coffey, L. A. DeWerd, C. Liu, R. Nath, S. M. Seltzer, and J. P. Seuntjens. (2001). "AAPM protocol for 40-300 kV x-ray beam dosimetry in radiotherapy and radiobiology." *Med. Phys.* 28(8):868–93.

Nath, R., L. L. Anderson, J. A. Meli, A. J. Olch, J. A. Stitt, and J. F. Williamson. (1997). "Code of practice for brachytherapy physics: Report of the AAPM Radiation Therapy Committee Task Group No. 56." *Med. Phys.* 24(10):1557-98. doi: 10.1118/1.597966.

Park, C. C., S. S. Yom, M. B. Podgorsak, E. Harris, R. A. Price, Jr., A. Bevan, J. Pouliot, A. A. Konski, and P. E. Wallner. (2008). "American Society for Therapeutic Radiology and Oncology (ASTRO) Emerging Technology Committee Report on Electronic Brachytherapy." *Int. J. Radiat. Oncol. Biol. Phys.* 76(5):963–72. doi: 10.1016/j.ijrobp.2009.10.068/

Potemin, S., I. Uvarov, and I. Vasilenko. (2015). "Intraoperative radiotherapy in locally-advanced and recurrent rectal cancer: retrospective review of 68 cases." Transl. Cancer Res. 4(2):189–95. doi:10.3978/j.issn.2218-676X.2015.04.07/

Rivard, M. J., B. M. Coursey, L. A. DeWerd, W. F. Hanson, M. S. Huq, G. S. Ibbott, M. G. Mitch, R. Nath, and J. F. Williamson. (2004) "Update of AAPM Task Group No. 43 Report: A revised AAPM protocol for brachytherapy dose calculations." *Med. Phys.* 31 (3):633–74. doi: 10.1118/1.1646040.

Rivard, M. J., S. D. Davis, L. A. DeWerd, T. W. Rusch, and S. Axelrod. (2006). "Calculated and measured brachytherapy dosimetry parameters in water for the Xoft Axxent X-Ray Source: An electronic brachytherapy source." *Med. Phys.* 33(11):4020–32. doi: 10.1118/1.2357021.

Rivard, M. J., C. S. Melhus, D. Granero, J. Perez-Calatayud, and F. Ballester. (2009). "An approach to using conventional brachytherapy software for clinical treatment planning of complex, Monte Carlo-based brachytherapy dose distributions." *Med. Phys.* 36(6):1968–75. doi: 10.1118/1.3121510.

Rusch T. W. and M. J. Rivard. (2004). "Application of the TG-43 dosimetry protocol to electronic brachytherapy sources." *Radiother. Oncol.* 71(S2):S84 (abstract).

Rusch, T. W. (2007). "Electronic Brachytherapy Sources." Technical Document. MC107 Rev 3. Xoft, Inc.

Schmidt, R. F. Wenz, T. Reis, K. Janik, F. Bludau, and U. Obertacke. (2012). "Kyphoplasty and intra-operative radiotheray, combination of kyphoplasty and intra-operative radiation for spinal metastases: Technical feasibility of a novel approach." *Intl. Orthopaedics* (SICOT) doi:10.1007/s00264-011-1470-9.

Schneider, F., H. Fuchs, F. Lorenz, V. Steil, F. Ziglio, U. Kraus-Tiefenbacher, F. Lohr, and F. Wenz. (2009). "A novel device for intravaginal electronic brachytherapy." *Int. J. Radiat. Oncol. Biol. Phys.* 74(4):1298–1305.

Schneider, F., S. Clausen, J. Thölking, F. Wenz, and Y. Abo-madyan. (2014). "A novel approach for superficial intraoperative radiotherapy (IORT) using a 50 kV X-ray source: A technical case report." *J. Appl. Clin. Med. Phys.* 15(1):167–76.

Seltzer, S. M., M. O'Brien, and M. G. Mitch. (2014). "New national air-kerma standard for low-energy electronic brachytherapy sources." *J. Res. Natl. Inst. Stand. Technol.* 119:554–74. doi: 10.6028/jres.119.022.

Simiele, S., B. Palmer, and L. DeWerd. (2016). "Experimental determination of modified TG-43 dosimetry parameters for the Xoft Axxent electronic brachytherapy source." *Med. Phys.* 43(6):3625 (abstract).

Thomadsen, B. R., P. J. Biggs, L. A. DeWerd, C. W. Coffey II, S.-T. Chiu-Tsao, M. S. Gossman, G. S. Ibbott, M. K. Islam, S. K. Jani, M. T. LaFrance, A. S. Meigooni, M. J. Rivard, V. Sehgal, R. J. Smith, D. J. Keys, M. Benker, and T. W. Rusch. "AAPM Report 152: The 2007 AAPM response to the CRCPD request for recommendations for the CRCPD's model regulations for electronic brachytherapy." College Park, MD:AAPM, 2009. ISBN 978-1-888340-78-5.

Turian, J., D. Bernard, Z. Hu, A. Dickler, and J. Chu. (2006). "Pre-clinical evaluation of the new Xoft Axxent^{TM} electronic brachytherapy system." *Med. Phys.* 33(6):2237 (abstract).

Udías, J. M., P. Ibáñez, M. Vidal, R. García-Marcos, G. Russo, C. Casarino, G. C. Candiano, G. Borasi, C. Messa, and M. C. Gilardi. (2014). "A fast Monte Carlo-based calculation algorithm for a Intra-Operative Radiation Therapy TPS: A validation study." *Radiother. Oncol.* 111(S1):S181–82 (abstract).

Vidal, M., P. Ibáñez, J. C. Gonzalez, P. Guerra, and J. M. Udías. (2014). "Hybrid Monte Carlo dose algorithm for low energy X-rays intra-operative radiation therapy." *Radiother. Oncol.* 111(S1):S106–S107 (abstract).

Weil, R., G. Mavinkurve, S. Chao, M. Vogelbaum, J. Suh, M. Kolar, and S. Toms. (2015) "Intraoperative radiotherapy to treat newly diagnosed solitary brain metastasis: initial experience and long term outcomes." *J. Neurosurg.* 122(4):825–32. doi:10.3171/2014.11.JNS1449.

Wenz, F., F. Schneider, C. Neumaier, U. Kraus-Tiefenbacher, T. Reis, R. Schmidt, and U. Obertacke. (2010a). "Kypho-IORT—a novel approach of intraoperative radiotherapy during kyphoplasty for vertebral metastases." *Radiat. Oncol.* 5:11. http://www.ro-journal.com/content/5/1/11.

Wenz, F., G. Welzel, E. Blank, B. Hermann, V. Steil, M. Sutterlin, and U. Kraus-Tiefenbacher. (2010b). "Intraoperative radiotherapy as a boost during breast-conserving surgery using low-kilovoltage x-rays: The first 5 years of experience with a novel approach." *Int. J. Radiat. Oncol. Biol. Phys.* 77(5):1309–14. doi:10.1016/j.ijrobp.2009.06.085.

Yanch, J. C. and K. J. Harte. (1996). "Monte Carlo simulation of a miniature, radiosurgery x-ray tube using the ITS 3.0 coupled electron-photon transport code." *Med. Phys.* 23(9):1551–58. doi: 10.1118/1.597885.

Yang, Y., C. S. Melhus, S. Sioshansi, and M. J. Rivard. (2011). "Treatment planning of a skin-sparing conical breast brachytherapy applicator using conventional brachytherapy software." *Med. Phys.* 38(3):1519–25. doi: 10.1118/1.3121510

Zeiss, Inc. *INTRABEAM Dosimetry.* User Manual EN_30_010_155II. Jena, Germany: Carl Zeiss Meditec, 2011.

Example Problems

(Answers are found at the end the book.)

1. Which of the following is not an attribute of electronic brachytherapy over radionuclide-based sources?
    a. ability to turn the source on and off
    b. less regulatory concerns
    c. higher electrical conductivity
    d. variable depth-dose
2. Which test is not part of commissioning the Axxent electronic brachytherapy system?
    a. source positioning accuracy using the manufacturer-provided test fixture
    b. system controller constancy of voltage and current applied to the x-ray tube
    c. measure of output constancy when reversing the voltage bias
    d. timer linearity over the range of expected patient treatments
3. What is the main difference for why the calculated dosimetry parameters differ between those determined by Rivard et al. (2006) and Hiatt et al. (2015)?
    a. the manufacturer changed the nomenclature for defining $\theta = 0°$
    b. the dosimetry formalism for high-energy sources was applied
    c. new applicators came to market
    d. the source design changed
4. Which item is of concern for commissioning Axxent dose calculations?
    a. matching the operating voltage and current
    b. setting a workaround for source strength
    c. keeping the dose grid constant in the TPS
    d. ensuring dose symmetry on the source transverse plane
5. How is the reference dose determined for the INTRABEAM system?
    a. TG-43
    b. TG-61
    c. TG-186
    d. TRS-398
6. How do shielding requirements for electronic brachytherapy treatment rooms typically differ from radionuclide-based HDR suites?
    a. EBT rooms have less shielding.
    b. EBT rooms have more shielding.
7. What is the inherent shape of the INTRABEAM source dose distribution?
    a. spherical
    b. donut (toroidal)
    c. egg (prolate ellipsoid)
    d. diamond

8. Dose output for Axxent surface applicator is significantly changed, requiring an additional calibration of dose rate at the exit window of the surface applicator because of the presence/absence of which item?
    a. flattening filter
    b. Tuohy–Borst adapter
    c. Plexiglas end-cap
    d. none of the above
9. Which item is not a part of the Axxent afterloader?
    a. touch screen electronic controller
    b. radioactive sealed source
    c. ADCL calibrated well chamber and electrometer
    d. none of the above (only the miniature x-ray tube assembly)
10. In the clinic, which methods would the QMP use in order to obtain corrections factors necessary to convert air kerma rate (measured at the exit window of the Axxent surface applicator) to dose in water?
    a. AAPM TG-61 report
    b. TRS-398 report
    c. MC simulations of the calibration geometry
    d. any of the above
11. In the treatment of breast IORT using the Axxent system, which item is not part of the equipment used in the operating theater?
    a. Axxent controller with a calibrated model S700 source
    b. additional calibrated model S700 source
    c. treatment planning workstation
    d. Flexi-shield drape

Chapter 4

LDR Prostate

Ronald S. Sloboda[1], Deidre L. Batchelar[2], Zhe (Jay) Chen[3], Wayne M. Butler[4], Luc Beaulieu[5], Dorin A. Todor[6], Mark J. Rivard[7], Frank-André Siebert[8], and Zoubir Ouhib[9]

[1]University of Alberta
Edmonton, Alberta

[2]British Columbia Cancer Agency
Kelowna, British Columbia

[3]Yale University School of Medicine
New Haven, Connecticut

[4]Schiffler Cancer Center
Wheeling, West Virginia

[5]Université Laval
Québec, Québec

[6]Virginia Commonwealth University Health System
Richmond, Virginia

[7]Tufts University School of Medicine
Boston, Massachusetts

[8]Universitätsklinikum Schleswig-Holstein
Kiel, Germany

[9]Lynn Regional Cancer Center
Delray Beach, Florida

4.1 Imaging and Targeting

4.1.1 Introduction

Medical images serve as a basis for treatment planning, radiation source placement guidance, and implant quality assessment in low-dose-rate (LDR) prostate brachytherapy. In current clinical practice involving transperineal source introduction (Aronowitz 2014), transrectal ultrasound (TRUS) is the predominant imaging modality used for treatment planning, with magnetic resonance imaging (MRI) recently having also been introduced for that purpose. The principal applications of imaging in the planning process are three dimensional (3D) anatomical structure identification and target delineation. In what follows, the discussion of targeting is limited to whole-gland prostate treatment as focal treatment guided by transperineal saturation biopsies (Merrick et al. 2007) or multiparametric MRI (mpMRI). Molecular imaging by positron emission tomography (Pouliot et al. 2012) is still investigational.

4.1.2 2D Mode Diagnostic Ultrasound

The core physics principles governing the operation of diagnostic ultrasound (US) systems are outlined in the report of Task Group 128 of the American Association of Physicists in Medicine (AAPM) (Pfeiffer et al. 2008) and references therein. In brief, an US transducer serves as both transmitter and receiver of a beam of pulsed, directionally steered, focused US energy emitted at a user-selectable frequency, f, in the 2 through 12 MHz range. The transmitted beam travels as a compression wave in materials with which the transducer is in acoustic contact, undergoes scattering in those materials, and is reflected from interfaces between constituent media having different acoustic impedance, $Z = c \times \rho$, where c is the velocity of sound in the medium and ρ is the density. The magnitude of the reflection increases with increasing impedance mismatch. The compression wave amplitude is attenuated exponentially as it travels through media, the attenuation coefficient being $\approx$0.5 dB cm^{-1} MHz^{-1} for typical clinical values of Z and f. Consequently, higher-frequency beams, although they provide better spatial resolution in the beam direction, are more highly attenuated than lower-frequency beams, so they yield reduced visibility of structures at depth. After transmission, the transducer "listens" for the reflected beam signal, which the system uses to build up a two-dimensional (2D) image via electronic directional scanning of the US beam. The time τ between beam transmission and subsequent signal reception for each beam direction is converted to a depth of reflection, r, in the image using the range equation $r = c \times \tau / 2$. The brightness of the displayed image at depth r is proportional to the amplitude of the received signal; hence, this mode of scanning was originally referred to as B-mode (brightness mode) scanning.

4.1.2.1 TRUS Equipment

For prostate brachytherapy, a monochrome B-mode scanner and biplane transrectal probe (transducer) such as those illustrated in Figure 4–1 comprise the basic imaging equipment. The probe incorporates two orthogonal arrays of transducer elements: a transverse array for axial imaging and a longitudinal array for base-apex imaging, as illustrated in Figure 4–2. It is important to note that because the transducer arrays for axial and longitudinal imaging occupy different physical locations within the probe, the axial and longitudinal images are offset. The probe manufacturer's literature should be consulted to determine the magnitude of the offset, and this information should be communicated to the brachytherapist, usually the radiation oncologist performing the brachytherapy procedure.

Also required are a probe stabilizer, stepper, and a needle template, as shown in Figure 4–3. The stabilizer fixes the probe with respect to either the floor or procedure bed and provides a solid mounting base for the stepper. The position and orientation of the stepper base can be precisely adjusted and then locked in position. The stepper holds the TRUS probe securely in a cradle and enables it to be manually advanced in 5-mm

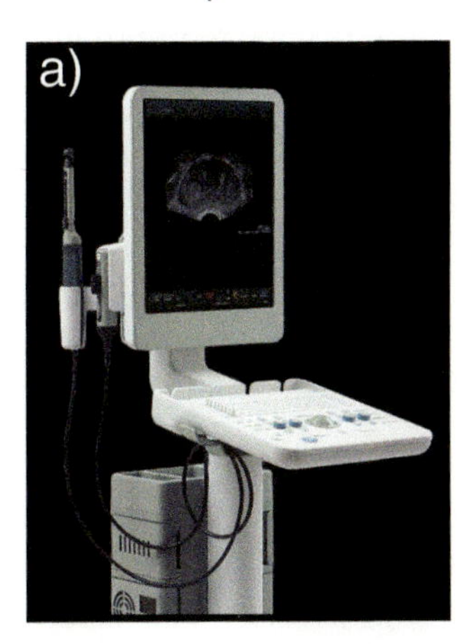

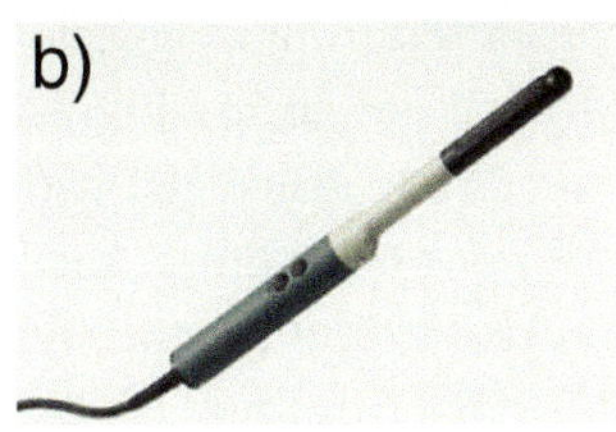

Figure 4–1 a) Monochrome B-mode US scanner and b) biplane transrectal probe.

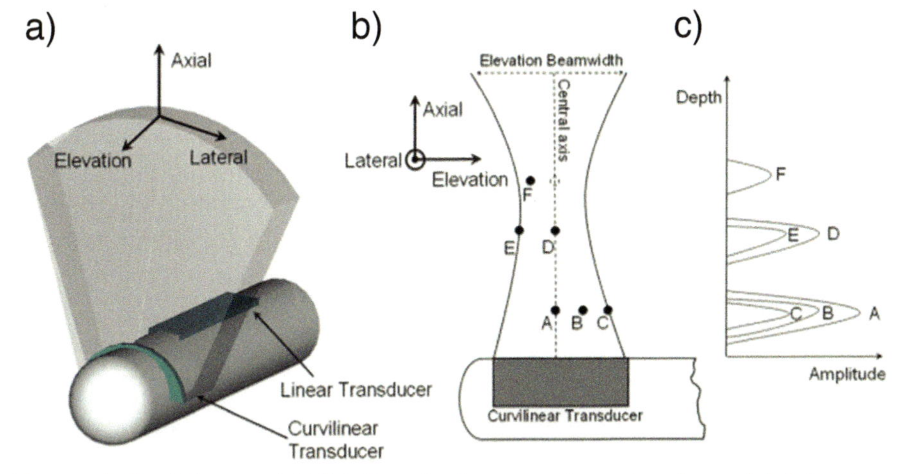

Figure 4–2 a) Transaxial TRUS probe beam pattern showing axial, lateral, and elevation axes (US coordinate system convention). b) and c) Beam pattern of a TRUS probe operating in B-mode with point reflectors A–F and their reflected US amplitudes. Amplitudes from all points within the elevation beam width are assigned to the central axis of the US beam. From (Peikari et al. 2011) with permission of John Wiley & Sons, Inc.

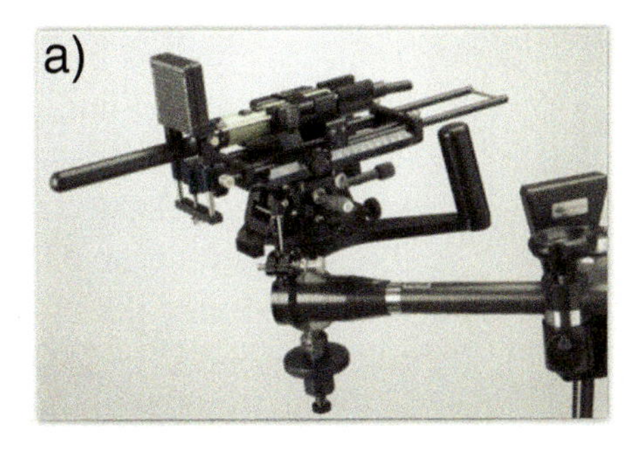

Figure 4–3 a) Assembled template, US probe stepper, and probe stabilizer. b) The large knobs on the stepper unit are used to advance the US probe in 5-mm increments.

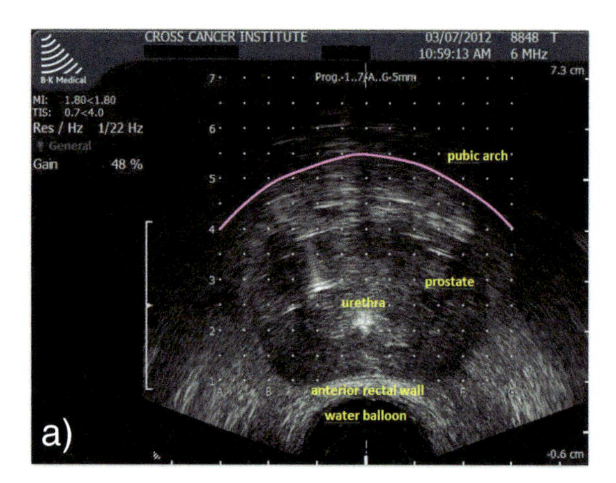

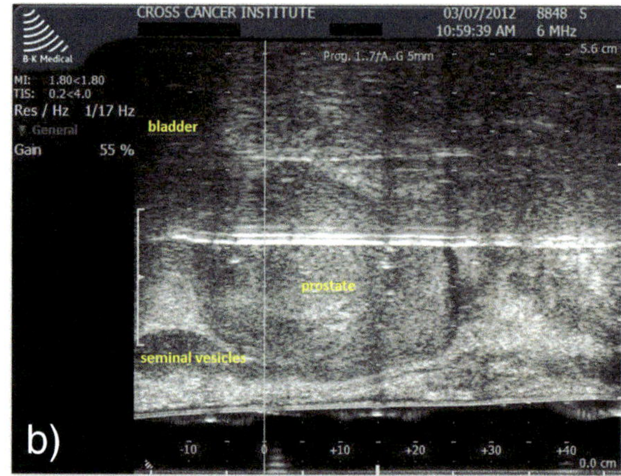

Figure 4–4 a) Transaxial and b) longitudinal TRUS images of a prostate implant in progress.

increments (for axial imaging), or manually rotated (for longitudinal imaging). It also holds the needle template in a fixed geometry relative to the probe. Optional motorized drive mechanisms for the probe and electronic indexers permit automated scanning and indexing of both axial and longitudinal 2D images to generate 3D image volumes (Fenster, Downey, and Cardinal 2001). Quality assurance requirements for prostate brachytherapy US systems are addressed in the report of Task Group (TG) 128 of the AAPM (Pfeiffer et al. 2008).

4.1.2.2 Image Formation and Interpretation

For transrectal prostate imaging, an US frequency near 5 MHz usually represents a good compromise between depth resolution and depth penetration. The operating frequency can be adjusted to accommodate small and large prostate sizes. In clinical practice, US gel, and a water balloon for some probes, are used to ensure good acoustic contact between the anterior surface of the probe and the anterior rectal wall. Figure 4–4 shows 2D axial and longitudinal images approximately bisecting a medium-sized prostate gland.

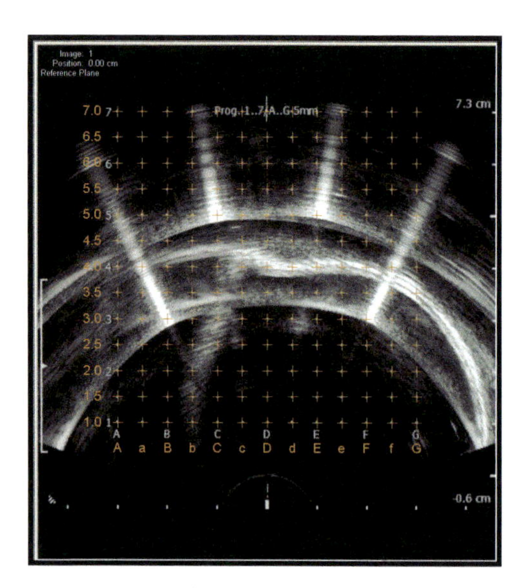

Figure 4–5 Alignment of the US electronic grid with the physical needle template grid is verified by suspending needles vertically from the template into a bucket of water. Here needles placed at grid locations B3.0, F3.0, C5.0, and E5.0 are identified by radial "comet tails."

Axial images are formed by acoustic beam scanning in a fan-line geometry; hence, spatial resolution is highest close to the probe and progressively worsens with distance away from it. Structures identified in Figure 4–4a include the prostate, urethra, pubic arch (projected from an inferior image on which some bony anatomy is seen), anterior rectal wall, and water balloon. Here the urethra is visualized with the aid of aerated gel injected into it prior to imaging; alternatively, a Foley catheter can be inserted to identify the urethra path. Longitudinal images are formed by acoustic beam scanning in a parallel-line geometry; hence, spatial resolution remains roughly constant with depth. Figure 4–4b identifies the prostate, bladder, and seminal vesicles.

The dark, hypoechoic "comet tail" (Figure 4–5) above the urethra is a ring-down artifact created by the large difference in acoustic impedance between the gel and surrounding tissue in combination with the urethra's small size; the same artifact can often be seen above implanted needles (Pfeiffer et al. 2008). Overlaid on the US image is an annotated electronic grid depicting the locations of the holes in the needle template. As such, it is **critically important** that (1) the correct grid be displayed (from the selection available on the US system) and that (2) the grid be properly aligned with the template as described in the AAPM TG-128 report (Pfeiffer et al. 2008) and illustrated in Figure 4–5.

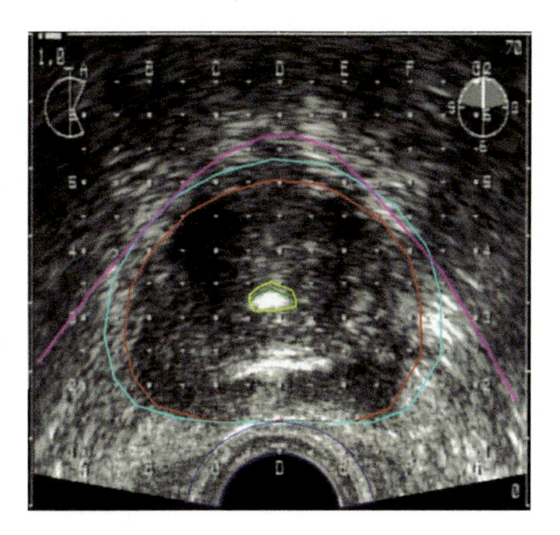

Figure 4–6 TRUS axial image of the prostate near mid-gland with overlaid anatomical structure contours. CTV (prostate) is red, PTV is light blue, urethra (infused with aerated gel) is yellow, rectal wall is dark blue, and pubic arch (projected from an image near the prostate apex) is magenta. The high-intensity feature below the urethra is a cluster of calcifications.

4.1.2.3 Target Definition

The 3D planning target volume (PTV) is defined by processing the stacked axial or swept-out longitudinal 2D images. For conventional whole-gland therapy, the PTV consists of the prostate as contoured by the brachytherapist on 2D images, i.e., the clinical target volume (CTV), plus a margin that is added to account for contouring and source placement uncertainties. Because of the multi-focal nature of most prostate cancer, the gross tumor volume (GTV), which is the visible or clinically demonstrable location and extent of the cancer, is not explicitly identified for most LDR prostate brachytherapy procedures. Typically, the margin on the CTV is asymmetric and extends 2 to 5 mm beyond the prostate in all directions, except at the posterior surface, where it is reduced to 0 mm to protect the adjacent rectum (Figure 4–6). Some common variations include (1) increasing lateral margins at the prostate apex and base to address the increased contouring uncertainty there and (2) reducing the anterosuperior margin to 0 mm to protect adjacent bladder. Suffice it to say that recommendations concerning margin size and placement vary, including those issued by professional societies (Ash et al. 2000; Salembier et al. 2007; Davis et al. 2012). The urethra and rectum are OARs that also are often delineated.

4.1.3 Magnetic Resonance Imaging (MRI)

As described by Bushberg et al. (Bushberg et al. 2011), nuclei containing an odd number of protons or neutrons possess a net spin and, hence, a non-zero magnetic moment m. The magnetic properties of materials containing such nuclei can be studied using magnetic resonance spectroscopy, which involves placing a sample of the material in a strong external magnetic field B_0. This causes the nuclear spins to precess with Larmor frequency $\omega = \gamma \times B_0$, where γ is the gyromagnetic ratio intrinsic to a particular nuclear species. A resonance energy coupling between the B_0 field and nuclei in the sample having non-zero values of m results in selective absorption of radiofrequency (RF) energy by the nuclei and subsequent energy release—the magnetic resonance (MR) signal—that is unique to each nuclear species and to its surrounding chemical environment. In the 1970s, magnetic field gradients were employed to localize the MR signal and thereby generate images containing clinically useful information. For any nuclear species, the strength of the MR signal depends on the magnitude of its magnetic moment, the physiologic concentration, and the isotopic abundance. Hydrogen, having the largest magnetic moment and greatest abundance, is a clear best choice for general medical imaging.

A 2D MR imaging plane is defined electronically, using a combination of the static B_0 field, three orthogonal transient magnetic gradients (G_x, G_y, G_z), and RF excitation pulses to alter the equilibrium distri-

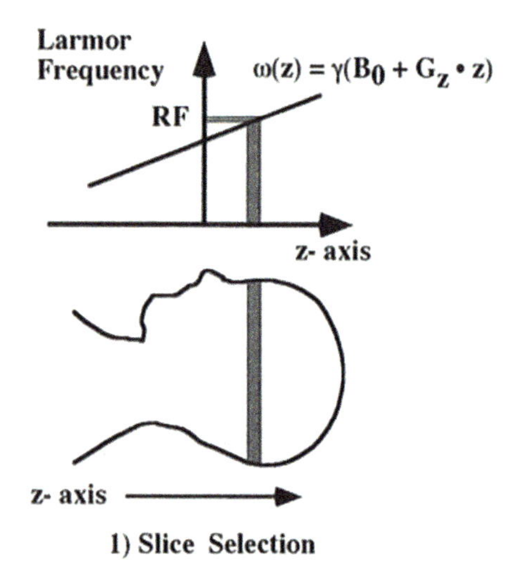

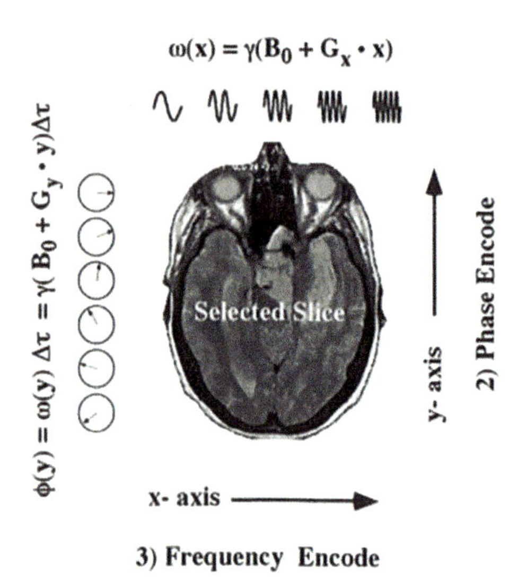

Figure 4–7 Spatial localization in MR imaging using gradients. From (Kessler and Ten Haken 1998) with permission of Medical Physics Publishing.

bution of nuclear spins in the volume of interest. Kessler and Ten Haken (1998) detail the basic steps involved: (1) slice selection using selective excitation; (2) phase encoding using a preparation gradient; (3) and frequency encoding using a readout gradient, as illustrated in Figure 4–7. To select a slice to image, one of the linear gradients (e.g., G_z) is turned on during the RF excitation pulse. Only spins precessing at a Larmor frequency equal to that of the RF pulse will be exited. The location and thickness of the slice depend on the gradient slope and RF pulse bandwidth. After slice selection, all the spins in the imaging plane are precessing with the same frequency and phase. Next, a *preparation* gradient (e.g., G_y) is applied to cause the spins to have a position-dependent phase, and then a *readout* gradient in the orthogonal direction (e.g., G_x) is applied during signal measurement to cause the spins to have a position-dependent frequency. Finally, the signal data are sorted and, through application of a Fourier transform, a 2D image is produced. Although 3D MRI acquisition techniques are available, it is common practice to create a volumetric image set by stacking 2D images.

4.1.3.1 Equipment and Accessories

A conventional diagnostic MRI scanner with a cylindrical bore can be used to acquire images for pre-operative planning (Section 4.2) or post-implant evaluation (also known as "post-planning") (see Section 4.9). A pelvic or cardiac RF coil is preferred over a whole-body coil as it yields superior spatial resolution and image quality. For pre-operative planning and post-implant evaluation, MR images are commonly registered with US (Section 4.1.3.3) and computed tomography (CT) (Section 4.7.4) images, respectively, to take advantage of MRI's superior soft tissue visualization capabilities.

In a very few centers, interventional MRI is used exclusively to create a treatment plan intra-operatively and guide needle placement (Rubens et al. 2006). In this setting, the implant needles and all operating room accessories must be MR-compatible. As this approach to prostate implantation is rare, it will not be addressed further here.

4.1.3.2 Image Formation and Target Definition

Although the main magnetic field strength B_0 influences image quality, field strengths in the range 0.5 T through 3.0 T provide images that are fully adequate for treatment planning purposes. T2-weighted fast spin-echo images display prostate anatomy clearly and unambiguously, as illustrated in Figure 4–8 (McLaughlin et al. 2005). Prostate cancer is more likely to be found in the peripheral zone than the transition zone or the central zone, which does not appear in Figure 4–8. Benign prostatic hypertrophy, which increases the prostate volume as men age, is primarily due to enlargement of the transition zone. When comparing MR and US images, it is important to keep in mind that patient setup differences—including pelvic tilt and the presence of the US probe in the rectum—generally lead to differences in prostate position, orientation, and deformations, although the latter have been shown to be relatively small (Liu et al. 2015).

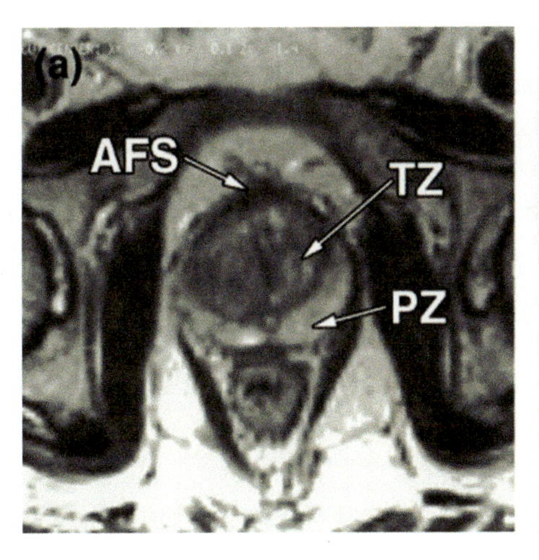

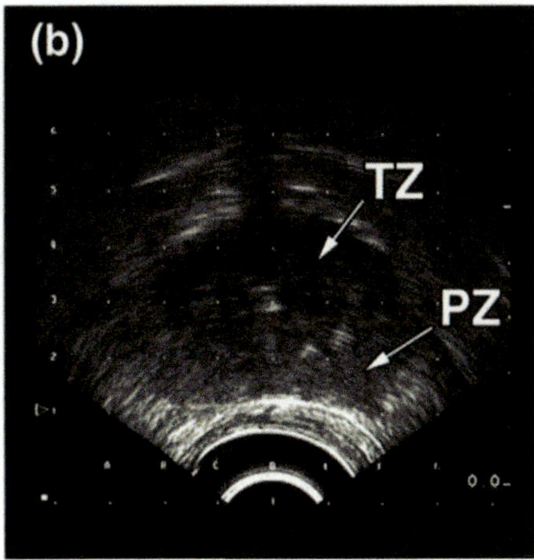

Figure 4–8 . Zonal anatomy of the prostate. Transition zone and peripheral zone on a) T2 MRI and b) US axial images. AFS = anterior fibromuscular stroma; PZ = peripheral zone; TZ = transition zone. From (McLaughlin et al. 2005) with permission from Elsevier.

The rationale for using MRI-based simulation for whole-gland LDR treatment as compared to US-based approaches is to reduce uncertainties and improve quality. As in US-based planning, 2D axial and sagittal images stacked to form a 3D dataset are useful in practice, as they facilitate visual comparison and anatomy-based registration with 3D US data. This is important because most prostate implants are performed using 2D US guidance.

Target definition for a 3D MRI dataset proceeds in exactly the same manner as described above for a 3D US dataset. A distinct advantage of MRI is better anatomic delineation of the apex, base, neurovascular bundles, external urinary sphincter, bladder neck, and intraprostatic ejaculatory ducts, as illustrated in Figure 4–9 (Tanderup et al. 2014). This must be weighed against the rapid real-time imaging processing and accessibility advantages of intra-operative US. Yet, as MRI evolves—with increasing acceptance for staging of dominant intraprostatic lesions and identification of extracapsular disease—interest in and investigation of MRI-based planning will expand.

4.1.3.3 MR–TRUS Registration

When MRI-based simulation is brought into the operating room virtually, either to refine or replace a TRUS-based treatment plan delivered using TRUS image guidance, 3D image registration is required. For prostate implants, systematic movement of the CTV relative to bony anatomy is common; hence, registration must be based on the discernible soft tissues of interest, namely the prostate and urethra.

Reynier and colleagues (2004) reported the application of both rigid registration and elastic octree-spline registration to prostate surface point sets for 11 patients, that were obtained by segmenting intra-operative TRUS images and pre-operative T2-weighted MR images acquired with an endorectal coil. The urethra position for four of the patients, measured by an indwelling catheter, was used to independently verify registration accuracy. The mean residual distance between TRUS and MRI surface point sets was found to be 1.6 ±0.4 mm and 1.1 ±0.4 mm for rigid and elastic registration, respectively, and between urethra lumen centers it was 1.3 ±0.6 mm and 1.6 ±0.6 mm, respectively. Thus, the two registration methods gave quite similar results. The authors concluded that TRUS-based prostate delineation at the apex and base could be reliably enhanced by the addition of MRI.

Tanaka and collaborators (2008) compared MRI-based and TRUS-based pre-plans in a prospective study involving 21 patients. T2-weighted MR images having a 5-mm slice thickness were acquired with a sense cardiac coil and used as the basis for treatment planning. TRUS images were then acquired within 2 hours and anatomical structures contoured on them. Finally, the MRI-based plan, including seed locations, was rigidly registered manually to the TRUS images, and dose-volume (DV) metrics were calculated for both the

MRI- and TRUS-based plans. The application of rigid registration to locate and orient the MRI dataset with respect to the TRUS dataset is a reasonable approach given that US probe-induced prostate deformation has been measured to be relatively small (Liu et al. 2015). The authors found that although MRI was performed in supine position and TRUS was performed in lithotomy position, there was no statistical difference in the prostate volume or anatomical structure DV metrics, with the exception of rectum $V_{100\%}$ (0.74 ±0.61 cm^3 for TRUS vs. 0.29 ±0.23 cm^3 for MRI). The latter finding is explained by the rectum taking on an unnaturally straight course as it follows the probe in TRUS imaging, whereas in MRI it maintains its natural shape. The authors concur with Reynier et al. that MRI enhances segmentation, and they further point out that MRI-based pre-procedure planning provides a clear indication of pubic arch interference and is able to minimize or eliminate intra-operative planning time.

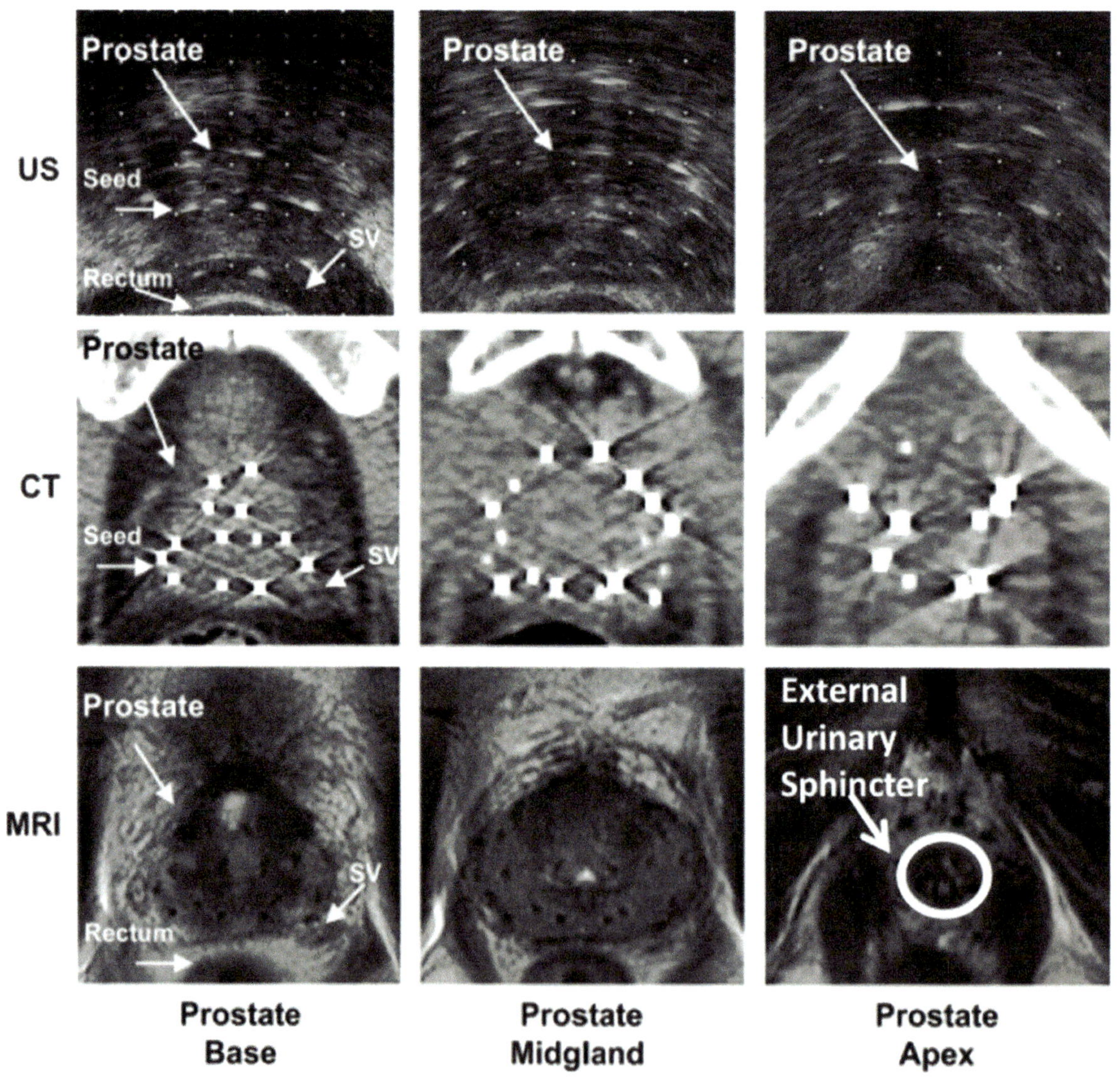

Figure 4–9 US, CT, and MRI images of the base, mid-gland, and apex of the prostate following LDR brachytherapy. MRI delineation of soft tissues is superior to US and CT. Better anatomic delineation of the apex, base, urinary sphincter (bottom right), neurovascular bundles, bladder neck, and intraprostatic ejaculatory ducts may improve disease outcomes and reduce treatment related morbidity. From (Tanderup et al. 2014), with permission of Elsevier.

4.2 Planning

Low-dose-rate brachytherapy for prostate can either be planned before the procedure—where a treatment plan is developed pre-operatively and that plan is delivered exactly during the implant—or it can be planned adaptively, where the plan is either formulated or modified substantially during the implant. This section discusses the most common clinical manifestation: pre-procedure planning for whole-gland implants. Adaptive planning is described in Section 4.5.

4.2.1 Target Volume Definitions

Target volumes for LDR prostate brachytherapy largely follow ICRU definitions (ICRU Report 58), although there is some heterogeneity in specific application of terms (Salambier et al. 2007; Davis et al. 2012). Histopathologic (Bott et al. 2010) and multi-parametric MRI (Riscke et al. 2013) studies have revealed most prostate cancer patients have one or more dominant intraprostatic lesions (DIL). When these can be identified on pre-operative MRI images, they can be contoured as gross tumor volumes (GTVs). See Chapter 1 for a discussion of target volumes.

Prostate cancer is not limited to the identifiable DILs. Even in early stage cancer, it is generally multifocal and diffuse. Thus, the clinical target volume (CTV) is at least the whole prostate gland. Additionally, the probability of having disease extending beyond the prostate capsule is at least 10%, even in T1c tumors (Salambier et al. 2007), making it necessary to provide a dosimetric margin of 3–5 mm around the prostate.

Planning target volume margins around the CTV are typically applied to compensate for organ motion and daily setup variations. These, obviously, do not apply to seed brachytherapy. There are, however, uncertainties in the placement of the seeds, which can range from 3 to 6 mm (Taschereau et al. 2000; Kaplan et al. 2004). This may be compensated for by applying a planning target volume (PTV) margin to the CTV.

The only group to provide detailed recommendations for a CTV definition is the PROBATE working group of Groupe European de Curietherapy and the European Society for Radiotherapy and Oncology (GEC ESTRO) (Salambier et al. 2007). They suggest that the CTV should correspond to the visible extent of the prostate, including any visible extracapsular extension, expanded uniformly by 3 mm. This expansion should exclude the rectal wall and bladder neck. This group considered the case where intra-operative planning was used and argued against the need for a further PTV margin. North American practice typically defines the prostate plus visible disease extension as the CTV and applies a PTV margin of 5 mm cranially, caudally, and anteriorly, with a 3 to 5 mm margin laterally and no posterior margin, although the cranial margin is frequently omitted (Crook 2011). Additionally, some brachytherapists implant the proximal seminal vesicles (Merrick 2005).

4.2.2 Organs at Risk

The two major organs at risk are the prostatic urethra and the anterior rectal wall. The anterior rectal wall is easily visualized on TRUS images and may be contoured. It is not, however, typical to assess dose to the rectum on a pre-procedural plan, and planners should just be mindful not to plan seeds adjacent to the rectum. The urethra is not visible on TRUS images and must either be opacified with aerated gel, or a catheter must be inserted if it is to be contoured. Although contouring of the urethra is recommended, it commonly is not. When not contoured, the planner should be cognizant of the region within which the urethra can be found, which is variable. The urethra mostly follows midline, but deviations of >5 mm are possible. Typically, the urethra enters the base in the upper third of the prostate and then moves toward the center of the gland until, at mid-gland, it ranges from the middle of the prostate in the anterior-posterior direction to the posterior third of the gland. It then curves upward until, at apex, it exits in the anterior third of the prostate. It is possible to limit dose to these regions, but greater precision is possible when the urethra is contoured.

4.2.3 Prescription Dose

Low-dose-rate seed brachytherapy can be used as a monotherapy for low-risk and low-tier, intermediate-risk patients. It may also be used in combination with external beam radiation therapy (EBRT) for high-tier, intermediate-risk and high-risk patients, with the EBRT typically preceding the brachytherapy (Davis et al. 2012).

4.2.3.1 Monotherapy

When used as the sole radiotherapy modality, the generally agreed upon prescribed doses are 145 Gy for ^{125}I and 125 Gy for ^{103}Pd, prescribed to the 100% isodose (Salambier et al. 2007; Rivard et al. 2007; Davis et al. 2012).

4.2.3.2 Combined Therapy

When brachytherapy is used as a boost to EBRT, 108 to 110 Gy is generally prescribed for ^{125}I or 90 to 100 Gy for ^{103}Pd. These follow EBRT delivery of 41.3 to 50.4 Gy to the prostate and pelvic nodes in 1.8 or 2.0 Gy/day fractions (Davis et al. 2012).

4.2.4 Pre-planning Dosimetric Goals

The goal of any plan is to produce a deliverable implant that will provide a curative dose to the prostate. Although there has long been evidence that having a post implant $D_{90\%}$ (the minimum dose covering 90% of the prostate volume) $\geq$ 140 Gy (Potters et al. 2003; Stock et al. 2006) will provide excellent biochemical control of prostate cancer, there is a dearth of evidence linking any particular pre-procedural plan criteria to achieving a $D_{90\%} \geq 140$ Gy. This situation arises from the multiple seed-planning philosophies that exist, making the development of definitive guidelines challenging. Regardless, it is important that any clinic adopting permanent prostate seed brachytherapy accept guidance from an experienced institution with a similar planning philosophy in setting plan dosimetric goals. Post-implant dosimetry should then be tracked and analyzed and the pre-procedural plan goals adjusted if necessary.

In a study performing a central review of pre-implant dosimetry for eight experienced brachytherapy practitioners, a wide range of dose volume histogram (DVH) values were found (Merrick et al. 2005). For ^{103}Pd monotherapy, prostate $V_{100\%}$ ranged from 97.6% to 100%, $V_{150\%}$ from 43.1% to 86%, and $V_{200\%}$ from 13.3% to 52.5%. For ^{125}I monotherapy, the ranges were: $V_{100\%}$, 95.7% to 100%; $V_{150\%}$, 32% to 92.1%; and $V_{200\%}$, 6.7% to 46.9%. These ranges encompass all published pre-implant DVH evaluation criteria. Mid-range values, such as those used by the British Columbia Cancer Agency (Table 4–1), seem reasonable.

Similarly, there is a lack of consensus on pre-implant dose constraints to the rectum and urethra. In the absence of a urethral contour, it is reasonable to keep the region through which the urethra passes to within 130% to 140%. If contoured, it should be possible to keep the urethra to less than 125% of prescription dose (Crook 2011). Some institutions do specify urethral DVH constraints, but there is no consensus on what parameters to use or what limits to apply. The basic principle, however, is to minimize the incursion of the 150% isodose into the urethra and to limit the dose to most of the urethra to less than 130% to 140%. Regarding the rectum, whether it is contoured or not, the overarching goal is to minimize the rectal $V_{100\%}$ in the pre-plan.

4.2.5 Seed Planning

4.2.5.1 Radionuclide Choice

Currently there are three radionuclides used clinically: ^{125}I, ^{103}Pd, and ^{131}Cs. The physical properties for each are summarized in Table 4–2.

Iodine and palladium have been used for more than 30 years and, as such, have robust long-term follow-up data demonstrating positive outcomes, both in terms of biochemical disease control and morbidity (Davis

Table 4–1 DVH criteria used to evaluate ^{125}I prostate implant pre-plans at the British Columbia Cancer Agency

DVH Parameter	Prostate (CTV)	PTV
$V_{100\%}$	≥99%	≥95%
$V_{150\%}$	56% to 65%	50% to 60%
$V_{200\%}$	<22%	<21%

Table 4–2 Radiological properties of radionuclides used for permanent seed prostate brachytherapy

Radionuclide	Half-life (d)	Average Energy (keV)
^{125}I	59.4	28.4
^{103}Pd	17.0	20.7
^{131}Cs	9.7	30.4

Table 4–3 Typical source strengths used in prostate seed implants

Radionuclide	Source Strength (U/seed)
^{125}I	59.4
^{103}Pd	17.0
^{131}Cs	9.7

et al. 2012). While there is evidence that treatment with ^{103}Pd leads to more rapid resolution of urinary morbidity (Wallner et al. 2002), there is no difference in efficacy for these radionuclides (Peschel et al. 2004). ^{131}Cs was introduced in 2004 (Bice et al. 2008) and shows promising early efficacy results, but follow-up is still short, and it does not currently show any advantage over the established radionuclide. A 2010 survey of 65 brachytherapy practitioners from 10 countries (Buyyounouski et al. 2012) revealed that ^{125}I was the most commonly used radionuclide for both monotherapy and boost treatments.

4.2.5.2 Seed Strength and Total Source Strength

The only traceable quantity for LDR seed brachytherapy source strength is air-kerma strength, S_K, in units of cGy cm² h⁻¹; for convenience those units often are referred to as U Typical source strengths for each radionuclide, and they are displayed in Table 4–3. There are no optimal values for air-kerma strength. Experienced practitioners often use a range of source strengths, choosing within that range to suit prostate anatomy and intent of treatment. Generally, lower strengths are used for boost implants. A choice of higher strength will, of course, lead to using fewer seeds per implant, but this does not necessarily result in a less robust plan (Beaulieu et al. 2004).

Equally, there is no consensus on the appropriate total source strength for a given prostate volume. A review of implants from three brachytherapy centers of excellence (Aronowitz et al. 2008) demonstrated that total source strength relative to prostate volume varied by 25% to 40%, with greater percentage variation for smaller prostates.

As seed and total implant source strength vary greatly with implant philosophy and still produce excellent clinical results, it is imperative for institutions to establish their own relationship between prostate volume, number of sources, and total source strength (Aronowitz et al. 2010). Determining this in the form of a nomogram or an equation permits preoperative calculation of the seed requirements for intra-operative planning. A nomogram can also be used to impose consistency among planners, acting as an important quality assurance tool to independently assess the total implanted source strength.

4.2.5.3 Seed Placement

Seed and needle placement can be determined either manually or using an optimization engine. In all cases, the aim is to develop a plan that meets dosimetric goals, is easily understood by the implant team, and that

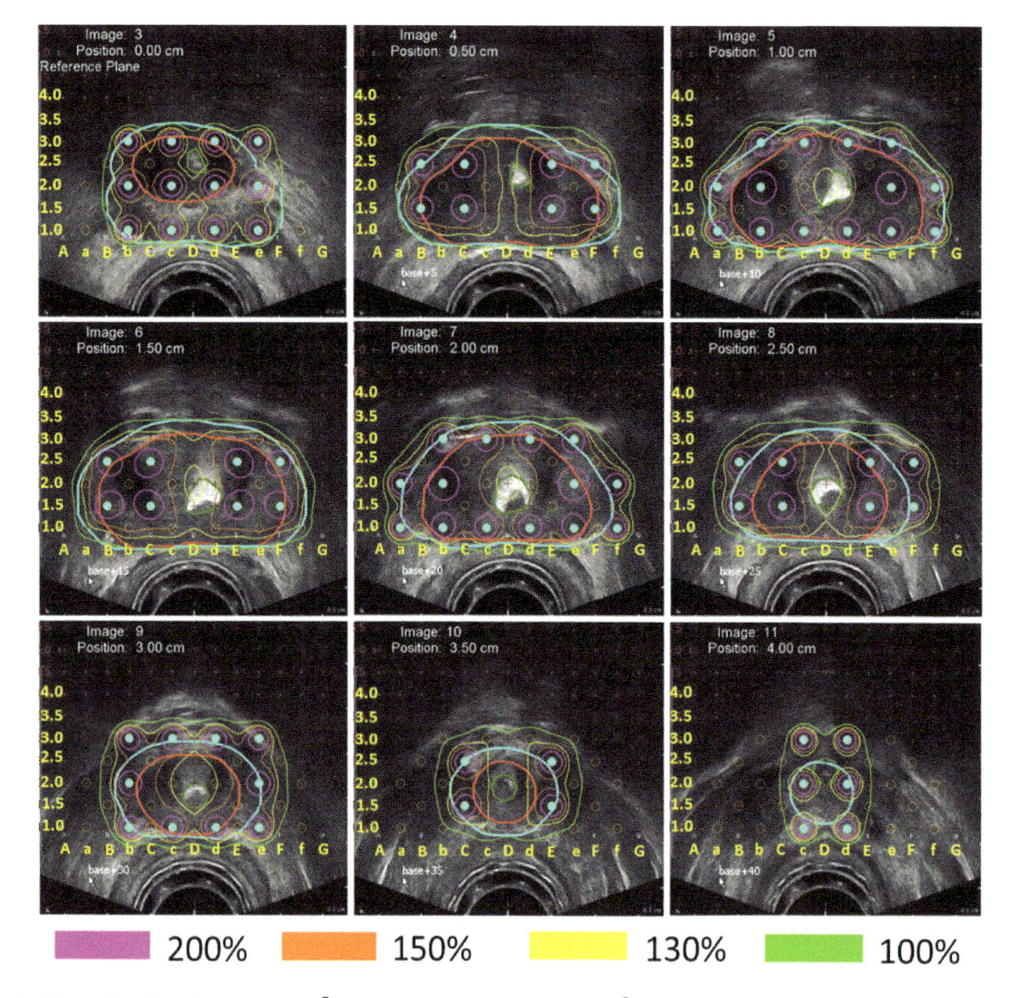

Figure 4–10 Sample plan for a 22 cm^3 prostate (red) and 36 cm^3 PTV (blue). The base of the prostate has been set lie on the 0.0 cm plane.

can be reproduced in the operating room. It should also be noted that, even without a formal adaptive planning process, the implanting physician can have a great deal of influence over the final result. Planners and physicians should work as a team so that the planners understand the physician's implant style and preferences. During the implant, needles may be steered so that producing complex loading patterns to spare the urethra, for example, may not be necessary. Although there are many strategies that can be employed to plan seed placement, they all result in some form of modified peripheral loading, wherein seeds are placed relatively uniformly throughout the outer portions of the prostate and are placed more sparsely in the interior in order to spare the urethra.

General Guidelines

- Avoid single-seed needles, as these increase trauma to the prostate without robust dosimetric benefit.
- Minimize two-seed needles. Needles with seeds only at the base and apex may be used centrally to spare the urethra, but peripherally they are not as robust as needles with three or more seeds, and they should only be used when more seeds are not needed.
- Avoid double loads (back-to-back seeds) in the most posterior row at apex, as implant needles are most often shifted caudally, and these can result in high rectal doses.

- The most posterior row of needles should be 3 to 5 mm inside the prostate to protect the rectum.
- Seeds on any slice should be separated >5 mm, regardless of the planning algorithm.
- Seeds and needles should be kept ≥5 mm from the urethra, or a strategy should be made to steer these needles away from the urethra in the OR.
- Extra-prostatic seeds should be within approximately 5 mm of the prostate, the exception being at the base, where posterior seeds may be needed to provide adequate coverage caudally.
- As the prostate is usually planned to lie centrally and symmetrically with respect to the template, the urethra most usually follows the central template column. No needles should be planned for this column unless the urethra has been contoured.

<u>Sample Plan</u>

As an illustration, a typical prostate plan has been generated for a 22-cm^3 prostate. Images of the prostate were acquired at 5-mm intervals, starting 5 mm cranial to the base. The prostate was positioned symmetrically on the grid with row 1 of the template lying 3 to 5 mm within the posterior capsule of the prostate. A PTV was then generated by expanding the prostate 5 mm laterally, 3 mm anteriorly, with no posterior margin. A PTV contour was placed 5 mm caudal to the apex, but there was no expansion cranial to the base (Figure 4–10). The PTV was modified at the base to match the prostate contour on the slice 5 mm below the base. The resulting PTV was 36 cm^3. The urethra was contoured based on the signal from aerated gel.

On the base slice, seeds were placed on the whole number template rows (1.0, 2.0, 3.0) and small letter columns (b, c, d, e) symmetrically and throughout the PTV. All seeds are 1.0 cm apart, and the urethra is receiving <130%. Other than the pair of needles at c2.0 and d2.0, these needles are then loaded at 1 cm intervals for as long as they lie within 5 mm of the PTV contour. At the slice 1.0 cm below the base, needles are added in the "a" and "f" columns to cover the lateral extent of the PTV. On the intermediate slices (0.5 cm, 1.5 cm, 2.5 cm, and 3.5 cm below the base), seeds are placed in the half-number rows (1.5 and 2.5) and capital letter columns (B, C, E, F) throughout the PTV, resulting in 8 more needles with seeds spaced every centimeter. On the final PTV slice below the apex, seeds are placed in the c2.0 and d2.0 needles to provide full dose coverage at the apex. The resulting planned dosimetry is found in Table 4–4, all of which meets the criteria found in Table 4–1.

On all slices, the bulk of the urethra is kept to <150%, although there is some overlap of the 150% isodose volume and urethra at mid-gland. Instead of modifying the seed-loading pattern to reduce urethral dose to <150% on all slices, the implanting physician would steer nearby needles away from the urethra (Figure 4–11).

Table 4–4 Dosimetric parameters for the sample plan

DVH Parameter	Prostate (CTV)	PTV
$V_{100\%}$	100%	99.4%
$V_{150\%}$	60%	57.6%

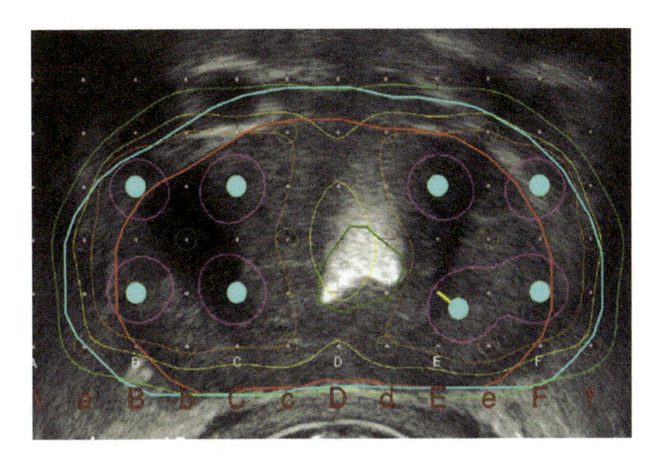

Figure 4–11 The needle in E1.5 is purposefully deviated from its template position to reduce urethral dose. The yellow line indicates the lateral-posterior shift of the actual needle insertion point (blue dot) from the planned insertion coordinate (E1.5).

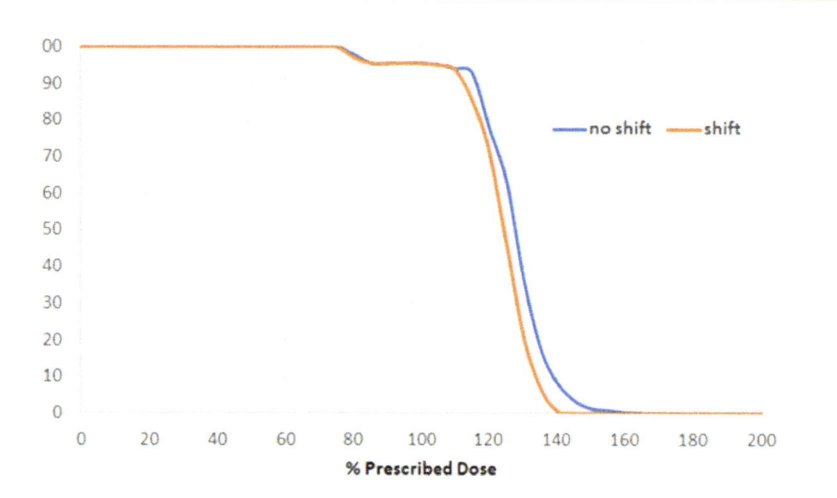

Figure 4–12 Urethral DVHs with and without needle deviation.

This small shift sufficiently reduces the urethral dose (Figure 4–12) without degrading the rest of the plan parameters.

4.3 Needle Loading Options

A *plan* in LDR prostate brachytherapy is a way of describing seed positions in 3D, as imagined in a pre-implantation plan. Two of the dimensions are implicitly given by position in the template, through which a needle will be inserted and consequently the seeds either contained in the needle or deployed with a Mick applicator (Eckert and Ziegler, Berlin, Germany). The third dimension is the *depth*, typically measured as an offset from a plane identified as zero. It is customary that seeds are planned in increments of 5 mm (even though the physical length of most seed models is 4.5 mm). A *loading* is, in a strict sense, simply the pattern of seeds and spacers within a needle. In a larger sense, one can talk about *needle loading* and understand the seed patterns within needles, but also the pattern of needles within the prostate (or CTV). In this section, needle loading is referred to only in the strict sense. In a pre-implantation plan, typically US-based, needles are created parallel to the US probe. In a real implant, needles can be steered and will likely result in being non-parallel to a probe and among themselves.

Section 4.2 states regarding planning objectives, "This situation arises from the multiple seed planning philosophies that exist, making the development of definitive guidelines challenging. Regardless, it is important that any clinic adopting permanent prostate seed brachytherapy accept guidance from an experienced institution with similar planning philosophy in setting pre-plan dosimetric goals." The same diverse field of philosophies (which, in this case, is really a nice way of saying habits and beliefs) affects needle loading. A very common, standard needle loading is a uniform pattern of seeds and spacers shown as needle 1 in Figure 4–13. While common—probably because of convenience and simplicity in planning and implant-

Figure 4–13 Examples of needle loading. Columns on the left indicate the sequential needle number, the retraction of the needle tip (in cm) from the base plane, the template insertion coordinates, and the number of seeds in the needle. Right: the needle contents, with the seeds colored gold and spacers white. Source trains usually end with a single spacer. Needle number 1 has a standard loading, while 4 and 14 (listed in this order since they have the same retraction from the zero plane) have "custom" loading.

ing (if, for example, a Mick applicator is used)—standard spacing is, for some groups (physicians and physicists) nothing short of a dogma. For these groups, a needle loading that contains two or three seeds, not separated by spacers, is the equivalent of blasphemy. In a somehow ironic twist, the LDR source market—previously monopolized by 4.5-mm and 5-mm long seeds of the radionuclides ^{125}I, ^{103}Pd and ^{131}Cs—recently accommodated a new source, the CivaString™ (CivaTech, Research Triangle Park, North Carolina), which is a long string (in increments of 1 cm) of uniformly deposited ^{103}Pd. Planning with CivaString showed the ability to obtain the same or better dosimetric parameters when compared to "discrete" seed plans, with up to 50% fewer needles (Rivard, Reed and DeWerd 2014)! In other words, a plan containing 24 needles using ^{103}Pd seeds could be delivered with 12 to 14 needles and produce the equivalent (or better) dose distribution and dose metrics. The price to pay is obliterating the "rule" that seeds cannot be allowed next to each other by making an implant in which each single seed (the 1-cm entity) is next to every other seed in the strand. Advantages and disadvantages of strands versus loose seeds are discussed in Section 4.5.

A non-standard, or custom, loading is simply a loading for which the seed spacing is irregular or larger than one spacer, such as a two-seed needle with the seeds separated by two or more spacers. While technically not part of the loading, an important part of executing the plan is inserting needles at various depths relative to plane zero. The depths are often called *retractions*, as the US probe is retracted in order to increase the depth relative to the base or reference plane. On brachytherapy stepper units, the template and US probe may be rigidly coupled to move in unison, or the two components may be uncoupled. If you measure the distance with a ruler from the template to the needle hub as a quality assurance check, that distance will depend on the stepper setup. With the probe and template rigidly coupled, that distance will be constant for all needles regardless of retraction when the needle tip is just visible on the US image. For non-coupled systems, that distance will increase by the retraction distance from the base plane. Some practitioners insert all the needles to the base plane and then retract the needle by a measured offset while leaving the US view fixed on the base plane. Others use various depths of retraction and view the plane where the needle tip is planned.

Correct configuration of seeds in the needles is essential in order to deliver the planned treatment. A comparison of an autoradiograph obtained by exposing film to the needles loaded with sources according to the patient treatment plan is a valuable quality assurance tool (Figure 4–14).

A decade ago or so, dosimetrists and physicists spent time loading sources and spacers into needles to prepare for prostate seed implants, using an average of 48 minutes for the loading. These days, most practices order needles loaded by the vendor or a third-party pharmacy (Bice et al. 2002). Once a plan is created, the loading pattern is send to the manufacturer or third party, which will calibrate the sources, load seeds according to the treatment plan, produce an autoradiograph, and sterilize the entire package, which will be ready to use in the OR. However, this service adds considerable cost to the implant compared to having a therapist load the needles. For practices using intra-operative technique, where a plan is created at the time of the implant, there are essentially two choices: either use seeds loaded in cartridges and a

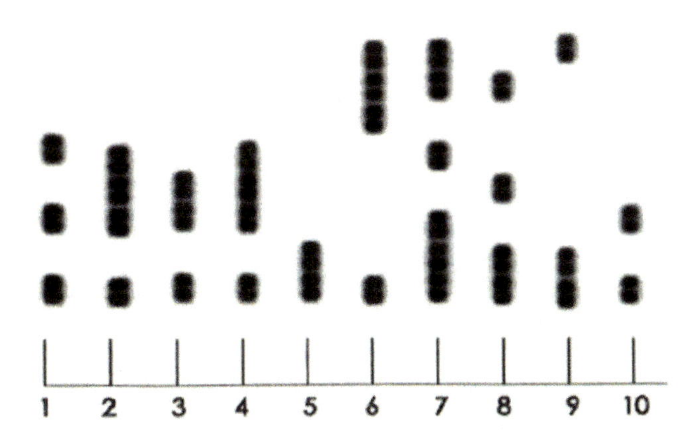

Figure 4–14 Example of an autoradiograph of the implant shown previously. Each seed of ^{103}Pd (Model 200, Theragenics, Buford, GA) appears as two adjacent dots because the seeds are manufactured by placing two palladium-coated graphite pellets separated by a lead pellet within the titanium shell. Needles number 1 and 4 can be compared with the plan shown in Figure 4–13.

Mick-type applicator, or load the seeds in needles on the spot, using a needle-loading device to accomplish that. An example is the QUICKLINK® Delivery System by Bard Medical, Covington, GA, which is a cartridge-based device designed to assemble brachytherapy seeds, spacers, and SOURCELINK® Connectors into seed trains of variable lengths and with variable seed-to-seed spacing as predetermined by the planner (physicist, dosimetrist, or physician). Various other needle-loading devices are commercially available, some of which are automated and combine needle loading and seed assay. The aim of the devices is to reduce the amount of time required for loading the needles, permit visual verification of the loading pattern, and reduce radiation exposure to personnel.

4.4 Seed Implanters for LDR Prostate Brachytherapy

In addition to manually delivered low-energy brachytherapy seeds for the treatment of prostate cancer, automated delivery may reduce radiation exposure to hospital staff and improve the consistency and reproducibility of seed delivery over manual delivery (Podder et al. 2014). In 2001, the Fully Integrated Real-time Seed Treatment (FIRST™) system was commercially introduced by Nucletron (Rivard et al. 2005). While much research is ongoing in university settings on about a dozen experimental devices, there are no other commercially available systems on the market (Podder et al. 2014). Since the introduction of the seedSelectron™, Nucletron has been acquired by Elekta AB (Stockholm, Sweden). The FIRST system was replaced in 2010 by the improved Oncentra Prostate™ (which includes Oncentra Seeds™) real-time prostate solution, under which the seedSelectron is available as an option.

Analogous to a high-dose-rate remote afterloading system, the seedSelectron builds in real-time any combination of seeds and spacers and then positions them automatically into needles, already implanted in the patient, using a digitally motorized and monitored drive-wire. The system verifies the approximate strength of each seed placed in the needle. The seeds and spacers are deposited in the prostate as the needle is robotically retracted with the drive wire acting as a stylet. The positioning of the most distal seed of each seed-spacer train can be planned for and delivered in 1-mm increments. The system also includes selectSeed and selectSpacer cartridges containing either 100 ^{125}I seeds (4.5 mm long) or 100 spacers (5.5 mm long) and can create a seed-spacer train up to 80 mm in length. Seed deployment is based on US image guidance, with possible comparisons of needle positions based on a pretreatment plan (Beaulieu et al. 2007). This approach permits real-time treatment planning while accounting for changes in prostate position due to mechanical insertion and shape due to needle pressure and edema (Lagerburg et al. 2005; Westendorp et al. 2016). Using the seedSelectron, improvements have been observed in the quality of the delivered treatment plan and efficiency of the involved medical staff (Van Gellekom et al. 2004; Beaulieu et al. 2007; Radford Evans 2007; Moerland et al. 2009).

In addition to the above-listed components (Figure 4–15), the seedSelectron is supported by an endocavity rotational mover (i.e., a rotational stage) for a bimodal TRUS probe. In this way, a fan of quasi-sagittal images are acquired for automatic 3D reconstruction to depict volumetric information. such as regions of interest (ROIs), needle and seed positions, and isodose distributions with ROI dose-volume histograms. In addition, the position of the needle is automatically shown on the live sagittal US view during seed delivery. Built into the seedSelectron is a diode array for detecting the presence or absence of ^{125}I seeds when creating the seed-spacer train. The diode array can be configured to further serve as an indication of source strength of each individual ^{125}I seed included in the train, although it cannot serve as a NIST-traceable measurement device for source strength (Perez-Calatayud et al. 2012). The medical physicist must still perform an independent check of the manufacturer certificate of ^{125}I air-kerma strength for a given selectSeed cartridge as required by the AAPM (Butler et al. 2008).

With new features beyond manual delivery of LDR seeds comes the medical physicist's responsibility to assure the safe delivery of new technologies, as recommended in the report of AAPM TG-167 (Nath et al.

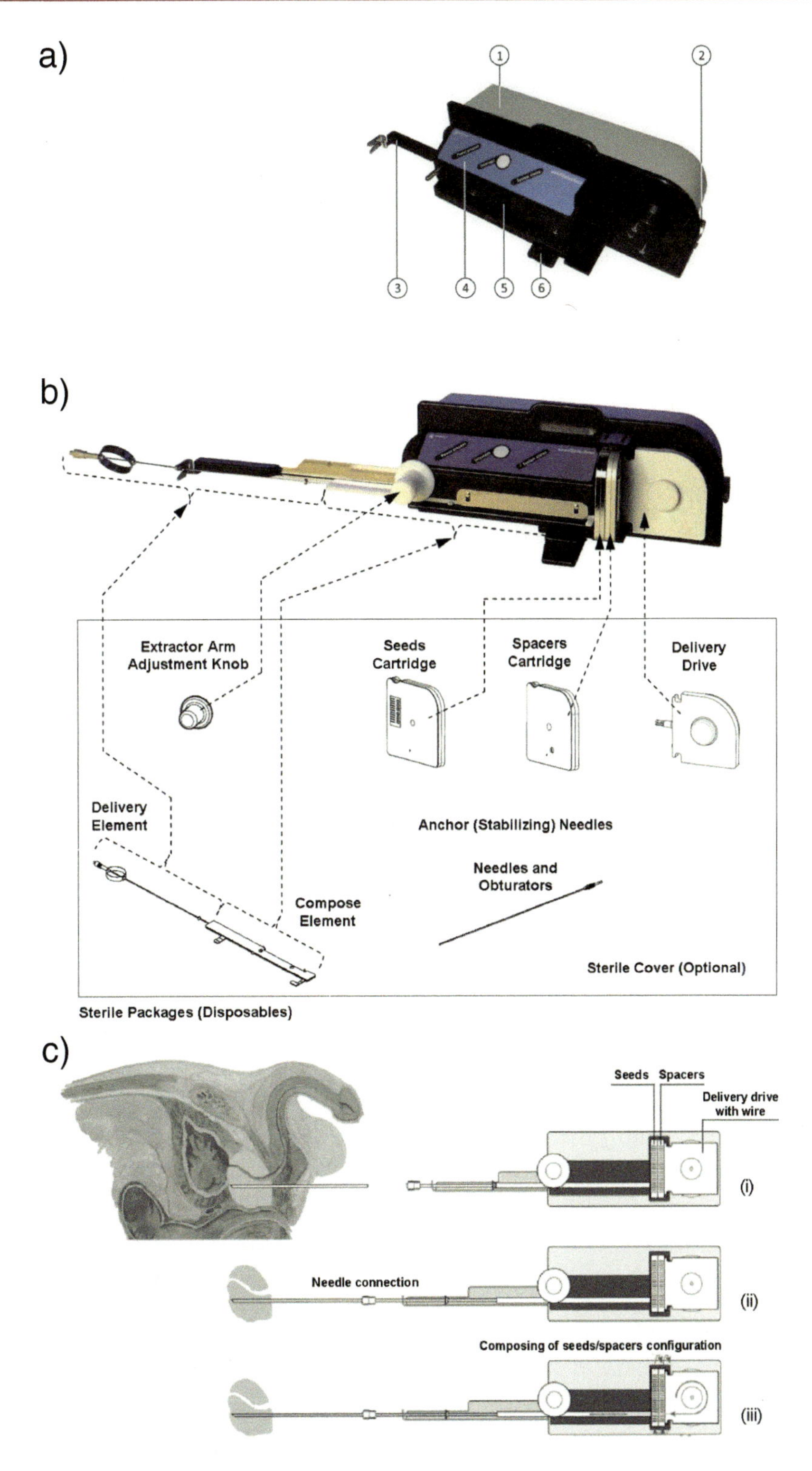

Figure 4–15 The elements of the seedSelectron are illustrated. a) The basic functions of the seedLoader are ① housing for electronics, ② cable connector, ③ extractor arm, ④ indicators and interrupt button, ⑤ radiation sensors (inside), and ⑥ attachment for fixation bracket. b) Configuration of various components to the seedLoader as they come together to create the seedSelectron. c) (i) The physician performs transperineal needle implantation of the prostate, guided by TRUS; (ii) the delivery element is connected to a needle according to the treatment plan composition; and (iii) the physicist or dosimetrist starts the needle loading process. Illustration courtesy of Elekta, Stockholm, Sweden

2016). Over a decade ago, concerns for quality assurance measures and quality checks specific to the seedSelectron were reported in the ESTRO Booklet #8 and by Rivard and colleagues (Venselaar and Perez-Calatayud 2004; Rivard et al. 2005). Venselaar and Perez-Calatayud identified the means of operation and the rationale for the device. Rivard and colleagues evaluated the system from the perspective of contemporaneous societal guidance for brachytherapy quality standards. In general, there were existing materials and documentation to satisfy the guidelines in several AAPM reports. Information or methods were provided where extant materials and documentation were lacking. This included suggestions for customer acceptance testing procedures where only an installation engineer checklist was previously available.

4.5 Stranding: Pros and Cons

4.5.1 Introduction

Permanent LDR prostate brachytherapy seeds do not always remain where they were deposited. The track left by the depositing needle is the most common path of movement, particularly noticeable if the needle is withdrawn too swiftly and the accompanying fluid flow sweeps the more distal seeds along the exit trail. Because the seeds are small—typically 0.8-mm diameter and 4.5- to 5-mm long and comparable to the dimensions of larger arterioles and venules of the circulatory system, as well as the larger glands and ducts of the prostate—some migration will inevitably occur along these more tortuous routes. Seeds that work their way into veins can sometimes be seen on fluoroscopy during the implant briskly departing the prostate.

Seed loss will reduce the dose to the target volume proportionate to the number of seeds lost, and seed migration will have adverse effects on dosimetry by potentially increasing the dose to OARs, such as the rectum and urethra. The frequency of loose seed loss and the magnitude of migration effects on dosimetry also depend on the experience of the brachytherapist and the implant philosophy applied. Operator dependence was observed between two brachytherapists at a single institution in Montreal, Canada, where any loss was seen in 19.4% of patients at a rate of 0.57% of the total number of seeds implanted in those patients who lost seeds (El-Bared et al. 2016). Analysis of the initial learning curve in the first 500 patients in groups of 100 at that institution found the loss rate of 0.31% in the last 300 patients to be one fourth that of the 1.25% for the first 100 patients (Taussky et al. 2012). The Montreal group determined seed loss 30 days after implant, and their implant philosophy placed virtually all seeds within the prostate. The Wheeling, WV, group places about 41% of seeds outside the prostate to treat likely extra prostatic extension of disease, and they found a logarithmic time dependence, with most seed loss occurring within 28 days (Merrick et al. 2000). The dosimetric effect of seed loss after two thirds of the ultimate brachytherapy dose is delivered becomes negligible. Two thirds of the prescribed dose is delivered within 28 days for a ^{103}Pd implant and over 90 days for an ^{125}I implant. Merrick et al. and found a seed loss rate of 1.4% for ^{103}Pd seeds at 28 days and 1.1% for ^{125}I seeds at 90 days. In their study, almost all peripheral and extra prostatic ^{125}I seeds were stranded, while all ^{103}Pd seeds were loose.

4.5.2 Loose Seeds

The purchase of loose seeds in non-sterile vials to be autoclaved in your facility and loaded into needles according to the plan is the least expensive way to acquire the necessary seeds.

4.5.2.1 Seeds Separated by Spacers

Although some needles are loaded with back-to-back seeds, most seeds are separated by one or more spacers. With the needle stylet held in position abutting the end of the source train, the brachytherapist withdraws the needle, leaving the source train with spacers within the surrounding tissue. Figure 4–16 is a schematic of part of the needle loading specified by a plan. In every case, a radioactive source occupies the initial position

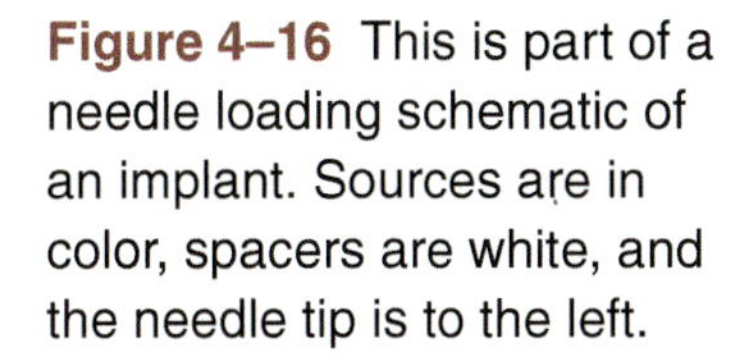

Figure 4–16 This is part of a needle loading schematic of an implant. Sources are in color, spacers are white, and the needle tip is to the left.

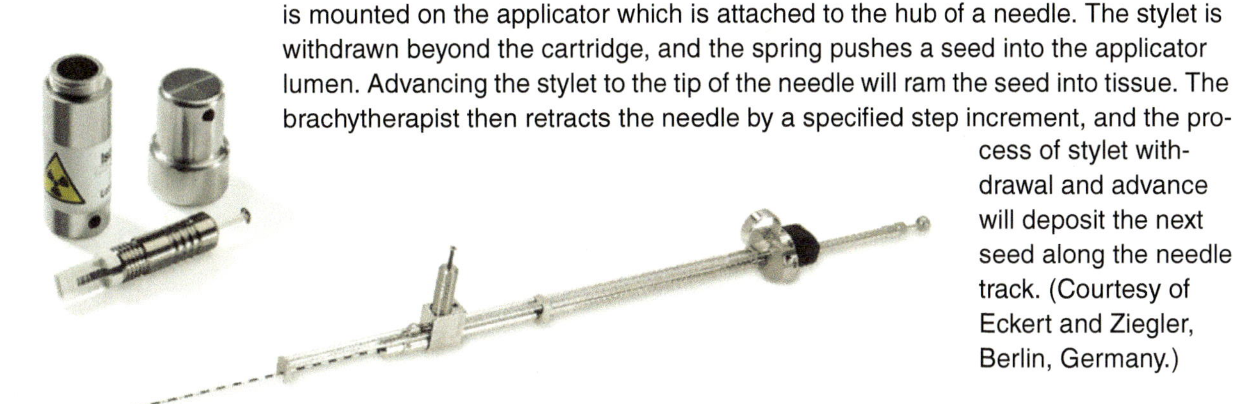

Figure 4–17 Mick applicator. A cartridge containing a spring-loaded stack of seeds is mounted on the applicator which is attached to the hub of a needle. The stylet is withdrawn beyond the cartridge, and the spring pushes a seed into the applicator lumen. Advancing the stylet to the tip of the needle will ram the seed into tissue. The brachytherapist then retracts the needle by a specified step increment, and the process of stylet withdrawal and advance will deposit the next seed along the needle track. (Courtesy of Eckert and Ziegler, Berlin, Germany.)

at the tip of the needle because the needle is inserted with the tip offset from the prostate base by a distance specified by the plan.

A spacer is added as the last item in every source train. Therefore, fluid flow along the needle track caused by removal of the needle and stylet after extruding the source train may displace the final spacer, but is less likely to affect more distant objects.

4.5.2.2 Seeds Deposited along Needle Track by Mick Applicator

Instead of intact source trains, the Mick applicator (Figure 4–17) requires the brachytherapist to deposit every seed individually.

Although the brachytherapist has greater control over the position of each seed and may account for intra-operative edema in real time, the process does take more operating room time. Because the seeds are not buttressed in position by spacers, every seed is susceptible to slippage effects as the implant needle is withdrawn stepwise to the next position. Nevertheless, in one report, only 2.0% of loose seeds were lost to pelvic imaging at a median of 133 days after the Mick-assisted implant (Kunos et al. 2004).

4.5.2.3 Seed Design Effects on Fixity

Although most seeds are hydro-dynamically streamlined, some seeds with squared-off or even cupped ends are expected to be resistant to movement along the needle track (Figure 4–18).

Another way to reduce migration of loose seeds is to coat them with a bioabsorbable polymer similar to that used to make strands. A randomized trial comparing implants with bare ^{125}I seeds and polymer-coated seeds found significantly fewer coated seeds lost from the prostate apex. However, the penile-bulb dose was significantly higher in the coated group, but there were dosimetric differences between the groups in prostate

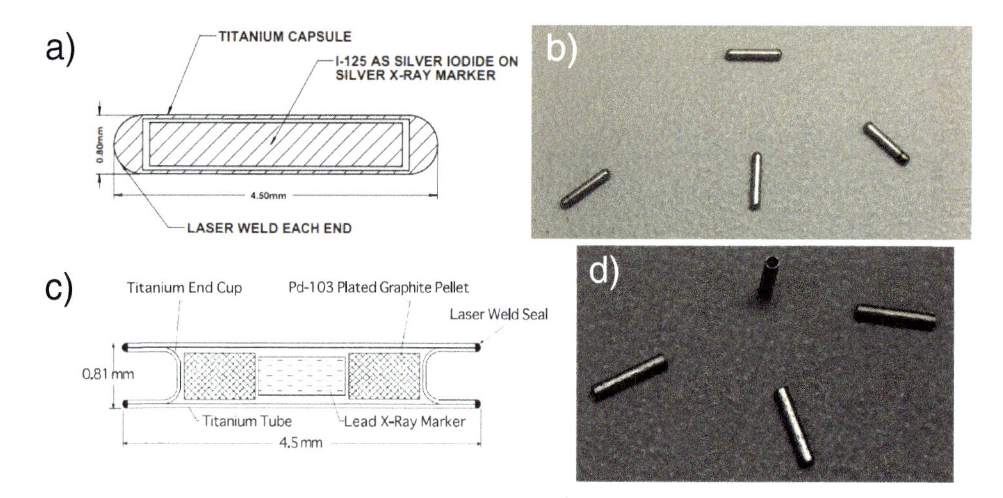

Figure 4–18 a) and b) depict streamlined ^{125}I seeds. c) and d) show ^{103}Pd seeds with cupped ends. Photos courtesy of Theragenics.

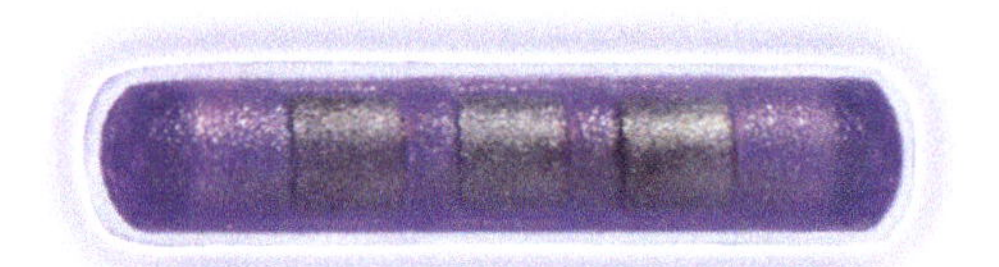

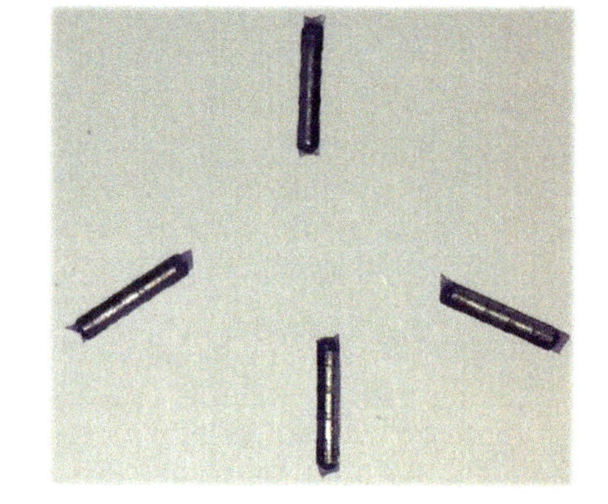

Figure 4–19 Anchor Seeds have a polymer coating that has an affinity for tissue. Photo courtesy of Theragenics.

$D_{90\%}$, prostate $V_{100\%}$, rectal $V_{100\%}$, and urethral $V_{150\%}$ (Bowes et al. 2013). Figure 4–19 illustrates the coating on Anchor Seeds® (Theragenics, Buford, GA).

4.5.3 Stranded Seeds

The first stranded-seed trains were introduce commercially in 1995 by Amersham as Rapid Strand. Their ^{125}I seeds were inserted at 1-cm intervals into a braided Vicryl sleeve, and the assembled sleeve was then roasted by the manufacturer to stiffen and embrittle the strand.

4.5.3.1 Stranding Materials and Source Spacing

The original Rapid Strand came with 10 seeds in a 10-cm long jig notched to facilitate cutting the strand into shorter, clinically relevant lengths. Other vendors followed with additional innovations, such as custom lengths and variable seed spacing, along with other bioabsorbable polymer materials. Users may also assemble strands on site using connectors that snugly fit over the seeds, such as SourceLink by Bard. Figure 4–20 shows some examples of strands.

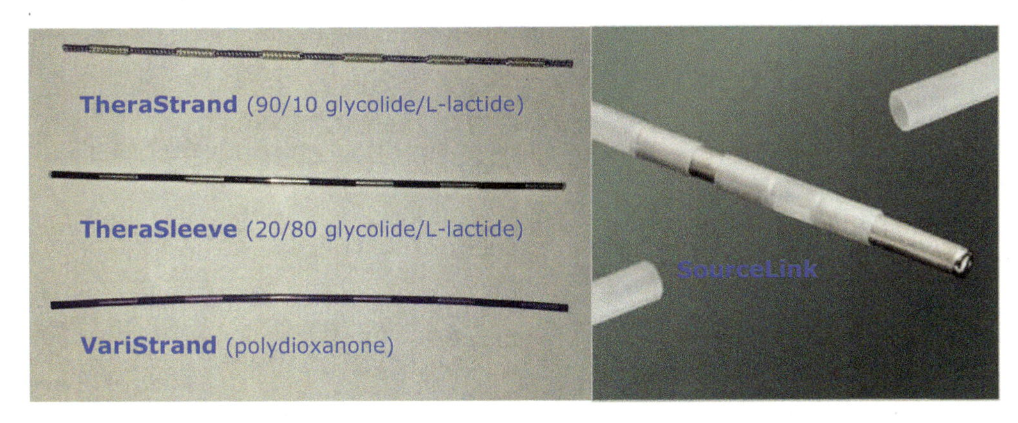

Figure 4–20 Strands made of various bioabsorbable materials with variable expected biomechanical integrity. Absorption of the stranding polymer will occur in 56 to 70 days for TheraStrand®, 140 to 180 days for TheraSleeve®, 182 to 238 days for VariStrand®, and ~170 days for SourceLink®. (Photos courtesy of Theragenics and Bard Medical.)

4.5.3.2 Stranded Seeds and Source Fixity

The first report of clinical results for stranded versus loose seeds found pulmonary seed embolization in 0.7% of patients implanted with stranded ^{125}I seeds 11% in loose ^{103}Pd seed patients, using chest radiographs on the day following implant (Tapen et al. 1998). In a series where lateral and AP chest radiographs were obtained at least 2 weeks post implant, Merrick et al. (2000) found lung embolization in 21.4% of stranded ^{125}I seed patients and 22.2% in loose ^{103}Pd seed patients.

Although stranding reduces the rate of seed migration to the lungs from the peri-prostatic vasculature, other pathways of seed loss remain viable. In two sequential cohorts of 20 men, 30% of the stranded-seed patients and 25% of the loose-seed patients experienced loss of seeds from the pelvis (Saibishkumar et al. 2009). Likewise, the overall seed loss in the stranded-seed cohort was greater than in the loose-seed group, with most of the loss via urine. These results were comparable to a review of a large implant population (n = 1,794) in which 29.7% of patients had seed loss via the urinary tract (Stutz et al. 2003). In the latter population, stranded ^{125}I seeds were employed at the brachytherapist's discretion, but rarely near the urethra.

4.5.4 Dosimetry Outcomes

The dosimetric outcome for any interstitial prostate brachytherapy procedure is dependent on the institutional implant philosophy, on where the brachytherapist is on the learning curve, and other factors. Unfortunately, there is no national registry for brachytherapy implants, so differentiation relies on single-institution or *ad hoc* multi-institutional assemblages. Most single institution studies of any avowed technological advancement consist of sequential cohorts—"We used to do it the old way, but then we adopted the new approach and our results are much better." The widespread adoption of permanent seed prostate brachytherapy occurred in the mid-1990s, so the move from loose to stranded seeds often occurred during the brachytherapy learning curve.

4.5.4.1 Randomized Trials and Prospective Cohorts

The first randomized trial comparing seed loss and dosimetry of stranded and loose seeds reported results for 64 patients (Reed et al. 2007). There was no significant difference in mean seed loss per patient, although the trend was as expected: 0.43 per patient implanted with strands vs. 1.09 in loose-seed implants, $p = 0.062$. Furthermore, the mean prostate $D_{90\%}$ and $V_{100\%}$ were lower in the stranded implants.

This dosimetric deficit was not seen in some sequential studies. For example, the mean prostate $D_{90\%}$ and $V_{100\%}$ of 103% and 89%, respectively, for 309 loose-seed ^{103}Pd implants before April, 2002 were significantly lower than in 136 stranded ^{125}I seed implants after that date (109% and 92% for $D_{90\%}$ and $V_{100\%}$, respectively) (Fagundes et al. 2004). The small differences, particularly between radionuclides, may be of no clinical consequence. In another sequential study using only ^{125}I seeds in 205 patients, statistically significant differences were seen between strands and loose seeds with $D_{90\%}$ at 119% vs. 114% and $V_{100\%}$ at 98% vs. 96% (Major et al. 2014). However, this latter study did find a better conformal index (see Chapter 1) for loose seeds, 0.70 vs. 0.63. In addition, four dosimetric quality measures each for the urethra and rectum found significantly better OAR sparing with loose seeds, where the mean improvement was more than 10% for every indicator.

There is one way to lessen the effect of confounding factors: randomly allocate the left and right halves of a treatment plan so one side is implanted with loose seeds and the other with stranded seeds. This was employed in a small trial of eight patients (Kaplan et al. 2004). At 4 to 6 weeks post implant, there was no seed loss, and there was no significant difference in the mean prostate $D_{90\%}$ (110% loose vs. 93% strand) or $V_{100\%}$ (89.8% loose vs. 89.6% strand).

4.5.4.2 Explanatory Model

Figure 4–21 illustrates schematically the cause of central dose escalation to the urethra and rectum and the loss of peripheral coverage of the prostate base and apex as edema waxes and wanes in a stranded-seed implant compared to a loose-seed implant. Using strands that are longer than the expected extent of edema along the needle path may prevent loss of coverage in a stranded implant, but the dose escalation is unavoidable.

Intact strands have also been observed moving both cranially and caudally, and a recent paper has provided an explanation for the phenomenon (Soni et al. 2016). Figure 4–22 is a schematic of cranial migration induced by bladder muscle contraction during micturition, while Figure 4–23 shows the caudal motion of a strand induced by contraction of the levator ani muscle during urination or defecation.

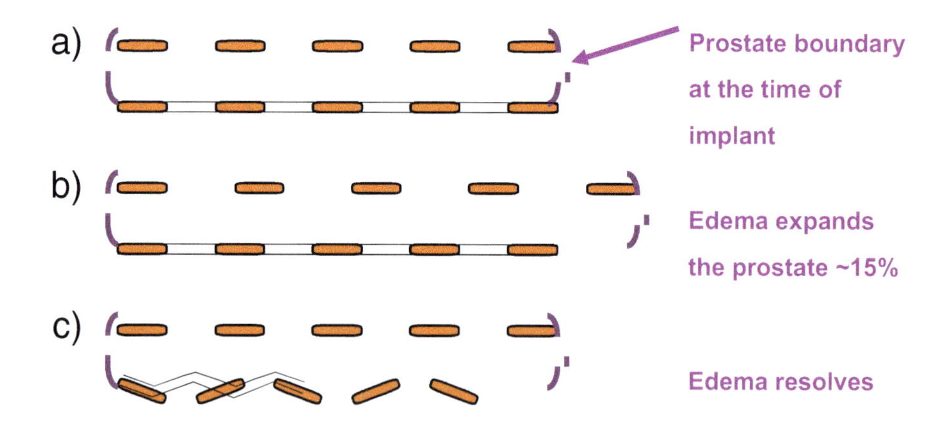

Figure 4–21 A schematic cartoon illustrating possibilities for prostatic edema. a) Source train with 5 loose seeds (top) and a 5-seed strand (bottom) extend 4.5 cm from the prostate base to the apex. b) Edema from trauma and radiation expands the prostate length by about 15%. Embedded loose seeds move with tissue and remain at the base and apex, but the strand does not stretch, so the base, the apex, or both will be uncovered and underdosed. c) As edema resolves, loose seeds follow tissue deformations and continue covering the entire length of the prostate as it returns to its original size. Strands have little compressive strength, and edema resolution draws the stranded seeds closer together increasing central dose to the urethra and rectum.

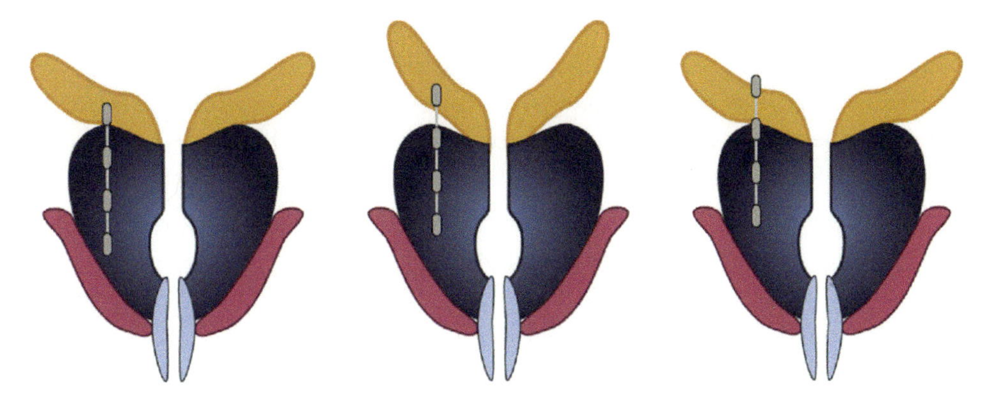

Figure 4–22 On the left, the leading seed of a strand is embedded in the bladder muscle. In the middle, bladder contraction during urination lifts the bladder and pulls the strand partially out of the prostate. On the right, relaxation may allow the next seed in the strand to be grabbed by the bladder. Illustration from (Soni et al. 2016).

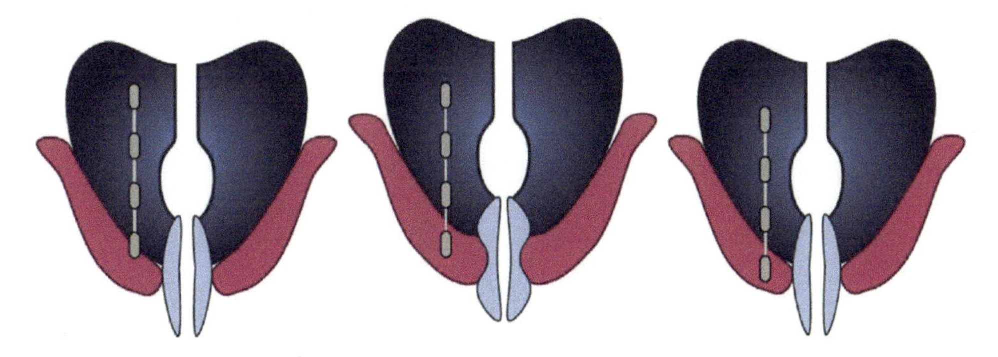

Figure 4–23 On the left, the leading seed of a strand is embedded in the levator ani muscle (fuchsia) just below the prostate apex. In the middle, levator ani contraction during urination or defecation grasps the seed and pulls the strand partially out of the prostate. On the right, relaxation may allow the next seed in the strand to be grabbed by the levator ani. Illustration from (Soni et al. 2016).

4.5.5 Biochemical Outcome Studies

Although dosimetric quality may be determined immediately after completion of the implant, determination of treatment failure requires frequent monitoring of a biochemical marker, prostate specific antigen (PSA), which is notably expressed by prostate cancer and, to a lesser extent, by normal prostate tissue. Whole-gland brachytherapy and radiation therapy of sufficiently high dose should ablate the prostate and gradually drive the PSA to zero. The ASTRO definition of failure—two successive PSA rises after a nadir—is still widely used, but in this era of high-quality brachytherapy and dose escalation, surgical-type definitions of PSA greater than some threshold are becoming more common. Violation of a PSA threshold of 0.4 ng/mL after a nadir or later than 48 months post implant constitute a biochemical failure. Because most prostate cancer is relatively slow growing and slow to undergo radiation cell death, PSA must be tracked for years. In a large implant population (n = 2,234) with some follow-up exceeding 17 years, the median time to failure in high-risk men was 2.4 years and in intermediate-risk patients, 2.7 years (Taira et al. 2013). Long-term actuarial failure rates at 15 years were 1.3% for low risk, 4.8% for intermediate risk, and 10.0% for high risk. All failures occurred prior to 9.2 years following implant, with no failures in any of 745 patients followed for more than 9.2 years regardless of their risk group. In that study, 88% of the patients were implanted with loose seeds only, but strands were used in 266 low-risk patients concurrently with over 500 loose-seed implants.

The British Columbia Cancer Agency initially used loose seeds, but then switched to strands. Their report on rates at 7 years for 327 patients implanted with loose seeds and 1,173 implanted with strands found no significant difference between biochemical progression-free survival rates of 91.3% and 91.9% for loose and strand implants, respectively, even though their early experience in the loose-seed era would have been during a learning curve (Herbert et al. 2011).

The Utrecht group reported a significant biochemical failure risk reduction of 43% where the 5-year survival rates were 86% for stranded-seed patients and 90% for loose-seed patients (Hinnen et al. 2010). The 896 implants were not randomized, but selection between strands or loose seeds was arbitrary, with a preference for strands in very small and very large prostates.

4.5.6 Conclusions

When implanting hydrodynamically shaped seeds outside the prostate, strands will reduce the relatively small incidence of lung embolization even further. The presence of mechanisms for whole strand migration makes the dosimetric benefits of strands problematic. Reports of less apical seed loss with strands compared to loose seeds may be due to learning curve effects where the last seed was inadvertently dragged below the apex. Excellent intermediate-term and long-term biochemical survival outcomes have been observed with both types of implants, but the one semi-randomized study found better results with loose seeds.

4.6 Adaptive Planning with TRUS or MRI

4.6.1 Introduction

In an ideal interstitial implant world, the sizes and shapes of undisturbed prostates would not change with time, brachytherapists would be capable of performing implants with seed distributions that perfectly match those in treatment plans, and implant needle-induced edema would be strictly forbidden by law. Alas, none of these propositions is true, and so the idea of performing implants adaptively in an attempt to mitigate the issues associated with their lack of veracity is conceptually appealing. In clinical practice, adaptive implantation is accomplished by treatment planning in the operating room (OR) (intra-operative planning), with the patient and imaging setups remaining fixed between initial imaging for treatment planning purposes and seed insertion. As succinctly described in the AAPM TG-137 report (Nath et al. 2009), intra-operative planning can take three different forms:

- *intra-operative pre-implantation planning*: creation of a plan in the OR just before the implant procedure, with immediate execution of the plan.
- *interactive planning*: stepwise refinement of the treatment plan using computerized dose calculations derived from image-based **needle-position feedback**.
- *dynamic dose calculation*: constant updating of dose distribution calculations using continuous **deposited-seed-position feedback**.

Intra-operative pre-implantation planning seeks to minimize uncertainties associated with changes in prostate volume over time and with replication of the patient setup in the OR for image-guided treatment. Note, however, that setup replication for conventional pre-implantation planning can be facilitated considerably by making a few key measurements, e.g., in TRUS imaging, using a digital level to measure the orientations of the US probe stepper and the patient stirrups in space. The only differences from conventional pre-implantation planning are the venue and the need to have a sufficient stock of seeds on hand. The additional cost of this approach, from both a financial and a procedural risk standpoint, is primarily related to the OR time spent in treatment planning. By comparison, both interactive planning and dynamic dose calculation utilize and execute conventional pre-implant planning and, hence, are discussed below.

4.6.2 Interactive Planning

In this approach, an intra-operative pre-plan is initially created and then refined as the implant proceeds. The first few needles are inserted as per the pre-plan, after which dosimetry is updated based on the **estimated positions of the implanted seeds as derived from the imaged needle positions**. The updated dosimetry is reviewed and inserted needles are repositioned in the plan, or subsequent needle positions are modified to maintain an acceptable level of implant quality. The dosimetry is then updated again. The needle insertion interval at which the seed distribution is updated and dosimetry recalculated is based on brachytherapist preference for uninterrupted flow of needle insertions. Changes in anatomic structures associated with implantation (e.g., prostate edema) are typically not accounted for, although they can be. Monitoring treatment plan modifications for quality is important as dosimetry is changing on the fly (Zaider et al. 2008).

Potential advantages of interactive planning as compared to intra-operative pre-implant planning include a shortening of the learning curve for inexperienced brachytherapists, reduced operator dependence of outcomes, and overall better dosimetry (Beaulieu et al. 2007). These advantages have proven somewhat elusive in practice (Meyer 2010), however, as the final seed positions frequently deviate from the imaged needle position due to mechanical needle bending on insertion caused by the needle's beveled tip, and to seed movement subsequent to needle and transrectal imaging probe (if used) withdrawal associated with relaxation of the prostatic and periprostatic tissues. Figure 4–24 (Chng et al. 2011) illustrates the variously curved paths along which stranded seeds in a TRUS-guided clinical implant ultimately lie as determined by post-implant CT imaging. Figure 4–25 (Waine et al. 2016) illustrates two types of needle bending that can occur during needle insertion. Figure 4–26 (Liu et al. 2015) shows a clinical example of the seed movement patterns associated with TRUS probe removal.

Figure 4–24 Views of a reconstructed implant delivered using 26 needles. Lines are drawn between segmented seeds and labeled with the planned needle number to identify a reconstructed strand during post-implant analysis. Transverse and sagittal views are shown on the right. From (Chng et al. 2011) with permission.

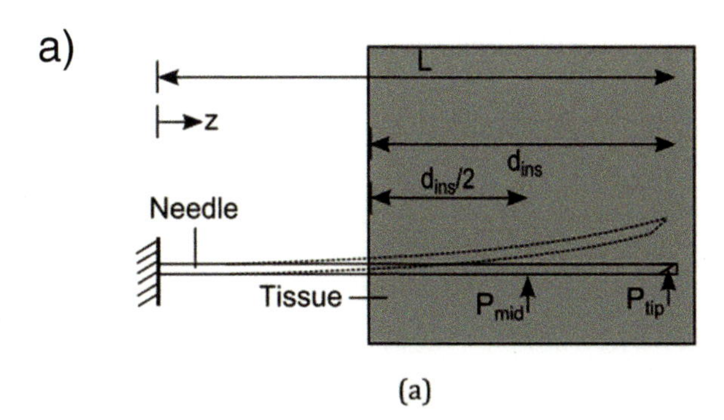

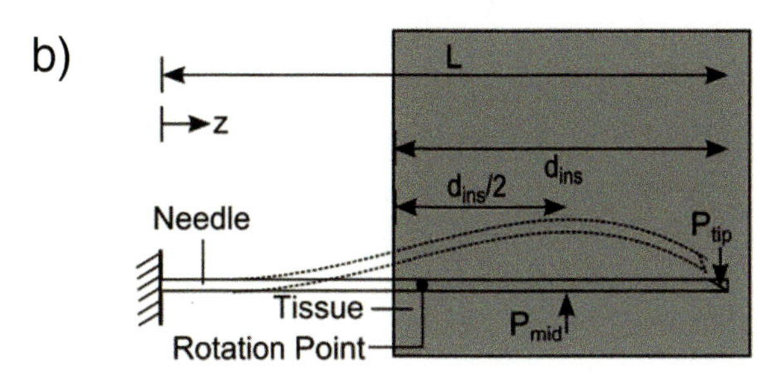

Figure 4–25 Implanted needle shape modeling based on point loads applied to a cantilever beam. a) Needle shape without rotation during insertion. b) Needle shape after 180° rotation at the indicated rotation point. P_{tip} represents the force applied to the needle tip. P_{mid} represents the force applied to the center of the inserted needle length. From (Waine et al. 2016) with permission of World Scientific Publishing.

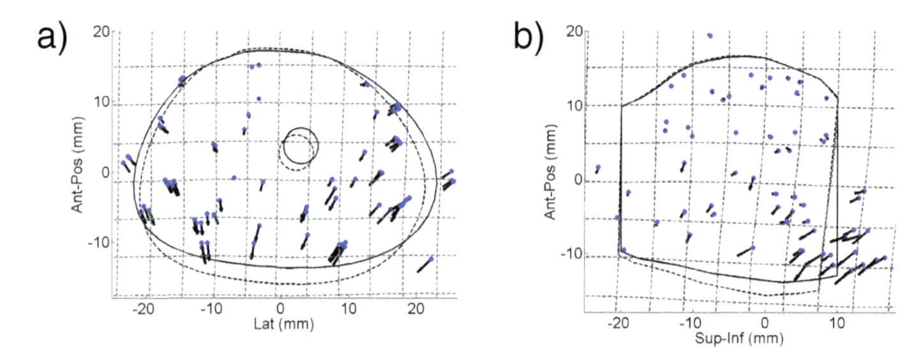

Figure 4–26 a) Transverse projection of implanted seed centroids under US probe pressure and subsequent movements on probe removal (arrows) for a representative patient showing elastic decompression movement. The probe is in for solid contours of the mid-gland prostate and urethra. The probe is out for dashed contours. b) Seed movement pattern and deformation mapping associated with superior-inferior shearing in the sagittal view. From (Liu et al. 2015) with permission from Elsevier.

Table 4–5 Conventional pre-planning vs. interactive planning

Consideration	Conventional Pre-planning	Interactive Planning
TRUS volume study location	Imaging room	Operating room
Pre-implantation planning location	Planning room	Operating room
Seed procurement	Based on pre-plan	Estimated or from seed bank
Implant dosimetry basis	Pre-implant structure contours, pre-planned seed positions	Real-time structure contour, real-time needle positions
Implant variations guidance	Brachytherapist alone	Real-time dosimetry, brachytherapist

An informative discussion of considerations associated with interactive planning vs. conventional pre-planning can be found in a pair of companion articles (Blasko 2006; Zelefsky and Zaider 2006). These considerations are summarized in Table 4–5.

It is important to note that the quality of an implant designed to treat the whole prostate is principally dependent on the details of the procedures followed at each individual clinic, no matter which general approach to planning is implemented (Bowes and Crook 2011). To date, clinical studies have demonstrated that conventional pre-planning yields nearly identical population-based outcomes as interactive pre-implantation planning in capable hands (Blasko 2006; Yoshida et al. 2013).

4.6.2.1 TRUS-based Implementation

A monochrome B-mode US scanner and biplane transrectal probe are used in conjunction with a commercial treatment planning system having the capacity to support intra-operative planning. Typically, the position of a needle is first identified in the transverse imaging plane and the probe then rotated so that the needle appears in the longitudinal imaging plane, as shown in Figure 4–27a. The needle's orientation in the longitudinal image is determined by locating its ends, and the positions of the sources it contains are subsequently identified by its loading pattern, as shown in Figure 4–27b.

A relatively recent development is the use of an MRI-based initial plan to save time in the operating room. In this approach, a set of volumetric T2-weighted MR images is acquired with the patient in the supine position, imported into the planning system, and reoriented so that the length of the anterior rectal wall underlying the prostate is perpendicular to the transverse slices, as illustrated in the top left panel of Figure

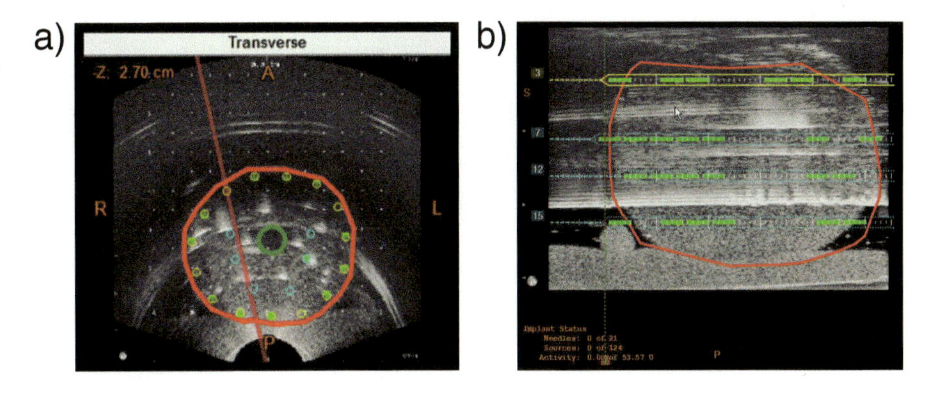

Figure 4–27 a) Identification in the axial image of the position of the longitudinal image plane (red line) containing the needle implanted near the 11 o'clock position in a prostate training phantom. b) Corresponding longitudinal image. The intensities of the displayed needle and seed positions are reduced with increasing distance from the imaging plane.

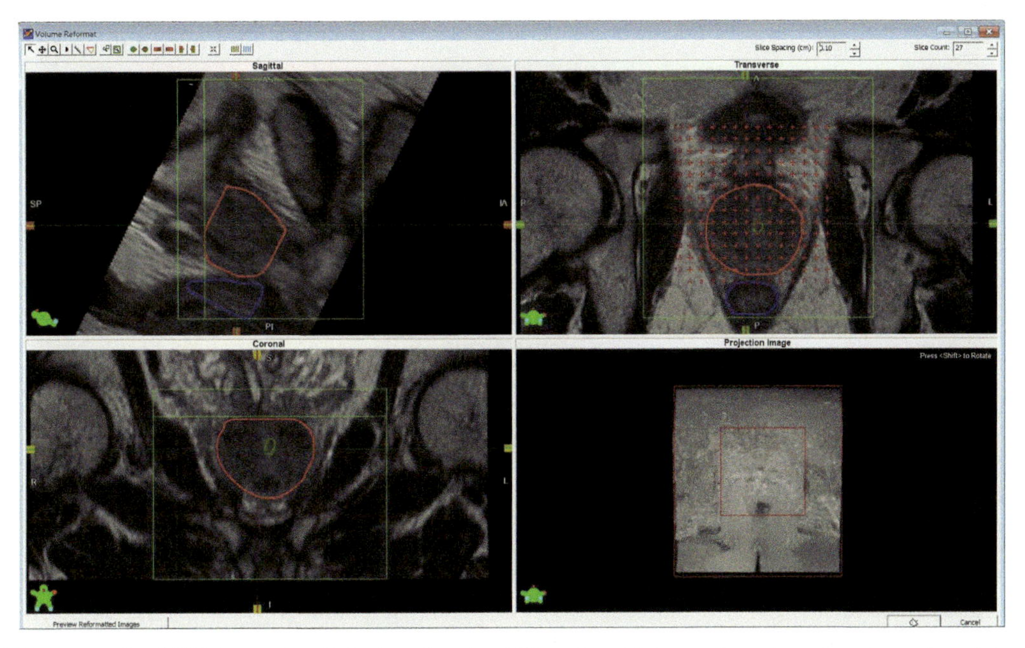

Figure 4–28 Setup for MRI-based initial planning for US-guided interactive planning, here involving reorientation of the 3D image data (top left), template grid placement (top right), and identification of the reference plane at the prostate base (top and bottom left).

4–28. Then an electronic template grid is positioned on the slices, as shown in the top right panel of Figure 4–28, and planning proceeds as for an US image-based treatment plan. In the operating room, the virtually introduced 3D MRI dataset is registered with a live 3D US dataset obtained with the patient in treatment position, using the positions of the reference plane and template grid to guide registration. At this point, anatomical contours previously drawn on MRI and initially planned seed locations can be adjusted as required based on the live US image information. This hybrid approach offers greater accuracy in estimating seed requirements, opportunity to create an initial plan without time pressure, and time savings in the operating room. This comes at a cost of the additional MRI scan and the initial planning time.

Some planning systems enable needle curvature to also be taken into consideration, which can improve the accuracy of source position determination. Nevertheless, non-negligible source position uncertainties remain unaccounted for, including the spatial accuracy with which a needle tip can be located in an US image (Siebert et al. 2009; Peikari et al. 2012), and the relaxation of prostatic and periprostatic tissues accompanying needle and TRUS probe withdrawal (Meyer 2010; Liu et al. 2015; Lian et al. 2016; Westendorp et al. 2016).

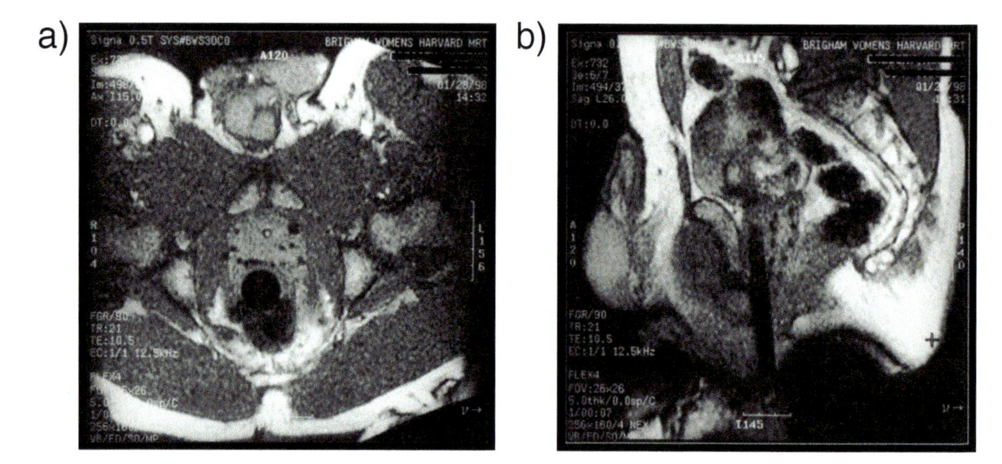

Figure 4–29 a) Real-time transaxial MR image of a prostate implant in progress. The inserted catheter carrying the seeds appears as a mid-sized circular void near the 2 o'clock position in the prostate, the rectal obturator as a large circular void below the prostate, and the implanted seeds as smaller circular voids within and adjacent to the prostate. b) Corresponding sagittal view. From (D'Amico et al. 1998) with permission of Elsevier.

4.6.2.2 MRI-based Implementation

Although TRUS-based treatment planning and implant guidance is currently the predominant standard of care for LDR prostate brachytherapy by a very wide margin, MRI-based planning and guidance is appealing because it offers substantially better anatomic definition. D'Amico and colleagues (1998) explored this latter approach in a 1998 study involving nine patients with early stage prostate cancer, using a 0.5-T, open-bore interventional MR imager. An MRI-compatible guidance template and implant catheters (needles) were used to enable simultaneous imaging with a pelvic coil during implantation. A rectal obturator affixed to the template was used to provide anatomical stability similar to that provided by a TRUS probe. The prostate peripheral zone was taken as the CTV. An in-house, real-time planning system was used to determine the seed loading for each catheter, which was subsequently placed using guidance from 5-mm thick, fast spin echo axial, sagittal, and coronal images, as shown in Figure 4–29. In the figure, both the implanted seeds and the inserted catheter appear as MRI signal voids that are noticeably larger than their actual size. Adjustments to account for prostate motion, edema, and catheter divergence were made as deemed necessary prior to seed deposition. This process was repeated until all catheters had been implanted. Dose-volume histogram parameters were calculated intra-operatively for the CTV, anterior rectal wall, and prostatic urethra based on the final source locations. The parameters reflected a high-quality implant in every case, although an uncertainty analysis of the parameter values was not done.

Major drawbacks of the MRI-based, real-time approach are its reliance on an expensive imaging resource with which the majority of brachytherapists have limited familiarity (the D'Amico study involved an expert MR radiologist who was present during the implant to interpret the images) and the time-intensive nature of the procedure (Stafford and Brezovich 2016). This may explain why the D'Amico study involved a small number of patients and has not been replicated elsewhere.

4.6.3 Dynamic Dose Calculation

This approach moves the concept of interactive planning two big strides forward by **basing updated dose calculations on deposited seed positions captured (ideally) in real time**, and by **accounting for prostate motion and intra-operative edema-induced prostate size and shape changes** during implantation. Thus, dynamic dose calculation involves adaptively "painting" a changing 3D target volume with an intended prescription dose. If a trans-rectal imaging probe is involved, the approach might also include accounting for the geometrical and dosimetric effects associated with probe removal.

4.6.3.1 TRUS-based Implementation

It is well known that identifying and locating all of the implanted seeds in a 3D US image dataset is a difficult task for a variety of reasons. Seeds distal to the US probe can be located in the shadow of proximal seeds. Some seeds are oriented such that they reflect the US beam away from the probe, thus rendering them invisible. Prostate calcifications and air bubbles in the rectum cause seed-mimicking artifacts. Manually identifying seeds in the longitudinal US image as they are ejected from the needle can overcome many of these challenges, but this cannot account for seed movement after deposition (Westendorp et al. 2016). Todor and colleagues described an approach for seed localization and dosimetry involving the integration of US and C-arm fluoroscopy (Todor et al. 2003) that has been further developed by Brunet-Benkoucha et al. (2009) and Dehghan et al. (2014). The latter system makes use of a radio-opaque fluoroscopic tracking fiducial affixed to the implant template to register the US and radiographic images, as illustrated in Figure 4–30. A fully automated image processing work flow is used to reconstruct the implanted seed locations on the US images, as shown in Figure 4–31. The system has been validated on a commercial phantom and on 16 clinical ^{103}Pd implant datasets. Compared with Day-1 CT-based dosimetry for the latter, the system yielded average $V_{100\%}$ and $D_{90\%}$ values that were 2.2% ±1.8% and 10.5% ±9.5% different, respectively. Further clinical evaluation of the system is required to adequately determine its potential to improve implant quality. Alternatives to fluoroscopy for seed localization include cone-beam CT (Westendorp et al. 2007) and real-time electromagnetic seed drop detection. The latter technology is currently under development (Racine et al. 2016) and involves a specially instrumented needle that is located with a commercial electromagnetic tracking system.

4.6.3.2 MRI-based Implementation

The current applications of MRI to brachytherapy for prostate treatment are described in overview in an accessible review article by Tanderup et al. (2014). Given the major drawbacks identified in the use of MRI for interactive planning, it is not well suited to support dynamic dose calculation. Indeed, no such approach has yet been demonstrated in the clinic. However, ongoing technical developments might make the approach not only feasible, but ultimately practically achievable as well. Work is ongoing by several groups to further

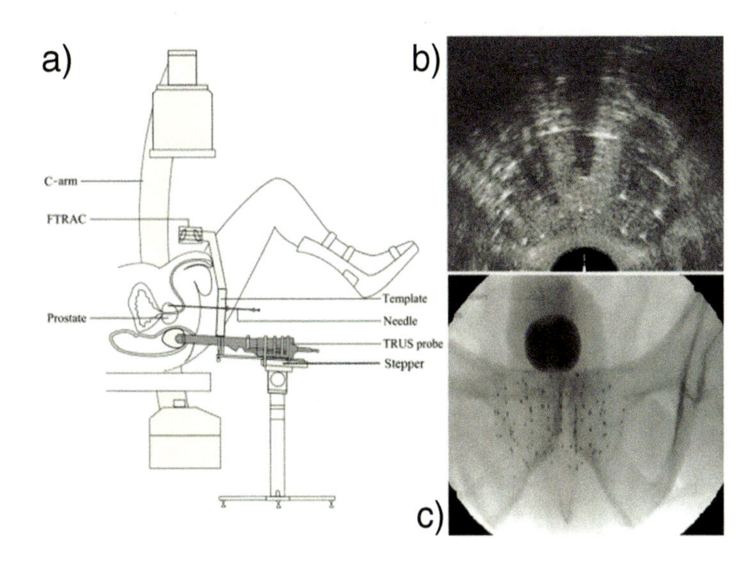

Figure 4–30 a) Prostate brachytherapy dynamic dose calculation setup showing the TRUS probe, guiding template, stepper, C-arm, and radio-opaque fluoroscopic tracking fiducial (FTRAC). b) A sample axial TRUS image of the prostate with implanted seeds. c) A sample radiographic image showing the seeds. From (Dehghan et al. 2014) with permission of Springer.

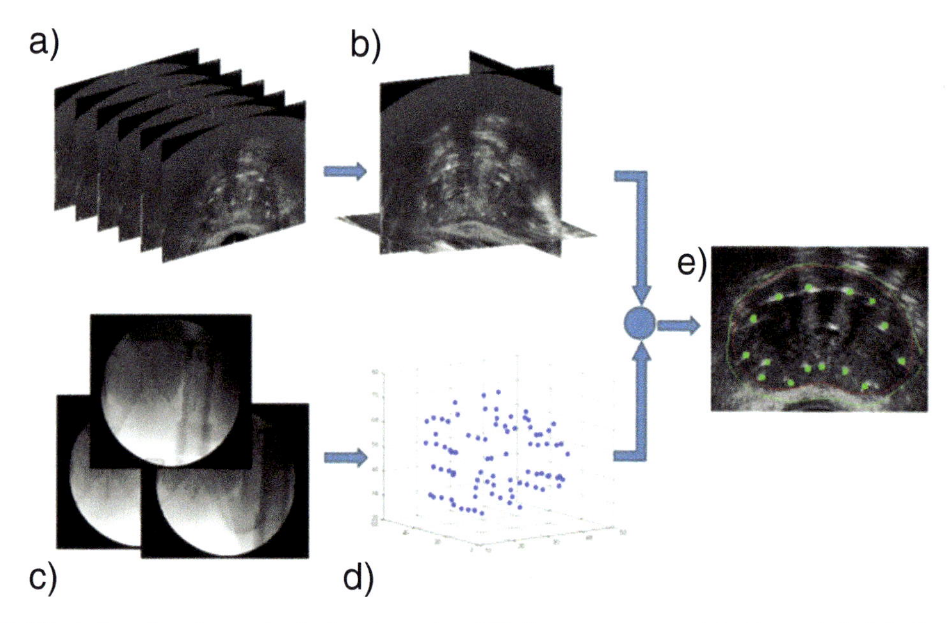

Figure 4–31 Work flow for dynamic dosimetry using US-fluoroscopy registration. a) Axial slices of TRUS images. b) Reconstructed TRUS volume. c) Radiographs acquired at three different view angles. d) Reconstructed seed centroids in 3D space. e) Sample axial slice of the TRUS volume overlaid with the registered seeds (green squares), prostate contour (red), and prescription isodose line (green). From (Dehghan et al. 2014) with permission of Springer.

develop the use of MRI for treatment planning (Tanaka et al. 2008; Albert et al. 2013) and seed localization (Frank et al. 2011; Seevinck et al. 2015), thereby continuing to enhance its attractiveness. With regard to MR imaging system cost and availability, technical innovations might one day enable MRI-based implant guidance to supplant TRUS-based guidance in the clinic.

4.7 Source Localization

4.7.1 The Timing of Source Localization

Source localization is, together with contouring (see Section 4.9), an important part of the post-planning procedure for LDR seeds. The reconstruction of the implanted sources requires high accuracy to obtain meaningful results. Because of edema after the implant due to needle trauma and an initially moderate dose rate of continuous radiation, the timing of post-implant imaging for dosimetric evaluation is important. Although some centers regard real-time intra-operative dosimetry to be adequate, most consider the post-implant CT to be definitive. Imaging shortly after the implant or within 24 hours (Day 0 or Day 1) is more convenient for the patient and encourages the timely completion of segmentation and calculations by the staff while interest in the quality of the procedure is still fresh. Because dosimetry is the brachytherapy equivalent of post-surgical pathology, early identification of dosimetric problems is valuable. Early evaluation also closes the learning loop to apply lessons learned to future implants while memory of the procedure is still recent. Day 0 or Day 1 dosimetry will underestimate dosimetric parameters because of the presence of edema, but imaging at that time provides a baseline of maximum edema for any subsequent scans.

Mathematical analysis of the effects of radionuclide decay and the dynamics of the individualized time course of edema resolution on deposited dose by Yue and Chen resulted in different optimal times for imaging of each radionuclide to minimize the dosimetric error inherent in calculating dosimetry at one point in time (Yue et al. 1999; Chen et al. 2000). They determined the error-minimizing time for a post-implant snapshot to be 10 ±2 days for ^{131}Cs, 16 ±4 days for ^{103}Pd, and 6 ±1 week for ^{125}I implants. Societal guidelines largely reflect these timing recommendations as an alternative to Day 0 or Day 1 dosimetry (Davis et al. 2012; Salembier et al. 2007; Nath et al. 2009).

4.7.2 Imaging Modalities

4.7.2.1 Fluoroscopy

Fluoroscopy for source reconstruction is not used very often in clinics any more. The main disadvantage is the inability to visualize soft tissue. The sources themselves can be seen very clearly in fluoroscopy. Images from two or more angles around the patient can be acquired and, using an appropriate reconstruction algorithm, the 3D source positions can be computed. A prerequisite is that the patient not move during imaging.

One mathematical challenge in the reconstruction of the sources is the assignment problem—in short, the identification of all the individual seeds in every fluoroscopy. In particular, overlapping seeds can cause problems for simple reconstruction algorithms. Hence, images from three or more fluoroscopy angles are recommended for a proper source reconstruction. Some algorithms offer the possibility of obtaining the 3D seed orientation (Tubic, Zaccarin, Pouliot et al. 2001; Tubic, Zaccarin, Beaulieu et al. 2001; Siebert, Srivastav et al. 2007), whereas others use TRUS imaging in parallel and thus allow for dynamic dosimetry in prostate implants (Todor et al. 2003).

Despite several advantages of fluoroscopic techniques for 3D seed reconstruction, those approaches are rarely used in clinics. Nevertheless, fluoroscopy is useful for quick assessment immediately after the implant of seeds. Moreover, it can be used to determine the number of implanted seeds. For this, two anterior fluoroscopy images are taken with angular spacing of about 5°. This allows counting the seeds, even if some seeds are overlapping in one image.

4.7.2.2 Computed Tomography

The standard imaging method for seed reconstruction is CT (Davis et al. 2012; Salembier et al. 2007). Implanted seeds on CT images can be identified either manually or with the aid of automatic identification software, or through a combination of both. Manual identification is tedious work, and automatic identification is not 100% reliable, so in practice they are combined. The imaging of seeds in the CT data is dependent on the seed design and the radiographic marker contained in the source—silver, lead, gold, or tungsten. This causes different seed models to have a different appearance in CT images. Additionally, the LDR source design, using different material composition, produces varying effects from x-ray attenuation and artifacts in CT.

The process begins by using the seed identification software to delineate a 3D search area in the CT dataset within which all seeds can be found and entering the number of seeds expected, as obtained from an independent source count (e.g., AP radiograph). The software is then asked to search for the seeds and return their locations, as shown in Figure 4–32.

When seeds are clustered, some seed models are easier to discern using the algorithm than others (Siebert et al. 2007). This work showed that the slice thickness for seed reconstruction should be $\leq$2 mm. Other CT parameters—such as CT tube current, scan field of view, and use of spiral or axial CT—have no impact on reconstruction accuracy. Bony structures should be excluded from the volume of interest to avoid misinterpretation between bone and sources by the algorithm. Care should be taken as well when calcifications are present in the prostate gland. In this case, calcifications may be labeled as seeds and must be corrected manually. For acceptable usage of automatic seed-finding algorithms, the exact number of implanted seeds must be known. Nevertheless, the user should always evaluate and correct, if necessary automatically determined seed locations and adjust the positions of the sources manually when necessary. Otherwise the algorithm might assign a calcification as a source and misinterpret or ignore one or more clustered seeds. Helpful tools

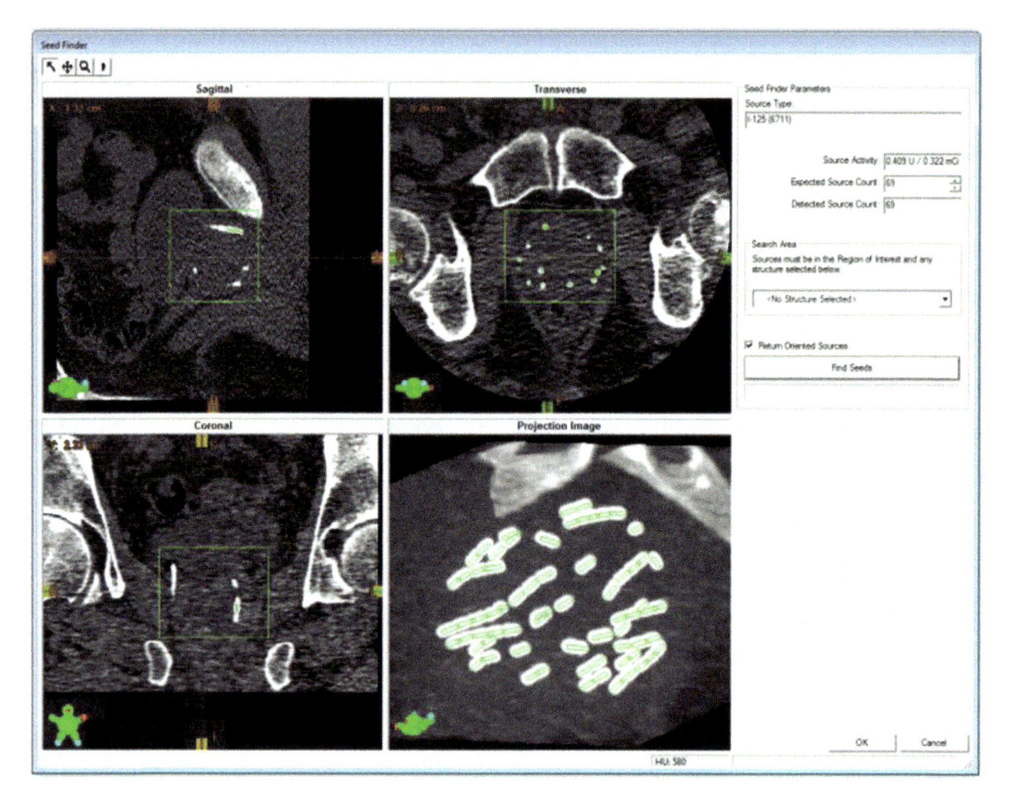

Figure 4–32 Automated seed localization in an operator-defined 3D region of interest (green box in top and bottom-left panels), showing results in a projection image (bottom right panel).

in this review process are the display window and level settings, which should be adjusted be provide the best images of the seeds, and the image pixel value query, which provides HU information useful for distinguishing a seed from bony anatomy. Some planning systems offer additional capabilities, such as a redundancy check, that can help determine whether a bright area on the image contains one or several seeds.

Some new approaches in post-procedural planning software allow for the detection of seed orientation when high-resolution CT scans are used and the seeds have a radiographic marker that is longer than the slice thickness. For sufficiently thin CT slices, obtaining seed orientations enables more accurate dose calculation when the line source model is used in the TG-43 formalism. When using the point-source approximation, the dose is not expected to be affected.

CT-based post planning of patients with hip prostheses is very difficult, and even impossible in double hip replacement patients, because of photon starvation in the area of the prostate. Metal artifact reduction technology on newer CTs helps clarify the images of the sources, even when the CT fan beam is attenuated by the overlapping artificial hips.

4.7.2.3 Magnetic Resonance Imaging

Post-procedural planning using MRI is also a common way to reconstruct the implanted sources. Typically, a combination of T1- and T2-weighted MRI scans is used. Although the sources appear as voids, the sources can be reconstructed in the T1 data set, whereas the delineation of the organs takes place in the T2-weighted images. Both scans, T1 and T2, should be conducted one after the other to minimize patient movement artifacts. For the post-procedural planning imaging, a zero registration between both data sets can often be performed. Nevertheless, warping effects between two scan protocols can occur, and a small affine transformation between both data sets may be necessary. Source reconstruction in post-procedural MRI-based planning software must be done manually, as automatic seed-finding algorithms are not yet on the market. The seed detection in MR images is more difficult than in CT. The sources give no signal in the MRI, but produce signal voids that are larger than the seed dimensions themselves (De Brabandere et al. 2006). Therefore, close or clustered seeds are difficult to discern. Knowledge of the number of seeds in the prostate is obviously important, and radiography might help as well. The accuracy of seed reconstruction in CT is higher than in MRI, as demonstrated in phantom cases with errors in localization of 0.9 mm in CT with 3 mm slice thickness, and 2.3 mm in MRI using 3 mm slice thickness in a 1.5-T scanner (De Brabandere et al. 2006).

4.8 Radiobiology

4.8.1 Introduction

As discussed in the previous sections, LDR prostate brachytherapy utilizes radioactive sources implanted permanently in the prostate gland to deliver a highly conformal radiation dose to the prostate with minimal dose to surrounding normal tissues. Due to the radioactive decay, cancer cells in LDR prostate brachytherapy are subjected to continuous LDR irradiation with instantaneous dose rate decreasing exponentially over time, very different from most other dose delivery techniques used in radiation therapy.

A number of radionuclides, including ^{131}Cs, ^{103}Pd, and ^{125}I, are used in current practice of LDR prostate brachytherapy. Their decay half-lives, however, vary widely, from 9.7 days to 59 days. As a result, the initial dose rate and the therapeutic treatment time (during which a major portion of the prescribed dose is delivered) are significantly different between patients treated with different sources. For example, the initial dose rates of ^{131}Cs, ^{103}Pd, and ^{125}I implants are approximately 0.30 Gy/h, 0.21 Gy/h, and 0.07 Gy/h, respectively. In addition, the time needed to deliver 90% of the total dose varies from approximately one month to nearly seven months between ^{131}Cs and ^{125}I implants.

Since the biologic effect of a given dose is known to depend on the dose rate and the type of tissues being irradiated (Hall 1972), the aforementioned variations in dose rate are expected to affect the treatment response of LDR prostate brachytherapy. In addition, since the effective treatment time can vary from one to seven months, tumor cell repopulation during treatment, if present, could also have a different effect on patients treated with different sources. Proper accounting of these variations on potential treatment response requires tools that are capable of describing the interactions between temporal dose delivery patterns and the tissues being irradiated. A number of mathematical models, based on the principles of radiobiology, have been developed for this purpose (e.g., see Chapter 1). Reviewed in this section are the radiobiological models suitable for LDR prostate brachytherapy, their use in elucidating the radiobiological features of LDR prostate brachytherapy, and potential implications on clinical practice.

4.8.2 Radiobiological Models for LDR Prostate Brachytherapy

4.8.2.1 Linear-quadratic (LQ) Model and Biologically Effective Dose (BED)

As discussed in Chapter 1 (Section 1.1), the LQ model for cell survival and the associated concept of BED has been used widely in elucidating the effects of changing dose fractionation in EBRT (Fowler 1989; 2010). Its application in brachytherapy took off in the late 1980s after Dale extended the model to protracted dose delivery under constant or exponentially decreasing dose rates (Dale 1985; 1989). Based on Dale's work, the BED for LDR prostate brachytherapy can be expressed in an analytic form as follows:

$$BED = D(T)RE(T) - \frac{\gamma}{\alpha}\max\{(T - T_k)\}, \tag{4.1}$$

where

$$RE(T) = 1 + \left(\frac{\beta}{\alpha}\right)\frac{\dot{D}_0}{(\mu-\lambda)} \times \frac{1}{1-e^{-\lambda T}}\left\{1 - e^{-2\lambda T} - \frac{2\lambda}{\mu+\lambda}(1-e^{-(\mu+\lambda)T})\right\}. \tag{4.2}$$

In the above, λ denotes the source's radioactive decay constant (inversely proportional to decay half-life), $\dot{D}$ the initial dose rate, T the elapsed treatment time, and $D(T)$ the cumulative dose delivered up to T; α and β are the LQ model parameters characterizing the relative importance of lethal damages caused by one-track and two-track interactions in a cell, respectively; and μ denotes the repair constant (inversely proportional to repair half-life) for DNA damage repairs following the mono-exponential kinetics. The last term in Equation (4.1) models the effect of cell proliferation during treatment, where γ represents the cell proliferation rate (inversely proportional to cell doubling time) and T_k the onset or lag time of cell proliferation.

Because the dose rate decreases exponentially with increasing T, the rate of cell-kill drops continuously over the treatment course. On the other hand, the rate of cell proliferation, if present, remains constant in Equation (4.1). At some time point, the rate of cell kill will fall to the level of cell proliferation rate and become lower than the proliferation rate beyond that time if any cancer stem cells survive, leading to treatment failure. This special time point, termed as *effective treatment time*, T_{eff}, by Dale (1989), is dependent on the decay constant (λ), cell proliferation rate (γ), dose to full decay (D), and radiosensitivity of the cell (α) as follows:

$$T_{eff} = \frac{1}{\lambda}\ln\left(\frac{\alpha D \lambda}{\gamma}\right). \tag{4.3}$$

Radiation delivered beyond T_{eff} will cease to produce a net gain in cell kill. As a result, BED will eventually become negative as $T \rightarrow \infty$ (when $\gamma \neq 0$). To alleviate this problem, BED calculated at T_{eff} has been used as *the* BED for permanent seed implants. While the definition of T_{eff} is physically intuitive, the need to use T_{eff} as the *time point* for computing the BED of permanent seen implants has remained as a point of concern when applying the BED concept to LDR prostate brachytherapy for proliferating tumors (Zaider and Hanin 2007). This situation is fundamentally linked to the simplistic nature of the LQ model, which assumes that radiation never completely eliminates all tumor cells and that cells proliferate at a uniform rate forever. Therefore, the use of BED for proliferating tumors in LDR prostate brachytherapy should always be viewed with these assumptions and the aforementioned technical maneuver (i.e., T_{eff}) in mind.

4.8.2.2 Isoeffective Dose (IED) Model

Zaider and Minerbo proposed an isoeffective dose model based on the theory of cell birth-and-death stochastic processes (Zaider and Minerbo 2000). It can also be used for brachytherapy with arbitrary temporal dose delivery patterns. When the LQ model is used for cell survival, the IED model was shown to be equivalent to the BED model when cell proliferation and spontaneous cell death can be neglected, while it is also mathematically well behaved in the limit of $T \rightarrow \infty$ for proliferating tumors (Carlson et al. 2013). In general,

$$IED(T) = -\frac{1}{\alpha}\log\left(\frac{S_0(T)e^{(b-d)T}}{1 + bS_0(T)e^{(b-d)T}\int_0^T \frac{du}{S_0(u)e^{(b-d)u}}}\right), \tag{4.4}$$

where $S_0(T)$ is the survival probability at time T in absence of cell proliferation; and b and d are the cell birth and spontaneous (non-radiation-induced) death rates, respectively. The factors b and d may be calculated from the more familiar quantities T_p (clonogenic doubling time), T_{pot} (potential doubling time), and φ (cell-loss factor) with $b - d = \ln(2) / T_p$, $T_p = T_{pot} / (1 - \varphi)$, and $d / b = \varphi$. In the $T \rightarrow \infty$ limit, IED is mathematically well behaved:

$$IED(\infty) = \frac{1}{\alpha}\log\left(b\int_{0}^{\infty}\frac{du}{S_0(u)e^{(b-d)u}}\right). \tag{4.5}$$

Historically, the IED model has not been used as often as the BED model. However, Zaider and Hanin showed that the numerical value of BED calculated at T_{eff} was lower than that of IED(∞) (Zaider and Hanin 2007). It has been shown in a recent systematic comparison between the BED and IED models for permanent brachytherapy with proliferating tumors that isoeffective prescription doses derived using the IED and BED models can also be different (Chen and Nath 2012). For example, using ^{125}I implant with 145 Gy as a reference, the isoeffective prescription doses derived from BED model was 2.7% and 3.5% lower than those from the IED model for ^{103}Pd and ^{131}Cs, respectively, for slow-growing tumors (e.g., at T_d of 42 days). The difference increased to 8.4% and 13.4%, respectively, for ^{103}Pd and ^{131}Cs implants for fast-growing tumors, e.g., at T_d of 5 days (Chen and Nath 2012). The clinical significance between the use of these two models remains to be tested.

4.8.2.3 Other Models

Another model, fundamentally equivalent to BED, is the EQD2 model. (Equivalent total dose when delivered in 2 Gy fractions would produce the same level of cell kill; see Chapter 1 Section 1.1.2.3 for details.) Unlike the BED and IED models, EQD2 is related directly to physical dose and is easily relatable to the clinical experience gained in conventional fractionated radiotherapy using 2 Gy per fraction. In addition, other models—such as equivalent uniform dose (EUD) and tumor control probability (TCP)—have also been used in LDR prostate brachytherapy evaluation (Niemierko 1997; Zaider et al. 2000). These models are not discussed further due to the limited scope of this section.

4.8.2.4 Model Parameters for Prostate Cancer

To use the BED or IED models for LDR prostate brachytherapy, numerical values of α, β, μ, γ, and T_k that are representative of prostate cancer cells are needed. At present, accurate estimates of these parameters for individual patients is still lacking. Indeed, the value of each parameter is likely to be different from patient to patient. In the absence of a reliable technique to accurately determine patient-specific model parameters, the numerical values of a calculated BED or IED cannot and should not be used as a predictor of treatment outcomes for individual patients.

In absence of patient-specific radiobiological parameters, a self-consistent set of parameters may be used for relative comparison of radiobiological impacts of different treatment strategies. The AAPM TG-137 report (Nath et al. 2009) has recommended a set of radiosensitivity parameters for prostate cancer (α = 0.15 Gy^{-1}, α/β = 3.0 Gy, T_d = 42 days, DNA repair half-life = 0.27 h, and T_k = 0 days) to increase the comparability of radiobiological indices reported by different institutions. This self-consistent set of parameter values will be used in the illustrative examples discussed below (unless otherwise noted). Obviously, these values should not be interpreted as definitive radiobiological parameters of prostate cancers.

The uncertainties associated with the existing radiobiological parameters for prostate cancer are quite large. For example, the α value estimated from the *in vitro* data ranged from 0.09 to 0.35 Gy^{-1}, and the α/β ratio ranged from 1.09 to 6.29 Gy (Carlson et al. 2004). The reported estimates of α/β range from 0.5 to 8.3 and 1.1 to 8.4 for *in vivo* and *in vitro* data, respectively (Dasu 2007; Carlson et al. 2004). The estimated *in vitro* repair half-time ranged from 5.8 to 8.9 h (with 95% confidence interval from 0.26 to 10.7 h) which appeared significantly larger than the estimates derived from clinical data (Carlson et al. 2004). The estimated cell-doubling time for prostate carcinoma also had a wide range, from 10 to over 60 days (Haustermans et al. 1997). It is, therefore, crucial in any modeling study to determine the sensitivity of the results to the assumed radiosensitivity parameters by using estimates sampled over their observed or expected uncertainty range.

4.8.3 Radiobiological Features of LDR Prostate Brachytherapy

The radiobiological models discussed in Section 4.8.2 can be used to compare the relative effectiveness of different technical variations using a fixed set of biological parameters. It can also be used to examine the influence of different biological process on a given technique by varying the values of model parameters. In this section, the BED model is used to elucidate the basic radiobiological features of LDR prostate brachytherapy. For simplicity and clarity in the presentation, the effect of heterogeneous dose/dose-rate distribution is discussed separately in Section 4.8.4.1.

4.8.3.1 Influence of Total Dose (or Initial Dose Rate)

In LDR prostate brachytherapy, the initial dose rate, total dose, and radioactive decay constant are interconnected:

$$\dot{D}_0 = \lambda D. \tag{4.6}$$

For a given radionuclide, increasing total dose leads to higher initial dose rate, while the characteristics of dose rate fall-off as a function of time remains the same (determined by λ). The influence of total dose (or initial dose rate) on BED is then as expected: BED increases with increasing total dose for a given source type.

4.8.3.2 Influence of Radioactive Decay Constant

The temporal variation in dose rate is determined by the source's decay constant λ. The influence of radioactive decay on BED is shown in Figure 4–33 where the BED per unit physical dose was plotted as a function decay half-life for a representative prostate cancer with a fixed total dose (solid line). Sources of different decay half-lives may emit different photon energies, which could have different relative biological effectiveness (RBE). This radiation quality effect is not considered in this example (see Section 4.8.3.3). The plots shows that the same total dose delivered by sources with shorter decay half-lives is biologically more effective than by sources with longer decay half-lives. In this illustration, a source with a decay half-life of 10 days (e.g., ^{131}Cs) is approximately 30% more effective than a source with a decay half-life of 60 days (e.g., ^{125}I) when given the same total dose.

This behavior is consistent with the fact that, for the same total dose, implants using sources with shorter decay half-lives have higher initial dose rate and shorter effective treatment time, which reduces the negative impacts of both DNA damage repair and cell proliferation, thereby increasing the BED. So, in theory, LDR prostate brachytherapy using ^{131}Cs source is expected to be biologically more effective than using ^{103}Pd or ^{125}I sources, if they were prescribed with the same total dose. While the same trend may exist, the impact of using sources with shorter decay half-lives on normal tissues can be further affected by the special radiobiological properties of and the dose/dose-rate received by normal tissues.

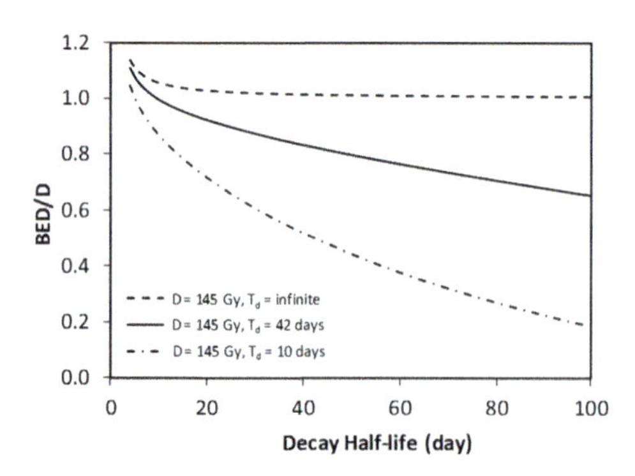

Figure 4–33 BED per unit dose as a function of decay half-life for the representative prostate cancer.

4.8.3.3 Influence of Photon Energy

Although not included in equations (4.1) and (4.2), the low-energy photons emitted by the sources used in LDR prostate brachytherapy can have different RBEs due to differences in linear energy transfer (LET). Several radiobiological studies using *in vitro* cell lines have shown that the RBE of continuous LDR irradiation using low-energy photons can be significantly different from that of high-energy photons typically used in EBRT. For example, relative to a high-energy photon reference source (e.g., 250 kVp or ^{60}Co gamma rays), the reported LET-dependent RBE for ^{125}I sources were between 1.2 and 1.5 and for ^{103}Pd sources between 1.6 and 1.9 (Ling, Li, and Anderson 1995; Nath et al. 2005). Because ^{131}Cs emits photons with similar energy as ^{125}I, it is reasonable to assume that ^{131}Cs has a similar LET-dependent RBE value as ^{125}I. Among the three sources used in current clinical practice, ^{103}Pd photons have greater RBE than photons from ^{125}I and ^{131}Cs sources. To accurately determine the equivalency between LDR implants and EBRT treatment, one needs to know the RBE specific to each irradiation condition and the tumor cells being irradiated. The effects of RBE may be evaluated using an approach proposed by Dale and his colleagues (Dale and Jones 1999; Antipas, Dale, and Coles 2001).

4.8.3.4 Influence of Cell Proliferation

Because the dose delivery time can be long in LDR prostate brachytherapy, the influence of cell proliferation, if present, could be more significant than in other dose delivery techniques. As shown in Equation (4.1), the presence of cell proliferation is expected to reduce the BED. The magnitude of this reduction is proportional to the cell proliferation rate.

Figure 4–34 plots the BED per unit dose as a function of tumor cell doubling time for three decay half-lives: 10, 20, and 60 days. At any given cell proliferation rate (i.e., cell doubling time), cell proliferation causes greater BED/D reduction for sources with longer decay half-lives. For a given decay half-life, BED/D reduces continuously with decreasing cell doubling time; the reduction become more rapid when cell doubling time falls below 20 days. Consistent with the observations in Section 4.8.3.2, sources with shorter decay half-lives are less susceptible to the negative effects of cell proliferation. Even at the cell doubling time of 42 days, sources such as ^{125}I (with the decay half-life of 59.4 days) can be significantly affected by the presence of cell proliferation compared to ^{103}Pd and ^{131}Cs sources.

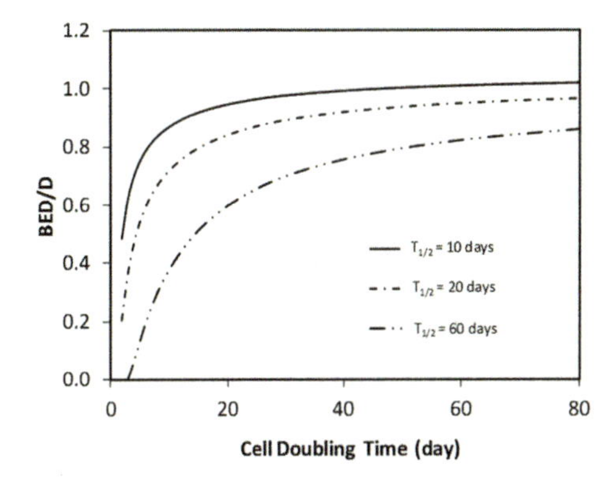

Figure 4–34 BED per unit dose as a function of cell doubling time for three different decay half-lives.

4.8.3.5 Influence of DNA Damage Repair

The presence of DNA damage repair reduces the effectiveness of LDR prostate brachytherapy. Intuitively, more efficient damage repair is expected to cause greater reduction in BED. The repair half-life is 0.27 h, as given in the AAPM TG-137 report recommended parameter set, which is relatively efficient among the mammalian cells. In addition, because the effective treatment time can be significantly different between sources of different decay half-lives, the impact of DNA damage repair is also expected to vary among implants using different sources.

Figure 4–35 compares the BED/D as a function of decay half-life for two different repair half-lives. The solid line, the same as in Figure 4–43, was calculated with the repair half-life of 0.27 h. The dash-dot line was calculated with a repair half-life of 1.5 h, while keeping all other parameters the same. It is apparent that the BED/D is greater for cells with longer repair half-life, i.e., less efficient in repair. The difference is relatively small for sources

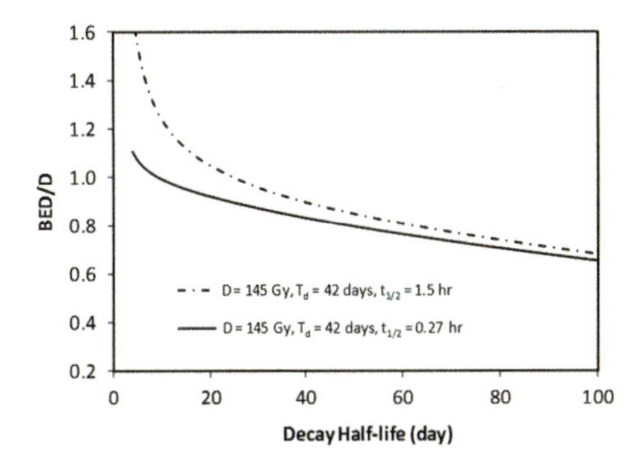

Figure 4–35 BED per unit dose as a function of decay half-life for two different DNA damage repair half-lives.

with long decay half-lives (approximately 5% at decay half-life of 60 days). But the difference keeps increasing with decreasing decay half-life, reaching approximately 13% and 24% at the decay half-lives of 20 days and 10 days, respectively.

4.8.3.6 *Influence of* α/β

As discussed in Chapter 1 (Section 1.1) and Chapter 5 (Section 5.9), the α/β ratio is an important radiobiological parameter that measures tissue radiosensitivity to dose fractionation change. It plays a key role in understanding the impact of different fractionation schemes used in EBRT and HDR brachytherapy for prostate cancer.

In LDR prostate brachytherapy, α/β should theoretically play the similar role. See Equation (4.2). This is easier to see in the absence of cell proliferation, in which case the BED (at $T = \infty$) becomes:

$$BED = D\left\{1+\frac{\lambda}{\mu+\lambda}\frac{D}{(\alpha/\beta)}\right\}. \tag{4.7}$$

For a given dose and radionuclide, tissues with a smaller α/β ratio would suffer greater biologic damage. In a typical situation where a tumor has a greater α/β ratio than late-reacting normal tissue, the continuous LDR irradiation over a relatively long period of time would be preferable. In the case of prostate cancer, however, the α/β ratio of tumors is, in fact, similar to (and possibly lower than) typical late-reacting normal tissue (e.g., 3 Gy). The advantage of LDR continuous irradiation becomes less apparent. Figure 4–36 plots the BED/D as a function of decay half-life for three different α/β values. For implants using sources with long decay half-lives, such as ^{125}I, the difference in biologic effect is small between tissues of different α/β values. The difference in biologic effect increases with decreasing decay half-life. Since implants with shorter decay half-lives deliver dose within shorter times and at higher initial dose rates, the increased difference in biologic effect is similar to situations observed in EBRT when changing dose fractionation toward hypofractionated schedules. With BED, one can perform detailed analysis of the effects of a prescribed dose on early- and late-responding normal tissue and assess potential complications associated with a particular LDR prostate brachytherapy treatment.

4.8.3.7 *Influence of Other Biological Factors*

Other biological factors—such as the cell cycle redistribution, reoxygenation, and spontaneous cell loss, etc. —are not modeled explicitly in the BED model. The influence of these factors on LDR prostate brachytherapy cannot be examined separately within this model.

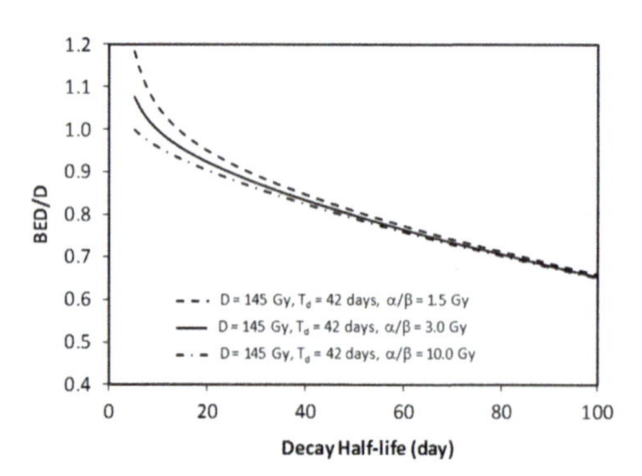

Figure 4–36 BED per unit dose as a function of decay half-life for three α/β values.

4.8.4 Radiobiological Comparison of ^{125}I, ^{103}Pd, and ^{131}Cs Implants

4.8.4.1 *Nominal Implants*

The discussions in Section 4.8.3 indicated that implants using radionuclides with shorter decay half-lives are biologically more potent on both the tumor and normal tissues when the same total dose is given. In clinical practice, the dose prescribed for implants using radionuclides with shorter half-life is, therefore, generally lower than those prescribed for longer-half-life radionuclides. For example, in LDR prostate brachytherapy, the prescribed dose for ^{131}Cs, ^{103}Pd, and ^{125}I implants is approximately 120 Gy, 125 Gy, and 145 Gy, respectively, aimed to achieve similar effects on tumors while maintaining acceptable normal tissue complications. A straightforward calculation of BED using the AAPM TG-137 model parameter set yielded BED values of 111 Gy, 115 Gy, and 117 Gy for ^{125}I, ^{103}Pd, and ^{131}C implants, respectively. One may conclude that implants using ^{125}I, ^{103}Pd, or ^{131}Cs sources under current dose prescriptions are biologically equivalent for prostate cancers characterized by the nominal radiobiological parameters.

Because the radiosensitivity and cellular processes can vary between patients, it is interesting to ask to what extent the equivalency would hold when a patient's radiobiological parameters differ from the nominal values. The question can be examined systematically using equations (4.1) and (4.2) with regard to cell proliferation, α/β ratio, and DNA damage repair. For example, the influence of cell doubling time is illustrated in Figure 4–37 where the BEDs of ^{125}I, ^{103}Pd, and ^{131}C implants are plotted as a function of cell doubling times. The equivalency holds reasonably well for tumors with cell doubling times between 40 and 100 days (with ^{125}I being slightly more effective at doubling times >60 days). However, for fast-growing tumors (e.g., cell doubling time <30 days), the equivalency begins to fall apart. Implants using ^{125}I are least effective compared to ^{103}Pd and ^{131}Cs implants for fast-growing tumors. Implants using ^{131}Cs, with a prescribed dose of 120 Gy, are more effective than either ^{103}Pd or ^{125}I over a wide range of tumor doubling times. However, this advantage could be offset by edema-induced source variations to be discussed in the next section.

4.8.4.2 Influence of Procedure-induced Prostate edema

In addition to radioactive decay, other dynamic processes—such as tumor shrinkage and the development and subsequent resolution of source-implant-induced prostate edema—may occur in LDR prostate brachytherapy, which can further increase the complexity of temporal dose delivery patterns. Source-implant-induced prostate edema is unique in LDR prostate brachytherapy. It causes the prostate gland to swell initially in response to the surgical trauma, followed by a gradual resolution of edema that can last weeks to over one month (Waterman et al. 1998; Moerland et al. 1997; Narayana et al. 1997). The quantitative characteristics of prostate edema vary widely from patient to patient (Moerland et al. 1997; Narayana et al. 1997; Waterman et al. 1998).

Since the presence of prostate edema forces the treatment volume and the location of the implanted sources to vary with time, the dose delivered by LDR prostate brachytherapy becomes dependent on the size and the resolution dynamics of edema (Leclerc et al. 2006). Implants using sources with shorter decay half-lives and lower effective photon energies were found to be more susceptible to the influence of prostate edema (Chen et al. 2006). Of the radioactive sources currently used in prostate brachytherapy, the ^{131}Cs and ^{125}I sources emit similar mean photon energies (~30.4 keV and 28.4 keV, respectively); both are greater than that of the ^{103}Pd source (~21 keV). On the other hand, the ^{131}Cs source has the shortest decay half-life (~9.7 days) compared to ^{103}Pd (~17 days) and ^{125}I (~60 days) sources. As a result, the magnitude of edema-induced dosimetry change is much greater in ^{131}Cs and ^{103}Pd implants compared to ^{125}I implants (Chen et al. 2006; Villeneuve et al. 2008).

The biologic effect of prostate edema on LDR prostate brachytherapy can be assessed systematically using the BED model (Chen et al. 2011). It has been shown that the presence of prostate edema could lead to increased cell survival and decreased probability of local control (Chen et al. 2011). The magnitude of edema-induced increase in cell survival becomes greater with increasing edema severity, decreasing decay half-life, or decreasing photon energy. At

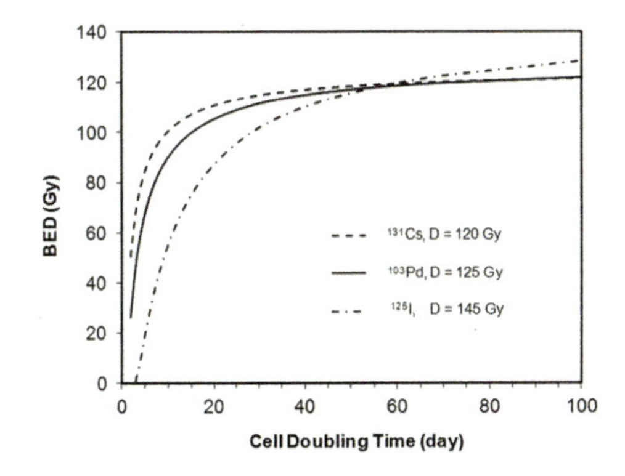

Figure 4–37 BED as a function of cell doubling time for nominal ^{131}Cs, ^{103}Pd, and ^{125}I implants.

the current level of dose prescriptions (for a "nominal" prostate cancer), implants using ^{125}I sources are less affected by edema than implants using ^{131}Cs or ^{103}Pd sources. Between ^{131}Cs and ^{103}Pd implants, the effect of edema on ^{103}Pd implants is slightly greater, even though the decay half-life of ^{103}Pd (17 days) is longer than that of ^{131}Cs (9.7 days) because the advantage of ^{103}Pd's longer decay half-life is negated by the lower effective energy of the photons it emits. The theoretical advantage of sources with shorter decay half-lives (e.g., ^{131}Cs) could, therefore, be offset by their greater susceptibility to edema-induced dose reductions in LDR prostate brachytherapy.

4.8.5 Further Considerations

4.8.5.1 Dose Heterogeneity

In addition to temporal variations, the instantaneous dose rates in LDR prostate brachytherapy also vary spatially due to the rapid dose fall-off around the implanted sources. It is not uncommon to have dose rates differ by more than a factor of two across the target volume. The BED for such an implant can be calculated by partitioning the tumor volume into small sub-volumes so that the dose rate distribution in each sub-volume can be considered uniform (Ling et al. 1994; Dale et al. 1997). The BED_i for a sub-volume *i* with initial dose rate of $\dot{D}_i(0)$ can then be calculated using the formulae discussed in 4.8.2.1. Mathematically, the BED for a clinical prostate implant, sometimes called as effective BED, can be calculated as:

$$BED = -\frac{1}{\alpha}\ln\left(\sum_i v_i e^{-\alpha \cdot BED_i}\right) \tag{4.8}$$

where v_i is the fractional volume receiving the dose rate $\dot{D}_i(0)$ with $\sum_i v_i = 1$. v_i is directly related to the differential dose (or initial dose rate) histogram of a permanent implant. The BED calculated with Equation (4.8) takes into account not only the time-dependent dose-rate variation, cell repopulation, and sublethal damage repair during the dose delivery, but also the spatial heterogeneity of dose rate distribution in permanent prostate brachytherapy. Ling has used Equation (4.8) in studying the effects of dose heterogeneity in permanent interstitial implants (Ling et al. 1994). It was found that dose heterogeneity in implants increased tumor cell kill and local control probability, although doses >20% higher than the prescription dose is wasted (Ling et al. 1994).

The BED calculated according to Equation (4.8) is preferentially weighted by low dose rates. To fully assess its significance, the three-dimensional distribution of BED within a permanent prostate implant can be calculated (Lee et al. 1995; Chen and Nath 2003). With 3D BED distributions, one can evaluate the biological significance of "hot" or "cold" dose-rate regions based on underlying anatomy. Calculation of 3D BED distribution with Equation (4.8) implicitly assumes that the tumor burden and radiosensitivity are spatially uniform. Nonetheless, it would be straightforward to incorporate the spatial distribution of tumor burden and radiosensitivity into the calculation of BED or 3D BED distribution when such information is accurately known.

4.8.5.2 Working with Models

In addition to investigating the effects of dose heterogeneity and prostate edema, the BED model has also been used to examine the possibility of radiobiologically optimal decay half-lives for permanent brachytherapy implants (Armpilia et al. 2003), the radiobiological effects of mixing sources with different decay half-lives (Chen and Nath 2003), the impact of tumor shrinkage during the implant (Dale, Jones, and Coles 1994), the probabilities of tumor control (Roy et al. 1993) and long-term normal tissue complication (Yaes 2001; Peschel et al. 1999), the possibility of dose escalation (Li et al. 2003), and the biological effect of combining prostate brachytherapy with EBRT (Jani et al. 2004). Stock et al. (2006) have also used BED as the implant quality index in a dose response study for ^{125}I implants.

Although the model predictions were largely consistent with general expectations, careful and rigorous model validations for brachytherapy applications are still lacking. Because of the simplistic nature of the models and the uncertainties associated with individual patient parameters, caution should be exercised in any modeling studies and in interpretation of model predictions. At present, the models are most useful in performing relative comparisons. They may also be used to check whether a new or alternate treatment regimen is plausible. Nonetheless, patient outcome data from carefully designed and controlled studies remains the gold standard for clinical development and optimization of LDR prostate brachytherapy.

4.8.5.3 Dose Calculation Algorithm

Until now, all dose parameters and DVHs used in this section have been based on the TG-43 dose calculation standard. The point was made in Chapter 2 that in the energy range of the radionuclides used in permanent seed implants, there is a significant departure from the Compton-dominated photon interaction process normally encountered in radiation therapy, including ^{192}Ir brachytherapy. As clearly stated in the TG-186 report, the average energy of all three seeds discussed in the section fall exactly in a range where tissue differentiation, from a physics interaction perspective, is maximum due to the photoelectric effect (Beaulieu et al. 2012). Thus, for LDR seed implants, the difference between calculations based on the TG-43 model and best-in-class dose calculation algorithms will yield significant differences in doses due to tissue heterogeneity and inter-seed attenuation (mainly lower doses as discussed in Chapter 2). Since all radiobiological models use either differential DVH or specific dose parameters (e.g., $D_{90\%}$) as input to calculate BED, equivalent uniform BED (EUBED), or other derived quantities, one should expect differences to appear once the field moves away from the TG-43 model.

Afsharpour and colleagues (2012) demonstrated the effect of a change in dose calculation algorithms in their study of permanent breast seed implants (PBSI). For this particular site, tissue heterogeneity can produce large differences in dose calculated with the TG-43 model versus a model-based approach such as Monte Carlo (MC). For the breast, they showed an underestimation of EUBED of about 10 Gy when using the TG-43 approach rather than MC.

While it is known that in prostate tissue radiation response, effects due to heterogeneity are not as sensitive as breast tissue in terms of dosimetric differences (Landry et al. 2010), the presence of localized calcifications can have a large impact on the whole prostate gland dose distribution (Collins Fekete et al. 2015). Figure 4–38a shows a voxel-by-voxel dose difference map for a prostate LDR implant in which a large local-

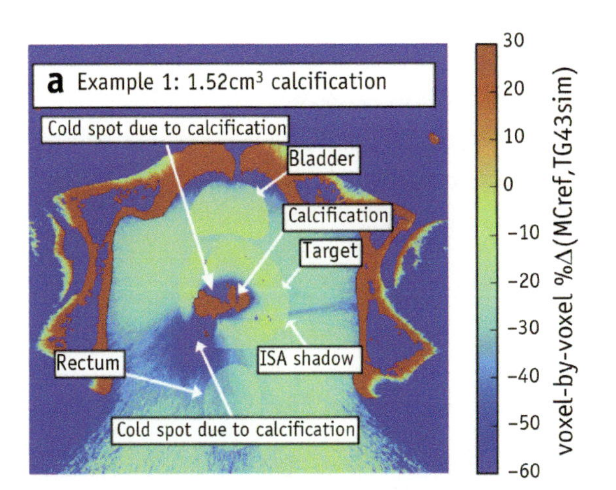

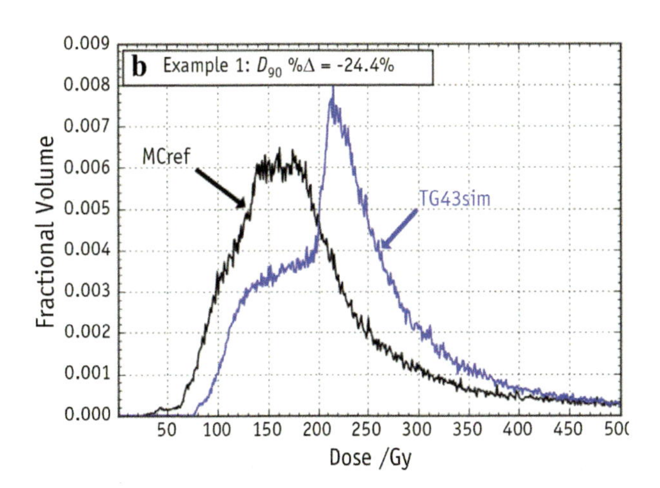

Figure 4–38 Dosimetric impact of dose heterogeneities in prostate LDR implants. a) Local dose difference map between TG-43-based calculations and MC for a prostate case with a large localized calcification. b) Differential DVHs for the prostate obtained using either TG-43-based calculations (blue curve) or MC (black curve). Adapted from Miksys et al. IJROBP 97 (2017) with permission from Elsevier.

ized calcification is identified (Miksys et al. 2017). The shielding effect of the calcification can clearly be seen in this specific CT slice and is furthermore localized in the anterior and posterior-lateral portions of the prostate, areas likely to contain cancer. It turns out that this localized calcification by itself decreases the prostate (whole gland) D_{90} by more than 24%. This decrease is, of course, even more pronounced locally, with a higher dose within the calcification and a lower dose beyond the calcification. The effect on the differential DVH is displayed in Figure 4–38b, where a clear shift toward lower values is seen when TG-43 is replaced with MC.

The shift toward lower values, populating the left part of the differential DVH, could lead to a large effect on radiobiological models, including potentially decreasing EUBED to very low values. Even for cases without visible calcification, Afsharpour and colleagues (2014) have demonstrated that the effect of tissue heterogeneity and inter-seed attenuation (or ISA) can lead to significant changes to extracted radiobiological values, such as the α/β ratio.

4.9 Post-implant Evaluation

4.9.1 Rationale and Dose-Volume Metrics

Post-implant assessment of dosimetry for a permanent prostate implant reflects a time-honored approach in interstitial brachytherapy, followed since the introduction of the Patterson–Parker rules for radium needle and radon seed placement (Meredith 1947), of using the actual source positions obtained by imaging to calculate the dose distribution in a patient. Its principal use in LDR prostate brachytherapy is to verify that the prescribed dose has been delivered to the clinical target volume and the organs at risk have received tolerable doses, so that the patient's ongoing medical management can proceed in an informed and appropriate manner. Other important uses of post-implant dosimetry include the identification and remediation (usually through a supplementary implant) of a clinically unsatisfactory implant (a relatively rare occurrence in a quality implant program), and the provision of objective feedback on treatment planning and implant techniques to the treatment team. Such feedback is valuable, not only when an implant program is first launched and the team is climbing the learning curve, but throughout the entire course of a program, as an important means to monitor and maintain quality. For these reasons, professional societies—including the AAPM, the American Brachytherapy Society (ABS), and GEC-ESTRO—strongly recommend that **post-implant dosimetry be performed for each and every implant** (Ash et al. 2000; Salembier et al. 2007; Nath et al. 2009; Davis et al. 2012).

Table 4–6 Recommended dose-volume reporting metrics from the AAPM TG-137 Report (Nath et al. 2009)

Anatomical Structure	Dose-Volume Metric	Units	Primary or Secondary Metric	Desired Value
Prostate	$V_{100\%}$	%SV	Primary	>95%
	$V_{150\%}$	%SV	Primary	≤50%
	$V_{200\%}$	%SV	Secondary	≤20%
	$D_{90\%}$	%PD	Primary	>90%
Prostatic Urethra	$D_{10\%}$	%PD	Primary	<150%
	$D_{30\%}$	%PD	Secondary	<130%
Rectum	D_{2cc}	%PD	Primary	<100%
	$D_{0.1cc}$	%PD	Secondary	<150%

prostate = post implant contour of the gland defined by the capsule on radiological examination
% SV = percentage of the anatomical structure volume
% PD = percentage of the prescribed treatment dose

In clinical practice, several dose-volume metrics derived from post-implant image-based dosimetry calculations are used to assess implant quality. These are commonly expressed for each anatomical structure of interest as V_x and D_y, where the former represents the volume of the structure receiving at least a dose of x (expressed in Gy or as a percentage of the prescribed dose), and the latter represents the minimum dose received by volume y of the structure (expressed in cm^3 or as a percentage of the total structure volume). Among professional societies, there is consensus regarding some, but not all, of the dose-volume metrics that should be reported and their acceptable values (Ash et al. 2000; Salembier et al. 2007; Nath et al. 2009; Davis et al. 2012). Table 4–6 summarizes selected post-implant reporting recommendations from AAPM TG-137 (Nath et al. 2009) which form a good starting point for developing a local operational set of reporting requirements. Every implant team should develop a set of initial reporting metrics that they feel best reflect local implant technique, and revise it as technique changes or peer-reviewed literature suggests that revision may be beneficial.

4.9.2 Limitations and Timing

Needle insertion causes prostatic and periprostatic tissues to swell, thereby changing the size of the prostate and the distances between implanted seeds. Resolution of the edema subsequently occurs over a span of about a month, after which the prostate size and seed geometry is presumed to be stable for dose calculation purposes, as shown in Figure 4–39.

The dynamics of edema resolution and radioactive decay alter dosimetry such that the dose distribution determined from imaging on the day of implant differs from that obtained one week or one month after the implant (Waterman et al. 1998; Chen et al. 2000; Yamada et al. 2003; Taussky et al. 2005; Leclerc et al. 2006; Chen et al. 2006; Sloboda et al. 2010; Monajemi et al. 2011), as illustrated for ^{125}I seeds in a reference phantom (Figure 4–40) in Figure 4–41. Note that the effect of edema will be greater for ^{103}Pd and ^{131}Cs seeds vs. ^{125}I seeds because of their shorter half-lives.

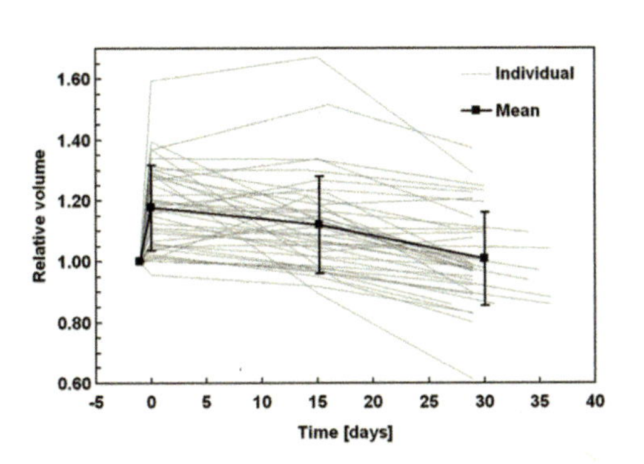

Figure 4–39 MRI contour-based relative prostate volume vs. time for individual patients (gray lines) and the group mean ± 1 S.D. (black line). From (Sloboda et al. 2010), with permission from Elsevier.

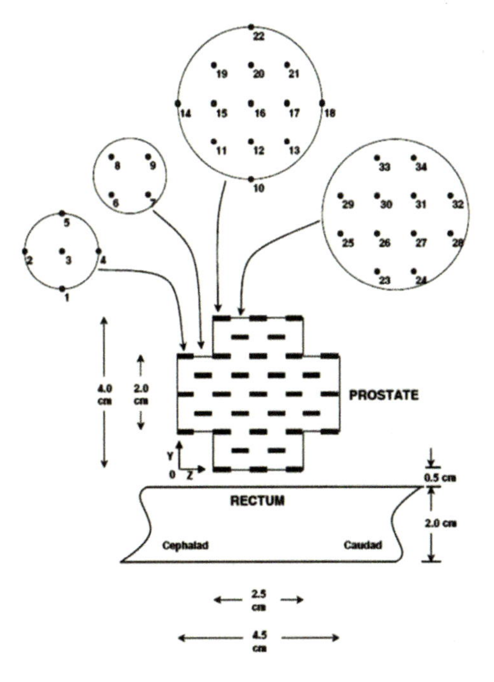

Figure 4–40 Reference case 2: a regular pattern of 81 seeds inside a virtual, cylindrically symmetric prostate phantom. From (Radiological Physics Center 2003), with permission.

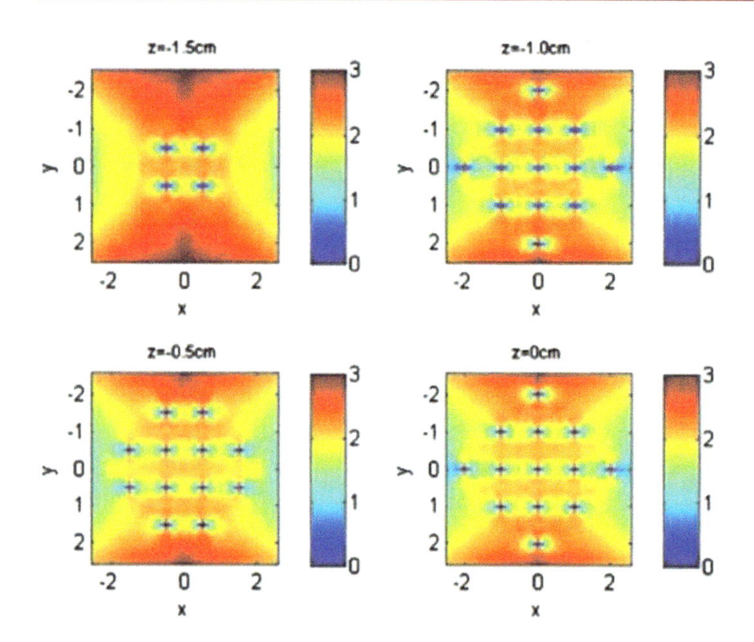

Figure 4–41 Percentage underdose maps associated with neglecting the effect of prostate edema for the ^{125}I seed arrangement in Figure 4–40, calculated using a clinically ased edema resolution model with parameters edema magnitude = 20% and edema resolution time = 28 days. The coordinate system origin is at the seed distribution center, and distances are in cm. From (Monajemi et al. 2011) with permission.

This results in the dose-volume metrics used for cumulative dose reporting changing with time of imaging (as edema is not usually considered in the dose calculation), and leads to the question, "What is the appropriate time to obtain images for post-implant dosimetry?" Interpreting the question scientifically, Chen and colleagues (Chen et al. 2000; Chen et al. 2006) determined nominal optimal dosimetry times that minimized the dosimetry errors associated with imaging at one particular time for ^{125}I, ^{103}Pd, and ^{131}Cs seeds based on the edema model of Waterman et al. (1998). Based on their results, the TG-137 report (Nath et al. 2009) recommends the following:

- ***For*** ^{125}***I:*** Because existing dose-response data were acquired one month after implantation, post-implant dosimetry should be performed at 1 month (±1 week)
- ***For*** ^{103}***Pd and*** ^{131}***Cs:*** Post-implant dosimetry should be performed at their respective nominal optimal times, 16 ±4 days and 10 ±2 days, respectively.

Alternatively, when interpreting the question clinically, other considerations, such as the timeliness of dosimetric feedback, come into play and suggest that an earlier time is preferable. Table 4–7 summarizes the scientific and clinical considerations for ^{125}I implants.

In practice, centers performing ^{125}I implants usually obtain post-implant dosimetry on Day 0 or1, or Day 30, but not both. Professional societies acknowledge and accommodate this dichotomy, but insist that the timing of post-implant dosimetry be reported, as several studies have demonstrated that Day 0 dosimetry cannot reliably predict dosimetry obtained at a later time (McLaughlin et al. 2006; Nag et al. 2008; Shaikh et al. 2015).

Table 4–7 Post-implant timing considerations for ^{125}I

Timing Consideration	Day 0/1	Day 30
Implant quality assessment without delay	✓	
Implant quality assessment based on dose-response data		✓
Patient convenience	✓	
Edema resolution mostly complete		✓

4.9.3 CT-based Evaluation

The use of CT imaging to evaluate an implant is currently the standard of care, and it is recommended by professional societies worldwide (Nath et al. 2009). As shown in Figure 4–42a, CT provides good source definition within the limits of axial slice spacing and partial volume artifact, but only moderate soft tissue contrast. If required, sinogram-based metal artifact reduction (Xu et al. 2011) or 3D adaptive median filtering of the reconstructed images (Basran et al. 2011) can reduce source streaking artifacts, as illustrated in Figure 4–43. A comparative study of various artifact reduction techniques has been conducted by Miksys and colleagues (2015). The AAPM TG-137 report recommends that contiguously acquired 2 to 3 mm thick axial images be used for post-implant evaluation. Although 5-mm thick slices provide better contrast and thereby enable easier delineation of anatomical structures, seeds are more accurately located using thinner slices, as demonstrated by the work of the European Brachytherapy Physics Quality Assurance Systems (BRAPHYQS) group (Siebert et al. 2007).

Other radiographic modalities that have been used in the intra-operative setting for post-implant evaluation include fluoroscopy and radiography (Todor et al. 2003; Dehghan 2014) and cone-beam CT (Westendorp et al. 2007).

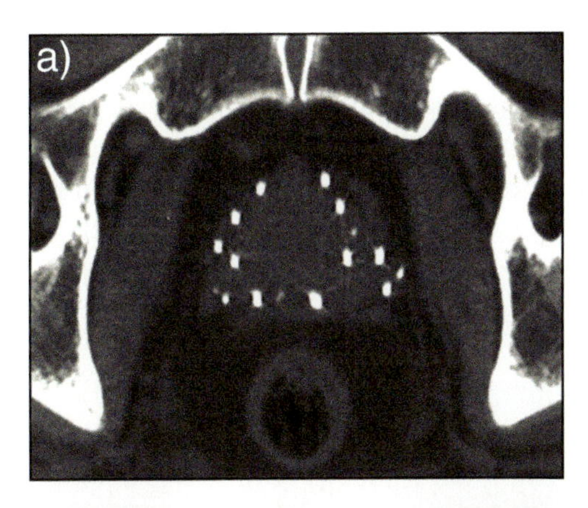

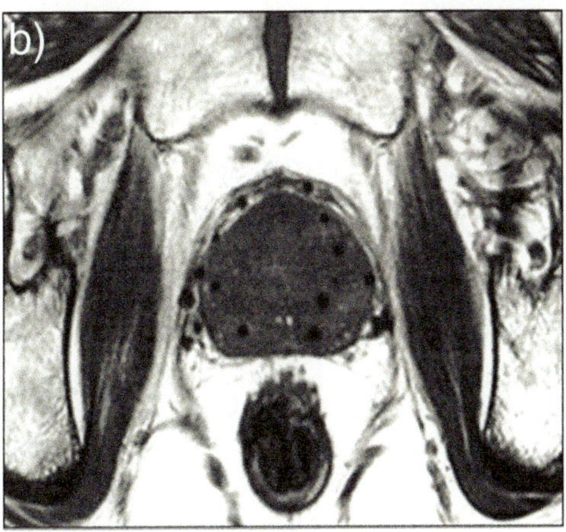

Figure 4–42 Corresponding post-implant a) CT and b) MR images of the pelvis near mid-prostate.

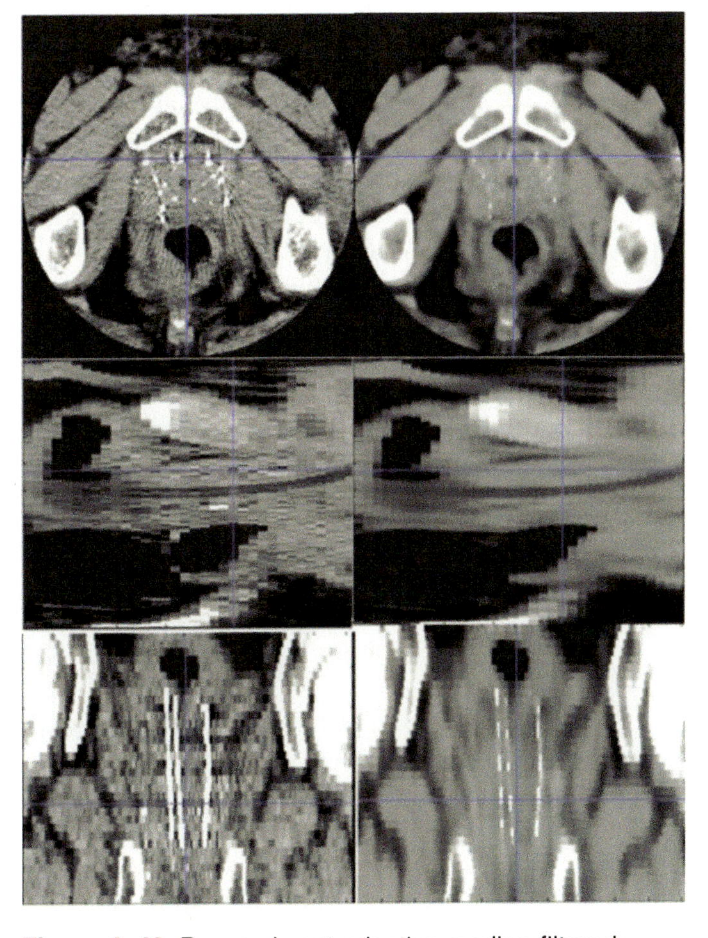

Figure 4–43 Pre- and post-adaptive-median-filtered images for a clinical case in transverse (upper), sagittal (middle), and coronal (lower) planes. The positions of the planes are indicated by the blue lines. From (Basran et al. 2011) with permission of John Wiley & Sons, Inc.

4.9.3.1 Contouring

Anatomical structures of interest for prostate brachytherapy include the prostate gland, urethra, rectum, and penile bulb. Although delineation of these structures is the primary responsibility of the brachytherapist performing the implant, the AAPM TG-137 report provides general guidance intended to promote more consistent and reproducible dose-volume parameter reporting. That guidance is replicated here for convenience:

1. The prostate should be contoured being mindful of the difficulties that are encountered at the prostate apex and base. In this regard, image interpretation advice from MRI, such as that related by McLaughlin et al. (2010) can be helpful.
2. The outer rectum should be contoured 1 cm superior and inferior to the prostate on CT. The rectum should not be distended when scanned.
3. The rectal wall on CT can be approximated by a 0.5-mm contraction of the outer rectal surface.
4. A Foley catheter should be used during Day 0 imaging, and the urethra should be contoured on all slices within the prostate (intraprostatic urethra). For post-implant dose evaluation at a later time, a Foley catheter is optional.
5. The penile bulb can be used as a surrogate for erectile tissues.

It has been well established that there is usually substantial inter- and intra-observer variability in post-implant contouring of the prostate on CT, which results in differences in prostate dose-volume parameter values (Crook et al. 2002; Lee et al. 2002). For this reason, and putting cost considerations aside, the ideal would be to obtain CT and MR images on the same day and register them for the purpose of post-implant evaluation, as discussed in Section 4.9.4 below.

Dose-volume parameters provide no spatial information regarding the distribution of dose within an anatomical structure. Consequently, they are of limited usefulness in reviewing post-implant dosimetry when the objective is to identify patterns of high or low dose in the prostate in an individual patient or a patient population. Bice and collaborators (2001) sought to address this issue by dividing the prostate into geometrical sectors or quadrants and performing dose-volume parameter analyses for each of the sectors individually, as illustrated in Figure 4–44. They used this technique to show that the same implant team working at two insti-

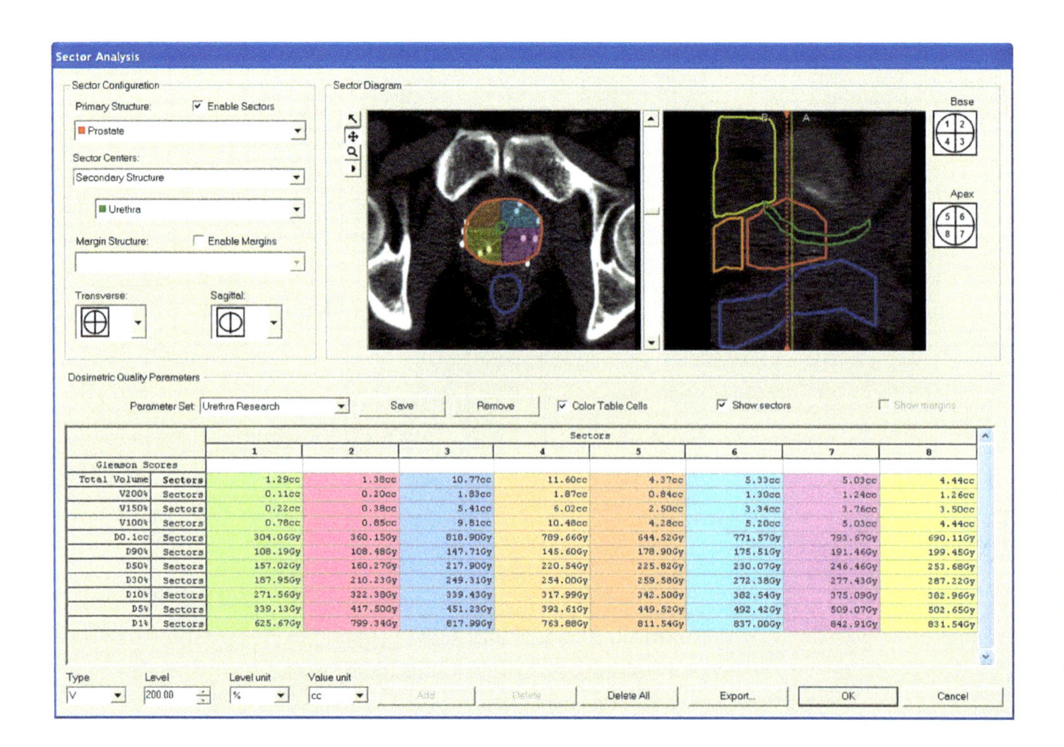

		Sectors 1	2	3	4	5	6	7	8
Gleason Scores									
Total Volume	Sectors	1.29cc	1.38cc	10.77cc	11.60cc	4.37cc	5.33cc	5.03cc	4.44cc
V200%	Sectors	0.11cc	0.20cc	1.83cc	1.87cc	0.84cc	1.30cc	1.24cc	1.26cc
V150%	Sectors	0.22cc	0.38cc	5.41cc	6.02cc	2.50cc	3.34cc	3.76cc	3.50cc
V100%	Sectors	0.78cc	0.85cc	9.81cc	10.48cc	4.28cc	5.20cc	5.03cc	4.44cc
D0.1cc	Sectors	304.06Gy	360.15Gy	818.90Gy	789.66Gy	644.52Gy	771.57Gy	793.67Gy	690.11Gy
D90%	Sectors	108.19Gy	108.48Gy	147.71Gy	145.60Gy	178.90Gy	175.51Gy	191.46Gy	199.45Gy
D50%	Sectors	157.02Gy	160.27Gy	217.90Gy	220.54Gy	225.82Gy	230.07Gy	246.46Gy	253.68Gy
D30%	Sectors	187.95Gy	210.23Gy	249.31Gy	254.00Gy	259.58Gy	272.38Gy	277.43Gy	287.22Gy
D10%	Sectors	271.56Gy	322.38Gy	339.43Gy	317.99Gy	342.50Gy	382.54Gy	375.09Gy	382.96Gy
D5%	Sectors	339.13Gy	417.50Gy	451.23Gy	392.61Gy	449.52Gy	492.42Gy	509.07Gy	502.65Gy
D1%	Sectors	625.67Gy	799.34Gy	817.99Gy	763.88Gy	811.54Gy	837.00Gy	842.91Gy	831.54Gy

Figure 4–44 Dose-volume analysis performed in eight geometrically-defined sectors, four in the base-most half and four in the apex-most half of the prostate, as shown in the schematic at the upper right.

tutions that used different seed introduction methods—a Mick applicator and loose seeds in needles—obtained better basal dose coverage with the latter method. Besides studying implant technique, sector analysis is useful to evaluate the dose distribution within individual implants. Ideally, this type of analysis should be based upon anatomically defined sectors instead of geometric ones, which could be accomplished via atlas matching of the contoured prostate volume.

4.9.4 MRI-CT Fusion-based Evaluation

The AAPM TG-137 report (Nath et al. 2009) describes the use of fused MR and CT images for post-implant evaluation as "ideal" because anatomical structures are best visualized on MRI and seeds on CT (see Figure 4–42). The report recommends that axial, coronal, and sagittal 3-mm thick T2-weighted contiguous images be obtained immediately before or immediately after the CT. More recently, 3D spatially isotropic MR imaging has proven to be a preferred choice for pelvic imaging, including prostate imaging (Lim et al. 2014). No matter what imaging techniques are used, patient body position should be as nearly as possible identical for the MR and CT scans (e.g., supine, with under-knee support) to facilitate good image registration.

Provided the image datasets are acquired closely in time using the same patient setups, MR-CT fusion can be accomplished using rigid registration, as illustrated in Figure 4–45. The registration should be based on a specified volume immediately surrounding the prostate and not the entire pelvis. Axial CT images should be used to determine the source positions, as described in Section 4.7.2.2, and the dose distribution should be displayed on an axial MRI dataset if 2D MRI slices are acquired. An advantage of 3D isotropic MR images is that good spatial resolution (1 mm cubic voxels) can be obtained, so the dose distribution can be displayed equally well in any chosen plane.

4.9.4.1 Contouring

Specific guidance taken from the AAPM TG-137 report (Nath et al. 2009) for contouring MR images is as follows:

1. The prostate should be contoured on MRI using the information from axial, coronal, and sagittal scans (for 2D MRI) or views (for 3D MRI).
2. The outer and inner rectum should be contoured on the axial MRI 1 cm above and below the prostate. The rectum should not be distended when scanned (i.e., do not use an endorectal coil).
3. The bladder should be contoured on the axial MRI.
4. The axial and sagittal MRI should be used to contour the urethra.
5. Other normal tissues responsible for erectile function should be contoured on MRI.

As with CT-based post-implant evaluation, sector analysis can be performed if desired.

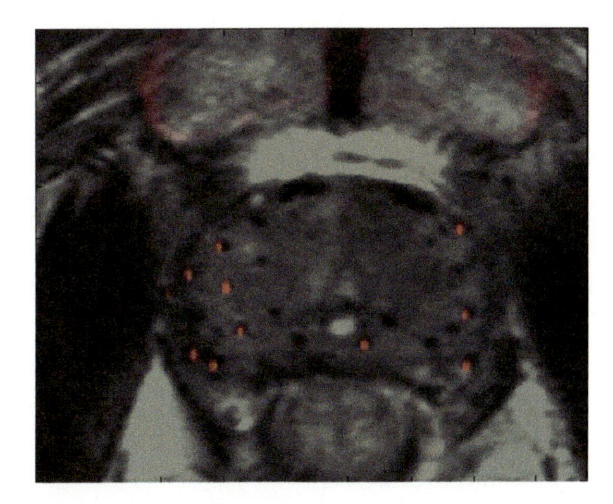

Figure 5–45 Prostate-only MR-CT 3D rigid registration. Gray scale image = MRI. Red overlay = seed centroids and bony anatomy from CT.

4.9.5 MRI-based Evaluation

Methods in which MRI alone is used not only for anatomical structure delineation but also for implanted seed localization are considered investigational at the present time (Tanderup et al. 2014). Potentially viable means to accomplish the latter task include the use of contrast-enhanced spacers placed between conventional sources (Frank et al. 2011), and new MR imaging protocols such as RASOR (RAdial Sampling with Off-resonance Reconstruction) that have been specifically designed to enhance the signal voids present at seed locations (Seevinck et al. 2015).

4.9.6 Post-implant Evaluation Summary

Evaluation of the completed implant is an essential part of any LDR prostate implant program, and it must be done for every patient to assess implant quality. In both the pre-implant planning and intra-operative settings, AAPM TG-137 identifies CT-based evaluation as the standard of care and recommends it be done at 1 month ±1 week for ^{125}I, 16 ±4 days for ^{103}Pd, and 10 ±2 days for ^{131}Cs. If available, MR imaging should be done immediately before or afterwards and MRI-CT registration performed to enable structure contouring on MRI and seed localization on CT, with the resulting dose distribution overlaid on MRI for detailed evaluation. In practice, some experienced brachytherapists with access to MRI use it solely to reassess poor-quality implants and plan remedial implants in order to manage costs and human resources.

In the intra-operative setting—where obtaining implant quality feedback before the patient leaves the operating room and maximizing patient convenience afterwards are overarching considerations—Day 0 post-implant dosimetry can be obtained using US alone (not recommended) or in combination with fluoroscopy, radiography, or cone-beam CT for seed localization. However, it should be noted that dose-volume metric values obtained on Day 0 are not predictive of those obtained at a later time, as the values can be affected substantially by prostate edema and seed movement after implantation, and so should be interpreted accordingly if used.

Lastly, it is important to appreciate that the dose-volume metrics used to assess implant quality each have an inherent uncertainty related to the time of imaging (as discussed in Section 4.9.2) and to structure contouring and seed localization. These latter operations are dependent on both the operator and the imaging equipment, making the assignment of generic uncertainties to them problematic. That said, some insight into the magnitude of dose-volume metric uncertainties can be gleaned from peer-reviewed literature reports from single institutions. Lee and colleagues (2002) reported variations in prostate $V_{100\%}$ and $D_{90\%}$ among five brachytherapists who independently contoured the prostate on Day 1 CT datasets for 10 patients. Population-averaged values for $V_{100\%}$ varied from 77.7 ±6.3% to 89.9 ±5.4%, and for $D_{90\%}$ from 75.1 ±11.3% to 102.6 ±16.7%. More recently, Mashouf et al. (2016) examined the sensitivity of prostate $V_{100\%}$ and $D_{90\%}$ to systematically scaled brachytherapist-drawn contours on 1 month CT images for 43 patients. They discovered that the number of post-plans having $V_{100\%}$ <90% increased from 7.0 ±0.1% for the original contours to 30.2 ±0.1% for a contour expansion of only 1 mm! Finally, in comparing post-implant dosimetry obtained using 1 month CT contours drawn by three oncologists for 10 patients vs. MR-CT fusion-based reference dosimetry, Crook et al. (2002) found that population-averaged $V_{100\%}$ values differed by 2.4% to 9.1% and $D_{90\%}$ values by 6.4% to 20.9%.

4.10 Medical Events in LDR Prostate Brachytherapy

4.10.1 Definition

Previously termed as "misadministration," the medical event (ME) defined by the Nuclear Regulatory Commission and agreement states is based on total dose delivered to the target. If the resulting dose differs from the prescribed dose by 20% or more, the case is considered as an ME and is to be reported to the authorities in accordance to the regulatory guidelines (code of federal regulations (CFR) 10 Part 35) or bureau of radiation control for agreement states.

4.10.2 Causes of MEs

Several components can go wrong before, during, and after the LDR prostate implant that can lead to poor results.

4.10.2.1 Pre-implant

- Poor volumetry. Based on this, the wrong number of seeds is ordered. In particular, if the prostate was measured smaller in the volumetry, too few seeds may be available at the implant.
- Some MEs can be associated with inadequate training and knowledge in brachytherapy, in general, and in the specifics of prostate brachytherapy, in particular.
- Commissioning of specific devices, such as imaging (US unit), the template being used, the Mick applicator, the treatment planning system (algorithm, source modeling).
- The resulting volume study had unacceptably poor resolution or was simply invalid due to erroneous probe alignment or inadequate training of staff on the imaging device.
- For pre-implant loaded needles, the pre-implantation planning process can have several flaws: wrong technique being used in seed placement during treatment planning, target volume with or without margin defined contrary to institutional protocol, wrong prescribed dose or source strength, failure to perform source leakage test upon their arrival, no source strength verification according to recommendations (Butler et al. 2008).
- Damage to seeds when cutting stranded-seed systems.
- Wrong source strength with an ordering or planning error using antiquated "mCi" https://www.nrc.gov/docs/ML0807/ML080710054.pdf.
- Wrong anatomy identified as the prostate. The penile bulb can look like a small prostate on US.

4.10.2.2 During the Implant

- Procedure, confirmation of target matching the volume study.
- Poor needle placement without accurate US verification.
- Seed delivery technique can provide poor results.
- Wrong needle at the wrong location.
- Poor handling of seeds during cystoscopy and seed recovery in the bladder or bladder wall that can lead to leaking source.
- Damaged source due to jamming of a Mick applicator. The seed must be removed carefully from the applicator or another applicator should be used to continue the implant.

4.10.2.3 Post Implant

- Lack of post-implant dosimetry
- Timing when post implant dosimetry was performed (could lead to an unfavorable dose distribution as discussed above).
- Lost or missing sources discovered during verification of the number of implanted seeds, performed in the OR by taking a scout film to account for implanted seeds and by surveying the OR to locate any seeds that may have been dropped or misplaced.

4.10.3 Role of the Qualified Medical Physicist in ME

Similar to any other treatment in radiotherapy, the qualified medical physicist (QMP) has an important role in ensuring that the intended dose as prescribed by the radiation oncologist is delivered to the patient in the safest possible way. While the QMP is not the authorized user, he or she should be able to question any abnormal or unacceptable situation that can lead to poor implantation. This includes poor needle placement that can lead to poor resulting dosimetry, inadequate probe placement leading to unreliable geometry, etc.

The QMP should also perform and provide such documentation of all pre-implant requirements (as stated in section b) to ensure a reliable outcome.

4.10.4 Steps to Avoid MEs

The QMP and the institution should have in place policies and procedures that address all steps previously stated. These policies should be used as intended and not created to pass a regulatory inspection. They also should be updated on a regular basis to be up to date with software and seed manufacturers' updates.

- Seed count should be performed at every step of this procedure: pre-loading, implant, and post implant.
- Seed source strength verification should be performed and not relying on a third-party assay (Butler et al. 2008).
- For pre-loaded needles, seed loading pattern is verified (imaging) against the planned one.
- Target and orientation of US images during the implant should be verified and should match images during the volume study.
- Needle placement verification using portal imaging (C-arm fluoroscopy as an example) should be used to ensure needle orientation and its location.
- After the implant, the sources not used must be secured. All equipment, e.g., stepper, needles, wipes, and bandages, must be checked with a Geiger counter or other detector for radioactivity.

4.10.5 Future Definition of ME in LDR Prostate Brachytherapy

The current definition of a medical event for prostate brachytherapy, a dose-based definition, has led to several medical events and was found inappropriate by the medical community. It has had an impact on the modality, which has seen a decrease in the number of LDR prostate brachytherapy cases in the past few years. The prostate volume (and therefore the resulting prostate dose) is dependent on the timing of the imaging. Also, the user has no control over possible seed migration. Therefore, a proposed ME definition using a source strength-based criterion has been identified as being more appropriate.

The Draft Final Rule containing the changes to the definition for permanent brachytherapy ME reporting (along with over 30 other significant changes) was submitted to the NRC Commission for their vote in June 2016. No final vote has been announced as of March, 2017.

References

Al-Halabi, H., L. Portelance, M. Duclos, B. Reniers, B. Bahoric, and L. Souhami. (2010). "Cone Beam CT-based three-dimensional planning in high-dose-rate brachytherapy for cervical cancer." *Int. J. Radiat. Oncol. Biol. Phys.* 77 (4):1092–97.

Alterovitz, R., E. Lessard, J. Pouliot, I. C. J. Hsu, J. F. O'Brien, and K. Goldberg. (2006). "Optimization of HDR Brachytherapy Dose Distributions Using Linear Programming with Penalty Costs." *Med. Phys.* 33: 4012–19. doi:10.1118/1.2349685.

Anacak, Y., M. Esassolak, A. Aydin, A. Aras, I. Olacak, and A. Haydaroğlu. (1997). "Effect of geometrical optimization on the treatment volumes and the dose homogeneity of biplane interstitial brachytherapy implants." *Radiother. Oncol.* 45:(71–76).

Anderson, L. L. (1986). "A 'natural' volume-dose histogram for brachytherapy." *Med. Phys.* 13(6):898–903.

Badiozamani, K. R., K. Wallner, S. Sutlief, W. Ellis, J. Blasko, and K. Russell. (1999). "Anticipating prostatic volume changes due to prostate brachytherapy." *Radiat. Oncol. Investig.* 7(6):360–64.

Baltas, D. and I-K. K. Kolkman-Deurloo. "Optimization and Evaluation." In *Comprehensive Brachytherapy: Physical and Clinical Aspects*, D. Baltas, J. L. M. Venselaar, A. S. Meigooni, and P. J. Hoskin, Eds. Boca Raton, FL: Taylor & Francis Group, 2013.

Barendsen, G. W. (1982). "Dose fractionation, dose rate and iso-effect relationships for normal tissue responses." *Int. J. Radiat. Oncol. Biol. Phys.* 8(11):1981–97.

Batchelar, D., M. Gaztañaga, M. Schmid, C. Araujo, F. Bachand, and J. Crook. (2014). "Validation study of ultrasound-based high-dose-rate prostate brachytherapy planning compared with CT-based planning." *Brachytherapy* 13(1):75–79.

Bentzen, S. M., W. Dorr, R. Gahbauer, R. W. Howell, M. C. Joiner, B. Jones, D. T. Jones, A. J. van der Kogel, A. Wambersie, and G. Whitmore. (2012). "Bioeffect modeling and equieffective dose concepts in radiation oncology—terminology, quantities and units." *Radiother. Oncol.* 15(2):266–68.

Beriwal, S., D. J. Demanes, B. Erickson, E. Jones, J. F. De Los Santos, R. A. Cormack, C. Yashar, J. J. Rownd, and A. N. Viswanathan. (2012). "American Brachytherapy Society consensus guidelines for interstitial brachytherapy for vaginal cancers." *Brachytherapy* 11:68–75.

Berns, C., P. Fritz, F. W. Hensley, and M. Wannenmacher. (1997). "Consequences of optimization in PDF brachytherapy—is a routine geometrical optimization recommendable?" *Int. J. Radiat. Oncol. Biol. Phys.* 37:1171–80.

Bice, W. S. Jr., B. R. Prestidge, and M. F. Sarosdy. (2001). "Sector analysis of prostate implants." *Med. Phys.* 28(12):2561–67.

Bittner, N., G. S. Merrick, A. Bennett, W. M. Butler, H. J. Andreini, W. Taubenslag, and E. Adamovich. (2015). "Diagnostic performance of initial transperineal template-guided mapping biopsy of the prostate gland." *Am. J. Clin. Oncol.* 38(3):300-03. doi: 10.1097/COC.0b013e31829a2954.

Brenner, D. J. (1997). "Radiation biology in brachytherapy." *J. Surg. Oncol.* 65(1):66–70.

Brenner, D. J. and E. J. Hall. (1999). "Fractionation and protraction for radiotherapy of prostate carcinoma." *Int. J. Radiat. Oncol. Biol. Phys.* (5):1095–101.

Brenner, D. J. and E. J. Hall. (1991). "Fractionated high dose rate versus low dose rate regimens for intracavitary brachytherapy of the cervix. I. General considerations based on radiobiology." *Br. J. Radiol.* 64(758):133–41.

Brenner, D. J., L. R. Hlatky, P. J. Hahnfeldt, Y. Huang, and R. K. Sachs. (1998). "The linear-quadratic model and most other common radiobiological models result in similar predictions of time-dose relationships." *Radiat. Res.* 150(1):83–91.

Bretschneider T, K. Mohnike, P. Hass, R. Seidensticker, D. Göppner, O. Dudeck, F. Streitparth, and J. Ricke. (2015). "Efficacy and safety of image-guided interstitial single fraction high-dose-rate brachytherapy in the management of metastatic malignant melanoma." *J. Contemp. Brachytherapy* 7(2):154–60.

Butler, W. M., G. S. Merrick, A. T. Dorsey, and B. M. Hagedorn. (2000). "Comparison of dose length, area, and volume histograms as quantifiers of urethral dose in prostate brachytherapy." *Int. J. Radiat. Oncol. Biol. Phys.* 48(5):1575–82.

Carlson, D. J., R. D. Stewart, V. A. Semenenko, and G. A. Sandison. (2008). "Combined use of Monte Carlo DNA damage simulations and deterministic repair models to examine putative mechanisms of cell killing." *Radiat. Res.* 169(4):447–59.

Carlson, D. J., Z. J. Chen, P. J. Hoskin, Z. Ouhib, and M. Zaider. "Radiobiology for Brachytherapy." In *Comprehensive Brachytherapy: Physics and Clinical Aspects*. D. Baltas, J.L.M. Venselaar, P. Hoskin, and A.S. Meigooni, Eds. Boca Roton, FL: CRC Press, 2013.

Chadwick, K. H. and H. P. Leenhouts. (1973). "A molecular theory of cell survival." Phys. Med. Biol 18 (1):78-87.

Chadwick, K. H. and H. P. Leenhouts. *The Molecular Theory of Radiation Biology*. Heidelberg: Springer-Verlag, 1981.

Chao, K. K., N. S. Goldstein, D. Yan, C. E. Vargas, M. I. Ghilezan, H. J. Korman, K. M. Kernen, J. B. Hollander, J. A. Gonzalez, A. A. Martinez, F. A. Vicini, and L. L. Kestin. (2006). "Clinicopathologic analysis of extracapsular extension in prostate cancer: should the clinical target volume be expanded posterolaterally to account for microscopic extension?" *Int. J. Radiat. Oncol. Biol. Phys.* 65(4):999–1007. doi: S0360-3016(06)00365-8 [pii].

Chaswal, V., B. Thomadsen, and D. Henderson. (2009). "Use of the Adjoint Analysis Based Greedy Heuristic Algorithms in Treatment Planning for LDR Brachytherapy of the Prostate and HDR Brachytherapy Using Multicatheter Breast Implant Technique." *Med. Phys.* 36(6):2423–24.

Chaswal, V., B. R. Thomadsen, and D. L. Henderson. (2012). "Development of an adjoint sensitivity field-based treatment-planning technique for the use of newly designed directional LDR sources in brachytherapy." *Phys. Med. Biol.* 57(4):963–82.

Chaswal, V., S. Yoo, B. R. Thomadsen, and D. L. Henderson. (2007). "Multi-species prostate implant treatment plans incorporating ^{192}Ir and ^{125}I using a Greedy Heuristic based 3D optimization algorithm." *Med. Phys.* 34(2):436–44.

Chen, Z. and R Nath. (2012). "On the use of biologically effective dose (BED) and iso-effective dose (IED) in radiobiological evaluations of permanent brachytherapy with proliferating tumors." *Int. J. Radiat. Oncol. Biol. Phys.* 84:S755.

Chen, Z., N. Yue, X. Wang, K. B. Roberts, R. Peschel, and R. Nath. (2000). "Dosimetric effects of edema in permanent prostate seed implants: a rigorous solution." *Int. J. Radiat. Oncol. Biol. Phys.* 47(5):1405–19. doi: S0360-3016(00)00549-6 [pii].

Coutard, H. (1932). "Roentgentherapy of epitheliomas of the tonsillar region, hypopharynx, and larynx, from 1920 to 1926." *AJR* 28(313–31):343–48.

Cunha, A., T. Siauw, I-C. Chow, and J. Pouliot. (2016). "A method for restricting intracatheter dwell time variance in high-dose-rate brachytherapy plan optimization." *Brachytherapy* 15:246–51.

Cunha, J. A. M., B. Pikett, and J. Pouliot. (2010). "Inverse planning optimization for hybrid prostate permanent-seed implant brachytherapy plans using two source strengths" *J. Appl. Clin. Med. Phys.* 11(3):64–77.

Curtis, S. B. (1986). "Lethal and potentially lethal lesions induced by radiation—a unified repair model." *Radiat. Res.* 106(2):252–70.

Dale, R. G. (1989). "Radiobiological assessment of permanent implants using tumour repopulation factors in the linear-quadratic model." *Br. J. Radiol.* 62(748):290–94.

Dale, R. G. (1990). "The use of small fraction numbers in high dose-rate gynaecological afterloading: some radiobiological considerations." *Br. J. Radiol.* 63(748):290–94.

Dale, R. G. (2010). "The BJR and progress in radiobiological modelling." *Br. J. Radiol.* 83(991):544–45. doi: 10.1259/bjr/52885245.

Dale, R. G. and B. Jones. (1998). "The clinical radiobiology of brachytherapy." *Br. J. Radiol.* 71(845):465–83.

Dale, R. G. (1985). "The application of the linear-quadratic dose-effect equation to fractionated and protracted radiotherapy." *Br. J. Radiol.* 58(690):515–28.

Davidson, M. T., J. Yuen, D. P. D'Souza, J. S. Radwan, J. A.Hammond, and D. L. Batchelar. (2008). "Optimization of high-dose-rate cervix brachytherapy applicator placement: the benefits of intraoperative ultrasound guidance." *Brachytherapy* 3(248–53).

De Boeck, L., J. Beliën, and W. Egyed. (2014). "Dose optimization in high-dose-rate brachytherapy: A literature review of quantitative models from 1990 to 2010." *Oper. Res. Health Care* 3(7):1157–61.

Deasy, J. O. (1997). "Multiple local minima in radiotherapy optimization problems with dose-volume constraints." *Med. Phys.* 24(7):1157–61.

Deufel, C. L. and K. M. Furutani. (2014). "Quality assurance for high dose rate brachytherapy treatment planning optimization: using a simple optimization to verify a complex optimization." *Phys. Med. Biol.* 59(3):525–40.

Dimopoulos, J. C., C. Kirisits, P. Petric, P. Georg, S. Lang, D. Berger, and R. Pötter. (2006). "The Vienna applicator for combined intracavitary and interstitial brachytherapy of cervical cancer: clinical feasibility and preliminary results." *Int. J. Radiat. Oncol. Biol. Phys.* 66(1):83–90.

Dinkla, A. M., R. van der Laarse, K. Koedooder, K. H. Petra, N. van Wieringen, B. R. Pieters, and A. Bel. (2014). "Novel tools for stepping source brachytherapy treatment planning: Enhanced geometrical optimization and interactive inverse planning." *Med. Phys.* 42(1):348–53.

Dowling, J. A., J. Lambert, J. Parker, O. Salvado, J. Fripp, A. Capp, C. Wratten, J. W. Denham, and P. B. Greer. (2012). "An atlas-based electron density mapping method for magnetic resonance imaging (MRI)-alone treatment planning and adaptive MRI-based prostate radiation therapy." *Int. J. Radiat. Oncol. Biol. Phys.* 83(1):e5–11. doi: 10.1016/j.ijrobp.2011.11.056.

Dresen, R. C., G. L. Beets, H. J. Rutten, S. M. Engelen, M. J. Lahaye, R. F. Vliegen, A. P. de Bruine, A. G. Kessels, G. Lammering, and R. G. Beets-Tan. (2009). "Locally advanced rectal cancer: MR imaging for restaging after neoadjuvant radiation therapy with concomitant chemotherapy. Part I. Are we able to predict tumor confined to the rectal wall?" *Radiology* 252(1):71–80. doi: 10.1148/radiol.2521081200.

Dyk, P. T., S. Richardson, S. N. Badiyan, J. K. Schwarz, J. Esthappan, J. L. Garcia-Ramirez, and P. Grigsby. (2015). "Outpatient-based high-dose-rate interstitial brachytherapy for gynecologic malignancies." *Brachytherapy* 14(2):231–37.

Edmundson, G. K. "Geometry Based Optimization for Stepping Source Implants." In *Brachytherapy HDR and LDR.* C. G. Orton, A. A. Martinez, and R. F. Mould, Eds. Columbia, MD: Nucletron Corporation, 1990.

Erickson, B., K. Albano, and M. Gillin. (1996). "CT-guided interstitial implantation of gynecologic malignancies." *Int. J. Radiat. Oncol. Biol. Phys.* 36(3):699–709.

Ezzell, G. "Optimization in Brachytherapy." In *Brachytherapy Physics,* 2nd edition. B. R. Thomadsen, M. J. Rivard, and W. M. Butler, Eds. Madison, WI: Medical Physics Publishing, 2005.

Ezzell, G., and R. W. Luthmann. "Clinical Implementation of Dwell Time Optimization Techniques for Single-Stepping Source Remote Applicators." In *Brachytherapy Physics*. J. F. Williamson, B. R. Thomadsen, and R. Nath, Eds. Madison, WI: Medical Physics Publishing, 1995.

Feder, B. H., A. M. Nisar Syed, and D. Neblett. (1978). "Treatment of extensive carcinoma of the cervix with the 'transperineal parametrial butterfly': a preliminary report on the revival of Waterman's approach." *Int. J. Radiat. Oncol. Biol. Phys.* 4(7–8):735–42.

Fleming, P., A. M. Nisar Syed, and D. Neblett. (1980). "Description of an afterloading ^{192}Ir interstitial-intracavitary technique in the treatment of carcinoma of the vagina." *Obstet. Gynecol.* 55(4):525–30.

Fowler, J. F. (2010). "21 years of biologically effective dose." *Br. J. Radiol.* 83(991):554–68.

Fowler, J. F. (1989). "The linear-quadratic formula and progress in fractionated radiotherapy." *Br. J. Radiol.* 62(740):679–94.

Fraass, B., K. Doppke, M. Hunt, G. Kutcher, G. Starkschall, R. Stern, and J. Van Dyke. (1998). "American Association of Physicists in Medicine Radiation Therapy Committee Task Group 53: Quality assurance for clinical radiotherapy treatment planning." *Med. Phys.* 25(10):1773–1829.

Guerrero, M. and X. A. Li. (2003). "Analysis of a large number of clinical studies for breast cancer radiotherapy: estimation of radiobiological parameters for treatment planning." *Phys. Med. Biol.* 48(20):3307–26.

Hall, E. J. (1991). "Weiss lecture. The dose-rate factor in radiation biology." *Int. J. Radiat. Oncol. Biol. Phys.* 59(3):595–610.

Hall, E. J. and D. J. Brenner. (1996). "Pulsed dose rate brachytherapy: can we take advantage of new technology?" *Int. J. Radiat. Oncol. Biol. Phys.* 34(2):511-12.

Hall, E. J. and J. S. Bedford. (1964). "Dose rate: its effect on the survival of Hela cells irradiated with gamma rays." *Radiat. Res.* 22:305-15.

Hellebust, T., C. Kirisits, D. Berger, J. Perez-Calatayud, M. De Brabandere, A. De Leeuw, I. Dumas, R. Hudej, G. Lowe, R. Wills, and K. Tanderup. (2010). "Recommendations from Gynaecological (GYN) GEC-ESTRO Working Group: Considerations and pitfalls in commissioning and applicator reconstruction in 3D image-based treatment planning of cervix cancer brachytherapy." *Radiother. Oncol.* 96:153–60.

Herron, B., D. Chin, and J. Pollock. (2008). "CT Imaging for gynecological HDR: Tools and Tricks." *Med. Dosim.* 33(4):310–14.

Hilts, M., H. Halperin,, D. Morton, D. Batchelar, F. Bachand, R. Chowdhury, and J. Crook. (2015). "Skin dose in breast brachytherapy:Defining a robust metric." *Brachytherapy* 14:970–78.

Holloway, C. L., T. F. DeLaney, K. M. Alektiar, P. M. Devlin, D. A. O'Farrell, and D. J. Demanes. (2013). "American Brachytherapy Society (ABS) consensus statement for sarcoma brachytherapy." *Brachytherapy* 3:179–90.

Holly, R., G. C. Morton, R. Sankreacha, N. Law, T. Cisecki, D. A. Loblaw, and H. T. Chung. (2011). "Use of cone-beam imaging to correct for catheter displacement in high-dose-rate prostate brachytherapy." *Brachytherapy* 10:299–305.

Holm, Å., T. Larsson, and Å. Carlsson Tedgren. (2013). "A linear programming model for optimizing HDR brachytherapy dosedistributions with respect to mean dose in the DVH-tail." *Med. Phys.* 40(8):081705–1.

Hoskin, P., P. Bownes, P. Ostler, K. Walker, and L. Bryant. (2003). "High dose rate afterloading brachytherapy for prostate cancer: catheter and gland movement between fractions." *Radiother. Oncol.* 68:285–88.

Hu, Y., J. Esthappan, S. Mutic, S. Richardson, H. A. Gay, J. K. Schwarz, and P. W. Grigsby. (2013). "Improve definition of titanium tandems in MR-guided high dose rate brachytherapy for cervical cancer using proton density weighted MRI." *Radiat. Oncol.* 8(1):16.

Huang, W. C., K. Kuroiwa, A. M. Serio, F. J. Bianco Jr., S. W. Fine, B. Shayegan, P. T. Scardino, and J. A. Eastham. (2007). "The anatomical and pathological characteristics of irradiated prostate cancers may influence the oncological efficacy of salvage ablative therapies." *J. Urol.* 177(4):1324–29; quiz 1591. doi: S0022-5347(06)03110-7 [pii].

ICRU. 1993. "Prescribing, recording and reporting photon beam therapy. ICRU Report 50." Bethesda, MD: ICRU.

ICRU. 1997. "Dose and volume specification for reporting interstitial therapy. ICRU Report 58." Bethesda, MD: ICRU.

ICRU. 1999. "Prescribing, recording and reporting photon beam therapy (Supplement to ICRU Report 50). ICRU Report 62." Bethesda, MD: ICRU.

ICRU. 2016. "Prescribing, Recording, and Reporting Brachytherapy for Cancer of the Cervix, ICRU Report 89." *J. ICRU* 13(1–2):Np. doi: 10.1093/jicru/ndw027.

Jaffray, D. A., M. C. Carlone, M. F. Milosevic, S. L. Breen, T. Stanescu, A. Rink, H. Alasti, A. Simeonov, M. S. Sweitzer, and J. D. Winter. (2014). "A Facility for Magnetic Resonance-Guided Radiation Therapy." *Sem. Rad. Oncol.* 24:193–95.

Kim, J., K. Garbarino, L. Schultz, K. Levin, B. Movsas, M. S. Siddiqui, I. J. Chetty, and C. Glide-Hurst. (2015). "Dosimetric evaluation of synthetic CT relative to bulk density assignment-based magnetic resonance-only approaches for prostate radiotherapy." *Radiat. Oncol.* 10:239. doi: 10.1186/s13014-015-0549-7.

Kim, T., T. N. Showalter, W. T. Watkins, D. M. Trifiletti, and B. Libby. (2015). "Parallelized patient-specific quality assurance for high-dose-rate image-guided brachytherapy in an integrated computed tomographyeon-rails brachytherapy suite." *Brachytherapy* 14:834–39.

Kirisits, C., M. Rivard, D. Baltas, F. Ballester, M. De Brabandere, R. van der Laarse, Y. Niatsetski, P. Papagiannis, T. P. Hellebust, J. Perez-Calatayud, K. Tanderup, J. L. M. Venselaar, and F. A. Siebert. (2014). "Review of clinical brachytherapy uncertainties: Analysis guidelines. *Radiother. Oncol.* 110(1):199–212. doi: 10.1016/j.radonc.2013.11.002. Epub 2013 Nov 30.

Kolkman-Deurloo, I-K. K., A. G. Visser, C. G. J. H. Niël, N. Driver, and P. C. Levendag. (1994). "Optimization of interstitial volume implants." *Radiother. Oncol.* 31:229-39.

Kubo, H., G. P. Glasgow, T. D. Pethel, B. R. Thomadsen, and J. F. Williamson. (1998). "High dose-rate brachytherapy treatment delivery: Report of the AAPM Radiation Therapy Committee Task Group No. 59." *Med. Phys.* 25(4):375–403.

Kuske, R. and R. Patel. (2007). "Breast Brachytheapy and Breast Augmentation: Breast Conservation with Capsular Contracutre." *Sem. Breast Disease* 10(1):42–49.

Kwan, D., R. Kagan, A. Olch, P. Chan, B. Hintz, and M. Wollin. (1983). "Single- and double-plan iridium-192 interstitial implants: Implantation guidelines and dosimetry." *Med. Phys.* 10(4):456–61.

Lahanas, M., D. Baltas, and N. Zamboglou. (1999). "Anatomy-based three-dimensional dose optimisation in brachytherapy using multiobjective genetic algorithms." *Med. Phys.* 26:1904–18.

Lambin, P., A. Gerbaulet, A. Kramar, P. Scalliet, C. Haie-Meder, E. P. Malaise, and D. Chassagne. (1993). "Phase III trial comparing two low dose rates in brachytherapy of cervix carcinoma: report at two years." *Int. J. Radiat. Oncol. Biol. Phys.* 25(3):405–12.

Lapuz, C., C. Dempsey, A. Capp, and P. C. O'Brien. (2013). "Dosimetric comparison of optimization methods for multichannel intracavitary brachytherapy for superficial vaginal tumors." *Brachytherapy* 12:637–44.

Lea, D. E. and D. G. Catcheside. (1942). "The mechanism of the induction by radiation of chromosome aberrations in Tradescantia." *J. of Genetics* 44:216–45.

Lee, C. D. (2014). "Recent developments and best practice in brachytherapy treatment planning." *Br. J. Radiol.* 87(1041):20140146.

Lessard, É. and J. Pouliot. (2001). "Inverse planning anatomy-based dose optimization for HDR-brachytherapy of the prostate using fast simulated annealing algorithm and dedicated objective function." *Med. Phys.* 28(5):773–79.

Lim, K. K., G. Noe, E. Hornsey, and R. P. Lim. (2014). "Clinical applications of 3D T2-weighted MRI in pelvic imaging." *Abdom. Imaging* 39(5):1052–62. doi: 10.1007/s00261-014-0124-y

Lindegaard, J. C., M. L. Madsen, A. Traberg, B. Meisner, S. K. Nielsen, K. Tanderup, H. Spejlborg, L. U. Fokdal, and O. Norrevang. (2016). "Individualised 3D printed vaginal template for MRI guided brachytherapy in locally advanced cervical cancer." *Radiother. Oncol.* 118:173–76.

Maspero, M., P. R. Seevinck, G. Schubert, M. A. Hoesl, B. van Asselen, M. A. Viergever, J. J. Lagendijk, G. J. Meijer, and C. A. van den Berg. (2017). "Quantification of confounding factors in MRI-based dose calculations as applied to prostate IMRT." *Phys. Med. Biol.* 62(3):948–65. doi: 10.1088/1361-6560/aa4fe7.

Mazeron, J. J., J. M. Simon, J. Crook, E. Calitchi, Y. Otmezguine, J. P. Le Bourgeois, and B. Pierquin. (1991). "Influence of dose rate on local control of breast carcinoma treated by external beam irradiation plus iridium 192 implant." *Int. J. Radiat. Oncol. Biol. Phys.* 21(5):1173–77.

McGeachy, P., J. Madamesila, A. Beauchamp, and R. Khan. (2015). "An open-source genetic algorithm for determining optimal seed distributions for low-dose-rate prostate brachytherapy." *Brachytherapy* 14:692–702.

Meertens, H., J. Borger, M. Steggerda, and A. Blom. "Evaluation and Optimization of Interstitial Brachytherapy Dose Distributions." In *Brachytherapy from Radium to Optimization*. J. J. Battermann, R. F. Mould, A. A. Martinez, and B. L. Speiser, Eds. Veenendaal, The Netherlands: Nucletron BV, 1994.

Meltsner, M. A., N. J. Ferrier, and B. R. Thomadsen. (2007). "Observations on rotating needle insertions using a brachytherapy robot." *Phys. Med. Biol.* 52(19):6027-37.

Merrick, G. S., W. M. Butler, A. T. Dorsey, and H. L. Walbert. (1998). "Influence of timing on the dosimetric analysis of transperineal ultrasound-guided, prostatic conformal brachytherapy." *Radiat. Oncol. Investig.* 6(4):182–90.

Merrick, G. S., W. M. Butler, A. T. Dorsey, and J. H. Lief. (1999). "The dependence of prostate postimplant dosimetric quality on CT volume determination." *Int. J. Radiat. Oncol. Biol. Phys.* 44(5):1111–17.

Merrick, G. S., W. M. Butler, K. E. Wallner, J. C. Blasko, J. Michalski, J. Aronowitz, P. Grimm, B. J. Moran, P. W. McLaughlin, J. Usher, J. H. Lief, and Z. A. Allen. (2005). "Variability of prostate brachytherapy pre-implant dosimetry: a multi-institutional analysis." *Brachytherapy* 4(4):241–51.

Merrick, G. S., W. M. Butler, P. Grimm, M. Morris, J. H. Lief, A. Bennett, and R. Fiano. (2014). "Multisector prostate dosimetric quality: Analysis of a large community database." *Brachytherapy* 13(2):146–51. doi: 10.1016/j.brachy.2013.08.003.

Moerland, M. A., R. van der Laarse, R. W. Luthmann, H. K. Wijrdeman, and J. J. Battermann. (2000). "The combined use of the natural and the cumulative dose-volume histograms in planning and evaluation of permanent prostatic seed implants." *Radiother. Oncol.* 57(3):279–84. doi: S0167814000002899 [pii].

Morton, D., D. Batchelar, M. Hilts, T. Berrang, and J. Crook. (2016). "Incorporating three-dimensional ultrasound into permanent breast seed implant brachytherapy treatment planning." *Brachytherapy* 16(1):167–73.

Nag, S., D. Shasha, N. Janjan, I. Petersen, and M. Zaider. (2001). "The American Brachytherapy Society recommendations for brachytherapy of soft tissue sarcomas." *Int. J. Radiat. Oncol. Biol. Phys.* 49(4):1033–43.

Nag, S., H. Cardenes, S. Chang, I. J. Das, B. Erickson, G. S. Ibbott, J. Lowenstein, J. Roll, B. Thomadsen, and M. Varia. (2004). "Proposed guidelines for image-based intracavitary brachytherapy for cervical carcinoma: report from Image-Guided Brachytherapy Working Group." *Int. J. Radiat. Oncol. Biol. Phys.* 60(4):1160–72.

Nath, R., L. L. Anderson, J. A. Meli, A. J. Olch, J. A. Stitt, and J. F. Williamson. (1997). "Code of practice for brachytherapy physics: Report of the AAPM Radiation Therapy Committee Task Group No. 56." *Med. Phys.* 24(10):1157–98.

Nath, R., W. S. Bice, W. M. Butler, Z. Chen, A. S. Meigooni, V. Narayana, M. J. Rivard, and Y. Yu. (2009). "AAPM recommendations on dose prescription and reporting methods for permanent interstitial brachytherapy for prostate cancer: report of Task Group 137." *Med. Phys.* 36(11):5310–22.

Niël, C. G. J. H., P. C. M. Koper, A. G. Viser, D. Simpkema, and P. C. Levendag. (1994). "Optimizing brachytherapy for locally advanced cervical cancer." *Int. J. Radiat. Oncol. Biol. Phys.* 29:873–77.

Niemierko, A. (1997). "Reporting and analyzing dose distributions: a concept of equivalent uniform dose." *Med. Phys.* 24(1):103–10.

Orcutt, K.P., B. Libby, L. L. Handsfield, G. Moyer, and T. N. Showalter. (2014). "CT-on-rails guided HDR brachytherapy: single-room, rapid-workflow treatment delivery with integrated image guidance." *Future Oncol.* 10(4):569–75.

Paley, P., W. J. Koh, K. J. Stelzer, B. A. Goff, H. K. Tamimi, and B. E. Greer. (1999). "A new technique for performing syed template interstitial implants of anterior vaginal tumors using an open retropubic approach." *Gynecol. Oncol.* 73:121–25.

Potter, R., C. Haie-Meder, E. Van Limbergen, I. Barillot, M. de Bragandere, J. Dimopoulos, I. Dumas, B. Erickson, S. Lang, A. Nulens, P. Petrow, J. Rownd, and C. Kirisits. (2006). "Recommendations from gynaecological (GYN) GEC ESTRO working group (II): Concepts and terms in 3D image-based treatment plannning in cervix-ancer brachytherapy-3D dose volume parameters and aspects of 3D image-based anatomy, radiation physics, radiobiology." *Radiother. Oncol.* 78:67–77.

Pouliot, J., D. Trembley, J. Roy, and S. Filice. (1996). "Optimizaiton of permanent 125-I prostate implants using fast simulated annealing." *Int. J. Radiat. Oncol. Biol. Phys.* 36(3):711–20.

Pouliot, J., É. Lessard, and I.-C. Hu. "Advanced 3-D Planning." In *Brachytherapy Physics,* 2nd edition. M. J. Rivard, B. R. Thomadsen, and W. M. Butler, Eds. Madison, WI: Medical Physics Publishing, 2005.

Ritter, M., S. Shahabi, M. Gehring, T. Shanahan, B. Thomadsen, and T. Kinsella. (1989). American Endocurietherapy Society, Hilton Head SC.

Roberson, P. L., V. Narayana, D. L. McShan, R. J. Winfield, and P. W. McLaughlin. (1997). "Source placement error for permanent implant of the prostate." *Med. Phys.* 24(2):251-57.

Roy, J. N., K. E. Wallner, S. T. Chiu-Tsao, L. L. Anderson, and C. C. Ling. (1991). "CT-based optimized planning for transperineal prostate implant with customized template." *Int. J. Radiat. Oncol. Biol. Phys.* 21:483–89.

Sachs, R. K. and D. J. Brenner. (1998). "The mechanistic basis of the linear-quadratic formalism." *Med. Phys.* 25(10):2071–73.

Sachs, R. K., P. Hahnfeld, and D. J. Brenner. (1997). "The link between low-LET dose-response relations and the underlying kinetics of damage production/repair/misrepair." *Int. J. Radiat. Oncol. Biol. Phys.* 72(4):351–74.

Shah, C., F. Vicini, D. E. Wazer, D. Arthur, and R. R. Patel. (2013). "The American Brachytherapy Society consensus statement for accelerated partial breast irradiation." *Brachytherapy* 12:267–77.

Shah, C., J. V. Antonucci, J. B. Wilkinson, M. Wallace, M. Ghilezan, P. Chen, K. Lewis, C. Mitchell, and F. Vicini. (2011). "Twelve-year clinical outcomes and patterns of failure with accelerated partial breast irradiation versus whole-breast irradiation: Results of a matched-pair analysis." *Radiother. Oncol.* 100:210–14.

Siauw, T., A. Cunha, A. Atamtürk, I.-C. Hsu, J. Pouliot, and K. Goldgerg. (2011). "IPIP: A new approach to inverse planning for HDR brachytherapy by directly optimizing dosimetric indices." *Med. Phys.* 38(7):4045–51.

Siauw, T., A. Cunha, D. Berenson, A. Atamtürk, I.-C. Hsu, K. Goldgerg, and J. Pouliot. (2012). "NPIP: A skew line needle configuration optimization system for HDR brachytherapy." *Med. Phys.* 39(7):4339–46. doi: 10.1118/1.4728226.

Slessinger, E. (2010). "Practical considerations for prostate HDR brachytherapy." *Brachytherapy* 9:282–87.

Sloboda, R. S. (1992). "Optimization of brachytherapy dose distribution by simulated annealing." *Med. Phys.* 19:234–44.

Stock, R. G., K. Chan, M. Terk, J. K. Dewyngaert, N. N. Stone, and P. Dottino. (1997). "A new technique for performing Syed-Neblett template interstitial implants for gynecologic malignancies using transrectal-ultrasound guidance." *Int. J. Radiat. Oncol. Biol. Phys.* 37(4):819–25.

Tagliaferri, L., S. Manfrida, B. Barbaro, M. M. Colangione, V. Masiello, G. C. Mattiucci, E. Placidi, R. Autorino, M. A. Gambacorta, S. Chiesa, G. Mantini, G. Kovác, and V. Valentini. (2015). "MITHRA—multiparametric MR/CT image adapted brachytherapy (MR/CT-IABT) in anal canal cancer: a feasibility study." *J. Contemp. Brachytherapy* 7(5):336–45.

Tanderup, K., R. Potter, J. C. Lindegaard, D. Berger, A. Wambersie, and C. Kirisits. (2010). "PTV margins should not be used to compensate for uncertainties in 3D image guided intracavitary brachytherapy." *Radiother. Oncol.* 97(3):495–500. doi: 10.1016/j.radonc.2010.08.021.

Taussky, D., L. Austen, A. Toi, I. Yeung, T. Williams, S. Pearson, M. McLean, G. Pond, and J. Crook. (2005). "Sequential evaluation of prostate edema after permanent seed prostate brachytherapy using CT-MRI fusion." *Int. J. Radiat. Oncol. Biol. Phys.* 62(4):974–80. doi: S0360-3016(04)03040-8 [pii].
Thames, H. D., Jr., H. R. Withers, L. J. Peters, and G. H. Fletcher. (1982). "Changes in early and late radiation responses with altered dose fractionation: implications for dose-survival relationships." *Int. J. Radiat. Oncol. Biol. Phys.* 8(2):219–26.
Thomadsen, B. R. "Achieving quality in brachytherapy." In *Achieving Quality in Brachytherapy.* Philadelphia: Institute of Physics Publishing, 1999.
Thomadsen, B. R., P. V. Houdek, G. Edmundson, R. van der Laarse, I-K. K. Kolkman-Deurloo, and A. G. Visser. "Treatment Planning and Optimization." In *High Dose Rate (HDR) Brachytherapy: A Textbook.* S. Nag, Ed. Armonk, NY: Futura Publishing Company, 1994.
Tobias, C. A. (1985). "The repair-misrepair model in radiobiology: comparison to other models." *Radiat. Res.* Supplement 8:S77–95.
van der Laarse, R. and R. W. De Boer. "Computerized High Dose Rate Brachytherapy Treatment Planning." In *Brachytherapy HDR and LDR.* C. G. Orton, A. A. Martinez, and R. F. Mould, Eds. Columbia, MD: Nucletron Corporation, 1990.
Venselaar, J. L. M., L. Beaulieu, Z. Chen, and R. L. Smith. "Advances in Brachytherapy Physics." In *Advances in Medical Physics.* J. Van Dyk, D. J. Godfrey, S. K. Das, B. H. Curran, and A. B. Wolbarst, Eds. Madison, Wisconsin: Medical Physics Publishing, 2016.
Viswanathan, A., S. Beriwal, J. De Los Santos, D. J. Demanes, D. Gaffney, J. Hansen, E. Jones, C. Kirisits, B. Thomadsen, and B. Erickson. (2012). "American Brachytherapy Society consensus guidelines for locally advanced carcinoma of the cervis. Part II: High-dose-rate brachytherapy." *Brachytherapy* 11:47–52.
Weitmann, H. D., T. H. Knocke, C. Waldhäusl, and R. Pötter. (2006). "Ultrasound-guided interstitial brachytherapy in the treatment of advanced vaginal recurrences from cervical and endometrial carcinoma." *Strahlenther. Onkol.* 182(2):86–95.
Williamson, J., P. Dunscombe, M. Sharpe, B. R. Thomadsen, J. A. Purdy, and J. A. Deye. (2008). "Quality Assurance Needs for Modern Image-Based Radiotherapy: Recommendations From 2007 Interorganizational Symposium on 'Quality Assurance of Radiation Therapy: Challenges of Advanced Technology.'" *Int. J. Radiat. Oncol. Biol. Phys.* 71(1, Supplement):S2–S12.
Wybranski, C., B. Eberhardt, K. Fischbach, F. Fischbach, M. Walke, P. Hass, F. W. Röhl, O. Kosiek, M. Kaiser, M. Pech, L. Lüdemann, and J. Ricke. (2015). "Accuracy of applicator tip reconstruction in MRI-guided interstitial ^{192}Ir-high-dose-rate brachytherapy of liver tumors." *Radiother. Oncol.* 115(1):72–77.
Yang, G., L. E. Reinstein, S. Pai, Z. Xu, and D. L. Carroll. (1998). "A new genetic algorithm technique for in optimization of permanent 125I prostate implants." *Med. Phys.* 25:2308–15.
Yoo, S., M. E. Kowalok, B. R. Thomadsen, and D. L. Henderson. (2003). "Treatment planning for prostate brachytherapy using region of interest adjoint functions and a greedy heuristic." *Phys. Med. Biol.* 48(24):4077–90.
Yoo, S., M. E. Kowalok, B. R. Thomadsen, and D. L. Henderson. (2007). "A greedy heuristic using adjoint functions for the optimization of seed and needle configurations in prostate seed implant." *Phys. Med. Biol.* 52(3):815–28.
Yu, Y. and M. C. Schell. (1996). "A genetic algorithm for the optimization of prostate implants." *Med. Phys.* 23:2085–91.
Zaider, M. (1998a). "Sequel to the discussion concerning the mechanistic basis of the linear quadratic formalism." *Med. Phys.* 25(10):2074–75.
Zaider, M. (1998b). "There is no mechanistic basis for the use of the linear-quadratic expression in cellular survival analysis." *Med. Phys.* 25(5):791–92.
Zaider, M. and G. N. Minerbo. (2000). "Tumour control probability: a formulation applicable to any temporal protocol of dose delivery." *Phys. Med. Biol.* 45(2):279–93.
Zaider, M. and L. Hanin. (2007). "Biologically-equivalent dose and long-term survival time in radiation treatments." *Phys. Med. Biol.* 52(20):6355–62.
Zannis, V. J., L. C. Walker, B. Barclay-White, and C. A. Quiet. (2003). "Postoperative ultrasound-guided percutaneous placement of a new breast brachytherapy balloon catheter." *Am. J. Surg.* 186(4):383–85.
Zoberi, J. E., J. Garcia-Ramirez, Y. Hu, B. Sun, C. G. Bertelsman, P. Dyk, J. K. Schwarz, and P. W. Grigsby. (2016). "Clinical implementation of multisequence MRI-based adaptive intracavitary brachytherapy for cervix cancer." *J. Appl. Clin. Med. Phys.* 17(1):121–31.
Zwicker, R. D. and R. Schmidt-Ulrich. (1995). "Dose uniformity in a planar interstitial implant system." *Int. J. Radiat. Oncol. Biol. Phys.* 31(1):149–55.
Zwicker, R. D., D. W. Arthur, B. D. Kavanagh, R. Mohan, and R. Schmidt-Ulrich. (1999). "Optimization of planar high-dose-rate implants." *Int. J. Radiat. Oncol. Biol. Phys.* 44(5):1171–77.

Example Problems

(Answers are found at the end of the book.)

1. For 2D B-mode ultrasound imaging with a transrectal probe, which of the following best describes how image resolution varies with increasing distance from the probe in the axial imaging plane?

   a. remains constant

   b. becomes better

   c. becomes worse

   d. varies in a complex manner

2. For whole gland permanent prostate implant therapy, a variably sized treatment margin is typically added to the contoured prostate to create the planning target volume. For which of the following uncertainties is this treatment margin intended to account?

   a. prostate contouring variability

   b. extraprostatic extension of disease

   c. seed placement

   d. All of the above are true.

3. Typical North American PTV margins for pre-planning are:

   a. 5 mm uniform margin

   b. 5 mm cranial, caudal, and anterior; 3–5 mm lateral; 0 mm posterior

   c. 5 mm cranial, 10 mm caudal, 5 mm anterior, 5 mm posterior

   d. 3–5 mm cranial, caudal; 5 mm anterior and lateral; 0 mm posterior

4. Which of the following statements are true.

   a. The prescription dose for ^{125}I monotherapy is 145 Gy.

   b. The prescription dose for ^{103}Pd boost therapy is 90-100 Gy.

   c. The prescription for ^{103}Pd monotherapy is 125 Gy.

   d. All of the above are true.

5. Which of the following is not a function of the selectSeed diode array?

   a. Validate the planned seed/spacer train configuration.

   b. Independently measure (NIST-traceably) individual source strengths.

   c. Position the most distally placed seed within 1 mm.

   d. Confirm implanted spacers are not ^{125}I seeds.

6. Comparing loose seeds versus stranded seeds for LDR permanent seed prostate implants, which of the following statements is true.

   a. Loose-seed implants provide better coverage of the prostate base and apex.

   b. Stranded-seed implants provide better urethral dose sparing.

   c. Stranded-seed implants result in better biochemical survival.

   d. The rate of seed loss is independent of brachytherapist experience.

7. What imaging modalities can be used for LDR source localization?
   a. positron emission tomography, PET
   b. anterior-posterior (AP) fluoroscopy
   c. computed tomography, CT
   d. T2-weighted magnetic resonance imaging, MRI
8. A limited number of clinical studies have compared population-based treatment outcomes achieved using conventional pre-planning vs. interactive planning in implant programs demonstrating high quality. What was the finding?
   a. Outcomes are nearly identical.
   b. Conventional pre-planning is better.
   c. Interactive planning is better.
   d. Results are indeterminate.
9. In TRUS-based implementations of dynamic dose calculation, which imaging technique has been successfully used clinically for seed localization?
   a. fluoroscopy
   b. cone-beam CT
   c. electromagnetic seed drop detection
   d. choices (a) and (b)
   e. choices (b) and (c)
10. The total prescription dose for a permanent prostate implant using a radionuclide with a shorter decay half-life like ^{131}Cs should be _____ compared to the prescription dose for a radionuclide with a longer decay half-life like ^{125}I.
    a. about the same
    b. slightly larger (by ~5%)
    c. much larger (by ~20%)
    d. slightly smaller (by ~5%)
    e. much smaller (by ~20%)
11. From the physical perspective of minimizing dosimetry errors, what is the recommended time after an ^{125}I implant to perform post-implant dosimetry?
    a. 0 days
    b. 16 ±4 days
    c. 1 month ±1 week
    d. anytime
12. What is the ideal imaging technique for post-implant dosimetry?
    a. CT alone
    b. MRI alone
    c. MRI co-registered with CT
    d. CT co-registered with pre-implant ultrasound

Chapter 5

High-dose-rate Brachytherapy for Prostate

Deidre Batchelar[1], Zhe (Jay) Chen[2], Bruce Libby[3],
Firas Mourtada[4], Susan L. Richardson[5], and Frank-André Siebert[6]

[1]British Columbia Cancer Agency
Kelowna, British Columbia Canada

[2]Yale University School of Medicine
New Haven, Connecticut

[3] University of Virginia
Charlottesville, Virginia

[4] Christiana Care Hospital
Newark, Delaware

[5]Swedish Medical Center
Seattle, Washington

[6]Universitätsklinikum Schleswig-Holstein
Kiel, Germany

5.1 Introduction

High-dose-rate (HDR) brachytherapy for prostate was introduced in 1976 (Martinez et al. 1985) as a boost to external-beam radiotherapy (EBRT) for locally advanced or recurrent prostate cancer. In the intervening 40 years, HDR prostate brachytherapy has proven to be a valuable and versatile part of the armamentarium for controlling prostate cancer. HDR brachytherapy as a boost is an established standard of care (Yamada et al. 2012; Hoskin et al. 2013), with level 1 evidence for its efficacy demonstrated via a randomized trial (Hoskin et al. 2012) as well as being supported by a large volume of clinical evidence (Morton and Hoskin 2013). There is also mounting evidence for using HDR brachytherapy as monotherapy, largely for low- and intermediate-risk patients, but in some cases for high-risk patients (Demanes and Ghilzean 2014; Yoshioka et al. 2013). More recently, HDR brachytherapy has been explored as a salvage therapy or to provide a focal boost to dominant intraprostatic lesions (Lee et al. 2007; Yamada et al. 2014; Chen et al. 2013; Crook et al. 2014; Banerjee et al. 2015).

5.1.1 Rationale for HDR BT for Prostate

There are manifold reasons that HDR brachytherapy is an attractive treatment option for prostate cancer, whether such cancers are low-, intermediate-, or high-risk disease. It permits a high biologically effective dose to the prostate (Yoshioka et al. 2014), with the greatest possible conformity, while exploiting the radiobiological advantage provided by prostate cancer cells having a lower α/β than the normal tissues that surround them (Morton 2014). By determining the treatment dosimetry prospectively, based on actual needle positions as placed in the prostate, HDR brachytherapy reduces operator dependence relative to low-dose-rate (LDR) brachytherapy and obviates concerns regarding inter- and intra-fraction motion that are associated with EBRT. HDR brachytherapy is a robust and successful treatment for prostate cancer as evidenced by the abundant clinical data generated over the past three decades supporting its use (Hoskin et al. 2013; Yoshioka et al. 2014; Morton 2014).

Conventional radiotherapy delivers doses of ≤70 Gy to the target volume. In the case of prostate cancer, several randomized controlled trials have demonstrated that doses in excess of 78 Gy are required to provide durable local control (Kuban et al. 2008; Zietman et al. 2010; Beckendorf et al. 2011; Al-Mamgani et al. 2011). For prostate cancer, local control is directly related to improved outcomes, both in terms of biochemical disease-free survival and the risk of distant metastasis (Zelefsky et al. 2012). Dose escalation has, therefore, become the standard of care for prostate cancer. EBRT doses of 76–80 Gy delivered with intensity modulated techniques in standard 2 Gy fractions have become routine (Pollack et al. 2013). Further dose escalation with EBRT is limited by the tolerance of surrounding normal organs (Pollack et al. 2013; Morton 2014). HDR brachytherapy provides a means of delivering significant dose escalation to the prostate and, as required, seminal vesicles, with a margin while maintaining low toxicity and providing a relatively easy means of boosting sub-volumes of the prostate. Yoshioka et al. found that, for monotherapy, HDR brachytherapy delivers EQD2 doses (i.e., doses biologically equivalent to 2 Gy fractions) ranging from 89 to 128 Gy with a median of 110 Gy (Yoshioka et al. 2014), while HDR brachytherapy boost fractionation schemes deliver 110 to 120 Gy in combination with EBRT.

Like permanent LDR brachytherapy, HDR brachytherapy delivers a higher dose more conformally than any EBRT technique (Skowronek 2013). Both LDR and HDR brachytherapy have been shown to provide a significantly better normal tissue sparing, including for the rectal and bladder walls, than any advanced external beam technique, including volumetric arc therapy, proton therapy, or carbon-ion therapy; HDR brachytherapy, however, provided the lowest overall normal tissue doses (Georg et al. 2014). In particular, HDR brachytherapy was shown to provide much better urethral sparing than LDR brachytherapy, with a mean EQD2 to the urethra of 62.9 ±13.3 Gy for HDR brachytherapy and 106.6 ±12.9 Gy for LDR brachytherapy. This is achieved by implanting needles within and around the prostate and using anatomy-based inverse planning to optimize the dwell positions and dwell times within the needles, with the needles in their actual treatment positions (Lessard and Pouliot 2001). With the energy of the ^{192}Ir source typically used for HDR brachytherapy and the degrees of freedom permitted by greater than 100 source-dwell positions per implant, HDR brachytherapy for prostate typically produces planned distributions that are finely sculpted around the organs at risk. Due to the high dose-per-fraction nature of this treatment, however, meticulous attention to detail is required to ensure that the dose is delivered as planned.

Beyond increasing the therapeutic ratio through good dosimetric geometry, HDR brachytherapy also exploits the radiobiological advantage provided by the low α/β ratio of prostate tissue, which is in the range of 1.2–3.1 and most likely lower than that of the surrounding normal tissue (Brenner and Hall 1999; Yoshioka et al. 2013). Prostate cancer cells are, therefore, more susceptible to large per-fraction radiation doses than are adjacent normal tissue cells, making HDR brachytherapy a very efficient means of escalating dose.

These factors combine to create a treatment that, despite considerable heterogeneity in dose and fractionation, produces consistently high rates of 5-year biochemical control, with 85% to 100% reported for low-risk disease, 83% to 98% for intermediate-risk disease, and 51% to 96% for high-risk disease (Yoshioka et al. 2014; Morton 2014; Yamada et al. 2012). These results are achieved with low rates of severe toxicity; late grade-3 rectal toxicity is rare, while reported rates of late grade-3 urinary toxicity are less than 15% (Morton 2014).

5.1.2 Review of Guidelines for HDR Brachytherapy for Prostate

Both the American Brachytherapy Society (ABS) and The Groupe Européen de Curiethérapie of the European Society for Radiotherapy & Oncology (GEC-ESTRO) have published consensus guidelines for HDR prostate brachytherapy (Yamada et al. 2012; Hoskin et al. 2013). These guidelines emphasize the need for an experienced multidisciplinary team to enable successful treatment of prostate cancer. The skill sets required of such teams (Hoskin et al. 2013; Thomadsen et al. 2014) include expertise in:

a. transrectal ultrasound (TRUS) imaging,
b. CT or MR interpretation,
c. transperineal procedures,
d. use of the treatment planning system (TPS),

e. use of the remote afterloader, and

f. patient care throughout the procedure.

The team should consist of a radiation oncologist, medical physicist, radiation therapists, nurses, as well as possibly an imaging specialist and urologist.

For physicists, the most relevant aspect of the medical guidelines are those used to assess the feasibility of prostate brachytherapy, including prostate size (generally <50 cm^3, although it is possible to successfully implant larger glands (Monroe et al. 2008; Vigneault et al. 2016), pubic arch interference, and prior radiation or surgery to the pelvis (for example, for colon cancer).

No particular dose and fractionation schedule is recommended given the heterogeneity of prescription doses reported in the literature, all of which are associated with excellent outcomes. HDR brachytherapy for prostate is typically delivered in 2 to 4 fractions of 4 to 11 Gy combined with EBRT delivering 40 to 50 Gy in 1.8 or 2 Gy fractions, although a single brachytherapy fraction of 15 Gy is becoming more common (Morton et al. 2011). Monotherapy HDR brachytherapy for prostate cancer is currently typically delivered in 2 to 6 fractions with doses per fraction ranging from 6.5 to 13.5 Gy.

The insertion of the needles is generally performed under US guidance, although some centers perform this task under MR or CT guidance. Dosimetry involves the accurate identification of prostate, with contouring of the critical structures—urethra, rectum, and bladder—also necessary to properly evaluate the treatment plan. It is good practice for the physicist to be present throughout the procedure to ensure that needles are properly placed, that all imaging and treatment equipment is functioning properly, and that the treatment proceeds according to plan.

5.1.3 Approaches to HDR Brachytherapy for Prostate

Just as there are many dose and fractionation schedules for HDR prostate brachytherapy, there are diverse approaches to implantation and planning. All techniques involve placing the patient under general or spinal anesthetic and then inserting needles transperineally, usually using a template for guidance, although a free-hand technique has been described (Kim et al. 2004). Needles are placed throughout the target volume using TRUS guidance. Use of MRI to guide placement is also possible, but currently most clinical implementations incorporate TRUS for intra-operative positioning. Once the needles are in place, images are acquired, the target and organ at risk (OAR) volumes are contoured, and the needles are reconstructed. If available, multiparametric MRI can be used to better delineate the target or identify dominant intraprostatic lesions (DIL). Anatomy-based inverse planning is then used to determine the dwell times for each dwell position. Following quality assessments of the treatment plan and the patient setup, treatment is delivered and needles are removed. At all steps, great care must be applied to ensure that any needle displacement is recognized and compensated for. All techniques produce excellent clinical results, and the choice of technique is a matter of institutional resources, logistics, experience, and personal evaluation of each technique's strengths and weaknesses.

5.1.3.1 CT-based Planning

CT-based HDR prostate brachytherapy (Martin et al. 1999) starts with transperineal TRUS-guided insertion of the needles with the patient in the dorsal lithotomy position. Following recovery from anesthetic, the patient is transferred to a CT scanner. Images are obtained and adjustments made to the needle positions to compensate for the displacement between needle tip and prostate that occur when the patient's legs are lowered (Holly et al. 2011). The physician then contours the target and OARs (bladder, rectum, urethra) before the physicist or dosimetrist identifies the needles and creates the treatment plan. The patient is then taken into the HDR brachytherapy treatment vault, needle positions are checked and corrected again, and treatment is delivered. It is not uncommon to deliver more than one fraction per implant. In these cases, the needle positions relative to both the prostate and the template should be confirmed before each fraction.

5.1.3.2 US-based Intraoperative Planning

First developed at Kiel University, US-based HDR prostate brachytherapy uses TRUS images to both guide needle insertion and plan the treatment (Bertermann and Brix 1990). Live 2D US images are used to ensure appropriate positioning of the needles and, once they are in place, a 3D US volume is acquired, in which the target and OARs are identified. Typically, only the urethra and anterior rectal wall are contoured. Needles are then reconstructed and a plan developed. Treatment is then delivered immediately, without moving the patient out of implant position. Following treatment, the needles are removed and the patient recovered from anesthetic. Significant changes in the relationship of the implant to the anatomy occur when the patient legs are lowered and the TRUS probe is removed (Seppenwoolde et al. 2008). These changes result in significant increases in urethral dose with a simultaneous decrease in PTV coverage. It is not, therefore, recommended to deliver more than one fraction per implant using a TRUS-based plan (Seppenwoolde et al. 2008).

5.1.3.3 MRI-based Intraoperative Planning

In MR-based planning, needle insertion and planning can both be performed under MR guidance (Menard et al. 2004). A rectal coil is placed, and a template is sutured to the perineum orthogonal to the coil, which is used to image the prostate. Needles are placed under MR guidance and, after all needles have been placed, a final MR image is taken to develop the treatment plan. Due to the excellent soft-tissue imaging, structures such as the penile bulb and neurovascular bundles can be contoured, along with the prostate, urethra, bladder, and rectum. If the patient is treated in the MR suite, pretreatment MR scans can be taken to ensure that the needles have not migrated during the planning process. If the patient is moved to the treatment room, imaging, such as orthogonal radiographs or CBCT, must be performed to ensure that the needles have not migrated during the patient transfer.

5.2 Equipment and Commissioning

Appropriate commissioning and staff training have been identified as key measures for the improvement of quality and safety in all forms of HDR brachytherapy (Thomadsen et al. 2014). A number of excellent guidance documents on the general practice of brachytherapy, and HDR for prostate specifically, are available to use as a basis for the development of commissioning standards (Erickson et al. 2011; Fraas et al. 1998; Kubo et al. 1998; Nath et al. 1997; Rivard et al. 2004; Thomadsen et al. 2014; Hoskin et al. 2013; Yamada et al. 2012; Perez-Calatayud et al. 2012). As with any commissioning process, the purpose is both to ensure functionality of the equipment and process while providing baseline data for ongoing quality assurance (QA). Care should always be taken to develop institutional-specific commissioning tests as warranted.

5.2.1 General

Regardless of imaging modality, all HDR prostate brachytherapy procedures involve an HDR brachytherapy afterloader, a TPS, needles, and guide tubes. A substantial majority of clinical work flows also involve a template, stepper, an US unit, and a trans-rectal probe.

5.2.1.1 Template, Stepper, and Ultrasound System

Geometric and volumetric fidelity of the US system should be checked with an appropriate phantom. Accuracy of the stepper motion should be confirmed. Registration between the template displayed on the US unit and the physical template mounted on the stepper should be assessed using a water bath or phantom as described in the AAPM TG-128 report (Pfeiffer et al. 2008).

5.2.1.2 Treatment Planning System

Specifically for prostate TPSs, care should be taken to ensure that images from US, CT, or other imaging modalities can be transferred into the TPS. Geometric and volumetric accuracy of the patient images should be validated by scanning phantoms of known dimensions (Pfeiffer et al. 2008). Phantom elements are then contoured and the dimensions and volume measured using tools in the TPS. The results are compared to actual measurements and volumes. Additionally, the geometric accuracy of the digital representations of the applicators—both needles and templates (if necessary)—and the source positions should be checked. Many planning systems come with source registry data already installed and commissioned; this data should be verified via source registry sites (e.g., the joint AAPM/IROC Houston QA Center Brachytherapy Source Registry at http://rpc.mdanderson.org/rpc/BrachySeeds/Source_Registry.htm). The correct brachytherapy dosimetry parameters can also be verified against published values. As described elsewhere, care should be taken to ensure the correct formalism is used (line vs point). Hand calculations should be performed to compute dose over an array of selected points or checked with a third party calculation system (Figure 5–1)

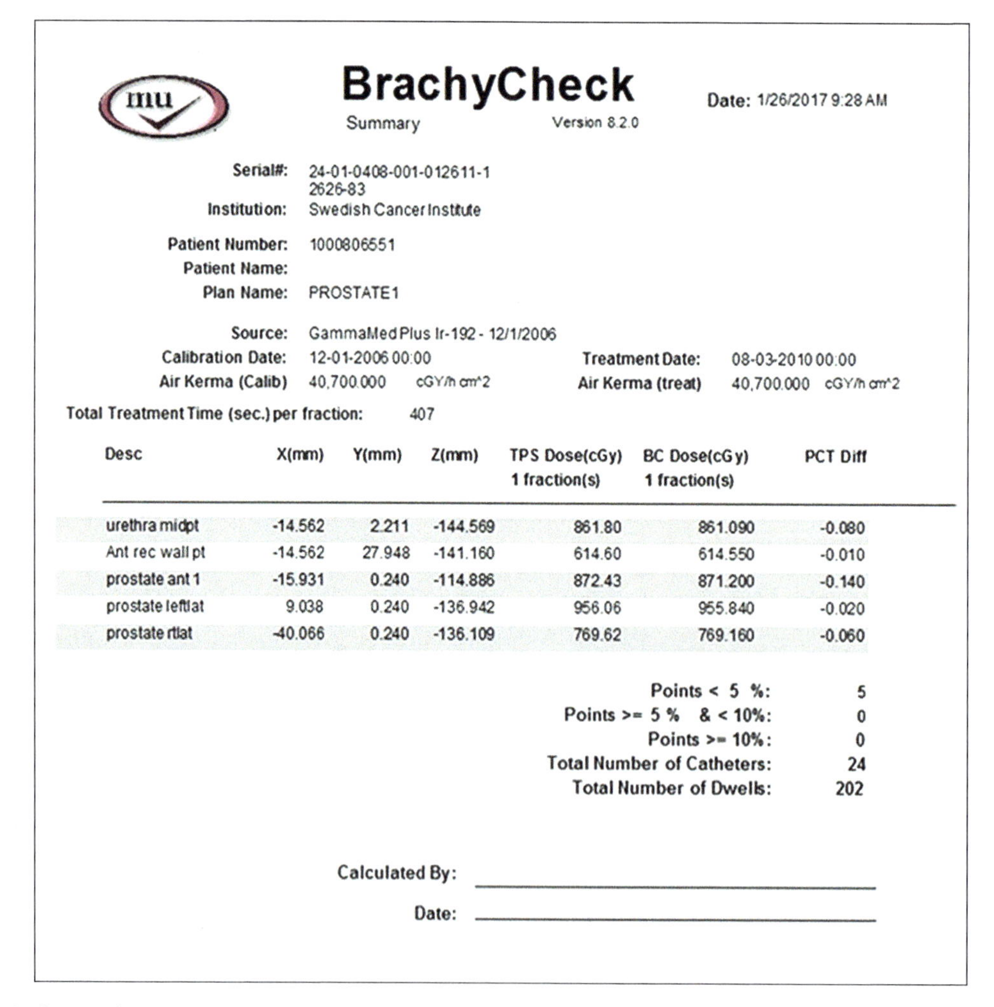

mu

BrachyCheck
Summary Version 8.2.0

Date: 1/26/2017 9:28 AM

Serial#: 24-01-0408-001-012611-1 2626-83
Institution: Swedish Cancer Institute

Patient Number: 1000806551
Patient Name:
Plan Name: PROSTATE1

Source: GammaMed Plus Ir-192 - 12/1/2006
Calibration Date: 12-01-2006 00:00 **Treatment Date:** 08-03-2010 00:00
Air Kerma (Calib) 40,700.000 cGY/h cm^2 **Air Kerma (treat)** 40,700.000 cGY/h cm^2

Total Treatment Time (sec.) per fraction: 407

Desc	X(mm)	Y(mm)	Z(mm)	TPS Dose(cGy) 1 fraction(s)	BC Dose(cGy) 1 fraction(s)	PCT Diff
urethra midpt	-14.562	2.211	-144.569	861.80	861.090	-0.080
Ant rec wall pt	-14.562	27.948	-141.160	614.60	614.550	-0.010
prostate ant 1	-15.931	0.240	-114.886	872.43	871.200	-0.140
prostate leftlat	9.038	0.240	-136.942	956.06	955.840	-0.020
prostate rtlat	-40.066	0.240	-136.109	769.62	769.160	-0.060

Points < 5 %:	5
Points >= 5 % & < 10%:	0
Points >= 10%:	0
Total Number of Catheters:	24
Total Number of Dwells:	202

Calculated By: ______________________

Date: ______________________

Figure 5–1 Secondary calculation performed for HDR prostate brachytherapy. The program independently performs a TG-43 calculation based on the DICOM coordinates of the dwell positions and times of the sources and calculation points.

Figure 5–2 Reusable needles are mounted in a phantom and a test plan is delivered to radiochomic film to verify that each needle delivers dose to the intended position along its length.

(Nathet al. 2016). The integrity of data transferred to the afterloader should be tested as well. These tests should be performed during initial implementation as well as whenever a TPS upgrade is performed (Kubo et al. 1998).

5.2.1.3 Needles and Guide Tubes

The mechanical integrity of the guide tubes should be confirmed via visual inspection to ensure they are free of cracks or breaks. The connector mechanism should also be tested for functionality, and the overall length of the guide tubes should be measured (Brown et al. 2016). Reusable needles should be visually inspected and their lengths and connections tested before being put into clinical use. Additionally, each new reusable needle should have its dosimetric length tested by running a test plan and creating an autoradiograph (Figure 5–2). For disposable needles, a set of needles should undergo these tests, and individual needles should be visually inspected before use. It is important to note that metal needles usually have a solid tip distal to the end of the source channel, resulting in some dead space at the tip of the needle. Manufacturers specify the position of the first dwell position relative to the physical tip, but this should be verified via autoradiography before use.

5.2.2 CT-based Planning

Needle identification in CT imaging is straightforward. Nevertheless, commissioning in a phantom should quantify the accuracy of the needle tip identification for each institution's CT imaging system to avoid geographical inaccuracies. Furthermore, individual institutions must carefully examine their clinical work flow to identify potential causes of needle displacement (e.g., patient transfers, changes in patient position, excessive elapsed time) and develop appropriate, yet clinically feasible, means of reducing, evaluating, and remedying needle migration. Prior to clinical implementation, devices to immobilize the patient or minimize the need for transfers should be investigated (Peddada et al. 2015). A robust technique for imaging the needle tips relative to a surrogate for the prostate base should be developed and tested, with attention paid to when the tip positions should be assessed (before imaging for planning and each treatment) and how to maintain the patient position consistently.

5.2.3 US-based Intraoperative Treatment Planning

Commissioning tasks specific to US-based intraoperative planning involve scrutiny of the reconstruction of the 3D US images to ensure the greatest degree of geometric accuracy possible. This will include validation of the encoded stepper used to determine where in physical space each individual US image was acquired. The imaging of string or bead phantoms can aid in assessing any distortions present in the images (Tong et al. 1996; Tong et al. 1998). Additionally, when using a biplane TRUS probe, the physical correlation between transversal and longitudinal imaging planes should be verified. It should also be noted that changing the frequency of the US transducer will change the scaling of the images. If the images are transferred via a video cable, the TPS scaling will need to be adapted manually.

Needle appearance in US images must be carefully examined in a phantom to fully understand the appearance of the needles before use in a patient (see Section 5.4.4). All commercial US-based TPSs incorporate some means of using a measure of the length of needle protruding from the template to increase needle tip identification accuracy (see Section 5.4.3). This process should be validated using phantoms, and users should expect to localize the tips within 1 mm (Siebert et al. 2009; Schmid et al. 2013; Batchelar et al. 2014). Accuracy of both image reconstruction and needle identification can be tested simultaneously by fusing US images with high-resolution CT scans of implanted phantoms (Figure 5–3) (Schmid et al. 2013).

5.2.4 MRI-based Planning

Use of MRI-based planning requires all equipment be MR compatible, including template, needles, and any ancillary equipment that is needed for the procedure (Murgic et al. 2016). Additional equipment needed includes a transrectal coil used to image the prostate and immobilization devices used for patient positioning on the MR couch. To properly use an MR for treatment planning, the distortions of patient anatomy must be measured. This requires the use of a specialized MR phantom to measure geometric distortion (Citrin et al. 2005).

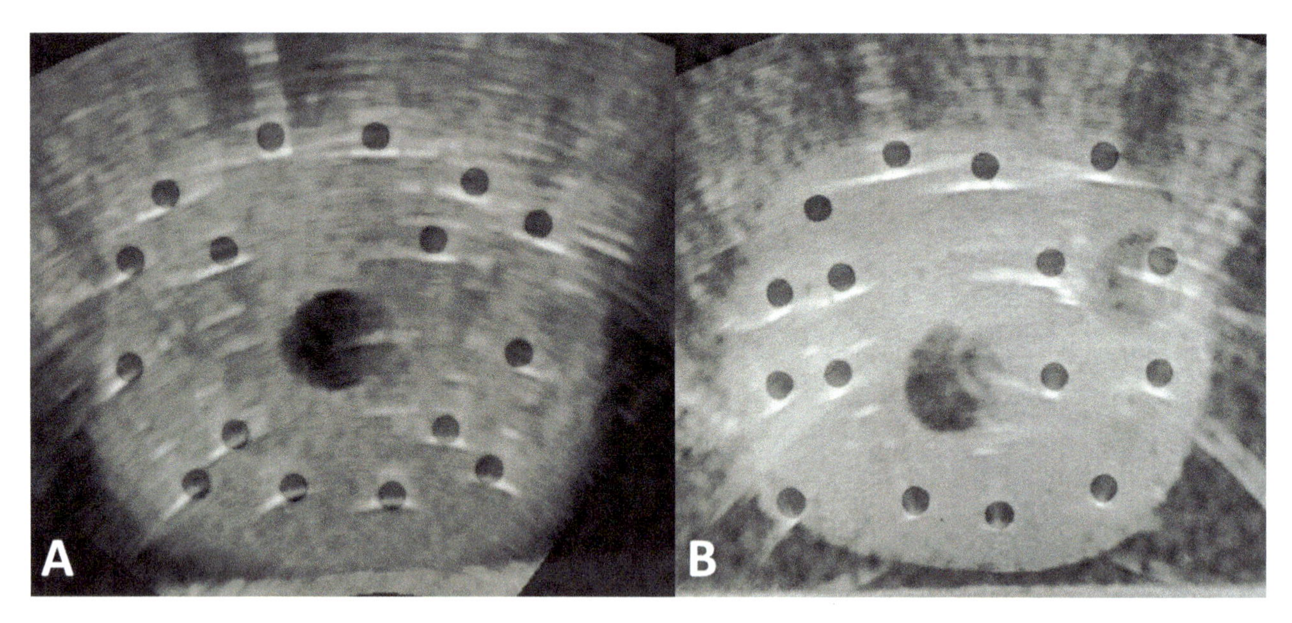

Figure 5–3 Fused CT-US images of a prostate phantom illustrating the relationship between the needle lumen on CT (dark circles) and the bright flash of the needle in US. It can be seen in both of these images that the flash from these needles is generated largely by echoes from the posterior surface of the needles. (A) Acquired using the sagittal TRUS crystals by rotating the TRUS probe, contains an US-reconstruction error, leading to the flash being located lateral to the actual needle locations for peripheral needles. (B) Acquired using the axial TRUS crystals by translating the probe and does not contain this error.

5.2.5 Technique Commissioning

Beyond commissioning of technical equipment, the implementation of any new brachytherapy technique must include the development of robust policies and procedures designed to address not only standard operating processes, but also to provide guidance in instances where the standard procedure cannot be followed (Brown et al. 2016). It has been found that the majority of errors in HDR brachytherapy arise from process-related issues (Richardson 2012). To decrease the likelihood of these occurring, it is valuable for facilities developing a new program to collaborate with an institution experienced in the delivery of prostate HDR brachytherapy to aid in formulating an effective treatment process. Each institution will need to account for local variations in infrastructure, human resources, and work flow, among other factors, as they develop and commission an effective treatment process. This should, as often as possible, be done collaboratively by a multi-disciplinary implementation team.

HDR brachytherapy for prostate involves all standard radiation therapy professions (radiation oncologists, medical physicists, radiation therapists, and medical dosimetrists) as well as surgical professionals (anesthetists and nurses) and potentially MRI technologists. The roles, responsibilities, and qualifications for each profession have been laid out for brachytherapy, in general, in an ASTRO white paper (Thomadsen et al. 2014). It is not possible to assign specific roles within HDR prostate brachytherapy procedures, as it will depend on local factors as to whether, for instance, a physicist or dosimetrist creates the treatment plans. What should be true for all institutions is that each profession's role should be clearly defined and documented.

Once a comprehensive policy and procedure document has been developed, representatives from all professional groups involved should review it and agree that the process is robust and complete from their point of view. The implementation team should develop a series of safety checks (e.g., time outs, independent checks, checklists, etc.) throughout the procedure (Huq et al. 2016). Effort should be made to have the completed procedure document peer reviewed by an institution with active experience in HDR brachytherapy for prostate (Brown et al. 2016).

Before any treatments are undertaken, an end-to-end test run with all staff involved should be performed. That is, the entire procedure should be performed with a phantom standing in for the patient. All decisions regarding equipment (including which surgical drapes are required, etc.) should be finalized before treatments commence. It should be a goal that all personnel involved in the procedure partake in the procedure test run. In practice, this can be challenging to realize. It should at least be ensured that a representative staff member from each profession be present for both the test run and the initial procedure, for example, if only one scrub nurse can attend the test run, that nurse should be present for the initial procedure.

It is encouraged that staff responsible for needle insertion and subsequent needle identification perform additional phantom implants in advance of clinical implementation. This allows the team to better understand what is required of them by other members of the team, helps establish communication channels between the disciplines, and increases familiarity with the imaging equipment and planning software. This is of particular importance for US- and MRI-based planning, where needle identification in planning images is not as straightforward as in CT-based planning (Schmid et al. 2013).

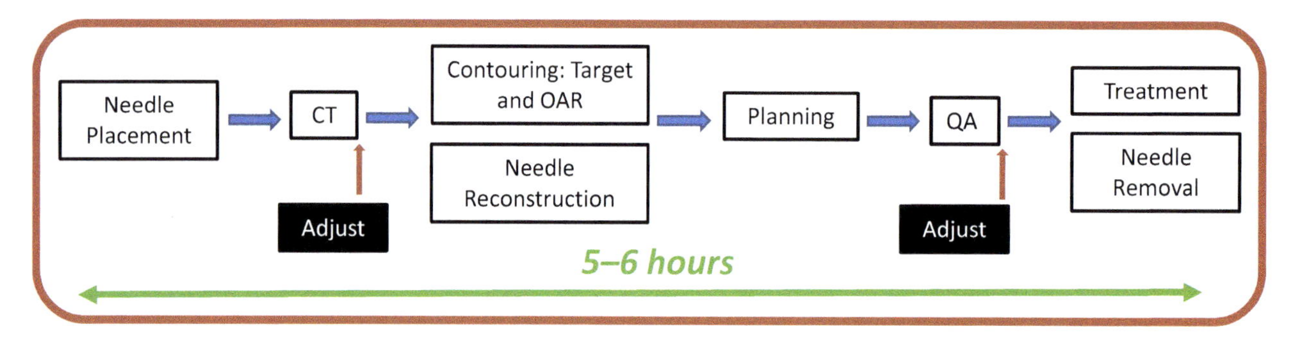

Figure 5–4 An overview of the CT-based procedure. (Courtesy G. Morton, University of Toronto.)

5.3 CT-based HDR Brachytherapy for Prostate

5.3.1 Process Overview

The work flow for CT-based prostate HDR brachytherapy is shown in Figure 5–4. The procedure consists of: 1) placing needles within the target volume under image guidance; 2) CT imaging with the needles in place; 3) contouring the relevant anatomical structures; 4) reconstructing the needles in the planning system; 5) optimizing the dwell times to achieve the dosimetric constraints; 6) performing pretreatment quality assurance, including verification and adjustment of needle positions to compensate for any shifts that occurred during planning; and 7) treatment delivery. If no subsequent fractions are to be delivered, the needles are then removed. This process typically takes 5–6 hours in total (Batchelar et al. 2016), although this time may be reduced if in-room CT is used to guide or plan the treatment.

5.3.2 Implant

Needles are typically inserted in an operating room (OR) with the patient under spinal or general anesthetic. Generating a list of equipment required in the OR (Figure 5–5) will ensure that the implant goes smoothly. This is particularly important if the OR used for implants is not in the radiation therapy department. The patient is positioned in dorsal lithotomy, and a Foley catheter is used to drain the bladder during the treatment procedure. Frequently, fiducial markers are inserted into the base and apex at the beginning of the implant. This aids both in judging the depth of needle tips and in delineating the prostate on CT images. Needle insertion is most commonly template-based and performed using TRUS image guidance, potentially augmented by fluoroscopy or C-arm CT imaging (Yamada et al. 2012; Hoskin et al. 2013; Morton 2014). Placement of the needles may be chosen individually for each patient, or a standard insertion pattern may be chosen as a starting point. In general, all needles are inserted to the base of the prostate.

When the implant is finished, the template should be sutured to the perineum and the legs lowered from lithotomy. As the change in leg position can alter the relative position of the prostate and needles, it is recommended that the depth of the tips be assessed before leaving the OR. This can be accomplished either via cystoscopy—in which an endoscope allows direct visualization of the needles tenting the bladder wall as they are advanced—or using C-arm fluoroscopy to visualize the needles relative to contrast in the bladder or the fiducial markers in the prostate base. Numbers may be assigned to the needles at this point; it is advisable to use the same numbering system for every implant (e.g., always increase numbering from bottom to top, left to right). Once the tips are satisfactorily positioned, the needles should be fixed to the template. This may be done using a template with a sliding plate (Figure 5–14), which locks all needles simultaneously, or by individually locking each needle using putty (Figure 5–6), locking collets (Figure 5–15), glue (Figure 5–35), or any other means devised by an individual institution. After fixation, the needles should be marked to make it easy to assess changes in the needle position relative to the template. These marks may be a simple line marking the point at which the needle exits the template or may be a series of marks at regular intervals as seen in Figure 5–6. Additionally, the template locations used should be mapped on a form, such as in figures 5–8 and 5–18, and photographs taken. It is also advisable to acquire radiographs recording the intended posi-

Effective date: 9/12/2016

HDR Prostate Supply Prep Checklist

We Bring to OR:

- ☐ Brachy Prostate Pack
- ☐ Prostate Cart
- ☐ Prolene Sutures x2
- ☐ Gold Seed Markers/Acculocs
  - ☐ Check for prior placement
- ☐ US manipulating arm
- ☐ US arm base
- ☐ US unit(w/prostate probe)
- ☐ Endocavity water standoff kit:
  - Tape for water channel
- ☐ Alcohol Gauze Jar
- ☐ 16 ISI Sterile Leur lock Catheter needles
- ☐ 4 pillows
- ☐ Dental putty in separate containers:
  - 1 scoop putty base
  - 1 scoop putty activator
- ☐ Paint pen
- ☐ Sterile marker & ruler
- ☐ Red catheter numbers
- ☐ Aquaplast grids (3 x 4.5cm aprox.)
- ☐ Striker stretcher ordered for OR by 4pm the day before.
- ☐ Patient's chart
  - Copy of brachy consent
  - Copy of procedure consent
  - HDR Prostate yellow Rx sheet
  - Radiation Oncology Admission History
  - Pathology
  - Volume Study Notes
  - Prior Acculocs?
- ☐ Sterilized friction cuffs.
- ☐ Copy of Updated Procedure
- ☐ Camera
- ☐ Catheter cutter/extra blades
- ☐ Stepper Drape
- ☐ Syringe and saline
- ☐ Zip sealed bag for patient supplies
- ☐ Pink styrofoam
- ☐ Hovermat w/ pump
- ☐ Betadine paint brush
- ☐ DVI cable

OR will Supply:

- ☐ Sterile gloves for Dr.
- ☐ Sterile instrument pack
  - Scissors
  - Needle driver
  - Pickups
  - Towel clamps
- ☐ Sterile H2O/ Saline for probe cover
- ☐ OR table
- ☐ OR Fluoro (If needed)
- ☐ Foley Catheter
- ☐ Contrast
- ☐ Cystoscope w/ 180 degree rotation

Urologist: ____________________

Number: ____________________

Date of Marker Placement: ____________________

Prior external radiation: Y or N

Therapist #1 Initials: ____________________ Date: ____________________

Therapist #2 Initials: ____________________ Date: ____________________

Figure 5–5 Generating an equipment list for the OR is useful. Lists for US- and MRI-based planning would be similar. (Courtesy of J Zoberi, Washington University.)

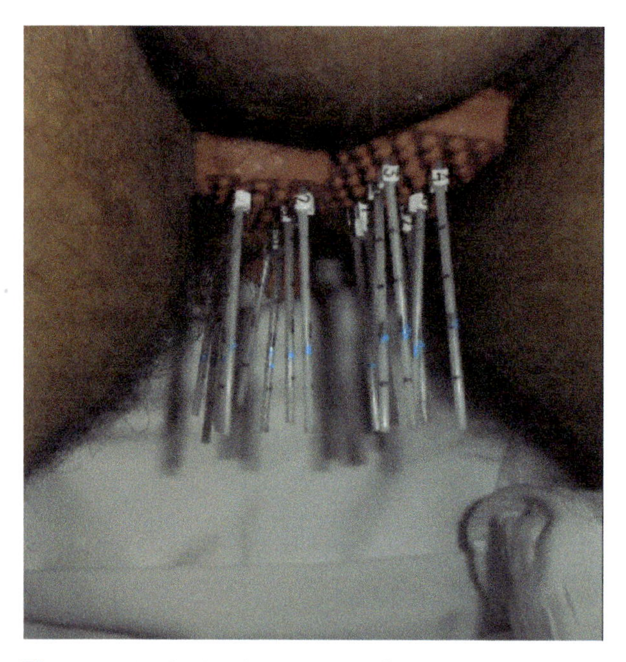

Figure 5–6 An implant using a Syed-type template illustrating the needle fixation and marking.

tion of the needle tips at the end of implant (Figure 5–7). The patient is then transferred to a stretcher for recovery from anesthetic.

5.3.3 CT Localization

Following recovery, the patient is transferred to obtain CT scans for treatment planning, and again the marks are checked to verify that the needles have not moved within the template. As the needle tips will shift caudally within the prostate even if they have not shifted relative to the template (Holly at al. 2011), it is imperative that the positions of the tips be adjusted before the planning CT scan is obtained. Needle displacement can be assessed relative to any fiducial markers placed at the prostate base (Figure 5–7), but most commonly the tip positions are judged relative to the bladder based on radiopaque contrast infused into the bladder. To avoid CT artifact and masking of anatomy and needles, the contrast-to-saline ratio (mixed by volume) should not exceed 10%. It is important to train the CT staff and

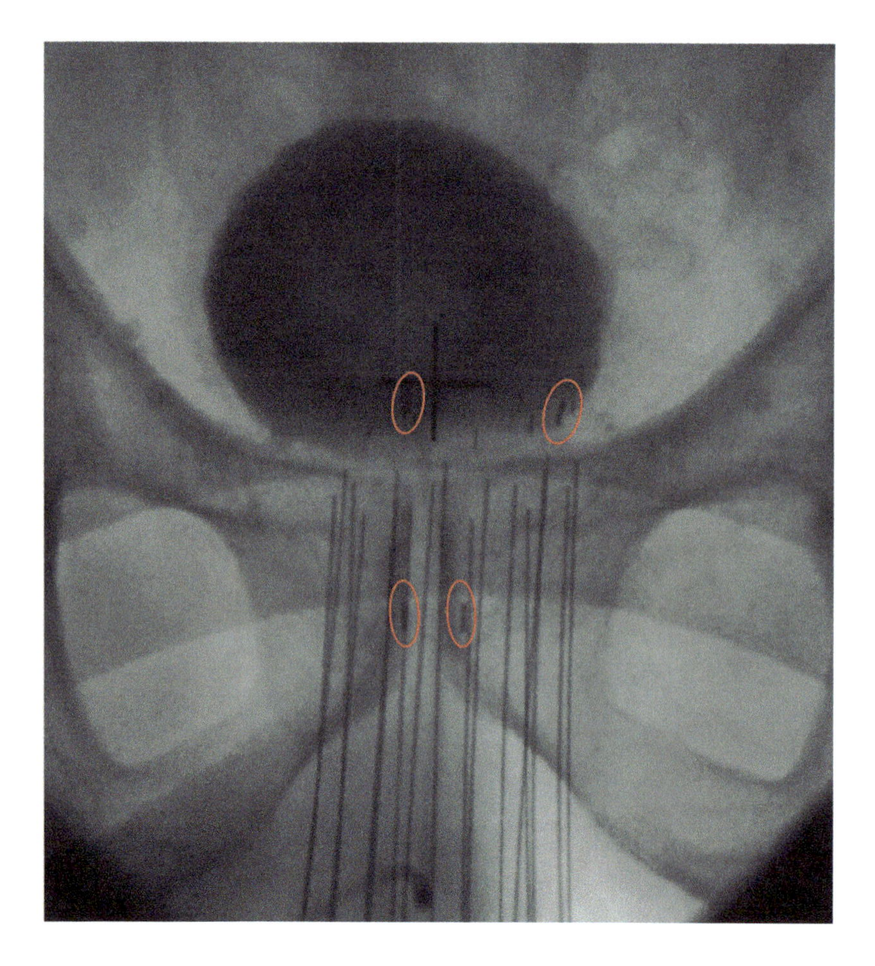

Figure 5–7 Maintaining the intended position of the needle tips with respect to the prostate base is challenging in any HDR brachytherapy work flow where the patient is moved or there is a significant period between insertion, planning, and treatment. An AP radiograph obtained at the end of the implant records the intended position of the needle tips relative to both contrast in the bladder and fiducial markers in the base and apex (circled in red).

document this protocol. Antero-posterior (AP) and lateral scout views of the pelvis are first obtained to verify needle location. If flexible plastic needles are used, metal stylets or CT markers should be used for the scout images to judge the tip positions. Any needle displacement observed should be adjusted and the scouts repeated until the needle positions are satisfactory. If necessary, needles should be re-affixed to the template, re-marked, and new photographs taken.

If used, metal stylets are then removed from the plastic needles and CT images are obtained from a level superior to the base to several centimeters inferior to the apex. The CT protocol should use slices of no more than 3 mm overlapping intervals (Hoskin et al. 2013). Smaller slice thicknesses will, however, reduce uncertainty when localizing the crainial-caudal position of the needle tips during CT-based HDR brachytherapy planning. In general, whatever imaging modality is used, the slice thickness should be minimized as much as is practical without degrading image quality (Kirisits et al. 2014).

The use of a CT simulation checklist, as shown in Figure 5–8, will help control the implant quality.

BJH/WUSM/SCC HDR Prostate Medical Physics Consultation

Patient: Patient ID:
Radiation Oncologist: Rx : 15 Gy or 19 Gy ? Implant Date:

Typical Needle Configuration

RT		LT	
1	2	3	4
5	6	7	8
9	10	11	12
13	14	15	16

**RT/LT = Patient's right and left

Actual Needle Configuration at Time of CT-sim

RT LT

Yes No

☐ ☐ **Record needle positions on diagram above?**
☐ ☐ **O.R. needle marks in blue aligned w/r.t. corresponding template?** If no, needles may be loose.
☐ ☐ **Needles tight within putty?** If no, may have to glue needles ***after*** MD has verified needle placement.
☐ ☐ **Exiting needle length at least 5 cm from template (min 3 cm)?** If no, alert MD that these needles cannot be pushed in further, or may not be usable for treatment. Maybe needle(s) can be pulled out.

☐ ☐ **Acquire CT using HDR Prostate Protocol with imaging extents set for 2mm and 1 mm recon?**
☐ ☐ **CT: is template flush against skin?** If no, have MD build up putty thickness adjacent to skin.
☐ ☐ **CT: is template thickness ≥ 3 cm (min 2cm)?** If no, have MD build up putty thickness.
☐ ☐ **CT: do needles reach prostate base?** If no, have MD adjust needle placement.

☐ ☐ **Adjustment performed by MD to push needles towards prostate base?** Re-check exit needle length
☐ ☐ **If adjusted, are needles still tight within putty?**
☐ ☐ **Needles glued into place by Physics (if necessary)?**
☐ ☐ **Needles re-marked by Physics in red (if necessary)?**
☐ ☐ **Are Aluminum markers required in certain needles due to air pockets on CT?** May need re-scan.
☐ ☐ **Photos of implant and marks acquired?** Please print photos for verification at time of treatment
☐ ☐ **Correct CT exported for planning?** CT study no./no. of images ________ / ______

Description of Medical Physics Consult: **(1) CT Sim**: Assists the MD by verifying integrity of implant and adequacy of images for planning; recommends adjustments to the implant if needed (see above checklist). **(2) Performs catheter length measurements** (see measurement sheets). **(3) Treatment Planning:** Assists MD and Dosimetry in plan generation and optimization; performs plan QA checks. **(4) Performs an independent calculation check of planned treatment time** (see Paterson-Parker Calc). **(7) Pre-treatment QA:** Assists RTT in treatment connections and treatment setup.
Notes:________________________________

Medical Physicist: ______________________ Date: ______________

Radiation Oncologist: ______________________ Date: ______________

jz created 10.26.15, revised 02.17.17

Figure 5–8 An example of a CT-Sim checklist used for prostate HDR brachytherapy physics consults. (Courtesy of J Zoberi, Washington University.)

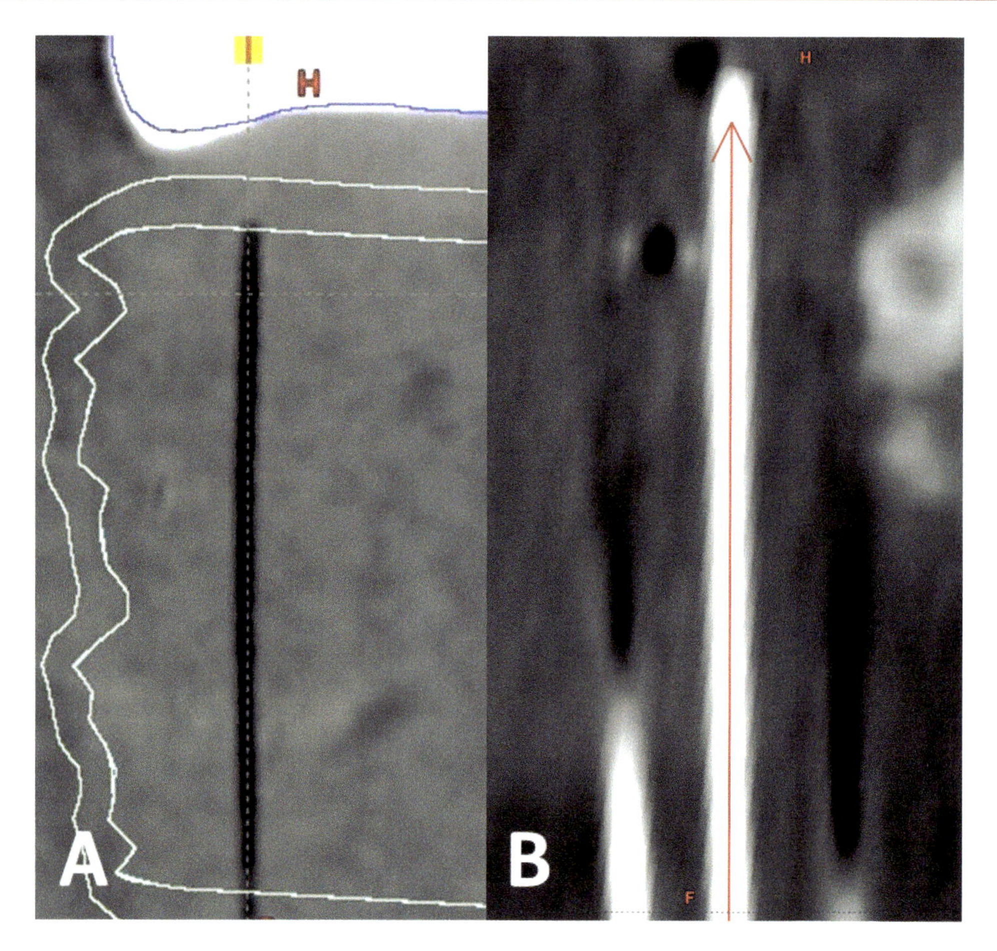

Figure 5–9 The appearance of plastic A) and steel B) needles in CT. In image B, the applicator reconstructed in the planning system is displayed showing that the tip of the applicator is offset from the tip of the needle as visualized in the images. This is to account for the solid, beveled needle tip, which is not part of the source channel.

5.3.4 Needle Reconstruction

In general, needle reconstruction based on CT images is very straightforward. If plastic needles are used, they will appear as air-filled holes within the gland (Figure 5–9). Steel needles may also be used; little obscuring artifact is generated by these thin-walled needles. In this case, the physical tip of the needle is easily located, but the dead space at the tip of the needle must be appropriately accounted for when reconstructing the needles (Figure 5–9). Digitally reconstructed radiographs from the planning system can be compared to CT scout images and radiographs taken in the OR to validate the reconstructed needles. If the needles were numbered at the time of implant, the reconstructed applicators in the planning system should be numbered identically. The map of the template locations used should be the guide for placing and numbering the applicators to match the implanted needle pattern and numbering. It is also possible to number the needles as they are reconstructed in the planning system. In this case, it should be ensured that the physical needles are assigned the same numbers.

5.3.5 Pretreatment Needle-positioning QA

Beyond the pretreatment QA discussed in Section 5.7, needle displacement is the most important element to scrutinize prior to treatment in CT-based prostate HDR brachytherapy. It is important to understand that there

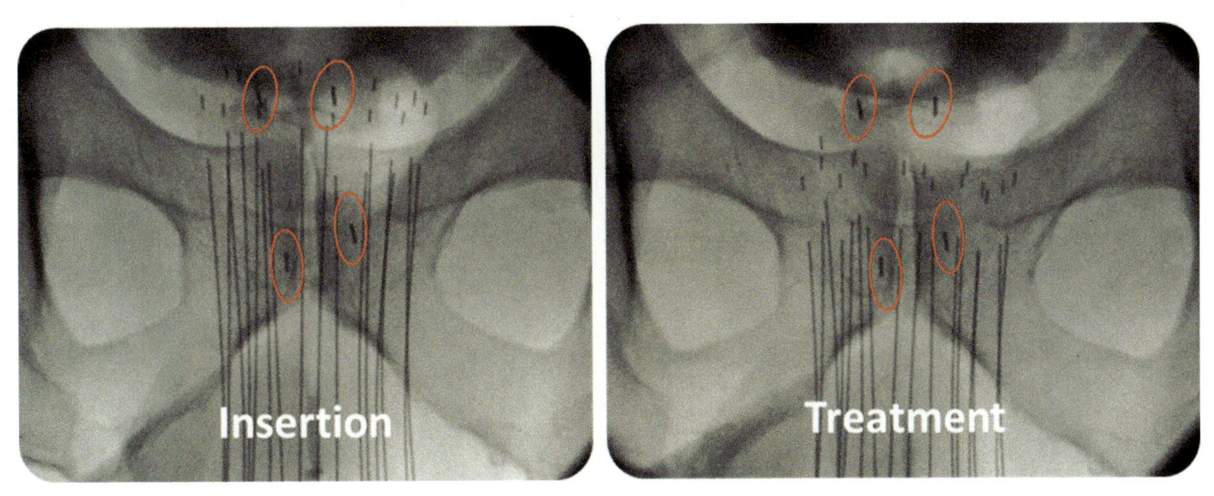

Figure 5–10 An illustration of the caudal shift of the needles relative to the prostate base that occurs between insertion and first fraction of treatment. Both images are AP radiographs of the same patient. Visible are the CT markers (dark vertical lines with a 1-cm, radio-transparent gap before a marker indicating the first dwell position) placed in plastic needles and four fiducial markers (circled in red), two at the base and two at the apex of the prostate. It is easily seen that significant caudal shift occurs, which must be corrected before delivery if the delivered dose is to match the planned dose.

are two components to this displacement (Hoskin et al. 2003). The first is that there may be movement of the needles within the template. This is easily detected by looking at the marks made on the needles themselves or by checking the length of needle protruding from the template. More challenging to assess and correct is ensuring that the needle tips are correctly placed relative to the prostate base. It is well known that such assessment should occur for subsequent fractions delivered with a single implant (Kim et al. 2007). It is, however, necessary even for the first, or only, fraction delivered with an implant. It has been demonstrated (Holly et al. 2011) that, even for a single-fraction delivery scheme, there is a mean caudal shift of the needle tips away from the prostate base of 1.1 ±0.8 cm (Figure 5–10), without any increase in the protruding length of the needle beyond the template. Left uncorrected, this shift can reduce the prostate V_{100} by 20% and the D_{90} by 37%. Fluoroscopy or cone-beam CT can aid in repositioning the needle tips relative to fiducial markers or with respect to contrast in the bladder. The use of specially designed patient positioning devices and transferring the patient with these devices can reduce the magnitude of observed needle shift (Peddada et al. 2015).

5.3.6 Tips and Tricks

- Many centers purposefully implant a needle between the urethra and the rectum, in the central row of the template, to provide more control over the dose in this region between OARs.
- All HDR brachytherapy units can be operated inside the CT room if the shielding requirements have been met. Thus, the patient can be treated after planning without being moved if the HDR afterloader shares a room with the CT scanner there. This is referred to as an in-room CT work flow, which saves time and reduces the chance of needle displacement. It is also possible to have both the CT scanner and the HDR afterloader in the brachytherapy procedure room. In either case, it is still necessary to assess the relative positions of the base and the needle tips after the legs have been lowered for imaging.
- Patient movement should be minimized while the needles are implanted. Procedures should be developed to make patient transfers between the OR bed, CT couch, and stretcher easy. This will reduce the possibility of needle displacement, and the number of such transfers should be reduced as much as possi-

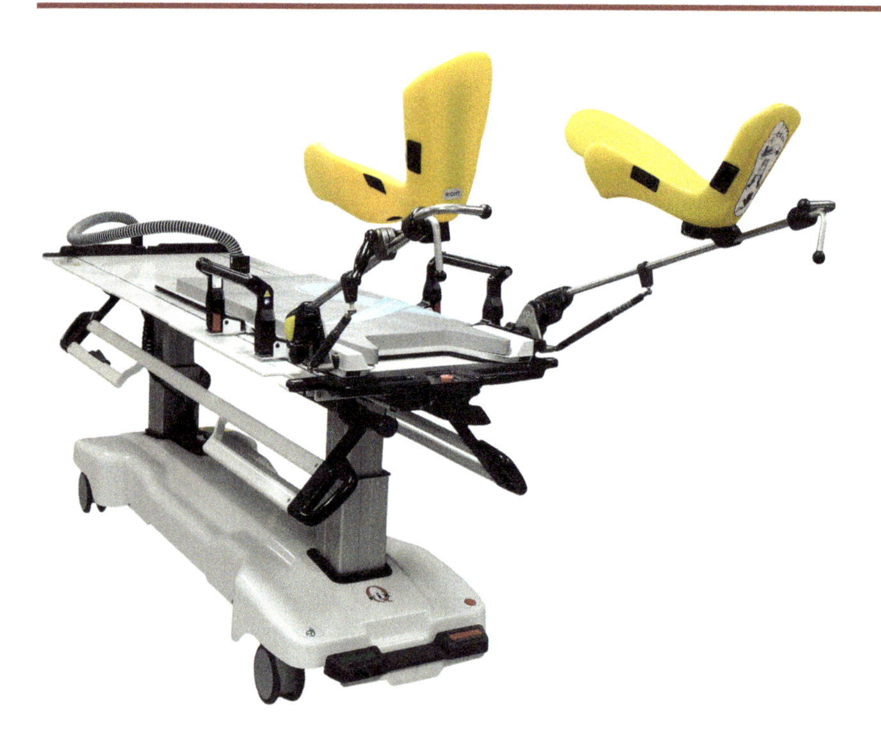

Figure 5–11 A patient trolley with MR/CT compatible transfer top that would permit the patient to be implanted, imaged, and treated without being transferred off the transfer top. Prototype is a work in progress at the time of this writing from Qfix (Avondale, PA).

ble. Commercial devices to reduce or eliminate the need to transfer patients, regardless of department work flow, have been developed (Figure 5–11). All procedures and equipment should be thoroughly validated to assess their impact on needle position.

5.4 US-based Intraoperative HDR Brachytherapy for Prostate

5.4.1 Process Overview

US-based intraoperative HDR brachytherapy is a one-step process that uses TRUS imaging to both guide the implantation of the needles and for treatment planning. Needles are implanted using real-time 2D US imaging, and then a 3D US image is acquired for needle reconstruction and contouring of the relevant anatomy. Dwell positions and dwell times are then determined using this image. Once the plan is complete, the patient is treated and the needles are removed. This is all accomplished with the patient under anesthetic and in the same position as for implant, providing no opportunities for the needles to shift. The work flow for US-based intraoperative prostate HDR brachytherapy, shown in Figure 5–12, takes between 1.5–2.0 hours to complete.

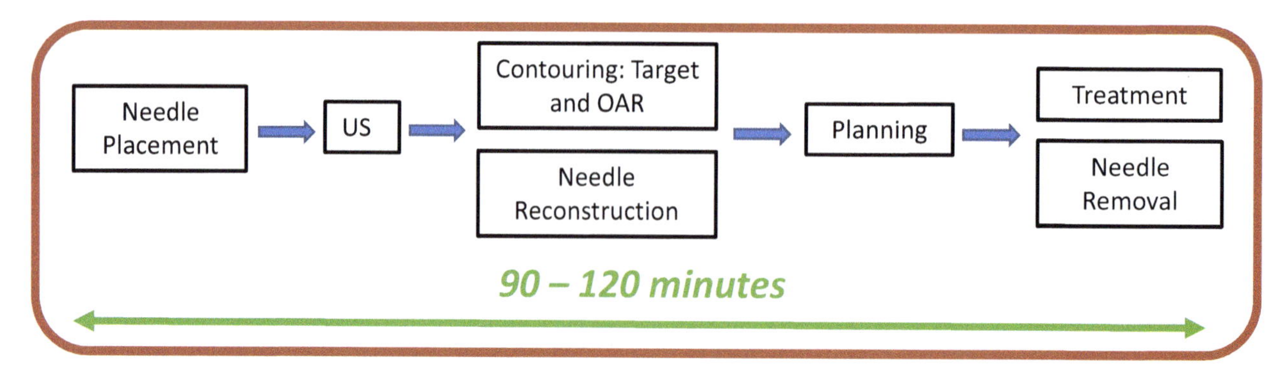

Figure 5–12 An overview of the US-based procedure. (Courtesy G. Morton, University of Toronto.)

5.4.2 Implant

As with CT-based planning, patients undergoing HDR brachytherapy using intraoperative TRUS-based planning have needles placed in an OR while under anesthetic and in the dorsal lithotomy position. It is preferable, when medically possible, to perform the procedure using general anesthetic as this provides a greater degree of patient immobilization than spinal anesthesia. A Foley catheter is inserted to drain the bladder and to aid in visualizing the urethra. This may be further enhanced by filling the catheter with aerated gel. The whole procedure is conducted using a bi-plane TRUS transducer mounted to a stepper equipped with an encoder for tracking the translational and rotational position of the transducer crystals. Three-dimensional US images are acquired either sagitally, by rotating the probe, or longitudinally, by translating it. Images are acquired at narrow intervals, and a 3D image set is reconstructed based on the encoded position the probe for each image (Tong et al. 1996). Needles are inserted using a template that is mounted to the stepper.

Commercial systems available for US-based planning incorporate acquisition of an initial image volume before needles are inserted. The intention is that the prostate and OAR would be contoured on this image set. These contours could either be used for inverse planning of needle positions or to aid in delineating the target following needle implantation. In practice, the current iteration of commercial optimizers for determining needle placement do not tend produce intuitive or robust needle arrangements. Since one of the strengths of using HDR brachytherapy is that you can prospectively determine the dose based on implanted needle positions, it currently makes more sense to manually develop a needle arrangement. This may be based on individual anatomy, or institutions may develop a standard pattern to use as a starting point (Batchelar et al. 2016). In the absence of optimization, a robust standard pattern (Figure 5–13) streamlines the implantation

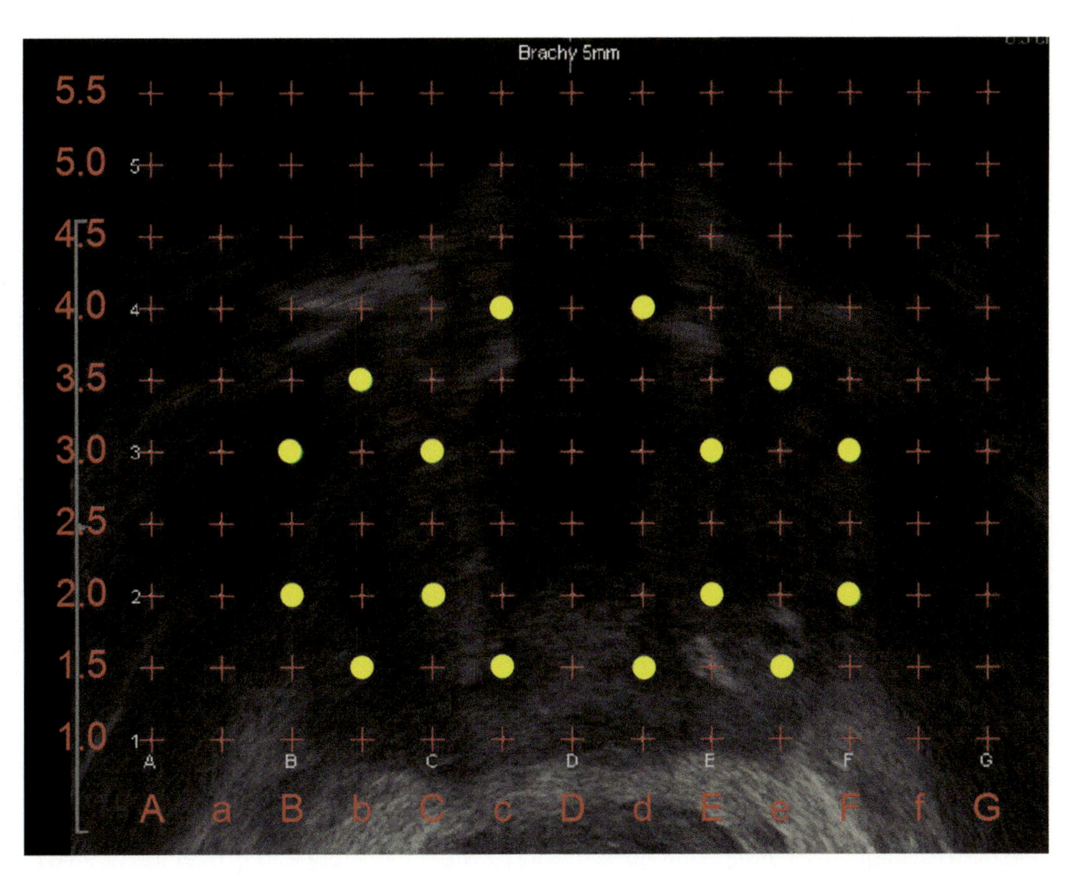

Figure 5–13 The standard pattern used at the BC Cancer Agency, derived from G Morton's pattern at the Odette Cancer Centre (Morton 2015). This pattern can be used as a starting point for any planning modality.

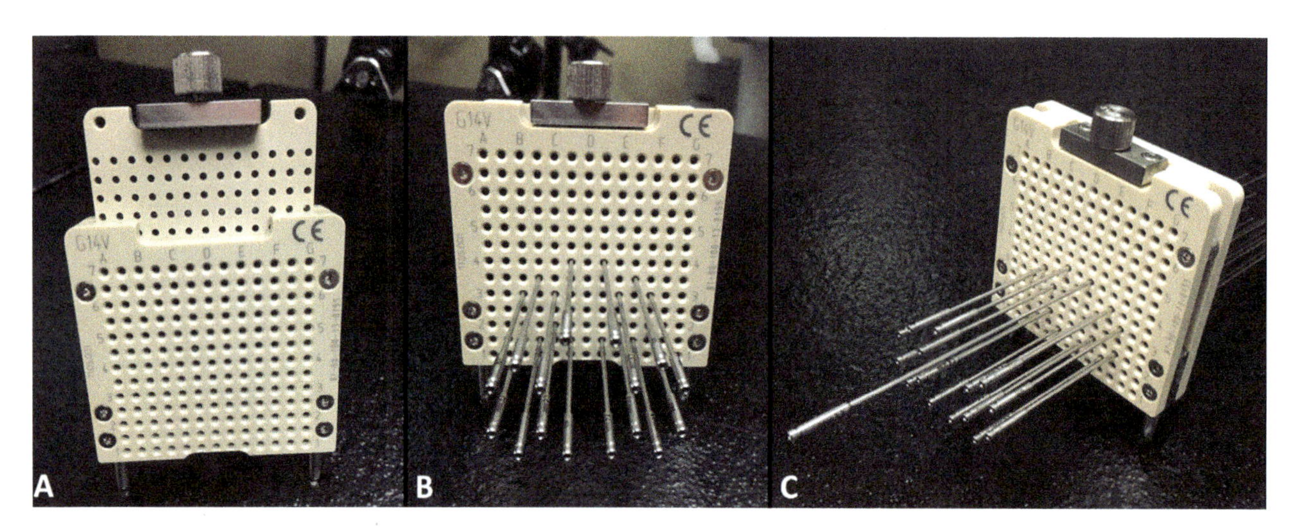

Figure 5–14 A sandwich style template. Templates of this style can be used with any imaging modality. This is a reusable template that fixes the cranial-caudal position of the needle by adjusting a sliding plate A) between the two main plates of the template. This quickly and simply locks all needles at the same time B). Care must be taken, however, because not all needles are fixed to the same degree. The needles are held in place by pressure applied by the sliding plate. The greatest pressure is applied in line with the adjustment screw in the center of the template, and it is possible to manually shift individual needles in the periphery, even when "locked" C) (Batchelar et al. 2014). Needles can be repositioned as intended, relative to the template, by marking on the needles the point at which they exit the face plate of the template and ensuring they have not shifted before acquiring a TRUS data set for planning and before treatment.

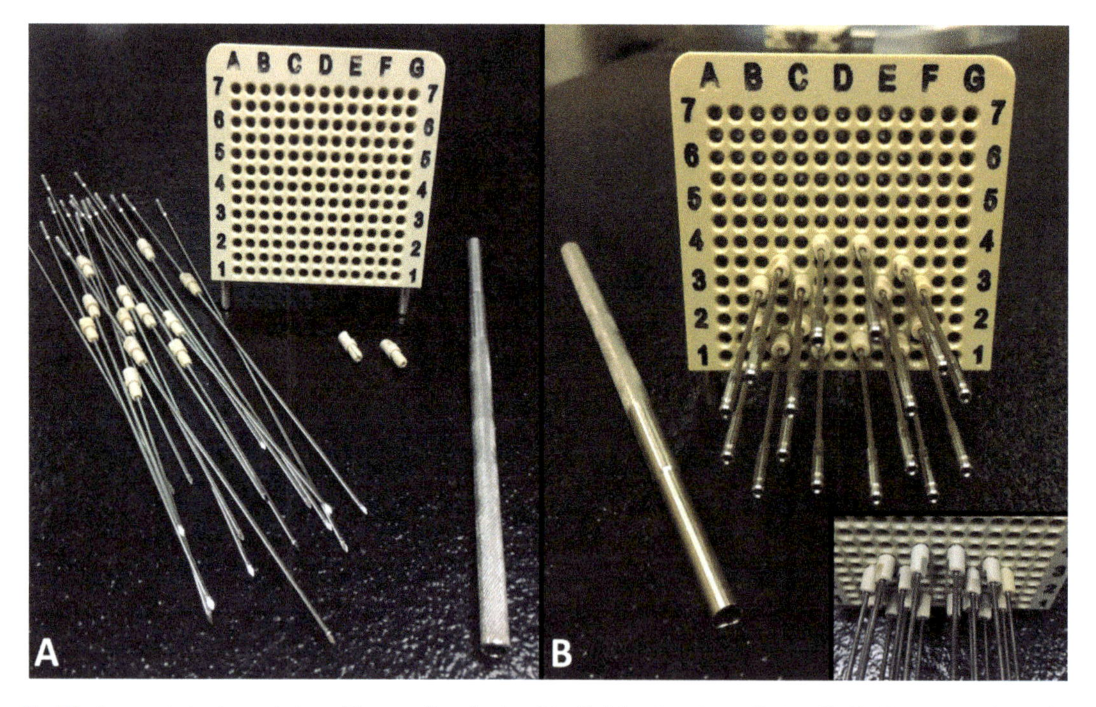

Figure 5–15 A prostate template with needles locked individually via collets. A) Collets are placed over the needles before implantation. B) As each needle is advanced to the prostate base, the collet is threaded into the template hole and tightened around the needle using a wrench designed to fit over the needle protruding from the template. No needle shift relative to the template is possible with this type of template. If equipped with the means of being stitched to the perineum, this template could be used with any imaging modality.

procedure by removing the need to start from scratch for each patient. It is not required by current clinical systems that these initial images or contours be used. Due to changes in the prostate position and volume induced by the insertion of needles, it is preferable to use the 3D US data acquired after implanting the needles for contouring and dose planning.

Another important factor to keep in mind is that posterior needles may shadow more anterior needles, obscuring them in the final image. It is helpful to insert all needles to mid-gland and ensure that all are visible before advancing them to their final depth. If any needles do fall within a shadow, retracting and reinserting either needle shifted by a small amount (~2 mm) usually brings them into clear view. Figure 5–13 shows the standard pattern of 16 needles used at the British Columbia Cancer Agency. This pattern generally prevents shadowing needles by other needles and leaves enough flexibility that needles can easily be shifted to improve visibility or to accommodate a wide range of prostate volumes. Needles may be added or removed (±2) to accommodate individual anatomical requirements. It should be noted that not all needles are necessarily inserted to the same depth in the patient. In particular, the anterior needles are often not as deep as the posterior ones. This can be verified in the longitudinal US view.

Once the needles are satisfactorily implanted, they are usually fixed to the template (Hoskin et al. 2013), either via a sliding plate within the template (Figure 5–14) or using individual locking collets (Figure 5–15). Marks may be made on the needles to easily monitor their cranial-caudal position. The template remains mounted to the stepper for TRUS-based HDR brachytherapy.

5.4.3 US Localization

Because the patient and applicators have already been positioned for treatment, simulation for TRUS-based HDR prostate planning consists of acquiring a final 3D US image and measuring the length of the needles protruding from the template. The planning image is acquired exactly as for the initial image. Care should be taken to ensure that all needle tips are included in the image. Visualization of the urethra is simplified by instilling a small volume of aerated gel in the Foley catheter (Figure 5–16).

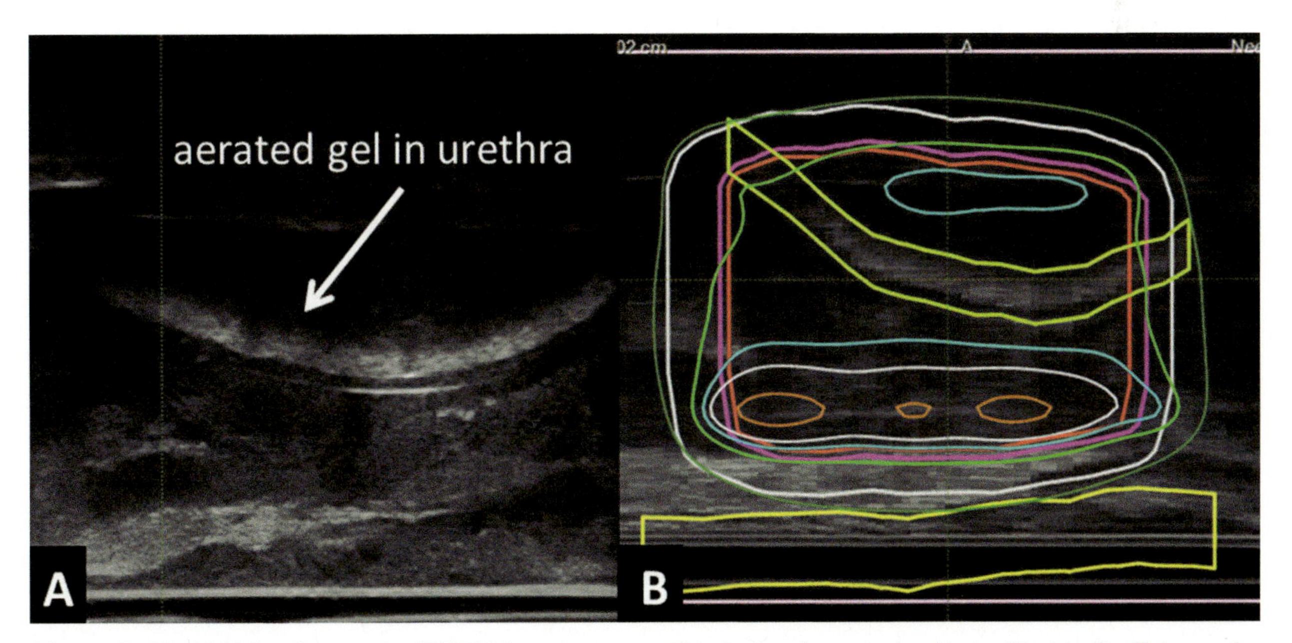

Figure 5–16 A) Using live sagittal TRUS imaging, a small volume of aerated gel is instilled in the Foley catheter just until the gel can be seen in the Foley balloon. B) This allows easy delineation of the urethra for planning without obscuring any anterior needles.

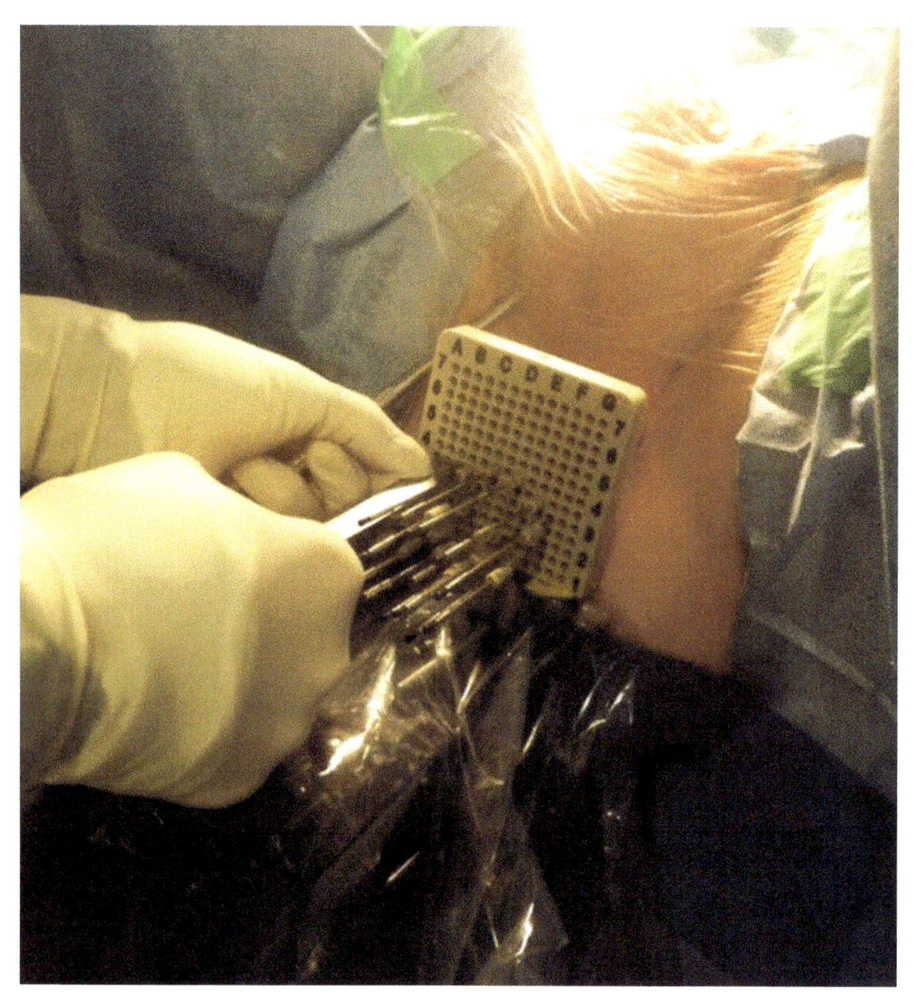

Figure 5–17 Measuring the protrusion length of a needle that has been implanted to the base of the prostate.

It is a recognized limitation of US-based planning that identifying needle tips based solely on the 3D TRUS images is challenging (Siebert et al. 2009; Schmid et al. 2013). This can be ameliorated by using the measured lengths of the needles that protrude from the implant template to adjust the cranial-caudal position of the needle tip. The ability to use the protrusion lengths for this purpose is incorporated into currently available commercial software, making needle tip localization in US as accurate as CT (Batchelar et al. 2014). Thus, the protrusion length of each needle should be measured (Figure 5–17), entered in the planning system as required, and independently recorded before planning. This independent record is useful in case of software issues; if anything happens to the digital record of the protrusion lengths, they can quickly be reconstructed without remeasuring. It is also good practice to independently record the template positions for each needle for QA purposes. These two records can be combined into a single, easily visualized check sheet (Figure 5–18).

HDR Prostate Needle Positions

Date: ____________

Initial reference needle number: _______ **Reference protrusion length:** ______

- Circle the channels to identify the location of implanted needles
- Record the protruding length for each needle
- Numbering is left to right, bottom to top. Physicist will verify the correct assignment in Vitesse

	A	a	B	b	C	c	D	d	E	e	F	f	G
6													
5.5													
5													
4.5													
4													
3.5													
3													
2.5													
2													
1.5													
1													

Figure 5–18 Needle position and protrusion length tracking document used at the BC Cancer Agency. As needles are inserted, their template positions can be marked in the correct locations on the sheet. When the protrusion length is measured, it can be entered in the corresponding square on the sheet.

5.4.4 Needle Reconstruction

Once the second 3D TRUS image set is acquired, the target and OARs are contoured (see sections 5.6.2 and 5.6.3). The needles are then reconstructed using tools integrated in the planning system. For TRUS-based planning, it is imperative that the appearance of the needles in US images is fully understood (Figure 5–19) to ensure correct interpretation of the clinical images. Moreover, the intended relationship between the physical and virtual needles must be understood to be able to judge that the lumen of the needle and the center of the first dwell position are correctly positioned relative to the flash of the needles in the TRUS images (Figure 5–20) (Schmid et al. 2013). Some TPSs offer the possibility to perform needle reconstruction in the live US image. The benefit of this method is that needle visualization can be increased due to moving interactions with the needles (e.g., by rotating or tipping the needles). Also, tipping or rotating the implanted needles can help to identify them in the reconstruction step. The resulting needle positions identified in the live images are then transferred automatically to the previously acquired 3D US image. If this feature is not available, it is still possible to use information from the live image to inform needle reconstruction in the static image set. The needle under consideration can be rotated, and by scrolling through the live image and looking at the position of the flash relative to the template grid, the position of the needle in the acquired image set can be elucidated.

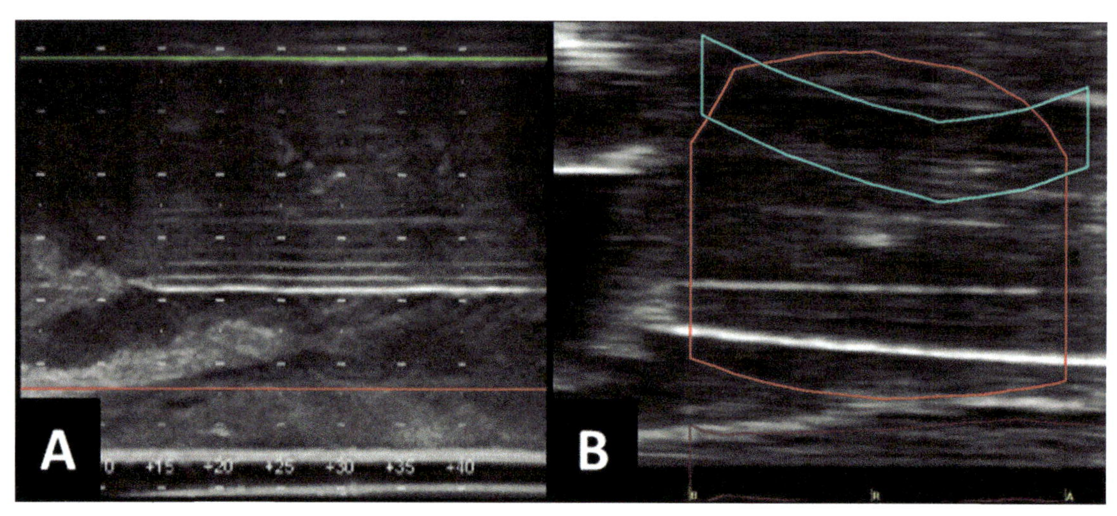

Figure 5–19 Images of two different needles in TRUS images: A is a reusable steel needle and B is a disposable plastic needle.

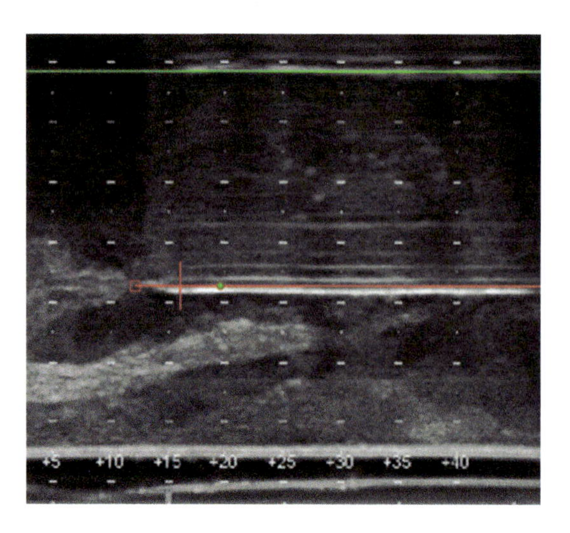

Figure 5–20 A correctly aligned virtual needle. The vertical line indicates the end of the active needle channel, the leftmost point of the horizontal line is the physical tip of the needle, and the green dot is the center of the first dwell position.

5.4.5 Pretreatment QA

All general pretreatment QA outlined in Section 5.7 applies for TRUS-based planned cases. As for any brachytherapy procedure, a crucial part of ensuring safe and accurate treatment is confirming that the applicators have been reconstructed appropriately. For most multi-needle interstitial procedures, it is challenging to perform this at the end of treatment planning and, therefore, beneficial to have a real-time independent check as the needles are identified by the planner. In TRUS-based HDR prostate brachytherapy, this is essential. Identifying the tips and trajectories of the needles depends too closely on information only available and verifiable during the implant for the needles to be reviewed at the end of planning.

It should be ensured that the US probe is inserted as it was for imaging to maintain the planned relationship between the rectum and the prostate. The cranial-caudal position of the needles should be checked after the guide tubes are connected. As treatment proceeds immediately after planning, with no change in patient position, it is not necessary to verify needle positions with respect to the prostate.

5.4.6 Tips and Tricks

- As this is a real-time technique, it is helpful to have a brief procedure checklist available for easy reference (Figure 5–21) to keep the process on track.
- Having the most posterior needles implanted 3–5 mm within the prostate capsule gives the best result with respect to rectal dose.
- Do not use a needle in the central row of the template, particularly posterior to the urethra. This will cause shadowing and obscure the urethra.
- The prostate will shift cranially as the needles are inserted. Make sure the probe can be inserted up to 3 cm farther than the initial position of the base. Also, the position of the anterior-most needles may need to be adjusted at the end of the implant.

HDR PROCEDURE CHECKLIST

BEFORE NEEDLES ARE INSERTED:

Confirm: Date=Today; Unit=HDRLake; Sk=40700; Prescription&isodoses match
Set stepper angle to zero and Zero stepper angle readout
Ensure that needle number assignment is correct (Screen L to R, bottom to top)
Set the probe to 2-3 cm SUP of base and record stepper position (stepper > 20)
Acquire a 3D US image set

AS NEEDLES ARE INSERTED TO MID-GLAND:

Turn on the embedded US overlay
Drag needle representation to US flash
Ensure can see all needles - not shadowed by other needles
Ensure needles have been assigned to correct template hole

AS NEEDLES ARE ADVANCED TO THE BASE:

Maunally align virtual needle one of the posterior needles
Enter measured protrusion length and set needle as reference
Enter each protrusion length in cm and click 'adjust'
Ensure that at least one other needle tip is clearly visualized
Lock all needles
Ensure that green reference line has not been moved
RO may review some of the needles and readjust depth

BEFORE ACQUIRING FINAL IMAGE SET

Ensure urethra is visible - aerated gel
Turn off the embedded US overlay
Set stepper to original position and begin acquire, step back above needle tips

AFTER IMAGE ACQUISITION

WARNING - DO NOT SAVE WHEN IN ACQUISITION TAB
Check that virtual needles align with the live and the acquired image set
Switch to contouring and select brush tool

AFTER CONTOURING

Print channel connection diagram for hook up
Check:StepSize=0.3cm;DeadSpace=0.4cm;Offset=0cm;NeedleDiam=1.47mm
Finalize the needle paths
Check 3D view to make sure everythign looks reasonable

AFTER NEEDLE TRACKING

Check that dwell position range for peripheral needles is reasonable
Load appropriate planning constraints
Manually adjust to meet planning goals

AFTER PLANNING

Ensure at least 1 dwell position in each needle; Check for dwell times < 0.3s
Planning Approve (RO) and Treatment Approve (MP)
Probe IN for treatment

Figure 5–21 Procedure checklist used at the BC Cancer Agency. The goal of this list is to ensure that no crucial steps are missed that would require repeating a portion of the procedure. It is not intended for plan checking.

- Rotation of the prostate during needle insertion can be minimized by having the first two needles inserted on the left and right of the prostate. It is particularly helpful to insert the central needles (e.g., those in columns C and E in Figure 5–13) first.
- If the prostate is too wide to be covered by template, an experienced operator can implant the needles "free-hand." Care should be taken in the labeling of the needles because no template coordinates can be used for identification of the needles.

5.5 MRI-based HDR Brachytherapy for Prostate

5.5.1 Process

MRI can be incorporated into HDR prostate brachytherapy in three main ways. Pretreatment multi-parametric MR images can be fused to planning images, whether CT, US, or MR (Crook et al. 2014; Hosni et al. 2016), to define a sub-prostatic volume for focal dose escalation. MRI can also replace CT as the post-implant imaging modality used for treatment planning in a process analogous to that shown in Figure 5–4 (Buus et al. 2016). If the MR scanner is in the brachytherapy suite, the number of patient transfers are reduced or eliminated. Finally, MRI can be used both to guide needle insertion and for planning (Murgic et al. 2016), analogous to the TRUS-based process seen in Figure 5–12. Especially during the learning phase for the team, the entire procedure can take much longer than in CT- or US-based work flows (Buus et al. 2016; Murgic et al. 2016), with initial procedure time reported >7 hours, even with MRI in suite. This makes it imperative that the entire process should be simulated prior to implementation of an MR-based work flow in order to anticipate areas of difficulty that may be encountered (Buus et al. 2016). A detailed study of all of the steps required for MR-based treatment planning for HDR prostate brachytherapy has been published. (Buus et al. 2016). This study has several steps that are required if needle placement is performed under US guidance, but with MR-based planning. The US steps can be replaced with similar tasks if only MR is used for needle insertion. There are additional complicating factors for the process of using MR-based insertion of needles that require careful consideration and planning: a) all equipment that is used must be MR compatible, b) there is often limited workspace for placement of the needles due to the size of the MR bore, c) the needle insertion may be in a different position than is normally encountered within the clinic, and d) the patient most likely would need to be moved to the treatment area without the implanted needles being affected.

5.5.2 Implant

When MRI replaces CT or US only as the planning modality, needles are implanted under TRUS guidance as described in Section 5.4.2, with the caveat that MRI-compatible needles must be used. For MRI used also for needle-insertion guidance, the patient must be immobilized for the implant, either in lithotomy, decubitus position (Menard et al. 2004; Laloksi et al. 2011), or in a frog-leg position (Murgic et al. 2016). For implantation in the decubitus position, Aquaplast® sheets can be used over the patient's pelvis in order to immobilize the patient in the MR unit (Lakosi et al. 2011). Needles should be placed under MR guidance, either with the patient within the MR field or with the MR table withdrawn from the bore and then moved back into the bore for imaging. As experience is gained by the operative team, several needles can be placed at a time, with imaging after placement of each group of needles. Final imaging is performed after all treatment needles have been placed.

For patients treated in a lithotomy position, the needle insertion procedure can be expedited. The first needle can be placed via manual guidance, in which a finger in the rectum guides a needle into place between the rectum and the urethra. The position of this needle is verified by MR, and it then serves as a reference for all other needles. A template is slid over this needle with all other needle placement positions predetermined and marked on the template (Ares et al. 2009).

In cases in which the institution would like to treat a DIL within the prostate, either solely or as an integrated boost, a pretreatment multiparametric MR can be performed in which the boost region has already been contoured. These pretreatment images are then registered to the treatment MR and the contour copied onto the registered image, ensuring that the DIL is properly treated (Hosni et al. 2016). A similar approach can be used to perform a focal boost in CT- or TRUS-based planning (Crook et al. 2014; Andrzejewski et al. 2015; Peach et al. 2016).

5.5.3 MRI Localization

The use of MR for localization and treatment planning can be performed using either closed bore or open (low field) (Ares et al. 2009) systems. While open systems allow increased access to the patient for the placement of needles, the low field strength leads to a lower signal-to-noise ratio and less soft tissue contrast than the high-field systems (Lakosi et al. 2011).

As stated above, the placement of needles and localization has been reported in the literature in a variety of patient positions. Prior to implantation, a rectal coil is placed and an MR-compatible perineal template is fixed perpendicular to the coil and immobilized against the perineum (Menard et al. 2004; Laloksi et al. 2011; Hosni et al. 2016). Figure 5–22 shows the patient set up in the decubitis position prior to implantation.

The patient is immobilized on an MR compatible couch which can be withdrawn from the scanner bore for needle placement. After each needle placement, either a single or small number of needles, the couch is placed back into the scanner for imaging (Lakosi et al. 2011). When all treatment needles have been inserted, a final treatment planning scan is performed. The patient must remain immobilized in the scan position

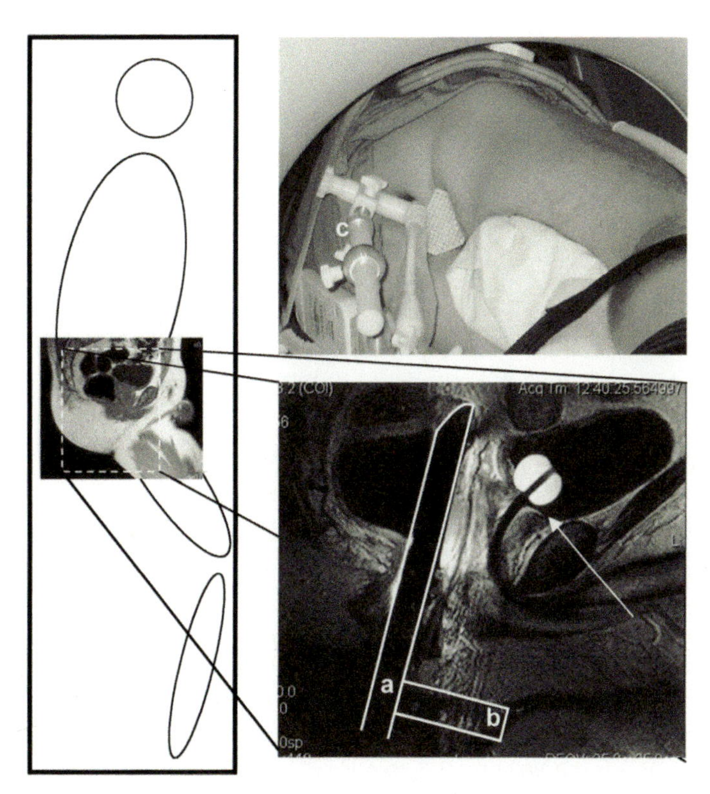

Figure 5–22 Patient set up in a MR scanner prior to implantation with a sagittal image showing the prostate, bladder, Foley catheter, and rectal coil (Menard et al. 2004)

throughout the entire procedure, from scanning through treatment, whether the treatment occurs in the MR suite or the patient is moved to the brachytherapy treatment room (Menard et al. 2004).

While MR can be used to guide needle insertion and treatment planning, it can also be used prior to an US-based work flow. For example, a pretreatment MRI can be performed and all structures can be contoured, along with any focus of disease within the prostate. This MRI scan is performed with a rectal coil the same diameter as the US probe in place. The MRI is then displayed on a split screen with the real-time US, and the MRI can then be rigidly registered to the US during the procedure. The contoured structures from the MR are transferred to the US set and adjusted until the contours are acceptable (Crook et al. 2014; Gomez-Iturriaga et al. 2016).

5.5.4 Needle Reconstruction

Needle reconstruction is performed in a similar method to other imaging modalities. The determination of the first dwell position is made by visualization of the signal void at the needle tip (Citrin et al. 2005; Buus et al. 2016), an example of which is shown in Figure 5–23 (see also Chapter 1). A study of the conspicuousness of brachytherapy needles on MRI showed that they could be readily visualized in most cases independent of the observer (Wybranski et al. 2015). A slice thickness no greater than 2.5 mm should be used to limit needle tip localization errors (Mason et al. 2014).

5.5.5 Pretreatment QA

Pretreatment QA is similar to other planning modalities. If required by vendor software, needle lengths must be measured and compared to the lengths in the TPS (or treatment console). As with all interstitial implants, special care must be taken to ensure that the correct transfer guide tube for each channel is connected to the correct needle. Additionally, verification of needle position relative to both the template and the prostate

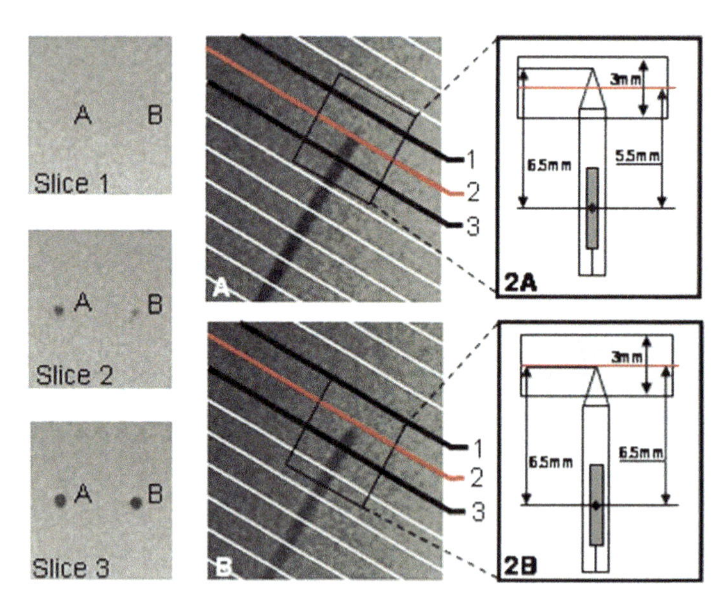

Figure 5–23 Needle A was advanced 1 mm beyond the center of slice 2 (red line). Needle A was clearly visible on slices 2 and 3 and not visible on slice 1 (left panel). Needle B was advanced to the exact center of slice 2 (red line); it was barely visible on slice 2 and was well visualized on slice 3, but not on slice 1 (left panel). The distance from the source center at the first dwell position (diamond) to the slice center was 5.5 mm for needle A (right upper corner) and 6.5 mm for needle B (right lower corner). Needle B represents the largest distance between the slice center of the visualized tip and the physical location of the first dwell position (Citrin et al. 2005).

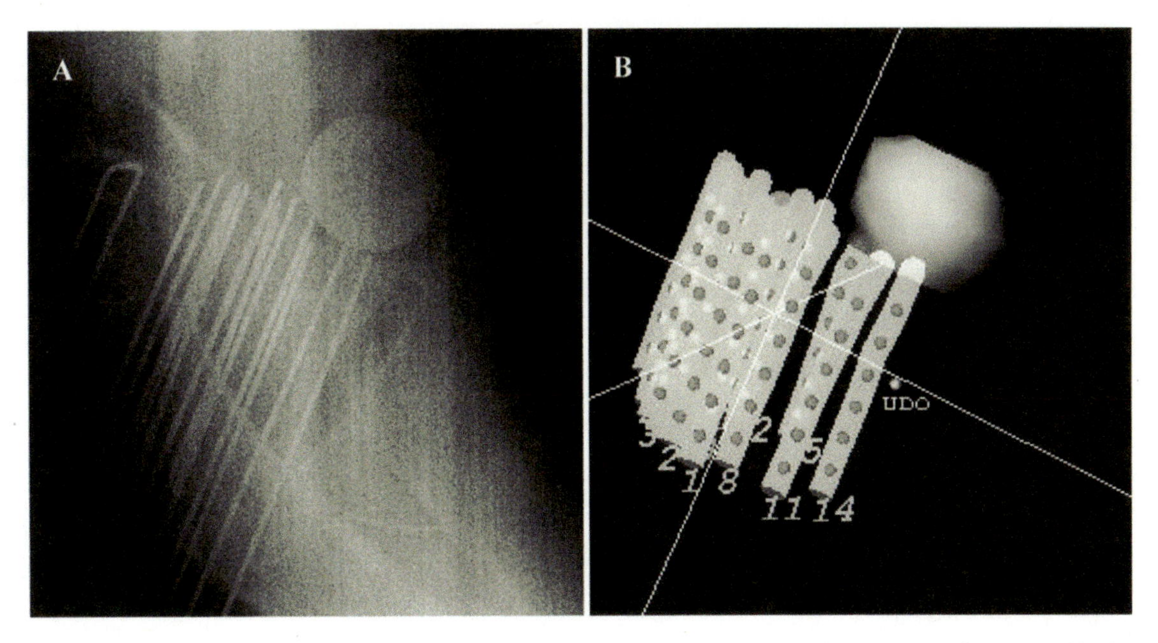

Figure 5–24 Example of comparison of lateral radiograph to the MR-based treatment plan to verify the position of the needles prior to treatment (Menard et al. 2004).

must be performed. If the patient is not moved from the MR scanner, a second MRI can be acquired immediately prior to treatment to ensure that the needles have not shifted (Lizondo Gisbert et al. 2016). If the patient needs to be moved from the MR scanner to the treatment room, the verification can be done via a variety of methods, including the use of orthogonal x-ray images to verify the position of the needle tips with respect to the Foley needle or anatomical landmarks (Menard et al. 2004). An example of this method of QA is shown in Figure 5–24.

Marking of the needle positions on the template should also be performed independently of where the patient is treated to provide a quick visual aid in determining whether any needles have shifted within the template.

5.5.6 Tips and Tricks

- A benefit of real-time MR imaging is the ability to visualize any extra-prostatic disease at the time of implantation, allowing the insertion of needles in specific locations to target this disease (Citrin et al. 2005).
- The soft tissue contrast of MR imaging allows the visualization of additional critical structures, such as the neurovascular bundles, which cannot be readily discerned by other imaging modalities.
- MR imaging may be superior to US guidance for patients with larger prostate volumes (>50 cm^3) or in patients with narrow pubic arches (Ares et al. 2009).
- An MR-only work flow eliminates the need for rigid or deformable registration of the MR to treatment planning CT or US, removing the inherent uncertainties of the registration process, but the geometric distortions of the MR system must be evaluated prior to implementation for brachytherapy treatment planning (Citrin et al. 2005).
- It may be more difficult to meet rectal dose constraints in MR-based planning if one uses the constraints developed for US- or CT-based HDR brachytherapy treatment. This may be due to the improved soft tis-

sue resolution of MR, allowing more accurate contouring of the rectal wall–prostate interface than is possible in CT or US (Murgic et al. 2016). It is also possible that a rectal coil displaces the anterior rectal wall closer to the prostate.

- If MR is used to determine any focus of disease within the prostate, interobserver variation in the delineation of the lesions can lead to significant variations in the DIL volume (Gomez-Itirriaga et al. 2016).

5.6 Treatment Planning

A variety of dose and fractionation schemes exist for HDR prostate brachytherapy. This is different from LDR brachytherapy, where the prescription dose follows narrow standards in most clinics. Moreover, HDR brachytherapy can be applied as a boost technique in combination with external beam or as a monotherapy. This also opens the possibility for a plurality of different dose and fractionation schemes (Kovács et al. 2005).

As is typical for brachytherapy, the applicator (here implant needles) geometry is paramount. With poor needle geometry, good planning results cannot be obtained in most cases. Not even modern, sophisticated inverse planning algorithms can compensate for a physically bad implant. But with good implant geometry and a modern TPS, HDR brachytherapy offers the possibility to create a very conformal treatment plan.

When using HDR brachytherapy afterloaders, the planner has two parameters to create and modify the dose distribution: dwell position and dwell time. Typically, the dwell positions are equidistant along the needle, e.g., in 5 mm steps. First, the dwell positions are defined in the TPS. The planner must ensure that the range of dwell positions is adequate to provide full coverage to the target. Typically, this will require some dwell positions outside the target to improve dose homogeneity within the prostate. Dwell positions should be within 5 mm of the target, and care must be taken to limit dwell positions adjacent to organs at risk.

In the next step, the dwell times are determined and the dose distribution calculated. Dwell times can be determined in several different ways. In a forward planning process, the medical physicist changes dwell positions manually and iteratively until satisfying the dosimetric criteria. Another semi-manual form is the use of dose shaping tools. Here, the planner can change the isodose shapes by "dragging" them in the images with the cursor while the dwell times change accordingly. Care should be taken when using this option as sometimes unintended large differences in dwell times from one dwell position to the next can occur. When using dose-shaping tools, the dwell times should be checked manually at the end of the planning process to verify that the change in time from one position to the next is reasonably smooth. A simple form of automated dwell time determination, geometrical optimization takes into account only the distances between dwell positions and ignores anatomical information. Typically, extensive manual intervention is required to obtain an acceptable dose distribution following geometrical optimization. In the absence of inverse planning, this can, however, produce a good starting point for manual forward planning. In inverse planning, the user defines dose constraints for the related organs, and the TPS computes the dose distribution via an optimization algorithm. It should be noted that different TPSs may produce different "optimal" solutions. The same is true when the software version changes, and the optimizing algorithm is altered. If standard dose constraints are used, the results of optimization should be scrutinized whenever such a change occurs.

Dose calculation for HDR prostate brachytherapy is usually conducted using the TG-43 formalism (Rivard et al. 2004, Perez-Calatayud et al. 2012). Tissue inhomogeneities do not play a significant role for high-energy photon-emitting brachytherapy sources in the region of the prostate (Rivard et al. 2009). Absorption effects of the needles are neglected using the TG-43 dose calculation formalism.

Table 5–1 Various dose/fractionation schemes for HDR prostate boost technique. (The list is not complete, but shows typical values.)

Institution	Dose and Fractionation EBRT (Gy/#)	Brachytherapy Fractions	Dose per Fraction Brachytherapy (Gy)	Target Volume
Sahlgrenska University Hospital (Borghede et al. 1997)	50/25	2	10	Prostate with boost to positive biopsy region
Kiel University (Galalae et al. 2002)	40/50	2	15	Peripheral zone (9 Gy to whole prostate)
William Beaumont Hospital (Vargas et al. 2006)	46/23	2 or 3	5.5–11.5	Prostate
California Endocurietherapy Cancer Centre (Demanes et al. 2009)	36/20–39.6/22	4	5.5–6	Prostate
Sunnybrook Hospital (Morton et al. 2011)	37.5/15	1	15	Prostate

Table 5–2 Typical dose and fractionation schemes for HDR prostate monotherapy

Institution	Total Brachytherapy Dose and Fractionation (Gy/#)
Osaka University (Yoshioka et al. 2016)	45.5/7 or 54/9
California Endocurietherapy Cancer Centre (Demanes et al. 2011)	42–43.5/6
Willia Beaumont Hospital (Demanes et al. 2011)	38/4
Mount Vernon Hospital (Hoskin et al. 2014)	26/2 or 19/1
GammaWest Cancer Services (Rogers et al. 2012)	39/6

5.6.1 Dose and Fractionation

5.6.1.1 Boost Dose Patterns

The most common use of HDR prostate brachytherapy is as a boost technique. The dose and fractionation for HDR brachytherapy boosts is not standardized. Several dose and fractionation schemes can be found in the literature (see Table 5–1).

In addition to the variety in total dose and number of fractions, there is also no consensus on the sequence of EBRT and brachytherapy. HDR brachytherapy is performed before, after, and interdigitated with EBRT. The only constant is that brachytherapy is not delivered the same day as EBRT. While it is still an open question as to whether there is an optimal dose and fractionation scheme for HDR brachytherapy treatment of prostate cancer—as well as which EBRT technique and dose should be chosen—the clinical outcomes of all dose and fractionation schemes are similar (Yamada et al. 2012).

There also exist differences in target volume definition. Although most commonly the whole prostate plus a margin is the target, some groups try to segment the tumor only, while others follow a two-CTV method (CTV1, CTV2), where the peripheral zone and the prostate are contoured and assigned to different prescription doses.

5.6.1.2 HDR Brachytherapy Monotherapy

In principle, HDR brachytherapy boost and monotherapy are very similar with regard to treatment planning. Nevertheless, using HDR brachytherapy as monotherapy for prostate makes the required meticulous attention to detail needed to obtain a good dose distribution even more crucial than for the boost technique because there is no EBRT dose to compensate for deficiencies in the dose distribution. HDR brachytherapy monotherapy is still under development (Hoskin et al. 2013) and is most commonly delivered under the auspices of a clinical trial. Typical monotherapy dose and fractionation schemes are listed in Table 5–2.

5.6.2 Target Volume Definition

The CTV in HDR prostate brachytherapy is the whole prostate capsule. In addition, macroscopic extracapsular extension or seminal vesicle involvement identified on diagnostic images, expanded by a 3-mm margin, are included. Margins are not typically added posteriorly (toward the anterior rectal wall) or superiorly (toward the base of the bladder) (Hoskin et al. 2013). Although the CTV definition is the same regardless of imaging modality, the prostate itself is visualized very differently in MRI, CT, and TRUS (Figure 5–25).

5.6.2.1 Target Volume Delineation on CT

Because of the poor soft tissue contrast in CT, delineation of the prostate gland is not as easy as when using T2-MRI or TRUS imaging (Smith et al. 2007). When contouring the prostate in CT, it is good practice to first delineate the rectum and the bladder, which are much easier to identify. Then the prostate gland can be delineated with more accuracy (Figure 5–26). Here it is important to check the contours in sagittal and coronal views. Knowledge of the prostate length and volume from MRI or US can increase the accuracy of the prostate contour. Some practitioners will place fiducial markers at the base and apex of the prostate under US

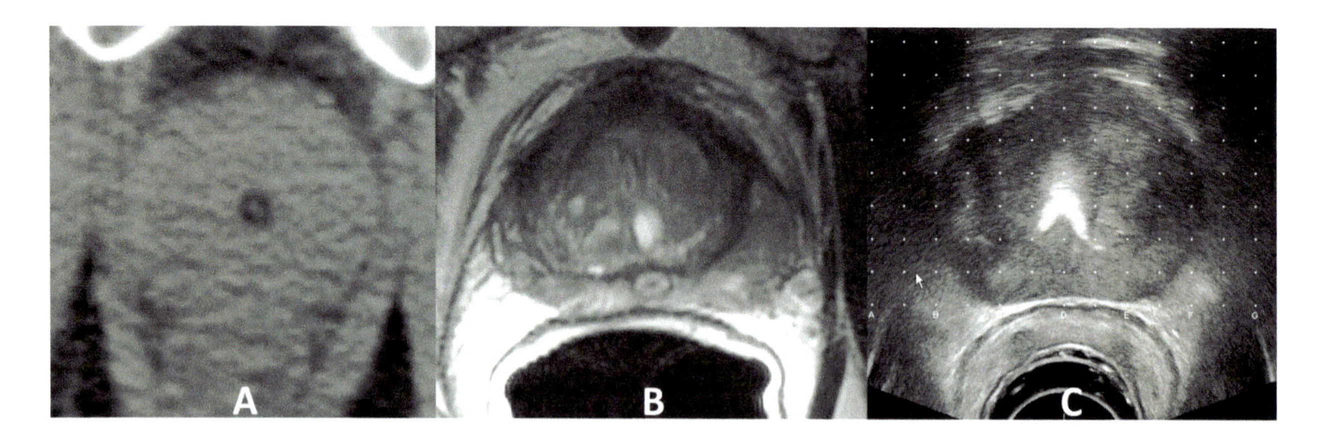

Figure 5–25 Mid-gland prostate images from the same patient acquired using A) CT standard pelvic protocol for EBRT simulation; B) MRI T2-weighted, 1.5T, endorectal coil; and C) 3D TRUS. This illustrates the differences in visualization of both the target and the adjacent organs at risk in each modality. Note that the patient was catheterized for the CT image, allowing for identification of the urethra as the low signal region in the center of the prostate. In the TRUS image, a urethrogram had been performed to identify the urethra (hyperechoic region in the center of the gland). The urethra is visible on its own in the T2-weighted image (enhanced signal). It is also possible to distinguish the peripheral zone in the T2 image.

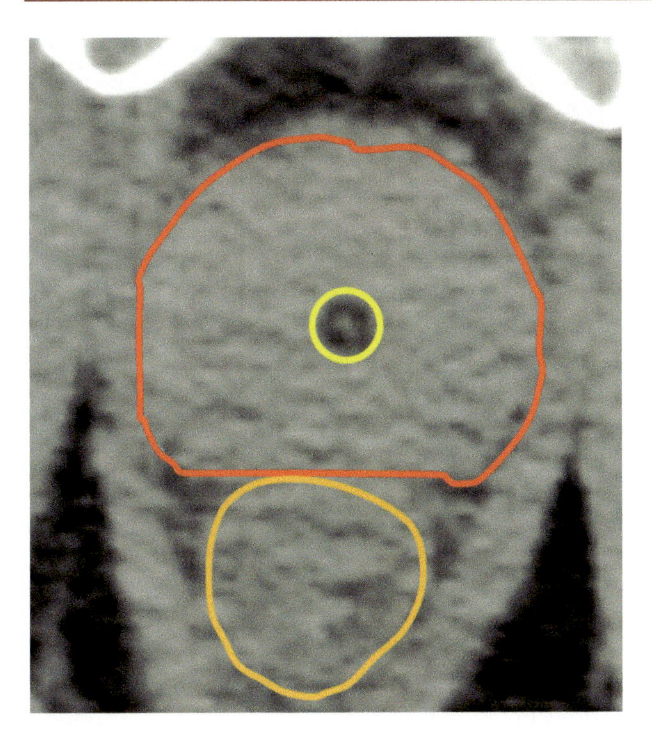

Figure 5–26 The CT image from Figure 5–25 with prostate (red), rectum (orange), and urethra (yellow) contoured.

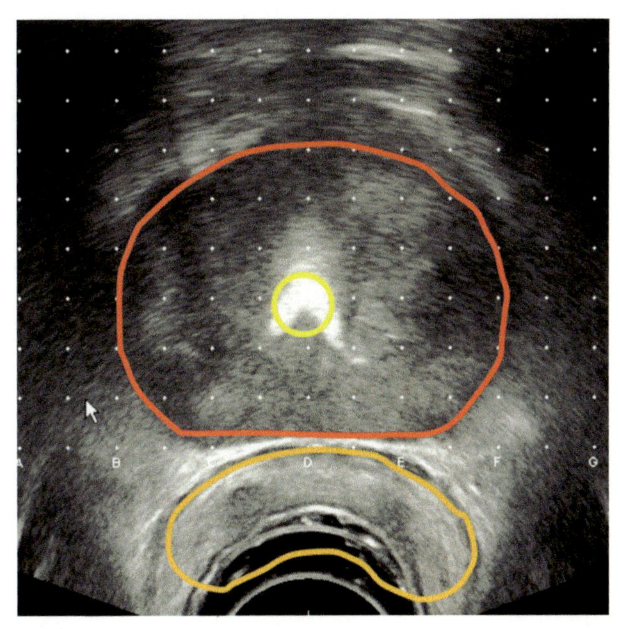

Figure 5–27 Transversal TRUS image from Figure 5–25 with delineated prostate (red), rectal wall (orange), and urethra (yellow), where the urethra has been defined as the central portion of the urethrogram.

guidance as an aid in defining the cranial-caudal extent of the prostate gland. It should be noted that contouring the prostate gland in CT usually overestimates prostate size.

5.6.2.2 Target Volume Delineation on US

TRUS is often used for planning HDR prostate treatments. The prostate gland is generally easy to delineate, and sometimes even the zonal structure can be visualized in TRUS (Figure 5–27). It must be considered that the image quality decreases after insertion of needles, and the quality of delineation of the ventral part of the prostate can be reduced. This can be ameliorated by noting where the anterior and peripheral needles have been placed with respect to the prostate capsule. Using a standard pattern of needles greatly helps in this respect. A biplane TRUS scanner can improve identification of the apical part of the prostate. TRUS acquisition settings, particularly acoustic frequencies and gains, should be adjusted to obtain the best possible imaging result.

5.6.2.3 Target Volume Delineation on MRI

MRI offers a very good soft tissue contrast, in particular when using T2 weighted scans. The central and the peripheral zone of the gland can be identified very well. Moreover, the anterior fibromuscular stroma is often well visualized. T2 MRI is used to define the local staging of prostate cancer (Kovacs and Hoskin 2013). Extracapsular spread of the tumor is often discovered by MRI imaging. The question of whether a rectal coil should be used or not depends on the magnetic field strength of the scanner and whether the scan is to be used for staging and diagnosis—which requires the highest possible image quality—or strictly for planning (Turkbey et al. 2016). For 3 T magnets, endorectal coils may only be necessary for multi-parametric MRI incorporating diffusion-weighted imaging, while they are commonly employed to improve image quality for

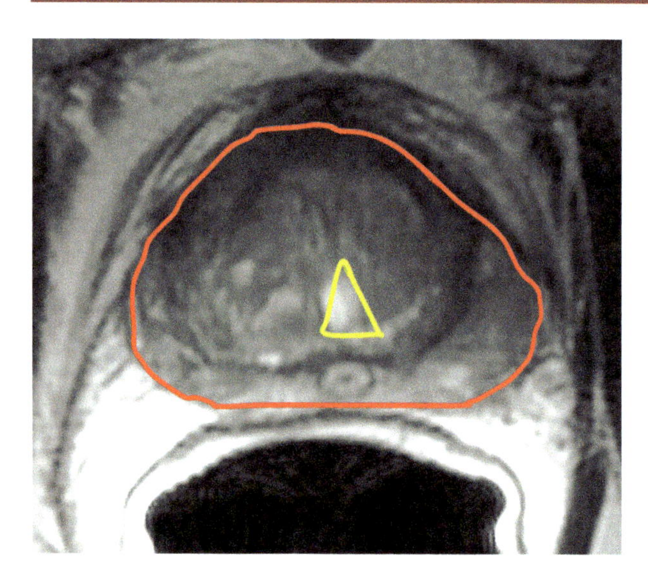

Figure 5–28. Prostate (red) and urethra (yellow) delineated in the T2-MRI image from Figure 5–25. The central gland and the peripheral zone are clearly visible in this image.

1.5 T or lower scanners. Thus, prostate delineation using MRI can be conducted with a high degree of accuracy (Figure 5–28). It is, however, important to note that the physician who performs the delineation must be trained in MR image interpretation as inter-observer variations in volume of up to 17% have been reported in MRI contouring studies (De Brabandere et al. 2012).

5.6.2.4 Multi-modality Imaging

Multi-modality imaging is accomplished by image fusion. Different image data sets, e.g., CT and MRI, can be registered to transfer information from one data set to another. This gives the possibility to add valuable image information unavailable in one image set.

Registration can be performed using different techniques. Very common is the use of landmarks. For this method, corresponding landmarks are set in both data sets, and the mean transformation between the two is calculated. When this method is used, care should be taken to use landmarks within and directly surrounding the prostate (e.g., fiducial markers, the urethra, and the rectal-prostate interface). Other methods include manual (freehand) image registration and automatic fusion algorithms which, for instance, use mutual information present in both image sets to determine the registration transformation. Before using automatic techniques, it is imperative to check if the technique is appropriate for the particular image sets; US images have insufficient signal-to-noise ratios to use intensity-based automated registration algorithms (Moradi et al. 2013), for example. Moreover, image data sets can be distorted, for example MRI data, or the organs may be compressed from the TRUS probe in US images. Some modern fusion algorithms allow for elastic, or deformable, matching of the data. Here the registration should be carefully checked and robust means of validating the results developed before they are used for clinical decision making.

It should be noted that image fusion has uncertainties. It has been shown (De Brabandere et al. 2012) that uncertainties in D_{90} arising from uncertainties in fusion between CT and MRI data in prostate brachytherapy can be on the same order as those arising from contouring uncertainties in CT (standard deviation in D_{90} of 16% and 23%, respectively). Thus, multimodality imaging should be used carefully and thoughtfully.

5.6.2.5 Defining Dominant Intra-prostatic Lesions

Image-based definition of dominant intra-prostatic lesions (DIL) is not a trivial task; investigations are ongoing to elucidate the most clinically useful method. US-based techniques under investigation are Doppler TRUS and sonoelastography. The tumor has limited elasticity and compressibility that yield to a possible detection in sonoelastography devices.

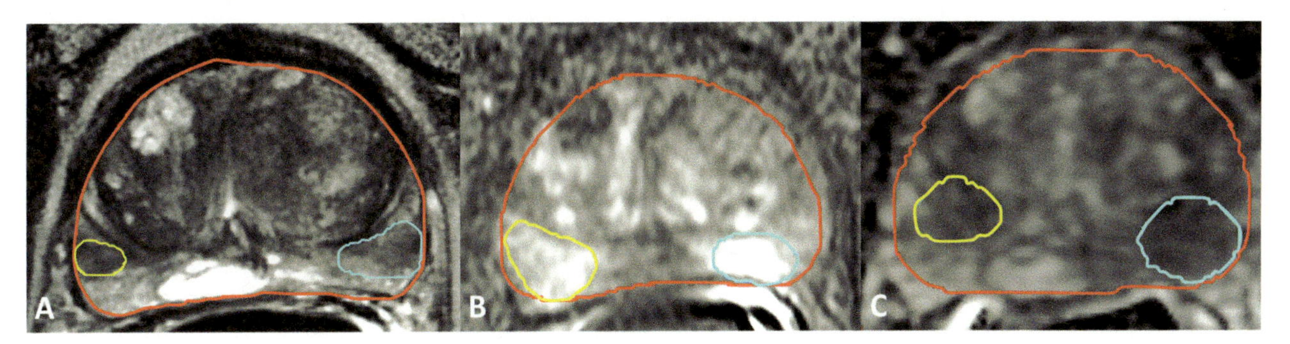

Figure 5–29 Multiparametric MRI images, acquired at 1.5T using an endorectal coil, displaying dominant intraprostatic lesions as identified on each sequence. A) In T2-weighted images, gross disease appears as a reduction in signal intensity. B) In DCE image, the DILs are the regions exhibiting high contrast earliest in the dynamic scan. C) In DWI, the DILs are again areas of reduced signal.

Perhaps a better candidate for defining intra-prostatic lesions is multiparametric MRI (mpMRI) (Mason et al. 2014). A combination of T2-weighted MR images, diffusion-weighted images (DWI), and a dynamic contrast-enhanced MRI (DCE) can be registered to fuse anatomical and functional regions and determine the location of any DIL (Figure 5–29) (Hegde et al. 2013; Turkbey et al. 2016). If MRI is not the modality used for treatment planning, the DIL defined on mpMRI can be transferred to the planning images via fusion based on the anatomy visible in the T2-weighted images. In addition, multiple core biopsies may complement the image-based information.

5.6.3 Organs at Risk

The primary organs at risk for HDR prostate brachytherapy are the urethra and the rectum, which can be clearly delineated in all planning modalities (figures 5–26 through 5–28). The rectal wall, consisting of the muscle layer overlaying the mucosa, is easy to contour in MRI and TRUS images (figures 5–27 and 5–28), although with TRUS only the anterior rectal wall can be contoured. Visualizing the rectal wall alone is less reliable in CT and, in practice, the whole rectum is usually contoured (Figure 5–26). In CT and US, the urethra is defined with the aid of the Foley catheter, which is often contoured as a surrogate to the urethra. A small volume of contrast medium can be injected in the catheter to aid in its identification. The urethra is straightforward to define based on T2-MR images, even without a urinary catheter in place (Figure 5–28). In CT and MR images, the bladder neck may be included, while MR permits delineation of the penile bulb and neurovascular bundles. Consistently following contouring concepts is advisable. Organs at risk should be delineated not only in slices where the prostate is visible but also in several adjacent slices.

5.6.4 Dosimetric Goals

There is no absolute consensus on dose constraints for normal tissues. The American Brachytherapy Society guidelines (Yamada et al. 2012) provide a summary of the limits used by experienced institutions. GEC/ESTRO proposes dose constraints for the organs at risk that use absolute EQD2 total doses, including the EBRT as necessary (Hoskin et al. 2013). The proposed constraints are:

- Rectum: $D_{2\text{cm}^3} \leq 75$ Gy EQD2
- Urethra: $D_{0.1\text{cm}^3} \leq 120$ Gy EQD2
  $D_{10} \leq 120$ Gy EQD2
  $D_{30} \leq 105$ Gy EQD2

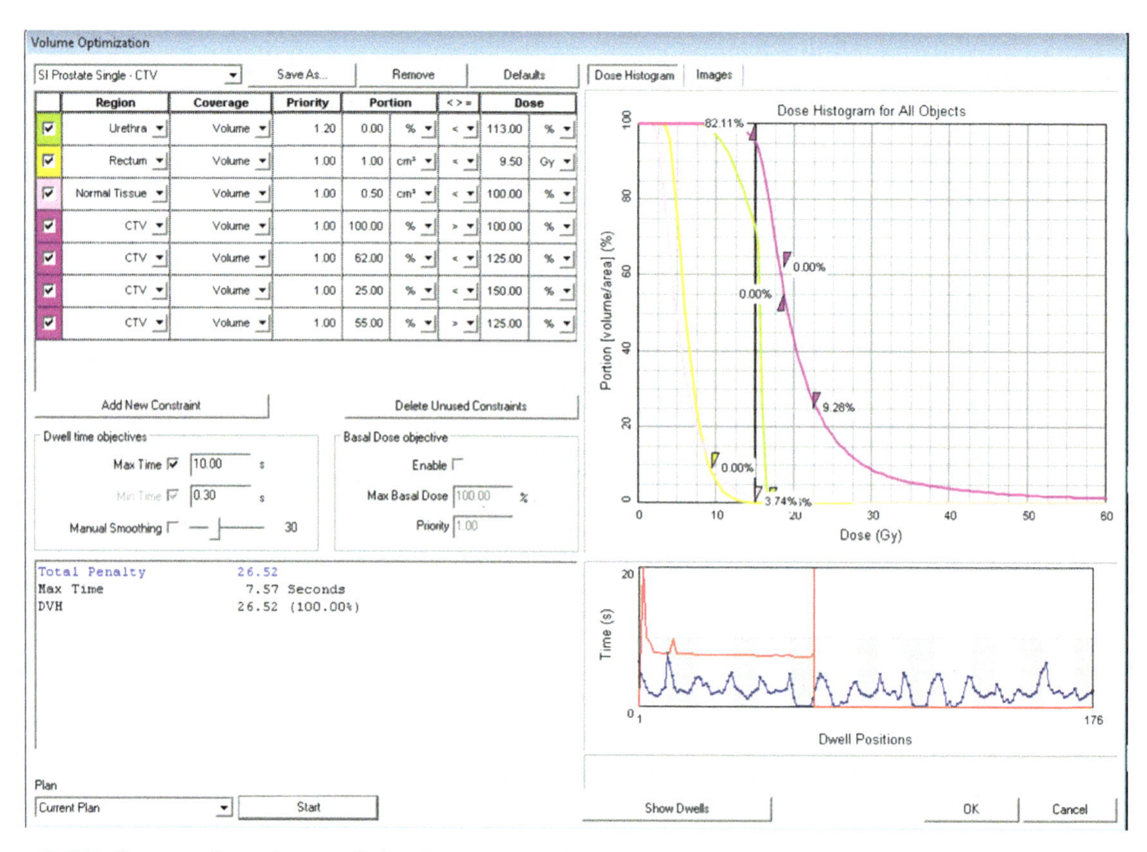

Figure 5–30 Screen shot of an optimization screen showing constraints for an US-based inverse plan. Constraints are applied to the urethra and rectum, the CTV, and normal tissue, which is the entire image volume minus the CTV. The actual DVH, dwell times (blue), and the cost function (red) are also displayed.

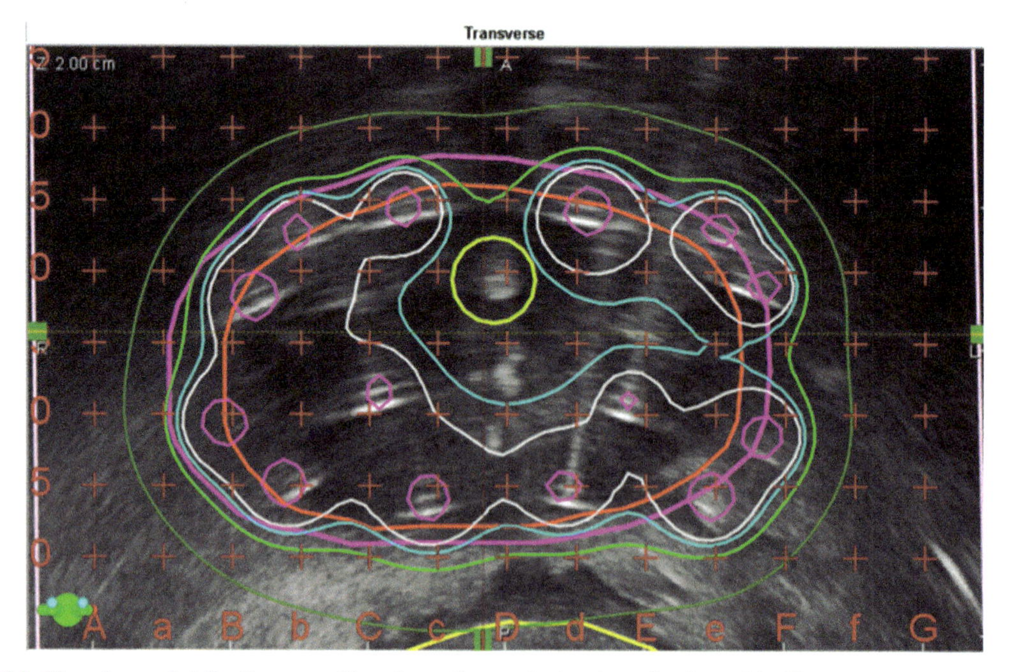

Figure 5–31 The dose distribution resulting from the optimization displayed in Figure 6–30. The prescription was 15 Gy. Isodoses displayed are 200% (magenta), 125% (white), 115% (cyan), 100% (light green), and 70% (dark green). The CTV V_{100} = 98% and the D_{90} is 108%. The urethra V_{115} = 0% and rectal D_{1cm^3} = 9.4 Gy. All institutional goals were achieved.

The D_{90} of the CTV should be >100%. Moreover, the V_{100} is recommended to be at least 95%. No clear recommendation can be given on dose homogeneity. But, dose non-homogeneity indices, such as the V_{150}/V_{100}, should be calculated (Hoskin et al. 2013).

5.6.5 Optimization

In dose optimization or inverse planning, the TPS calculates a proposed dose distribution. Dose constraints for CTV and organs at risk can be defined and saved. Depending on the TPS, further constraints, such as smoothness (the degree of change between adjacent dwell positions) of dwell times, minimum and maximum dwell times, and basal dose points can also be considered (Figure 5–30). The dose constraints are not necessarily the dosimetric goals listed in Section 5.6.4, but the results of the optimization should lead to the dosimetric aims (Figure 5–31). It is of high importance to check the results after completion of the optimization. In many cases, additional manual optimization may be needed to reach an acceptable plan. Several publications have demonstrated that inverse planning leads to similar results as forward planning (Siebert et al. 2014). The benefits of inverse planning are that the results are largely user-independent and the needed planning times are decreased (Lachance et al. 2002; Batchelar et al. 2016).

5.6.6 Tips and Tricks

In forward-dose planning, it is sometimes advisable to plan with only the peripheral needles at first. Once the peripheral dose is adequate, central needles can be loaded and afterwards, the peripheral ones optimized.

5.7 Pretreatment Verification

Patient-specific QA for HDR prostate brachytherapy is similar to any other HDR treatment planning check; the plan should be independently reviewed, the calculation verified, the needle and patient positioning confirmed, and the needles must be correctly connected to the afterloader. The order of these checks will largely be determined by local work flow demands.

5.7.1 Treatment Plan QA

The following documentation, generally available in the plan report, must be independently verified as correct:

- patient name and ID
- treatment site
- treatment unit and source model
- planned source strength
- treatment date and time
- prescription dose
- step size
- correct numbering of needles
- assignment of needles to the correct channel (a road map or printed template grid should be used to number and verify that the physical needle location and number is given the same number and channel in the planning system)
- any default settings used by the planning system, e.g., needle diameter, dead space
- required approvals are present (this should include checking that the plan matches the constraints of the prescription)

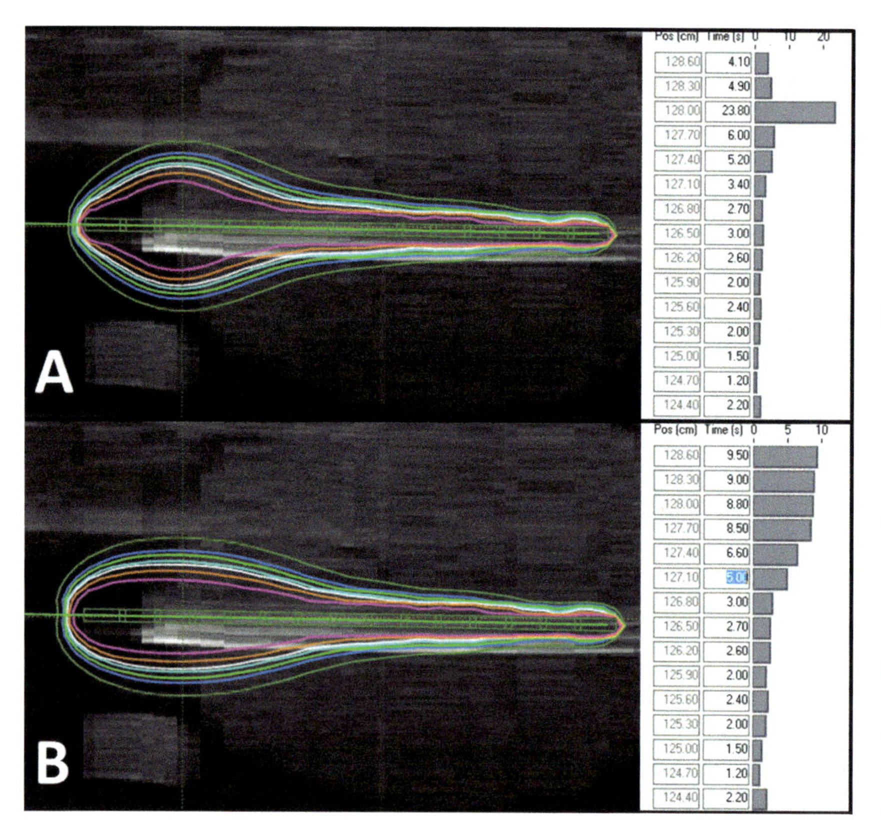

Figure 5–32 Illustration of improving the smoothness of dwell time distribution. This image focuses on one of 16 needles implanted during an HDR brachytherapy treatment. A) Following graphical dose-shaping, a single dwell position had a dwell time 4–5 times larger than its neighbors, causing a bulge in the dose sleeve around this needle. B) The dwell time has been manually redistributed along the needle. The total dwell time for the needle displayed is the same in both parts A and B, and the prostate V_{100} was equivalent (97.4% vs 97.2%). In B, the high dose sleeve is smoothed along the needle, making the overall dose distribution less sensitive to needle placement uncertainties.

HDR Prostate Checklist:	Checked?	Comments
Correct image sets (Series?, # of images?)		
Plan Properties:		
Rx Dose (Mosaiq vs Plan)		
Correct CT for planning		
Check all contours		
Channel:		
Channel numbering		
Digitization		
Correct machine assigned to applicators		
Catheter Lengths		
Step size = 0.5 cm?		
Offset >= 0.3 cm?		
Dwell Times:		
Distribution of dwell times reasonable?		
No 0.1s dwells		
No empty channels		
Isodoses		
Prostate DVH:		
V100≥ 90% (Min V100 > 85%)		
Urethra:		
V125 < 1cc		
V150 = 0%		
Bladder & Rectum:		
V75 < 1cc		
V150 = 0%		
Implant DVH:		
HI = (1- V150/V100) ≥ 0.5		
'P-P' Calc within 15%?		

Figure 5–33 An example of a checklist used to verify a CT-based HDR prostate brachytherapy plan prior to treatment. (Courtesy of J. Zoberi, Washington University.)

Much of the remaining plan-check tasks are best performed directly in the planning system. Before the dose distribution is scrutinized, it should be confirmed that the needles have been reconstructed correctly. This requires verifying that the virtual needle tips are properly aligned with the end of the source channel, that the path of the needle has been mapped correctly, and that the first dwell position is in a physically realizable position. For interstitial brachytherapy, the orientation and determination of each needle path can be difficult depending on how many needles are used and the straightness of the insertion. It is often more straightforward to have a real-time check of the needle reconstruction than to attempt it at the end of planning; this would consist of having the person responsible for the verification directly observe the planner as the needles are reconstructed. In US-based planning, it is necessary to perform this check in real time.

The dose distribution should be reviewed on a slice-by-slice basis, assessing both target coverage and organ sparing, as well as making sure high- and low-dose regions are in reasonable locations. The DVH parameters should be compared to institutional dosimetric goals for all targets, and OAR and any deficiencies should be justified. Each dwell position time should be inspected, both to ensure that any minimum dwell-time requirements are met and to be sure that the delivered dose is smooth along the needle (i.e., each dwell position time is similar to the next) rather than a few selected dwell positions delivering most of the dose (Figure 5–32). This check is particularly important if inverse planning has been used or if the dwell times have been manipulated graphically. A checklist, such as the one seen in Figure 5–33, should be developed to ensure no QA steps have been missed.

5.7.2 Independent Calculation

Following these checks, an independent check of the dose calculation should be performed. This can be accomplished by using an independent dose calculation (Figure 5–1) or by assessing the total time for a fixed source strength relative to the target volume (Figure 5–34). A global evaluation of the reasonableness of the plan can be accomplished with the treatment time per volume check. This is analogous to using nomograms in LDR brachytherapy to assess the number of sources planned for a given prostate volume. The range of reasonable treatment times can be determined based on treatment data from an experienced institution with similar implant philosophy and dosimetric goals or from historical tables converted to the radionuclide used for treatment.

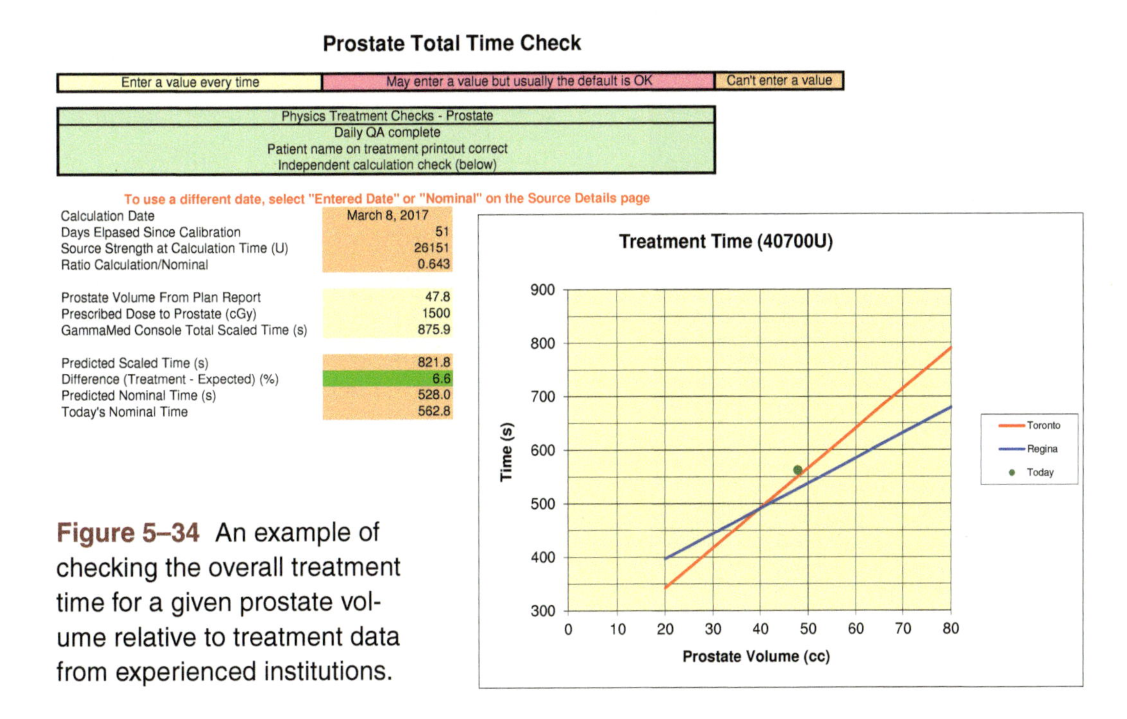

Figure 5–34 An example of checking the overall treatment time for a given prostate volume relative to treatment data from experienced institutions.

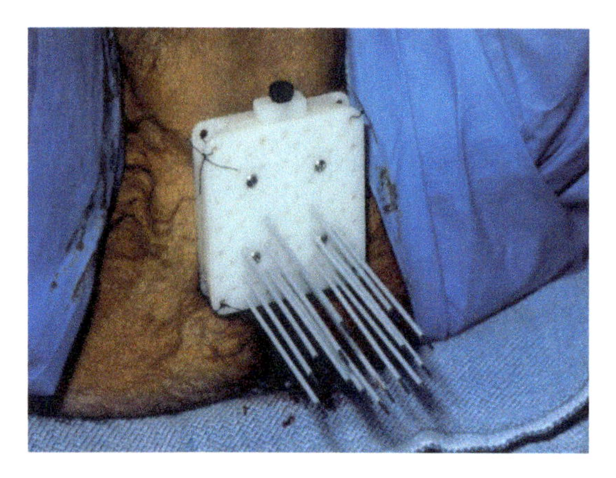

Figure 5–35 An illustration of the protrusion of needles beyond the template. Unintentional changes to this length should be prevented by a combination of fixing the needles to the template, marking their positions on the needles themselves, and checking the protrusion length prior to treatment.

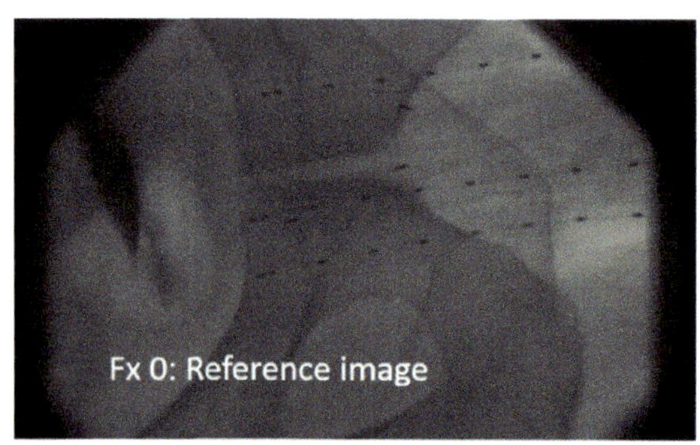

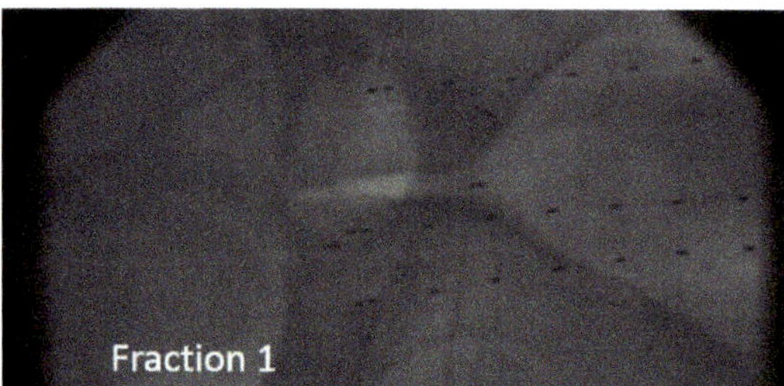

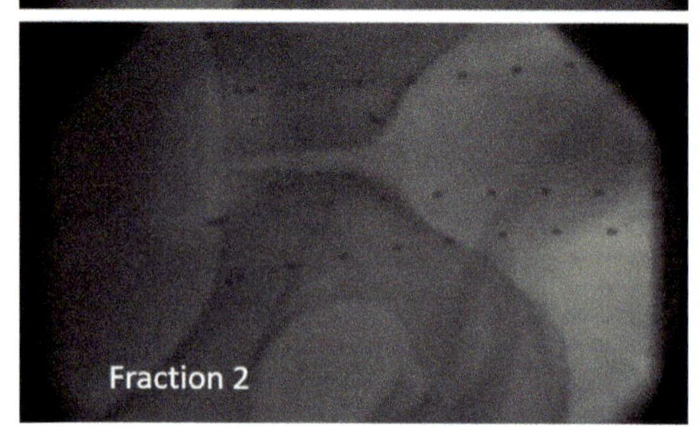

Figure 5–36 Needle tip position can be assessed before each fraction using radiographs and comparing the tip positions to gold markers in the prostate.

5.7.3 QA of Needle Position within the Template

Before the needles are connected to the afterloader via guide tubes, the position of the needles must be verified. This check must be performed before each fraction and consists of two parts: confirming the length of needle protruding from the template has not changed and confirming that the needles have not been displaced relative to the prostate internally. The second step is covered in the next section. The length of needle protruding from the template (Figure 5–35) should be constant throughout the procedure, unless the needles are intentionally adjusted by the physician. This consistency can be confirmed visually by checking that marks made on the needles have not changed position or by measurement of the needle protrusion length with a ruler (as in Figure 5–17). This can be cross verified against measurements made at the time of implant (or following adjustment) or using CT planning images to determine the implanted depth (implanted depth + template thickness + amount protruding = needle length). Each needle may be gently touched to ensure it is still fixed tightly to the template. Special care should be taken when handling needles in a sandwich template, whether for QA or while hooking up the guide tubes, as the peripheral needles will not be fixed as securely as the central needles. As necessitated by equipment or institutional standards, the length of each needle may also be assessed before the first fraction is delivered.

5.7.4 Patient and Needle Positioning

To ensure the treatment is delivered as planned, it is important to maintain the same patient position from planning through each treatment. For intra-operative planning, where a single fraction is delivered per implant, this is implicit. In all other cases, care should be taken to minimize patient movement using positioning and immobilization devices (Peddada et al. 2015). Consistency of patient positioning should be confirmed before each fraction.

Additionally, it is known that needles can be displaced internally between planning and treatment with negative impact on dosimetry (Holly et al. 2011). Thus, if the patient position changes between implant, planning, and treatment, measures must be put in place to ensure that the needles have been adjusted back to the planned position. For example, in Figure 5–36, the patient has had steel needles and two gold markers implanted in the operating room. Upon transfer to the HDR suite, the patient has an AP radiograph taken as a baseline (Figure 5–36 top) with a C-arm unit. Marker wires are placed in needles for visualization and distances between relative points measured (e.g., distance from any given needle tip to a gold marker tip). The patient proceeds to the CT and has treatment planning images acquired, where again, needle positioning relative to the gold markers can be assessed on scout images. After plan approval, the patient has another AP radiograph just prior to initiation of the first treatment to verify no needles have shifted or moved (Figure 5–36 middle). For multiple fractions delivered per implant, patients are transfer to their rooms for 6–24 hours, and then a subsequent fraction is given, again with an AP radiograph prior to treatment to verify the needle positions (Figure 5–36 bottom). Since the patient is not positioned rigidly each time in the brachytherapy suite bed, the needles may appear slightly rotated or moved relative to the patient anatomy. Agreement within 3 mm to planned locations should be achieved (Holly et al. 2011) to maintain planned dosimetry. If movement of the needles is greater than institutional standards, the needles should be repositioned before treatment.

5.7.5 Needle Connection

The connection of the patient's needles to the afterloader can be complicated due to the number of needles and connectors involved in HDR prostate brachytherapy. Between 10 and 20 needles may be used and the area in the vicinity where the connection takes place is small. Multiple individuals should be involved in verifying the connection at the turret end of the after-loader and at the template and needle end (Figure 5–37). If a sandwich template is used, care must be taken not to displace the needles within the template; marks on the needles at the template face provide visual confirmation that the needles have not shifted.

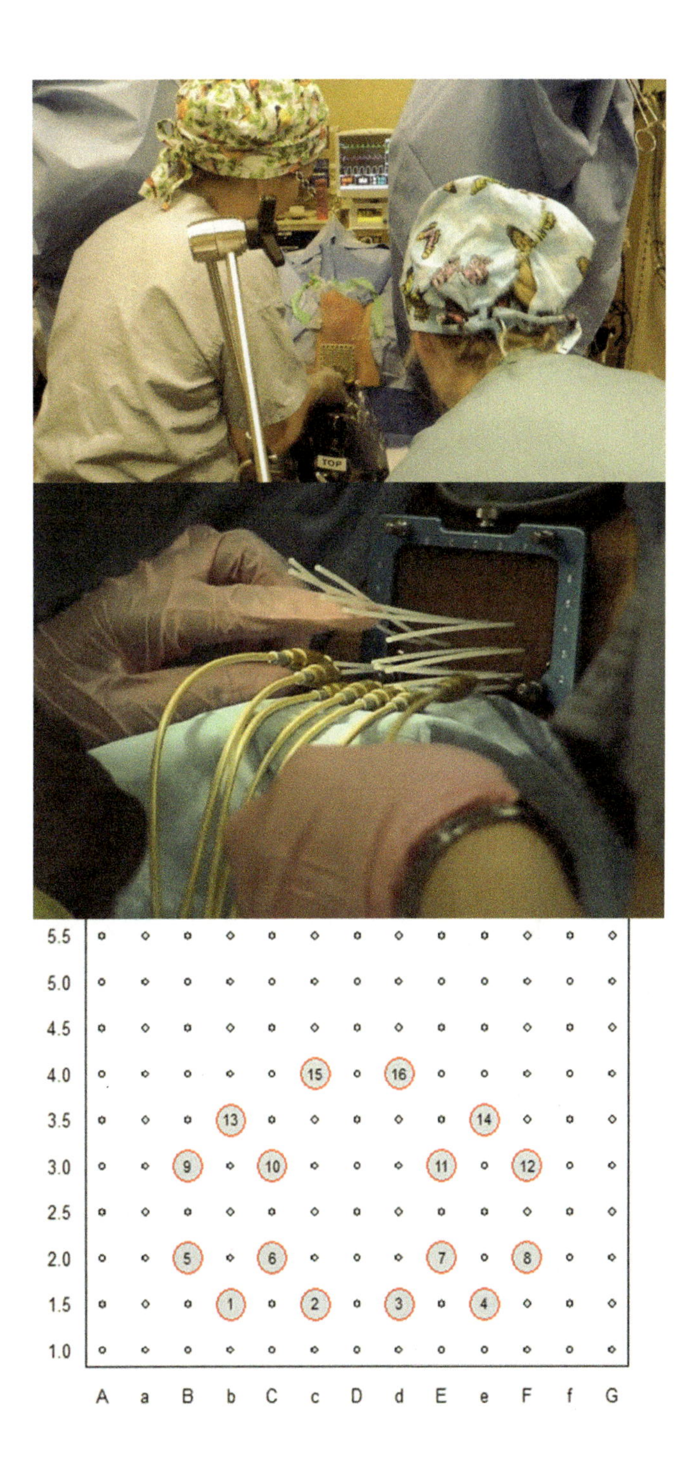

Figure 5–37 Connecting the needles to the afterloader. Top: two staff members should use a printed representation of the template (bottom) with the needle positions and channel numbers indicated to confirm the correct connection. The connection of each needle should be confirmed by both staff members (middle).

5.7.6 Treatment Console Checks

Once the plan has been transferred to the treatment console, the following final checks should be undertaken:

- patient name and ID
- treatment date and time
- correct current source strength
- correct current treatment time
- correct correspondence between template position for each needle, as noted on the plan report or check form, and channel number
- correct treatment time per channel
- correct dwell positions, in particular, most distal position and distances between dwell positions

5.8 Quality Assurance

5.8.1 HDR Unit

Several recommendations for HDR quality assurance have been reported (ACR 2015; CPQR 2015; DeWerd et al. 1995; Kutcher et al. 1994; Venselaar and Perez-Calatayud 2004). When considering prostate treatments, the HDR afterloader QA should follow regular HDR QA procedures (see Section 5.8.3). However, one item to note is that HDR prostate typically employs many channels and needles, possibly to or over the machine limit, depending on the type of afterloader used. In some systems, if the plan requires more needles than available on the turret of the afterloader, the plan is separated into two separate deliveries (e.g., for 25 needles, 20 needles on the first delivery and 5 on the second). Plan transfer checks should be performed to make sure this is handled smoothly by the afterloading console computer and the treatment machine itself; for example, no issues in duplication of the patient name and number, etc. A full plan autoradiograph with a multi-needle jig, such as the one shown in Figure 5–2, can verify the intended treatment length and relative dwell positions for each needle. At some institutions, this is done per patient; more commonly it would be part of daily or quarterly QA.

5.8.2 Templates, Needles, and Guide Tubes

Template and US QA should be performed as recommended by Task Group 128 (Pfieffer et al. 2008). In Figure 5–38, the TRUS probe is fixed into a phantom with the appropriate template on top. Needles are inserted in specific locations, and the depth and visualization of the needle placement can be verified.

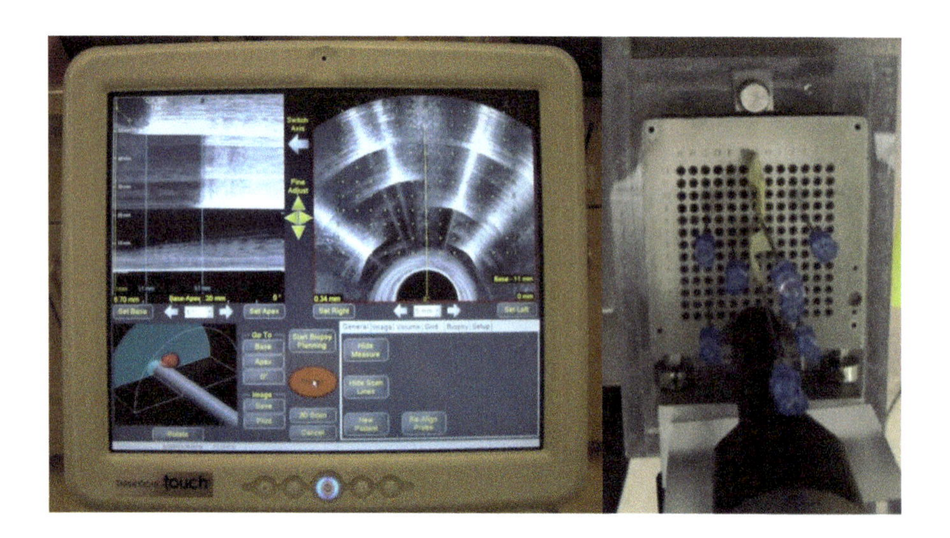

Figure 5–38 Verifying the location, depth, and visualization of the needles in US.

Prior to use, templates should be visually inspected to ensure the correct and anticipated type has been received, as several different types that look similar are sold by the same vendor. Reusable templates should be inspected for damage following sterilization and replaced as necessary. Needle lengths can be verified physically with a dummy wire and by performing audioradiography. Dummy, or representative, needles may be used in the case of single use needles. For reusable needles, their lengths should be checked using autoradiography before being put into clinical use. Usage should be tracked to prevent re-use beyond the manufacturer's number of sterilization cycles. Guide tubes should be visually inspected quarterly and their lengths checked annually.

5.8.3 Useful Forms

When developing institutional QA tests and procedures, all local or national regulatory requirements should be taken into consideration. Tables 5–3, 5–4, and 5–5 are comprehensive, but they may not cover all regulatory environments.

Table 5–3 Table of daily QA checks for a high-dose-rate afterloader based on that used at the British Columbia Cancer Agency–Centre for the Southern Interior and recommendations by the Canadian Partnership for Quality Radiotherapy (CPQR 2015). These checks are performed at the start of each treatment day.

Test	Action
CCTV/intercom	Functional
Survey meter	Functional
Emergency equipment	Available
Last person out button and timer delay	Functional
Primalarm	Functional
AC power alarm	Functional
Warning lights	Functional
Console displays (treatment status indicator) and key switch	Functional
Interrupt button	Functional
Door interlock	Functional
Battery backup	Functional
Console date, time, and source strength	Accurate
Source (and dummy) position check	2 mm
Dwell time accuracy	2%

Table 5–4 Table of checks for a high-dose-rate afterloader to be performed Quarterly or at Source Exchange. Based on that used at the British Columbia Cancer Agency–Centre for the Southern Interior and recommendations by the Canadian Partnership for Quality Radiotherapy (CPQR 2015). Some of these checks may be performed by the vendor's service engineer. A qualified medical physicist should verify that they are completed correctly.

Test	Action
Emergency stop buttons	Functional
Room radiation monitor	Functional
Internal battery power supply (power failure recovery)	Functional
Source/dummy transit interlocks	Functional
Radiation survey (relative to baseline readings)	±15%
Treatment interruption and delivery recovery	Functional
Dummy positional accuracy	1 mm*
Radiological source positioning accuracy	1 mm
Source strength calibration	±5%
Mechanical integrity of applicators, guide tubes, connectors	Functional
Source positioning all channels	±2%
Review of QA records	Complete

*This may be 3 mm for afterloaders where the dummy wire is not used to check catheter length.

Table 5–5 Table of Annual checks for a high-dose-rate afterloader based on that used at the British Columbia Cancer Agency–Centre for the Southern Interior and recommendations by the Canadian Partnership for Quality Radiotherapy (CPQR 2015). As with the quarterly checks, any tests performed by the vendor should be verified by a qualified medical physicist. Some institutions may choose to perform the institutional checks quarterly.

Test	Action
Hand crank operation	Functional
Leakage radiation	Baseline
Multi-channel indexer function	Functional
Dwell time accuracy	±1%
Timer linearity	±1%
Transit time/transit dose reproducibility	Baseline
Dosimetric length of applicators and guide tubes	±1 mm
Applicator/template dimensions	Baseline
Shield integrity for shielded applicators	Baseline
X-ray marker positional accuracy	±1 mm
Review of emergency response procedures for all staff	Complete
Independent review	Complete

5.9 Radiobiology

As discussed at the beginning of this chapter (Section 5.1), the rationale for using HDR brachytherapy for prostate cancer stems from both the technical capabilities of HDR brachytherapy and the radiobiological properties of prostate cancer. The technical aspects of this clinical application have been discussed systematically in the preceding sections. In this section, we review the radiobiological aspects of HDR brachytherapy for prostate cancer and their potential clinical implications.

HDR brachytherapy is typically delivered in multiple fractions, although the current trend is moving toward extreme hypofractionation with fewer fractions for prostate cancer (Yoshioka et al. 2013). The treatment fractions are usually separated from each other with enough time to allow complete damage repair between fractions. Each fraction is delivered within a relatively short time period at a nearly constant dose rate. The temporal pattern of HDR brachytherapy is, therefore, similar to fractionated dose delivery used in external beam radiotherapy. The basic radiobiology of HDR brachytherapy can, therefore, be analyzed using the radiobiological models for fractionated dose delivery, even though the spatial dose distribution of HDR brachytherapy is characteristically more heterogeneous compared to EBRT.

5.9.1 Radiobiological Model(s) for Fractionated Dose Delivery

As discussed in detail in Chapter 1 (Section 1.2), the linear-quadratic (LQ) cell survival model and, in particular, the concepts of biologically effective dose (BED) or equivalent total dose in 2 Gy fractions (EQD2), both derivable from the LQ model, are useful tools for analyzing the radiation effects of fractionated dose delivery. For a total dose of D given n fractions with dose-per-fraction of d, the BED and EQD2 are given, respectively, by

$$BED = D\left(1+\frac{g}{(\alpha/\beta)}d\right)-\frac{\gamma}{\alpha}(T-T_k) \quad (5.1)$$

and

$$EQD2 = BED/[1+2/(\alpha/\beta)]. \quad (5.2)$$

In the above expressions, $d = D/n$; α and β are two model parameters characterizing cellular radiosensitivity to lethal damages caused by one-track and two-track interactions, respectively; The ratio α/β provides an important measure of a cell's radiosensitivity to dose fractionation; g is the dose protraction factor characterizing the effects of intra-fractional sublethal damage repair during the dose delivery of each fraction, with a value bounded between 0 and 1. In general, g approaches 1 in the absence of repair or when the dose of each fraction is delivered much shorter than the damage repair half-life. The last term in Equation (5.1) models the effect of cell proliferation during the course of treatment, where γ represents the cell proliferation rate, T the elapsed time of treatment course, and T_k the onset or lag time of cell proliferation. In addition, Equation (5.2) assumes the equivalent total dose in 2 Gy fractions is delivered instantaneously at each fraction and the cell proliferation effect is either absent or the same as the treatment course under study (see Section 1.1.2.3).

By definition, different dose delivery schemes that have the same BED (or EQD2) would produce the same level of cell killing on a given tissue. The model is, therefore, useful for comparing the relative effectiveness of different dose fractionation schemes. It can also be used to determine alternate fractionation schedules to match the cell killing of a known fractionation scheme.

Several general observations can be drawn from Equation (5.1):

1. BED increases with increasing total dose D, i.e., more dose produces greater biological effect (which is quite obvious).
2. BED increases with increasing dose per fraction, d, i.e., dose delivered with larger dose per fraction (therefore. in fewer fractions) produces greater biological effect.
3. BED increases with decreasing α/β value, i.e., the biologic effect of a given dose and fractionation is greater on tissues having smaller α/β values.
4. BED decreases with increasing intra-fractional repair, i.e., more efficient repair or prolonged dose delivery increases the chance of damage repair and leads to decreased cell kill or biologic effect.

5. Cell proliferation during treatment course always reduces BED, hence, the effectiveness of radiation. These observations are independent of the use of BED; using EQD2 would reach the same conclusions.

The implications of these observations on HDR brachytherapy for prostate cancer are discussed further below.

5.9.2 Influence of α/β

As mentioned earlier, α/β is an important measure of cell's radiosensitivity to dose fractionation. This is easy to appreciate by examining the quantity in the angular brackets of Equation (5.1), also known as the relative effect (RE) because the product of RE and D yields BED in the absence of cell proliferation. As written in Equation (5.1), $(\alpha/\beta)^{-1}$ serves as the coefficient of fraction size d (when $g = 1$), thus measuring the impact of changing fraction size on RE. Tissues with larger α/β would be less responsive to fractionation change.

The influence of α/β on fractionated dose delivery stems primarily from the observation that the α/β values of early-reacting tissues are generally different than those of late-reacting tissues. Early-reacting tissues—including most tumors and highly proliferative normal tissues such as mucosa—typically have larger α/β values (~10 Gy) while late-reacting tissues—such as most normal organs and tissues—have generally lower α/β values (~3 Gy) (Hall and Giaccia 2012). Since the coefficient of fraction size is inversely proportional to α/β in RE, tumors are, therefore, typically less responsive to fraction size change than are late-reacting normal tissues.

The practical importance of this difference relates to the fact that tumors in radiation therapy are typically surrounded by normal tissues/organs. If the tumor and the surrounding normal tissues/organs were to receive the same dose and possess the same intrinsic radiosensitivity (α), the late-reacting normal tissues would suffer greater biological damage than tumors, and this differential response would increase with increasing fraction size. Therefore, dose delivered in larger dose-per-fraction is more detrimental on late reacting normal tissues. Due in part to these considerations, fraction sizes of 1.8 to 2.0 Gy have been used in conventional fractionation schemes. Other considerations include the differences in radiosensitivity (α) and damage repair capacity between tumors and normal tissues (Hall and Giaccia 2012).

It is interesting to note, the difference in α/β between typical tumor and late-reacting normal tissue also favored low dose rate (LDR) dose delivery in traditional brachytherapy applications. As discussed in Section 1.2.3.3, for the conditions described above—and assuming both tumor and normal tissue have a repair half-life of 1.5 h using cervix cancer as an example—it would need 11 fractions of HDR (i.e., smaller dose per fraction) to match both the early and late effects to a continuous LDR dose delivery of 48 Gy in 48 hours (Dale and Jones 1998).

The conclusions reached above would change completely if the α/β value of a tumor would be lower than that of surrounding normal tissue and critical organs. In this scenario, a gain in therapeutic ratio would be achieved by delivering dose in large dose per fraction. This favorable scenario was never thought clinically possible until 1999 when Brenner and Hall reported that the α/β of prostate cancer could be as low as 1.5 Gy based on a retrospective analysis of the biochemical long-term controls achieved using conventional EBRT and LDR permanent brachytherapy for prostate cancer (Brenner and Hall 1999). A series of studies were reported subsequently which seemed to confirm a low (~1.5 to 3 Gy) population-averaged α/β value for prostate cancer (Fowler et al. 2001; Miralbell et al. 2012; Wang et al. 2003; Pedicini et al. 2013). The low α/β value is one of the main rationales behind the move toward hypofractionation in EBRT and HDR brachytherapy for prostate cancer.

5.9.3 Influence of Normal Tissue Dose Sparing

The discussions in Section 5.9.2 tactically assumed the normal organ/tissue received the same dose as tumor. This need not be the case, especially with intensity-modulated radiotherapy (IMRT), stereotactic body radiation therapy, and, particularly, HDR brachytherapy. Depending on their proximity to the target volume, dose to late-reacting critical organs, such as rectum and bladder, can usually be reduced compared to tumor dose. The reduction in dose to normal organs would manifest in both total dose and dose-per-fraction, which can significantly reduce the biologically effective dose to these critical structures. In the case of HDR brachytherapy for prostate cancer, normal organ dose reductions could provide added physical benefit to the already low α/β of prostate cancer for therapeutic gain.

In the context of comparing with LDR brachytherapy as mentioned earlier, when the total dose (and dose per fraction) to the normal tissue is reduced by 20% from those received by tumor in HDR, the number of HDR fractions needed to match both the early and late effects to a continuous LDR dose delivery of 48 Gy in 48 hours could be reduced from 11 to 5 fractions (Dale and Jones 1998). Parenthetically, these considerations revealed that HDR brachytherapy may be more radiobiologically acceptable than initially thought, even for tumors having typical α/β values, because of the normal tissue dose reduction achievable in brachytherapy (Dale 1990; Brenner and Hall 1991).

5.9.4 Influence of Sub-lethal Damage Repair

Most radiobiological analysis of HDR brachytherapy for prostate cancer neglected the effect of intra-fractional damage repair and cell proliferation during the course of treatment. This is justifiable only when the times needed to deliver each fraction and the elapsed time of the treatment course are much shorter than sub-lethal damage repair half-life and tumor cell doubling time, respectively.

The reported damage repair half-life and effective tumor cell doubling time for prostate cancer were 0.27 h and 42 days, respectively, in one of the self-consistent estimates (Wang et al. 2003; Nath et al. 2009). It is safe to neglect the effect of cell proliferation. However, for fractionation schemes using very few fractions, especially the single fraction scheme (e.g., 19 Gy), the dose delivery time for a new ^{192}Ir source can approach that of repair half-life. In addition, due to radioactive decay (half-life ~74 days for ^{192}Ir), the time needed to deliver the same dose would become progressively longer. It would be more than twice as long when the source reaches 90 days old (just before a quarterly source exchange).

For a single 19-Gy scheme, accounting for intra-fractional repair could lead to a reduction of tumor BED by ~10% for patients treated with a new ^{192}Ir source. An additional ~10% reduction would be added for patients treated with a 90-day-old source. The reduction in tumor BED for patients treated in-between source exchanges could be anywhere between 10% and 20% (Tien et al. 2017). BED calculated without taking into account the effect of damage repair would overestimate the potential biologic effect by a variable amount, depending on when a patient is treated relative to source exchange. This could have a negative impact on patient outcome and analysis.

5.9.5 Potential Influence of Other Factors

The simplified LQ model—and, hence, the BED and EQD2 presented in equations (5.1) and (5.2)—does not include cellular processes such as reoxygenation and cell cycle redistribution, both of which are known to enhance the treatment effectiveness in fractionated treatments (Hall and Giaccia 2012, Carlson et al., 2011). Their impact on extremely hypofractionated HDR treatment schemes remains to be fully quantified. In theory, a single-fraction scheme will likely diminish the advantages of fractionation provided by these two factors, and its treatment effectiveness could be severely affected by the presence of hypoxic tumor cells (Carlson et al. 2011).

The heterogeneous dose distribution inherent in HDR brachytherapy is, in principle, radiobiologically advantageous for tumor control so long as all tumor cells receive at least the prescribed dose. With the information of 3D dose distribution, a corresponding 3D BED (or EQD2) distribution can be computed. The presence of radiobiological "cold" or "hot" spots and their impact on overall cell survival can be analyzed and quantified. Nonetheless, proper consideration of the overall effect of dose heterogeneity is warranted when comparing HDR brachytherapy with other treatment modalities such as EBRT.

5.9.6 Further Considerations

The favorable radiobiological properties of prostate cancer and HDR brachytherapy dose distribution have provided a unique opportunity in HDR brachytherapy for prostate cancer. A growing number of clinical studies have been initiated to examine the role of HDR monotherapy for prostate cancer (Yoshioka et al. 2013). Although mature, long-term follow-up data from these studies are yet to be reported, the overall biochemical control rates reported so far have been excellent, with generally acceptable toxicity levels (Yoshioka et al. 2013). Many institutions began using a 4-fraction regimen (e.g., 38 Gy in 4 fractions) in the 1990s and the early 2000s. Recently, the trend is moving toward even fewer fractions: 3-, 2-, and even 1-fraction regimens are being tested and adopted (Yoshioka et al. 2013). The clinical effectiveness and toxicity profiles of these extremely hypofractionated schemes can only be adjudicated by forthcoming clinical results.

It should be noted that the reported α/β values were derived from population-averaged outcome data. While the reported nominal values were consistently low (1.2 to 3.1 Gy), the uncertainties associated with the estimates were quite large, varying from 0.03 Gy up to 4.8 Gy (Fowler et al. 2001; Brenner et al. 2002; Williams et al. 2007). In addition, the α/β values of individual patients could be very different from the derived nominal value. If an individual patient happens to possess an α/β value greater than the nominal value, the favorable radiobiology condition discussed earlier would not be applicable for this patient. Since clinical data are currently analyzed using aggregated date, poor results in patients that could have been caused by their greater-than-nominal α/β value may not be readily appreciated and properly analyzed because of the present lack of a reliable and accurate method for determining patient-specific α/β ratios and other radiobiological parameters.

While the LQ model is very useful in elucidating the radiobiology of dose fractionation, it is well known that the model can over-predict the level of cell killing at large dose per fraction when compared to more sophisticated kinetic reaction rate models (Carlson et al. 2013). This is especially relevant as HDR brachytherapy for prostate cancer is trending toward extreme hypofractionation, with dose per fraction as large as 19 Gy in the single-fraction scheme. One should be cautious when using this model to derive iso-effective hypofractionated schemes. In addition, the LQ model does not include several biological processes that are known to modulate the radiation response of cells, such as tumor hypoxia, low-dose hyper-radiosensitivity, bystander effects, and the possibility of additional biological targets other than the cellular DNA of tumor cells. Therefore, all usual caveats of modeling analysis apply to any radiobiological exercises regarding HDR brachytherapy for prostate cancer.

5.10 Summary

Despite significant heterogeneity in dose and fractionation, as well as diverse planning modalities, HDR brachytherapy provides an effective means of treating prostate cancers of all risk strata. The treatment is well tolerated with low complication rates and excellent long-term disease control rates. Achievement of such results does, however, require meticulous attention to detail at all stages of planning and delivering the treatment. The robustness of HDR brachytherapy for prostate can be attributed to a combination of the precision with which dose distributions can be sculpted to avoid normal tissues and its exploitation of the radiobiology of prostate cancer with large fraction sizes.

References

Al-Mamgani, A., W. L. van Putten, G. J. van der Wielen, P. C. Levendag, and L. Incrocci. (2011). "Dose escalation and quality of life in patients with localized prostate cancer treated with radiotherapy: long-term results of the Dutch randomized dose-escalation trial (CKTO 96-10 trial)." *Int. J. Radiat. Oncol. Biol. Phys.* 79(4):1004–12.

American College of Radiology. 2015 "ACR-AAPM Technical Standard for the Performance of High-Dose-Rate Brachytherapy Physics, Technical Standard." Available at: http://www.acr.org/~/media/EF9F4CCED3C5426B915BC2C91974CE51.pdf. Accessed March 9, 2017.

Andrzejewski, P., P. Kuess, B. Knäusl, K. Pinker, P. Georg, J. Knoth, D. Berger, C. Kirisits, G. Goldner, T. Helbich, R. Pötter, and D. Georg. (2015). "Feasibility of dominant intraprostatic lesion boosting using advanced photon-, proton- or brachytherapy." *Radiother. Oncol.* 117(3):509–14.

Ares C., Y. Popowski, S. Pampallona, P. Nouet, G. Dipasquale, S. Bieri, O. Ozsoy, M. Rouzaud, H. Khan, and R. Miralbell. (2009). "Hypofractionated boost with high-dose-rate brachytherapy and open magnetic resonance imaging-guided implants for locally aggressive prostate cancer: a sequential dose-escalation pilot study." *Int. J. Radiat. Oncol. Biol. Phys.* 75(3):656–63.

Banerjee, R., S. J. Park, E. Anderson, D. J. Demanes, J. Wang, and M. Kamrava. (2015). "From whole gland to hemigland to ultra-focal high-dose-rate prostate brachytherapy: A dosimetric analysis." *Brachytherapy* 14(3):366–72.

Batchelar, D. L., H. T. Chung, A. Loblaw, N. Law, T. Cisecki, and G. C. Morton. (2016). "Intraoperative ultrasound-based planning can effectively replace postoperative CT-based planning for high-dose-rate brachytherapy for prostate cancer." *Brachytherapy* 15(4):399–405.

Batchelar, D., M. Gaztanaga, M. Schmid, C. Araujo, F. Bachand, and J. Crook. (2014). "Validation study of ultrasound-based high-dose-rate prostate brachytherapy planning compared with CT-based planning." *Brachytherapy* 13(1):75–79.

Beckendorf, V., S. Guerif, E. Le Prisé, J. M. Cosset, A. Bougnoux, B. Chauvet, N. Salem, O. Chapet, S. Bourdain, J. M. Bachaud, P. Maingon, J. M. Hannoun-Levi, L. Malissard, J. M. Simon, P. Pommier, M. Hay, B. Dubray, J. L. Lagrange, E. Luporsi, and P. Bey. (2011). "70 Gy versus 80 Gy in localized prostate cancer: 5-year results of GETUG 06 randomized trial." *Int. J. Radiat. Oncol. Biol. Phys.* 80(4):1056–63.

Bertermann, H. and F. Brix. "Ultrasonically guided interstitial high dose rate brachytherapy with Ir-192: Technique and preliminary results in locally confined prostate cancer." In *Brachytherapy HDR and LDR: Remote Afterloading State of the Art.* A. A. Martinez, C. F. Orton, and R. F. Mould, Eds. Leersum, The Netherlands: Nucletron International, 1990.

Borghede, G., H. Hedelin, S. Holmäng, K. A. Johansson, F. Aldenborg, S. Pettersson, G. Sernbo, A. Wallgren, and C. Mercke. (1997). "Combined treatment with temporary short-term high dose rate iridium-192 brachytherapy and external beam radiotherapy for irradiation of localized prostatic carcinoma." *Radiother. Oncol.* 44(3):237–44.

Brenner, D. J. and E. J. Hall. (1991). "Fractionated high dose rate versus low dose rate regimens for intracavitary brachytherapy of the cervix. I. General considerations based on radiobiology." *Br. J. Radiol.* 64(758):133–41/

Brenner, D. J. and E. J. Hall. (1999). "Fractionation and protraction for radiotherapy of prostate carcinoma." *Int. J. Radiat. Oncol. Biol. Phys.* 43(5):1095–1101.

Brenner, D. J., A. A. Martinez, G. K. Edmundson, C. Mitchell, H. D. Thames, and E. P. Armour. (2002). "Direct evidence that prostate tumors show high sensitivity to fractionation (low alpha/betaratio), similar to late-responding normal tissue." *Int. J. Radiat. Oncol. Biol. Phys.* 52(1):6–13.

Brown, D. W., A. L. Damato, S. Sutlief, S. Morcovescu, S. J. Park, J. Reiff, A. Shih, and D. J. Scanderbeg. (2016). "A consensus-based, process commissioning template for high-dose-rate gynecologic treatments." *Brachyhterapy* 15(5):570–77.

Buus, S., S. Rylander, S. Hokland, C. S. Søndergaard, E. M. Pedersen, K. Tanderup, and L. Bentzen. (2016). "Learning curve of MRI-based planning for high-dose-rate brachytherapy for prostate cancer." *Brachytherapy* 15(4):426–34.

Cahlon, O., M. J. Zelefsky, A. Shippy, H. Chan, Z. Fuks, Y. Yamada, M. Hunt, S. Greenstein, and H. Amols. (2008). "Ultra-high dose (86.4 Gy) IMRT for localized prostate cancer: toxicity and biochemical outcomes." *Int. J. Radiat. Oncol. Biol. Phys.* 71(2):330–37.

Canadian Partnership for Quality Radiotherapy (CPQR). (2015). "Technical Quality Control Guidelines for Brachytherapy Remote Afterloaders." Available at: http://www.cpqr.ca/wp-content/uploads/2016/07/BRA-2015-06-02.pdf. Accessed March 9, 2017.

Carlson, D. J., P. Keall, B. W. Loo, Z. Chen, and J. M. Brown. (2011). "Hypofractionation results in reduced tumor cell kill compared to conventional fractionation for tumors with regions of hypoxia." *Int. J. Radiat. Oncol. Biol. Phys.* 79(4):1188–95.

Carlson, D. J., Z. J. Chen, P. J. Hoskin, Z. Ouhib, and M. Zaider. "Radiobiology for Brachytherapy." In: *Comprehensive Brachytherapy: Physics and Clinical Aspects.* J. L. M. Venselaar, D. Baltas, P. Hoskin, and A. S. Meigooni, Eds. Boca Raton, FL: CRC Press, Taylor & Francis Group, 2013.

Chen, C. P., V. Weinberg, K. Shinohara, M. Roach III, M. Nash, A. Gottschalk, A. J. Chang, and I. C. Hsu. (2013). "Salvage HDR brachytherapy for recurrent prostate cancer after previous definitive radiation therapy: 5-year outcomes." *Int. J. Radiat. Oncol. Biol. Phys.* 86(2):324–29.

Citrin, D., H. Ning, P. Guion, G. Li, R. C. Susil, R. W. Miller, E. Lessard, J. Pouliot, X. Huchen, and J. Capala. (2005). "Inverse Treatment Planning based on MRI for HDR Prostate Brachytherapy." *Int. J. Radiat. Oncol. Biol. Phys.* 61(4):1267–75.

Crook, J., A. Ots, M. Gaztañaga, M. Schmid, C. Araujo, M. Hilts, D. Batchelar, B. Parker, F. Bachand, and M. P. Milette. (2014). "Ultrasound-planned high-dose-rate prostate brachytherapy: dose painting to the dominant intraprostatic lesion." *Brachytherapy* 13(5):433–41.

Dale, R. G. (1990). "The use of small fraction numbers in high dose-rate gynaecological afterloading: some radiobiological considerations." *Br. J. Radiol.* 63(748):290–94.

Dale, R. G. and B. Jones. (1998). "The clinical radiobiology of brachytherapy." *Br. J. Radiol.* 71(845):465–83.

De Brabandere, M., K. Haustermans, F. Van den Heuvel, .P Hoskin, and F. A. Siebert. (2012). "Prostate post-implant dosimetry: interobserver variability in seed localization, contouring, and fusion." *Radiother. Oncol.* 104(2)192–98.

Demanes, D.J., D. Brandt, L. Schour, and D. R. Hill. (2009). "Excellent results from high dose rate brachytherapy and external beam for prostate cancer are not improved by androgen deprivation." *Am. J. Clin. Oncol.* 32(4):342–47.

Demanes, D. J., A. A. Martinez, M. Ghilezan, D. R. Hill, L. Schour, D. Brandt, and G. Gustafson. (2011). "High-dose-rate monotherapy: safe and effective brachytherapy for patients with localized prostate cancer." *Int. J. Radiat. Oncol. Biol. Phys.* 81(5):1286–92.

Demanes, D. J. and M. I. Ghilezan. (2014). "High-dose-rate brachytherapy as monotherapy for prostate cancer." *Brachytherapy* 13(6):529–41.

DeWerd, L., P. Jursinic, R. Kitchen, and B. Thomadsen. (1995). "Quality assurance tool for high dose rate brachytherapy." *Med. Phys.* 22(4):435–40.

Erickson, B. A., D. J. Demanes, G. S. Ibbott, J. K. Hayes, I. C. Hsu, D. E. Morris, R. A. Rabinovitch, J. D. Tward, and S. A. Rosenthal. (2011). "American Society for Radiation Oncology (ASTRO) and American College of Radiology (ACR) practice guideline for the performance of high-dose-rate brachytherapy." *Int. J. Radiat. Oncol. Biol. Phys.* 79(3):641–49.

Fraass, B., K. Doppke, M. Hunt, G. Kutcher, G. Starkschall, R. Stern, and J. Van Dyke. (1998). "American Association of Physicists in Medicine Radiation Therapy Committee Task Group 53: quality assurance for clinical radiotherapy treatment planning." *Med. Phys.* 25(10):1773–1829.

Fowler, J., R. Chappell, and M. Ritter. (2001). "Is alpha/beta for prostate tumors really low?" *Int. J. Radiat. Oncol. Biol. Phys.* 50(4):1021–31.

Galalae, R. M., G. Kovács, J. Schultze, T. Loch, P. Rzehak, R. Wilhelm, H. Bertermann, B. Buschbeck, P. Kohr, and B. Kimmig. (2002). "Long-term outcome after elective irradiation of the pelvic lymphatics and local dose escalation using high-dose-rate brachytherapy for locally advanced prostate cancer." *Int. J. Radiat. Oncol. Biol. Phys.* 52(1):81–90.

Georg, D., J. Hopfgartner, J. Gòra, P. Kuess, G. Kragl, D. Berger, N. Hegazy, G. Goldner, and P. Georg. (2014). "Dosimetric considerations to determine the optimal technique for localized prostate cancer among external photon, proton, or carbon-ion therapy and high-dose-rate or low-dose-rate brachytherapy." *Int. J. Radiat. Oncol. Biol. Phys.* 88(3):715–22.

Gomez-Iturriaga, A., F. Casquero, A. Urresola, A. Ezquerro, J. I. Lopez, J. M. Espinosa, .P Minguez, R. Llarena, A. Irasarri, P. Bilbao, and J. Crook. (2016). "Dose escalation to dominant intraprostatic lesions with MRI-transrectal ultrasound fusion High-Dose-Rate prostate brachytherapy. Prospective phase II trial." *Radiother. Oncol.* 119(1):91–96.

Hall, E. J. and A. J. Giaccia. *Radiobiology for Radiologists.* 7th ed. Philadelphia: Lippincott Williams & Wilkins, 2012.

Hegde, J. V., R. V. Mulkern, L. P. Panych, F. M. Fennessy, A. Fedorov, S. E. Maier, and C. M. Tempany. (2013). "Multiparametric MRI of prostate cancer: an update on state-of-the-art techniques and their performance in detecting and localizing prostate cancer." *J. Magn. Reson. Imaging* 37(5):1035–54.

Holly, R., G. C. Morton, R. Sankreacha, N. Law, T. Cisecki, D. A. Loblaw, and H. T. Chung. (2011). "Use of cone-beam imaging to correct for catheter displacement in high dose-rate prostate brachytherapy." *Brachytherapy* 10(4):299–305.

Hoskin, P. J., P. J. Bownes, P. Ostler, K. Walker, and L. Bryant. (2003). "High dose rate afterloading brachytherapy for prostate cancer: catheter and gland movement between fractions." *Radiother. Oncol.* 68(3):285–88.

Hoskin, P. J., A. M. Rojas, P. J. Bownes, G. J. Lowe, P. J. Ostler, and L. Bryant. (2012). "Randomised trial of external beam radiotherapy alone or combined with high-dose-rate brachytherapy boost for localised prostate cancer." *Radiother. Oncol.* 103(2):217–22.

Hoskin, P. J., A. Colombo, A. Henry, P. Niehoff, T. Paulsen Hellebust, F. A. Siebert, and G. Kovacs. (2013). "GEC/ESTRO recommendations on high dose rate afterloading brachytherapy for localised prostate cancer: an update." *Radiother. Oncol.* 107(3):325–32.

Hoskin, P., A. Rojas, P. Ostler, R. Hughes, R. Alonzi, G. Lowe, and L. Bryant. (2014). "High-dose-rate brachytherapy alone given as two or one fraction to patients for locally advanced prostate cancer: acute toxicity." *Radiother. Oncol.* 110(2):268–71.

Hosni, A., M. Carlone, A. Rink, C. Ménard, P. Chung, and A. Berlin. (2017). "Dosimetric feasibility of ablative dose escalated focal monotherapy with MRI-guided high-dose-rate (HDR) brachytherapy for prostate cancer." *Radiother. Oncol.* 2017122 (1):103-108.

Huq, M. S., B. A. Fraass, P. B. Dunscombe, J. P. Gibbons, G. S. Ibbott, A. J. Mundt, S. Mutic, J. R. Palta, F. Rath, B. R. Thomadsen, J. F. Williamson, and E. D. Yorke. (2016). "The report of task group 100 of the AAPM: Application of risk analysis methods to radiation therapy quality management." *Med. Phys.* 43(7):4209.

Kim, Y., I. C. Hsu, E. Lessard, J. Vujic, and J. Pouliot. (2004). "Dosimetric impact of prostate volume change between CT-based HDR brachytherapy fractions." *Int. J. Radiat. Oncol. Biol. Phys.* 59(4):1208–16.

Kim, Y., I. C. Hsu, and J. Pouliot. 2007. "Measurement of craniocaudal catheter displacement between fractions in computed tomography-based high dose rate brachytherapy of prostate cancer." *J. Appl. Clin. Med. Phys.* 8(4):2415.

Kirisits, C., M. J. Rivard, D. Baltas, F. Ballester, M. De Brabandere, R. van der Laarse, Y. Niatsetski, P. Papagiannis, T. Paulsen Hellebust, J. Perez-Calatayud, K. Tanderup, J. L. M. Venselaar, and F. A. Siebert. (2014) "Review of clinical brachytherapy uncertainties: Analysis guidelines of GEC-ESTRO and the AAPM." *Radiother. Oncol.* 110:199–212.

Kovács, G., R. Galalae, T. Loch, H. Bertermann, P. Kohr, R. Schneider, and B. Kimming. (1999). "Prostate preservation by combined external beam and HDR brachytherapy in nodal negative prostate cancer." *Strahlenther. Onkol.* 175(S2):87–88.

Kovács, G. and P. Hoskin. *Interstitial Prostate Brachytherapy LDR-HDR-PDR.* Berlin Heidelberg: Springer-Verlag, 2013.

Kuban, D. A., S. L. Tucker, L. Dong, G. Starkschall, E. H. Huang, M. R. Cheung, A. K. Lee, and A. Pollack. (2008). "Long-term results of the M. D. Anderson randomized dose-escalation trial for prostate cancer." *Int. J. Radiat. Oncol. Biol. Phys.* 70(1):67–74.

Kubo, H. D., G. P. Glasgow, T. D. Pethel, B. R. Thomadsen, and J. F. Williamson. (1998). "High dose-rate brachytherapy treatment delivery: report of the AAPM Radiation Therapy Committee Task Group No. 59." *Med. Phys.* 25(4): 375–403.

Kutcher, G. J., L. Coia, M. Gillin, W. F. Hanson, S. Leibel, R. J. Morton, J. R. Palta, J. A. Purdy, L. E. Reinstein, G. K. Svensson, M. Weller, and L. Wingfield. (1994). "Comprehensive QA for radiation oncology: Report of AAPM Radiation Therapy Committee Task Group 40." *Med. Phys.* 21(4):581–618.

Lachance, B., D. Béliveau-Nadeau, E. Lessard, M. Chrétien, I. C. Hsu, J. Pouliot, L. Beaulieu, and E. Vigneault. (2002). "Early clinical experience with anatomy-based inverse planning dose optimization for high-dose-rate boost of the prostate." *Int. J. Radiat. Oncol. Biol. Phys.* 54(1):86–100.

Lakosi, F., G. Antal, C. Vandulek, A. Kovacs, G. L. Toller, I. Rakasz, G. Bajzik, J. Hadjiev, P. Bogner, and I. Repa. (2011). "Open MR-guided high-dose-rate (HDR) prostate brachytherapy: feasibility and initial experiences open MR-guided high-dose-rate (HDR) prostate brachytherapy." *Pathol. Oncol. Res.* 17(2):315–24.

Lee, B., K. Shinohara, V. Weinberg, A. R. Gottschalk, J. Pouliot, M. Roach, and I. C. Hsu. (2007). "Feasibility of high-dose-rate brachytherapy salvage for local prostate cancer recurrence after radiotherapy: The University of California–San Francisco experience." *Int. J. Radiat. Oncol. Biol. Phys.* 67(4):1106–12.

Lessard, E. and J. Pouliot. (2001). "Inverse planning anatomy-based dose optimization for HDR-brachytherapy of the prostate using fast simulated annealing algorithm and dedicated objective function." *Med. Phys.* 28(5):773–79.

Lizondo, Gisbert M., S. Buus, S. B. Hokland, S. Rylander, E. M. Pedersen, L. Bentzen, and K. Tanderup. (2016). "Dosimetric Impact of Needle Migration in Magnetic Resonance Imaging-Based High-Dose-Rate Brachytherapy" *Int. J. Radiat. Oncol. Biol. Phys.* 96(S2):E278.

Martin, T., C. Kolotas, T. Dannenberg, G. Strassmann, H. G. Vogt, R. Heyd, B. Rogge, D. Baltas, R. Kurek, U. Tunn, and N. Zamboglou. (1999). "New interstitial HDR brachytherapy technique for prostate cancer: CT based 3D planning after transrectal implantation." *Radiother. Oncol.* 52(3):257–60.

Martinez, A., G. K. Edmundson, R. S. Cox, L. L. Gunderson, and A. E. Howes. (1985). "Combination of external beam irradiation and multiple-site perineal applicator (MUPIT) for treatment of locally advanced or recurrent prostatic, anorectal, and gynecologic malignancies." *Int. J. Radiat. Oncol. Biol. Phys.* 11(2):391–98.

Mason, J., B. Al-Qaisieh, P. Bownes, D. Wilson, D. L. Buckley, D. Thwaites, B. Carey, and A. Henry. (2014). "Multi-parametric MRI-guided focal tumor boost using HDR prostate brachytherapy: a feasibility study." *Brachytherapy* 13(2):137–45.

Ménard, C., R. C. Susil, P. Choyke, G. S. Gustafson, W. Kammerer, H. Ning, R. W. Miller, K. L. Ullman, N. Sears Crouse, S. Smith, E. Lessard, J. Pouliot, V. Wright, E. McVeigh, C. N. Coleman, and K. Camphausen. (2004). "MRI-guided HDR prostate brachytherapy in standard 1.5T scanner." *Int. J. Radiat. Oncol. Biol. Phys.* 59(5):1414–23.

Miralbell, R., A. A. Roberts, E. Zubizarreta, and J. H. Hendry. (2012). "Dose-fractionation sensitivity of prostate cancer deduced from radiotherapy outcomes of 5,969 patients in seven international institutional datasets: $\alpha/\beta = 1.4$ (0.9–2.2) Gy." *Int. J. Radiat. Oncol. Biol. Phys.* 82(1):e17–e24.

Monroe, A. T., P. O. Faricy, S. B. Jennings, R. D. Biggers, G. L. Gibbs, and A. V. Peddada. (2008). "High-dose-rate brachytherapy for large prostate volumes (> or =50cc)—Uncompromised dosimetric coverage and acceptable toxicity." *Brachytherapy* 7(1):7–11.

Moradi, M., F. Janoos, A. Fedorov, P. Risholm, T. Kapur, L. D. Wolfsberger, P. L. Nguyen, C. M. Tempany, and W. M. Wells. (2012). "Two solutions for registration of ultrasound to MRI for image-guided prostate interventions." *Conf. Proc. IEEE Eng. Med. Biol. Soc.* 2012:1129–32.

Morton, G., A. Loblaw, P. Cheung, E. Szumacher, M. Chahal, C. Danjoux, H. T. Chung, A. Deabreu, A. Mamedov, L. Zhang, R. Sankreacha, E. Vigneault, and C. Springer. (2011). "Is single fraction 15 Gy the preferred high dose-rate brachytherapy boost dose for prostate cancer?" *Radiother. Oncol.* 100(3):463–67.

Morton, G. C. and P. J. Hoskin. (2013). "Brachytherapy: current status and future strategies—can high dose rate replace low dose rate and external beam radiotherapy?" *Clin. Oncol. (R. Coll. Radiol.)* 25(8):474–82.

Morton, G. C. (2014). "High-dose-rate brachytherapy boost for prostate cancer: rationale and technique." *J. Contemp. Brachytherapy* 6(3):323–30.

Morton, G. C. (2015). "Prostate high-dose-rate brachytherapy: Transrectal ultrasound based planning, a technical note." *Pract. Radiat. Oncol.* 5(4):238–40.

Murgic, J., P. Chung, A. Berlin, A. Bayley, P. Warde, C. Catton, A. Simeonov, J. Abed, G. O'Leary, A. Rink, and C. Menard. (2016). "Lessons learned using an MRI-only workflow during high-dose-rate brachytherapy for prostate cancer." *Brachytherapy* 15(2):147–55.

Nath, R., L. L. Anderson, J. A. Meli, A. J. Olch, J. A. Stitt, and J. F. Williamson. (1997). "Code of practice for brachytherapy physics: report of the AAPM Radiation Therapy Committee Task Group No. 56. American Association of Physicists in Medicine." *Med. Phys.* 24(10):1557–98.

Nath, R., W. S. Bice, W. M. Butler, Z, Chen, A. S. Meigooni, V. Narayana, M. J. Rivard, and Y. Yu. (2009). "AAPM recommendations on dose prescription and reporting methods for permanent interstitial brachytherapy for prostate cancer: Report of Task Group 137." *Med. Phys.* 36(11):5310–22.

Nath, R., M. J. Rivard, L. A. DeWerd, W. A. Dezarn, H. H. Thompson 2nd, G. S. Ibbott, A. S. Meigooni, Z. Ouhib, T. W. Rusçh, F. A. Siebert, and J. L. Venselaar. (2016). "Guidelines by the AAPM and GEC-ESTRO on the use of innovative brachytherapy devices and applications: Report of Task Group 167." *Med. Phys.* 43(6):3178–3206.

Peach, M., D. M. Trifiletti, and B. Libby. (2016). "Systematic review of focal prostate brachytherapy and the future implementation of image-guided prostate hdr brachytherapy using MR-ultrasound fusion." *Prostate Cancer* 2016:1–13.

Peddada, A. V., O. C. Blasi, G. A. White, A. T. Monroe, S. B. Jennings, and G. L. Gibbs. (2015). "Prevention of needle displacement in multifraction high-dose-rate prostate brachytherapy: A prospective volumetric analysis and technical considerations." *Pract. Radiat. Oncol.* 5(4):228–37.

Pedicini, P., L. Strigari, and M. Benassi. (2013). "Estimation of a self-consistent set of radiobiological parameters from hypofractionated versus standard radiation therapy of prostate cancer." *Int. J. Radiat. Oncol. Biol. Phys.* 85(5):e231–e237.

Pérez-Calatayud, J., F. Ballester, R. K. Das, L. A. DeWerd, G. S. Ibbott, A. S. Meigooni, Z. Ouhib, M. J. Rivard, R. S. Sloboda, and J. F. Williamson. (2012). "Dose calculation for photon-emitting brachytherapy sources with average energy higher than 50 keV: Report of the AAPM and ESTRO." *Med. Phys.* 39:2904–29.

Pfeiffer, D., S. Sutlief, W. Feng, H. M. Pierce, and J. Kofler. (2008). "AAPM Task Group 128: quality assurance tests for prostate brachytherapy ultrasound systems." *Med. Phys.* 35(12)5471–89.

Pollack, A., G. Walker, E. M. Horwitz, R. Price, S. Feigenberg, A. A. Konski, R. Stoyanova, B. Movsas, R. E. Greenberg, R. G. Uzzo, C. Ma, and M. K. Buyyounouski. (2013). "Randomized trial of hypofractionated external-beam radiotherapy for prostate cancer." *J. Clin. Oncol.* 31(31):3860–68.

Richardson, S. (2012). "A 2-year review of recent Nuclear Regulatory Commission events: what errors occur in the modern brachytherapy era?" *Pract. Radiat. Oncol.* 2(93):157–63.

Rivard, M. J., B. M. Coursey, L. A. DeWerd, W. F. Hanson, M. S. Huq, G. S. Ibbott, M. G. Mitch, R. Nath, and J. F. Williamson. (2004). "Update of AAPM Task Group No. 43 Report: A revised AAPM protocol for brachytherapy dose calculations." *Med. Phys.* 31(3):633–74.

Rivard, M. J., J. L. M. Venselaar, and L. Beaulieu. (2009). "The evolution of brachytherapy treatment planning." *Med. Phys.* 36:2135–53.

Rogers, C. L., S. C. Alder, R. L. Rogers, S. A. Hopkins, M. L. Platt, L. C. Childs, R. H. Crouch, R. S. Hansen, and J. K. Hayes. (2012). "High dose brachytherapy as monotherapy for intermediate risk prostate cancer." *J. Urol.* 187(1):109–16.

Schmid, M., J. M. Crook, D. Batchelar, C. Araujo, D. Petrik, D. Kim, and R. Halperin. (2013). "A phantom study to assess accuracy of needle identification in real-time planning of ultrasound-guided high-dose-rate prostate implants." *Brachytherapy* 12(1):56–64.

Seppenwoolde, Y., I. K. Kolkman-Deurloo, D. Sipkema, M. de Langen, J. Praag, P. Jansen, and B. Heijmen. (2008). "HDR prostate monotherapy: dosimetric effects of implant deformation due to posture change between TRUS- and CT-imaging." *Radiother. Oncol.* 86(1):114–19.

Siebert, F. A., M. Hirt, P. Niehoff, and G. Kovács. (2009). "Imaging of implant needles for real-time HDR-brachytherapy prostate treatment using biplane ultrasound transducers." *Med. Phys.* 36(8):3406–12.

Siebert, F. A., S. Wolf, H. Bertermann, N. Nürnberg, G. Bockelmann, and B. Kimmig. (2014). "Introduction of inverse dose optimization for ultrasound-based high-dose-rate boost brachytherapy—How we do it in Kiel." *Brachytherapy* 13(3):250–56.

Skowronek, J. (2013). "Low-dose-rate or high-dose-rate brachytherapy in treatment of prostate cancer—between options." *J. Contemp. Brachytherapy* 5(1):33–41.

Smith, W. L., C. Lewis, G. Bauman, G. Rodrigues, D. D'Souza, R. Ash, D. Ho, V. Venkatesan, D. Downey, and A. Fenster. (2007). *Int. J. Radiat. Oncol. Biol. Phys.* 67(4):1238–47.

Thomadsen, B. R., B. A. Erickson, P. J. Eife, I. C. Hsu, R. R. Patel, D. G. Petereit, B. A. Fraass, and M. J. Rivard. (2014). "A review of safety, quality management, and practice guidelines for high-dose-rate brachytherapy: executive summary." *Pract. Radiat. Oncol.* 4(2):65–70.

Tien, C. J., D. J. Carlson, R. Nath, and Z. Chen. (2017). "HDR monotherapy in prostate cancer: the impact of cellular repair and repopulation on biologically effective dose." Submitted for American Brachytherapy Society 2017 Annual Meeting, private communication.

Tong, S., D. B. Downey, H. N. Cardinal, and A. Fenster. (1996). "A three-dimensional ultrasound prostate imaging system." *Ultrasound Med. Biol.* 22(6):735–46.

Tong, S., H. N. Cardinal, D. B. Downey, and A. Fenster. (1998). "Analysis of linear, area and volume distortion in 3D ultrasound imaging."*Ultrasound Med. Biol.* 24(3):355–73.

Turkbey, B., A. M. Brown, S. Sankineni, B. J. Wood, P. A. Pinto, and P. L. Choyke. (2016). "Multiparametric prostate magnetic resonance imaging in the evaluation of prostate cancer." *CA Cancer J. Clin.* 66(4):326–36.

Vargas, C. E., A. A. Martinez, T. P. Boike, W. Spencer, N. Goldstein, G. S. Gustafson, D. J. Krauss, and J. Gonzalez. (2006). "High-dose irradiation for prostate cancer via a high-dose-rate brachytherapy boost: results of a phase I to II study." *Int. J. Radiat. Oncol. Biol. Phys.* 66(2):416–23.

Venselaar, J. and J. Pérez-Calatayud, Eds. *A Practical Guide to Quality Control of Brachytherapy Equipment, European Guidelines for Quality Assurance in Radiotherapy.* Booklet no. 8. Brussels, Belgium: ESTRO, 2004.

Vigneault, E., K. Mbodji, M. É. Beaudet, P. Després, M. C. Lavallée, A. G. Martin, W. Foster, S. Aubin, and L. Beaulieu. (2016). "Does Prostate Volume Has an Impact on Biochemical Failure in Patients with Localized Prostate Cancer Treated with HDR Boost?" *Radiother. Oncol.* 121(2):304–9.

Wang, J. Z., M. Guerrero, and X. A. Li. (2003). "How low is the alpha/beta ratio for prostate cancer?" *Int. J. Radiat. Oncol. Biol. Phys.* 55(1):194–203.

White, E. C., M. R. Kamrava, J. Demarco, S. J. Park, P. C. Wang, O. Kayode, M. L. Steinberg, and D. J. Demanes. (2013). "High-dose-rate prostate brachytherapy consistently results in high quality dosimetry." *Int. J. Radiat. Oncol. Biol. Phys.* 85(2):543–48.

Williams, S. G., J. M. Taylor, N. Liu, Y. Tra, G. M. Duchesne, L. L. Kestin, A. Martinez, G. R. Pratt, and H. Sandler. (2007). "Use of individual fraction size data from 3756 patients to directly determine the alpha/beta ratio of prostate cancer." *Int. J. Radiat. Oncol. Biol. Phys.* 68(1):24–33.

Wybranski, C., B. Eberhardt, K. Fischbach, F. Fischbach, M. Walke, P. Hass, F. W. Röhl, O. Kosiek, M. Kaiser, M. Pech, L. Lüdemann, and J. Ricke. (2015). "Accuracy of applicator tip reconstruction in MRI-guided interstitial ^{192}Ir-high-dose-rate brachytherapy of liver tumors." *Radiother. Oncol.* 115(1):72–77.

Yamada, Y., L. Rogers, D. J. Demanes, G. Morton, B. R. Prestidge, J. Pouliot, G. N. Cohen, M. Zaider, M. Ghilezan, and I. C. Hsu. (2012). "American Brachytherapy Society consensus guidelines for high-dose-rate prostate brachytherapy." *Brachytherapy* 11(1):20–32.

Yamada, Y., M. A. Kollmeier, X. Pei, C. C. Kan, G. N. Cohen, S. M. Donat, B. W. Cox, and M. J. Zelefsky. (2014). "A Phase II study of salvage high-dose-rate brachytherapy for the treatment of locally recurrent prostate cancer after definitive external beam radiotherapy." *Brachytherapy* 13(2):111–16.

Yoshioka, Y., K. Yoshida, H. Yamazaki, N. Nonomura, and K. Ogawa. (2013). "The emerging role of high-dose-rate (HDR) brachytherapy as monotherapy for prostate cancer." *J. Radiat. Res.* 54(5):781–88.

Yoshioka, Y., O. Suzuki, Y. Otani, K. Yoshida, T. Nose, and K. Ogawa. (2014). "High-dose-rate brachytherapy as monotherapy for prostate cancer: technique, rationale and perspective." *J. Contemp. Brachytherapy* 6(1):91–98.

Yoshioka, Y., O. Suzuki, F. Isohashi, Y. Seo, H. Okubo, H. Yamaguchi, M. Oda, Y. Otani, I. Sumida, M. Uemura, K. Fujita, A. Nagahara, T. Ujike, A. Kawashima, K. Yoshida, H. Yamazaki, N. Nonomura, and K. Ogawa. (2016). "High-Dose-Rate Brachytherapy as Monotherapy for Intermediate- and High-Risk Prostate Cancer: Clinical Results for a Median 8-Year Follow-Up." *Int. J. Radiat. Oncol. Biol. Phys.* 94(4):675–82.

Zelefsky, M. J., Y. Yamada, Z. Fuks, Z. Zhang, M. Hunt, O. Cahlon, J. Park, and A. Shippy. (2008). "Long-term results of conformal radiotherapy for prostate cancer: impact of dose escalation on biochemical tumor control and distant metastases-free survival outcomes." *Int. J. Radiat. Oncol. Biol. Phys.* 71(4):1028–33.

Zietman, A. L., K. Bae, J. D. Slater, W. U. Shipley, J. A. Efstathiou, J. J. Coen, D. A. Bush, M. Lunt, D. Y. Spiegel, R. Skowronski, B. R. Jabola, and C. J. Rossi. (2010). "Randomized trial comparing conventional-dose with high-dose conformal radiation therapy in early-stage adenocarcinoma of the prostate: long-term results from proton radiation oncology group/american college of radiology 95-09." *J. Clin. Oncol.* 28(7):1106–11.

Example Problems

1. True or False. The α/β value of prostate cancer can be lower than that of typical late-reacting normal tissues, making hypofractionated HDR brachytherapy a suitable treatment option for prostate cancer.
2. True or False. In MR based planning for prostate HDR, it is easier to meet the dosimetric constraints for the rectal wall than for other planning modalities
3. An advantage of MR planning for prostate HDR is:
   a. It can be performed more quickly than other planning modalities.
   b. It can be performed with larger slice thickness, decreasing the scanning time.
   c. Extra prostatic disease and other critical structures can be readily visualized.
   d. There is no imaging distortion with MR.
4. True or False. Dose calculation for HDR prostate brachytherapy should never be conducted using the TG-43 formalism.
5. What parameter should be >100% in an HDR brachytherapy prostate treatment plan?
   a. Urethra D_{10}
   b. CTV D_{90}
   c. CTV V_{150}
6. The main document that describes complete ultrasound and prostate brachytherapy QA processes is:
   a. TG-43
   b. TG-56
   c. TG-128
   d. TG-182
7. True or False. For a second fraction treatment, if the length of the catheter beyond the template has remained the same as fraction 1, the needles have not moved within the prostate.
8. Which of the following MRI sequences is NOT used to delineate a DIL?
   a. T2 weighted images
   b. perfusion-weighted images
   c. dynamic contrast-enhanced images
   d. diffusion-weighted images
9. True or False. Needle tip reconstruction accuracy is equivalent in TRUS- and CT-based planning.

10. Which of the following regarding HDR brachytherapy for prostate are true (list all)?
    a. It can be used to treat low-, intermediate-, and high-risk prostate cancer.
    b. It can be used as a boost, as monotherapy, or as a salvage treatment.
    c. There is consensus regarding the most appropriate dose and fractionation schedule.
    d. Clinical results are good, both in terms of outcome and toxicity.
    e. There is only one way to plan and treat.
    f. It provides higher dose escalation than EBRT.
11. True or False. An advantage of TRUS-based HDR brachytherapy for prostate is that imaging, planning, and treatment proceed without moving the patient.

Chapter 6

Brachytherapy for Gynecological Malignancies

Antonio L. Damato[1], Luc Beaulieu[2], Firas Mourtada[3], Sujatha Pai[4], Jason J. Rownd[5], Daniel J. Scanderberg[6], Amandeep Taggar[7], Bruce R. Thomadsen[8], and Christian Kirisits[9]

[1]Memorial Sloan Kettering Cancer Center
New York, New York

[2]Université Laval
Québec, Canada

[3] Christiana Care Hospital
Newark, Delaware

[4]Memorial Hermann Texas Medical Center, LMP
Houston, Texas

[5]Medical College of Wisconsin
Milwaukee, Wisconsin

[6]University of California–San Diego
La Jolla, California

[7]Memorial Sloan Kettering Cancer Center
New York, New York

[8]University of Wisconsin
Madison, Wisconsin

[9]Medical University of Vienna
Vienna, Austria

6.1 Introduction

Brachytherapy is an essential component of curative treatment of an array of gynecologic malignancies. The importance of brachytherapy in gynecologic cancers has been demonstrated by a large body of literature showing excellent local control, improvement in survival, and low toxicity (Syed et al. 2002; Nakano et al. 2005; Potter et al. 2007; Nout et al. 2010; Potter et al. 2011; Charra-Brunaud et al. 2012; Lee et al. 2013; Townamchai et al. 2013; Brown et al. 2015; Fallon et al. 2016; Manuel et al. 2016). The success of brachytherapy in gynecologic cancers also makes intuitive sense, given the relative accessibility of the treatment sites via intracavitary insertions. With applicators placed inside the tumor through body cavities such as the vagina and uterine canal, very high doses of radiation to the target can be delivered while sparing surrounding normal tissues. In most cases, these insertions do not require patient hospitalization. Sedation and pain management is provided depending on the type of application, clinical practice, and an individual patient's need. Brachytherapy has been used for the treatment of endometrial, cervical, vaginal, and vulvar cancers. A vast selection of applicators exists for each of these indications. In modern times, high-dose-rate (HDR) and pulsed-dose-rate (PDR) techniques have mostly replaced the temporary low-dose-rate (LDR) techniques. Permanent LDR implants are rarely performed for gynecologic malignancies, although reports of their use in modern times exist (Wooten et al. 2014). Given the variability in indications, applicators, clinical practices, and modalities, gynecologic brachytherapy encompasses a broad variety of treatment techniques and defies generalization.

This chapter focuses on the two most common indications of gynecologic brachytherapy: post-operative endometrial and cervix. The objective of this chapter is to provide practical guidance for the physicist in starting and maintaining a gynecologic brachytherapy service. Aspects to be covered include: an assessment of imaging needs for various procedures, the availability of equipment and applicators, treatment planning, and establishing a quality management program (QMP). In most cases, we refer to HDR techniques, which are most commonly practiced nowadays. The reader can also refer to the vast literature of recommendations and guidelines published by the Groupe European de Curietherapy and the European Society for Radiotherapy and Oncology (GEC-ESTRO) and by the American Brachytherapy Society (ABS) on endometrial (Small et al. 2012; Schwarz et al. 2015), cervical (Haie-Meder et al. 2005; Potter et al. 2006; Lee et al. 2012; Viswanathan et al. 2012; Viswanathan et al. 2012), and vaginal (Beriwal et al. 2012) brachytherapy. The most recent and comprehensive guidelines for cervical cancer are in ICRU/GEC-ESTRO Report 89.

6.2 Vaginal Cuff Brachytherapy

6.2.1 Introduction

Phase III randomized trials that established the importance of adjuvant radiotherapy post total abdominal hysterectomy (TAH) for endometrial cancer noted approximately 20% higher local-regional relapses in the observation arms without radiotherapy, and 75% of these relapses were at the vaginal cuff (Creutzberg et al. 2000; Keys et al. 2004). Pathological features associated with higher risk of relapse include: >50% myometrial invasion (MI), higher grade, lymphovascular space invasion (LVSI), or older age. These findings led to the post-operative radiation therapy in endometrial cancer trial (PORTEC 2) where women with risk features listed above and considered "intermediate risk" were randomized to pelvic external-beam radiation therapy (EBRT) or vaginal cuff brachytherapy (Nout et al. 2010). The primary endpoint of vaginal cuff recurrences was equivalent in both arms (1.6% vs 1.8%, p=0.7). However, a small but significant difference existed in pelvic recurrences (0.5% vs 3.8%, p=0.02). The most recent study that compared pelvic EBRT and vaginal cuff brachytherapy to vaginal cuff brachytherapy alone for "intermediate risk" patients confirmed the findings of PORTEC 2 trial—no significant difference exists in overall survival or vaginal relapses in either of the arms, but a small but significant difference was present in pelvic relapses (1.5% vs 5%, p=0.013) (Sorbe et al. 2012). Patients who underwent vaginal cuff brachytherapy alone, however, had significantly lower gastrointestinal (GI) and genitourinary (GU) toxicity, as well as better quality of life compared to those who received an EBRT component. Low pelvic recurrences in both arms with high GI and GU toxicity involved with pelvic EBRT led the authors of both studies to conclude that vaginal cuff brachytherapy is the preferable treatment modality for these patients. A panel of gynecological experts from the American Society for Radiation Oncology (ASTRO) convened and produced guidelines and indications for vaginal cuff brachytherapy (Klopp et al. 2014). Current indications for vaginal cuff brachytherapy based on high-quality evidence include: 1) patients with Grade 1 or 2 cancers with >50% myometrial invasion (MI), and 2) patients with Grade 3 cancers with <50% MI. However, vaginal cuff brachytherapy is also often recommended in older women due to its lower toxicity profile and better quality of life.

6.2.2 Applicators

The most commonly used applicator in vaginal cuff brachytherapy is the vaginal cylinder. Traditional HDR brachytherapy cylinders are composed of metal and plastic parts (Figure 6–1). Specifically, there are often metal rings embedded on the cylinder surfaces to help correctly identify cylinder size differences on radiographic images. CT- and MR-compatible HDR brachytherapy cylinders are generally composed of plastic and carbon fiber to minimize artifacts during CT and MRI scans (Figure 6–2). Radiographs of CT MR com-

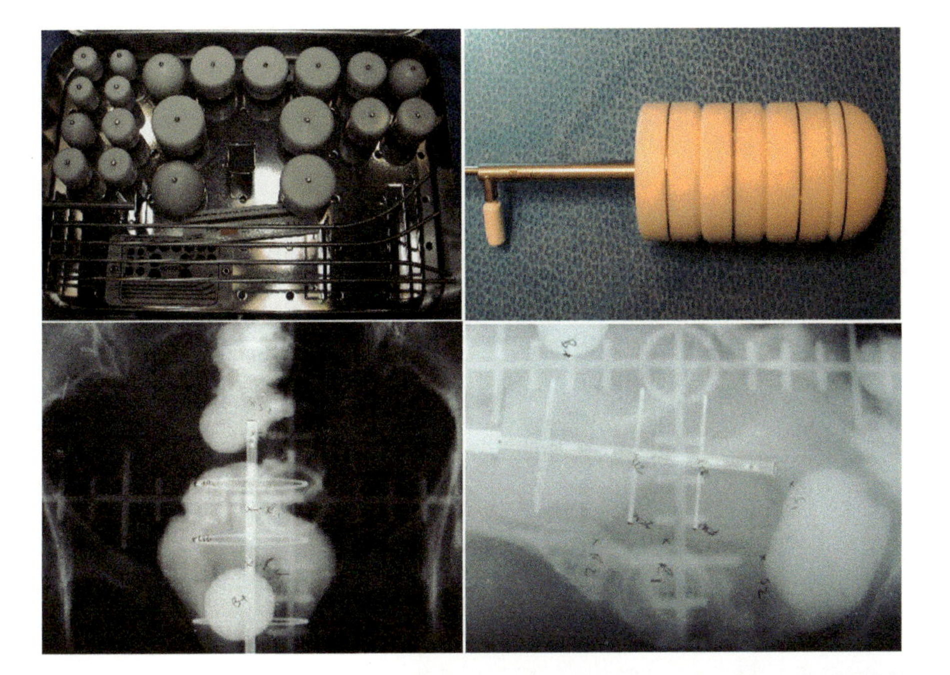

Figure 6–1 Traditional HDR ^{192}Ir brachytherapy cylinder applicator set. Set in a sterilization container (top left) and assembled applicator (top right). AP (bottom left) and lateral (bottom right) radiographic images show the central metallic catheter and the metallic rings marking the cylinder outer shape.

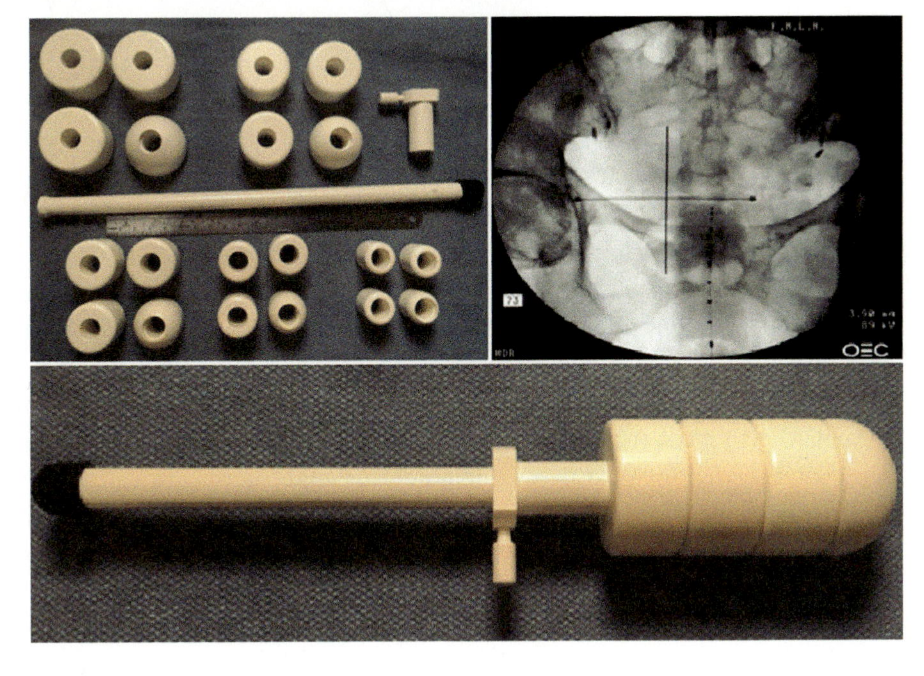

Figure 6–2 In the top left panel, a CT/MR-compatible HDR ^{192}Ir vaginal brachytherapy cylinder set is separated out into the various diameter groups, with a fully assembled cylinder shown in the bottom panel. The top right panel shows an AP fluoroscopic image that illustrates the applicator localization using added x-ray markers inserted into the central lumen of the applicator.

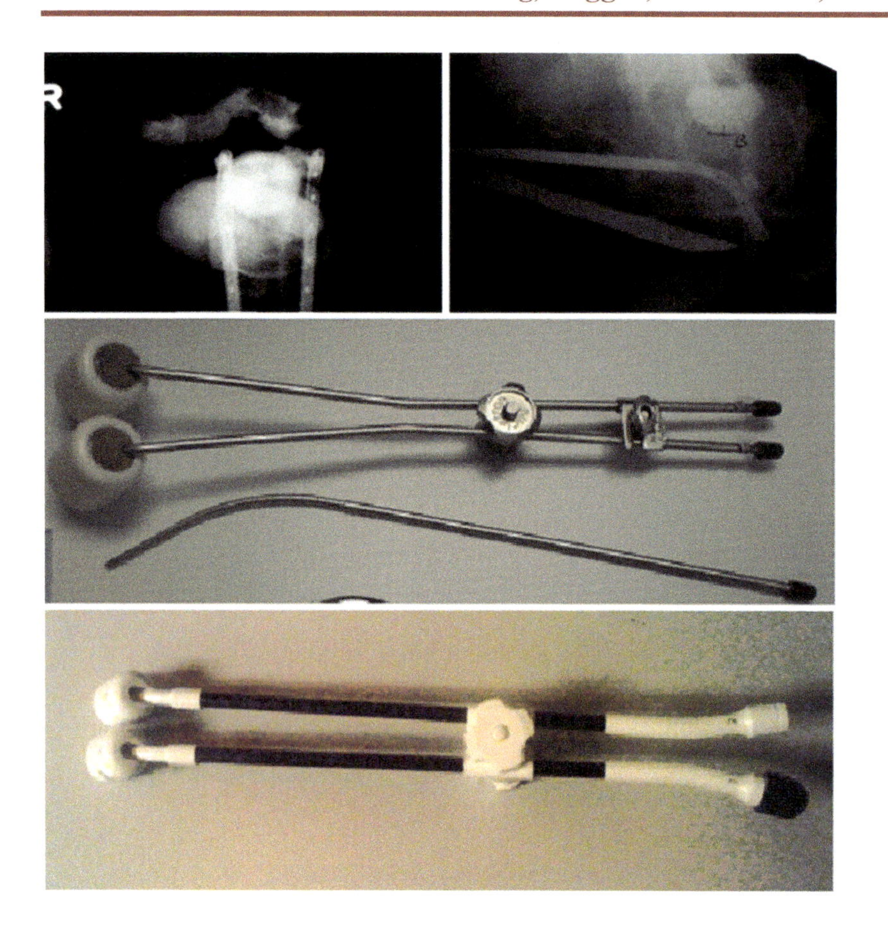

Figure 6–3 The top panel shows the orthogonal radiographic images of a traditional metal HDR ^{192}Ir colpostats, or ovoids, applicator with the actual applicator displayed in the middle panel. The bottom panel shows a CT/MR-compatible version of this applicator type for HDR ^{192}Ir brachytherapy.

patible applicators typically need x-ray opaque marker to help visualize the other transparent applicator. Colpostats, also called ovoids, have also been used (Figure 6–3). Ovoids can be advantageous in dosing the vaginal fornices but are, in general, considered harder to insert. Irradiation of the vagina longer than 3 cm requires the use of a cylinder applicator.

6.2.3 Vaginal Cylinder Planning

6.2.3.1 Role of Imaging

In radiotherapy, imaging is typically used for simulation, planning, and treatment verification. The adoption of 3D imaging for treatment planning of vaginal cuff brachytherapy in North America has increased from 18% in 2003 (Small et al. 2005) to 83% in 2016 (Harkenrider et al. 2016), although 73% of those performing image-based planning do so only on the first fraction. For vaginal cuff brachytherapy cases treated with a vaginal cylinder, it has been shown in one study that inter-fraction variation of organs-at-risks dose does not warrant re-planning at each insertion (Holloway et al. 2011). Despite the preference of most practitioners to perform 3D treatment planning for the first fraction, the clinical benefit for doing so instead of using a standard plan (that is, a plan generated based only on idealized geometry, not on patient specific anatomy) is unclear (Small et al. 2012; Harkenrider et al. 2017). Planning based on 3D imaging may not be required, and a standard plan given the specific cylinder dimension, treatment length, and dose specification can be used.

Volume-based imaging may be required for applicators with variable geometry, such as inflatable balloon applicators or applicators for which the dose is sensitive to rotational alignment of the applicator, such as multi-channel cylinders. Also, volume-based imaging can be used to verify that substantial air gaps are not present between the applicator and the vaginal surface (toward ensuring tissue contact with the applicator). Volume-based planning is performed in all vaginal cylinder cases where the attending physician requests to modify the dose distribution to accommodate specific anatomical considerations. CT- and MR-compatible applicators for use in vaginal cuff brachytherapy are commercially available. Imaging properties and MR safety need to be independently verified in phantom before clinical use of these applicators. CT is, in general, sufficient to contour organs at risk. The target is typically not contoured, and the vaginal cuff may be difficult to visualize on CT, particularly for patients with deep, undistended vaginal fornices (Chapman et al. 2016). Use of MRI may assist in contouring the vaginal tissue, but logistics, costs, and the difficulty in applicator reconstruction on MRI has limited its use in this setting.

6.2.3.2 Prescription Parameters

The prescription should contain all information necessary to plan a vaginal cuff brachytherapy case, that is: dose and number of fractions, dose specification, cylinder diameter, and treatment length. A broad range of fractionation schemes have been reported, and these are summarized in Table 6–1 (Harkenrider et al. 2017). The dose is usually specified either to the vaginal surface or to a 5-mm depth. When the dose is specified to the vaginal surface, the dose to 5 mm in tissue varies depending on the cylinder size (see example in Figure 6–4). Conversely, if the dose is specified to a 5-mm depth, the dose to the vaginal surface varies depending

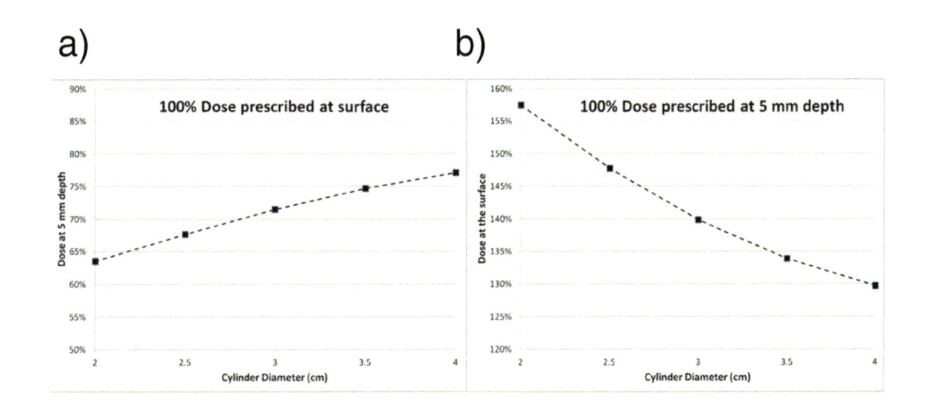

Figure 6–4 a) Variability of dose at 5 mm depth with cylinder size if dose prescription is specified at the vaginal. b) Variability of dose at the vaginal surface with cylinder size if dose prescription is specified at 5 mm depth. Actual doses will vary depending on specific optimization technique and clinical practice.

Table 6–1 Common fractionation schemes for vaginal cuff brachytherapy (Townamchai et al. 2012; Harkenride et al. 2016; Harkenrider et al. 2017)

	Dose per Fraction	Prescribed at
Brachytherapy Alone (monotherapy)	7.5 Gy × 3 fractions	surface
	7.0 Gy × 3 fractions	5 mm
	6.0 Gy × 5 fractions	surface
	6.0 Gy × 4 fractions	5 mm
	5.5 Gy × 4 fractions	5 mm
	5.0 Gy × 4 fractions	5 mm
	4.0 Gy × 6 fractions	surface
	3.0 Gy × 6 fractions	5 mm
Brachytherapy After EBRT	6.0 Gy × 3 fractions	surface
	5.5 Gy × 3 fractions	5 mm
	5.0 Gy × 3 fractions	5 mm
	5.0 Gy × 3 fractions	surface
	4.0 Gy × 3 fractions	0.5 cm
	4.0 Gy × 3 fractions	surface

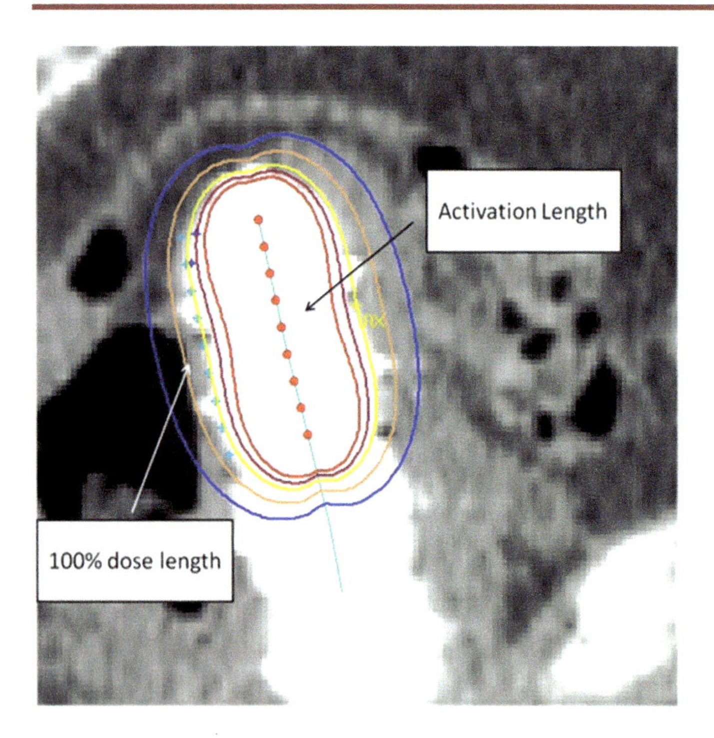

Figure 6–5 An illustration of the possible ambiguity of the term "treatment length," which may be used loosely to mean either the length of the source train ("activation length") or of the prescription isodose surface ("100% dose length").

on cylinder size. Note that as a function of cylinder size, the dose variation at the vaginal surface is higher than the dose variation at 5 mm. There is no recommendation for one or the other specification methodology, but the ABS recommended reporting the dose to both locations. The presence of two common dose specification methodologies presents two challenges when initiating a new brachytherapy service. First, the physicist needs to present a clear assessment of the uncertainty in the dose to the points not used for dose specification, given the clinic-specific planning practice. Second, the physicist needs to ensure that no confusion between prescription modalities can arise during clinical practice. The simplest methodology is to use only one dose specification methodology in a given practice. Interestingly, 53% of the clinics in North America report using both modalities: 33% adopt only a prescription at 5 mm depth, and 13% adopt only a prescription at the vaginal surface (<1% report using some other dose specification) (Harkenrider et al. 2016).

The geometrical parameters of the prescription are the vaginal cylinder diameter and the treatment length. The cylinder diameter is selected to be the largest that the patient can tolerate to ensure contact between the vaginal surface and the applicator. In some cases, this may not be possible, in particular for patients with an introitus that is smaller than the vaginal cavity and patients with large vaginal fornices. In these cases, colpostats or balloon applicators may provide better conformity to the tissue (Kim et al. 2002; Park et al. 2013; Damato et al. 2015). The treatment length should also be specified on the prescription. Practice varies on treatment lengths, with a majority of institutions prescribing a fixed length of 3 to 5 cm, or to the superior 1/3 or 2/3 of the vagina (Harkenrider et al. 2016). The exact meaning of treatment length in the planning process should be discussed when initiating a brachytherapy practice: some institutions interpret it as the activation length of the applicator, while other institutions define it as the length receiving the prescription dose at the prescribed depth (Figure 6–5). Treatments using ovoids usually cover approximately the upper 1/3 of the vagina.

6.2.3.3 Optimization Systems

Most commercial TPSs offer an array of optimization options. For vaginal cuff brachytherapy performed with a vaginal cylinder, the most common approach is to perform a point-based optimization (see Chapter 2).

Table 6–2 Example of distance from the central axis to the vaginal surface for a given cylinder size as used at the Medical College of Wisconsin for Elekta vaginal cylinders. These may change depending on manufacturer and model.

Dwell Position along +/- Lateral Axis	2.0 cm Diameter	2.5 cm Diameter	3.0 cm Diameter	3.5 cm Diameter	4.0 cm Diameter
1	9	11	12	13	15
2	10	12.5	14	16	18
3	10	12.5	15	17.5	20
4	10	12.5	15	17.5	20
5+ to last active dwell	10	12.5	15	17.5	20

This approach describes a series of points and requests that the treatment planning system (TPS) varies the dwell time distribution within the active length of the vaginal cylinder to ensure that the average dose to those points matches a given objective (typically, the prescription dose). The specific optimization process needs to be documented in the medical records, e.g., with a detailed printout of the plan. The simplest pattern is composed of points along a straight line at a fixed distance from the central channel. This configuration is simple to implement and produces symmetrical dwell time distributions that are insensitive to directional errors in catheter reconstruction. The downside of this approach is that the points do not represent the actual shape of a vaginal cylinder, which is domed on top. As a result, the dose to the apex of the vagina is typically higher than the prescription dose. Moreover, the difference between the apex dose and the prescription dose increases as the cylinder diameter increases. To avoid this feature, it is possible to laterally adjust the positions of the superior optimization points to conform to the expected shape of the vaginal cylinder dome. In this configuration, an asymmetrical dwell time distribution is obtained. In a 2003 survey, 83% of clinics in North America reported tapering the dose points to the dome of the cylinder (Small et al. 2005). Table 6–2 shows an example of lateral distances from each dwell position to the cylinder surface for typical vaginal cylinders.

While there is no clinical data suggesting that one approach is superior to the other, the ABS recommends the use of tapered dose points with an apex point positioned at the tip of the cylinder (Small et al. 2012). Other clinics use custom geometry for the location of the optimization points. Moreover, the optimization problem of finding a distribution of dwell times delivering a given dose on average to a set of points has, in general, multiple solutions. Most planning systems provide the opportunity to set a parameter indicating the maximum acceptable variability between dwell times, and this additional factor is used to obtain a unique solution. Some institutions adopt optimization strategies aimed at increasing the dose toward the apex, instead of tapering it to follow the cylinder dome. The variability in optimization strategies results in a broad range of possible dose distributions associated to the same prescription. This fact complicates the comparison of the results among institutions.

6.2.4 Quality Assurance

6.2.4.1 Potential Pitfalls

Reviewing the summaries compiled by the Advisory Committee on the Medical Uses of Isotopes (ACMUI) of the U.S. Nuclear Regulatory Commission (NRC) of medical events reported to the Nuclear Materials Event Database, vaginal cuff irradiation using cylinders forms one of the most common types of reports (Advisory Committee on the Medical Uses of Isotopes of the US Nuclear Regulatory Commission 2014). Given the prevalence of vaginal cuff brachytherapy, the higher number of reports may not indicate that vaginal cuff brachytherapy is more prone to errors than other brachytherapy techniques. The reports may seem surprising given vaginal cuff treatments are one of the simplest brachytherapy modalities, yet the simplicity may play an important part in these events. Because the treatments are not technologically challenging, facilities that perform no other type of brachytherapy treat the vaginal cuff. This may indicate that such facilities do not have a well-developed and robust brachytherapy quality management program (QMP). However, the events also occurred at about the same rate at major centers with busy brachytherapy practices. This procedure's simplicity may also play a role in the events, regardless of the size of the center: the procedure, and particularly the quality checks, may be rushed through, with staff members thinking that little could go wrong with such a simple procedure. Several of the reported events mention the omission of the normally performed quality checks for the treatment.

Looking at the events reported over the last decade can be instructive. Ignoring the events involving mistaken identities and source calibration errors, both of which are not specific to vaginal cuff treatments, almost all events in some way involved the dose being delivered to the incorrect location. The methods of error covered a considerable diversity, some of the more common being:

- the cylinder slipping from its intended position (by far the most common failure) or the source-path catheter slipping along the cylinder,
- the use of an unintended transfer tube,
- mis-measurement of the length of the source path,
- specification of the dose at the wrong distance from the surface of the cylinder, and
- placement of the cylinder in the wrong orifice.

Using FMEA terminology (Huq et al. 2016), the way that a procedure goes wrong leading to an event is a failure mode. A failure mode not seen in the reports is the prescription for a diameter cylinder different than that used. One would expect that this type of failure would be relatively frequent. It is possible that, once the error has been made, detection becomes unlikely. Most of these events had simple remedial solutions:

- Immobilize the cylinder and have someone check it to address cylinder slippage.
- Place the wrong set of transfer tubes in a secure location to address the use of the unintended transfer tube.
- Know the expected length limits of the applicator to prevent mis-measurement of the length of the source path.
- Have an independent check of the dose specification to prevent specification of the dose at the wrong distance from the surface of the cylinder.
- Institute a check of cylinder placement to avoid placement of the cylinder in the wrong orifice.

Each of these interventions would have prevented the event from reoccurring, but they do nothing to prevent the weakness that allowed and facilitated the event in the first place. A robust QMP could be designed following the methodology of the AAPM TG-100 report (Huq et al. 2016). This includes the assessment of the potential ways that procedures could fail at a given institution and the investigation of the failure pathways through a fault tree. This methodology allows institutions to not only prevent the events that actually happened, but to also intercept multiple other possible failures. Jumping at the simple, obvious solutions prevents looking deeper into the operations of a facility (Ostrom et al. 1994; Ostrom et al. 1996). Most of the events resulted from assuming that the procedure was simple and, as a result, unlikely to fail.

6.2.4.2 Role of Image Guidance in Treatment Delivery

In many centers, cylinder treatments are performed without image guidance, sometimes due to logistical problems, such as the lack of in-room imaging. The ABS recognizes these logistical difficulties and states that localization radiographs are not mandatory for fixed geometry cylinders (Small et al. 2012). Lack of

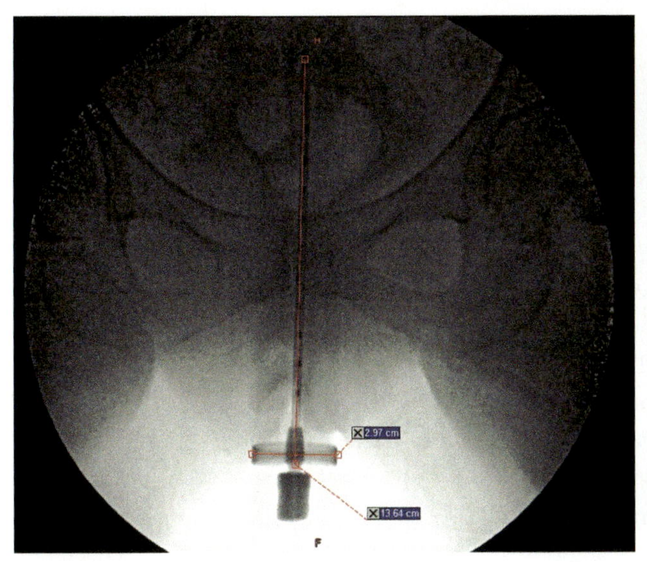

Figure 6–6 Radiograph of vaginal cylinder application. The image has been calibrated based on markers positioned in the central canal. Location of the cylinder compared to the bony anatomy is visible. Measurements have been performed to verify the correct mounting of the applicator and cylinder size.

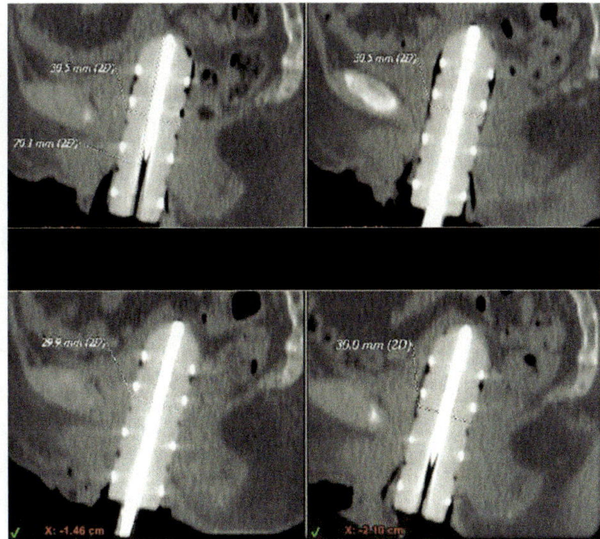

Figure 6–7 Examples of reconstructed sagittal images from a CT scan acquired in-room for four insertions of a vaginal cylinder. Measurements of cylinder diameter and vaginal length (for first fraction, top right) have been documented. Soft tissue anatomy is visible to evaluate applicator positioning.

imaging capabilities requires the implementation of alternative quality assurance procedures to ensure safety of treatment. Cylinder diameter should be confirmed immediately before insertion by measurement with a caliper. When applicable, correct assembly of the cylinder should be confirmed, in particular validating that the relative location of the central channel and the outside features of the applicators are correct and properly affixed. Finally, steps such as clinical photography or measurements should be considered to verify the reproducibility of the clinical insertions across treatment fractions.

The use of imaging is desirable, when available. Localization radiographs can be calibrated based on markers that are commercially available and used to confirm the correct assembly of the cylinder and the cylinder diameter (Figure 6–6). The location of the cylinder compared to the surrounding bony anatomy can be compared in between fractions to confirm reproducibility of the clinical insertions. Moreover, images can be approved by the physician and saved, providing documentation that quality checks were performed. Clinics with capabilities to perform volume-based imaging in the treatment room can scan the patient to perform the same verification and, furthermore, evaluate the conformity of the soft tissues to the applicator (Figure 6–7). Volume-based imaging (CT, MRI) can also be used to retrospectively calculate the dose received by soft tissues in case of re-treatments in the future. Despite its advantages, the use of volume-based imaging for vaginal cuff brachytherapy is not commonplace due to logistical difficulties, increased imaging dose to the patient, and small benefit compare to radiographic imaging.

6.3 Cervical Cancer Brachytherapy

6.3.1 Historical Background

6.3.1.1 Standard Systems

A brachytherapy dosimetry system refers to a comprehensive set of rules involving a specific applicator type and radionuclide with a defined distribution of sources in the applicator in order to deliver a prescribed dose to a designated point or volume of interest (Stitt et al. 1992). Since the early 1900s, ^{226}Ra was used as a gynecological brachytherapy source. Initially there was very little understanding of dosimetry and the mechanism of radiobiology. A standard approach was necessary and particularly useful at a time when absorbed-dose computation for the individual patient was limited. The three basic dosimetry systems developed during the first half of the 20th century were the Stockholm system, the Paris system, and the Manchester system. A combination of the Paris and Manchester systems later evolved into the Fletcher system (also referred to as the MD Anderson system).

In the Stockholm system, usually two or three implants were used, each lasting 20–30 hours. In this system, the applicators consisted of intravaginal applicators and a flexible, plastic intrauterine tube. The Stockholm system advocated an unequal loading for the uterine and vaginal radium sources, normally with 30 to 90 mg of radium placed inside the uterus and 60 to 80 mg placed in the vagina. Originally, the prescription for the treatment was specified as the product of the amount of source loading in milligrams of radium and the number of hours of treatment, i.e., mg_{Ra} h (Tod et al. 1953). Using this concept of dose prescription, a total application of 6500 to 7100 mg_{Ra} h was prescribed for the treatment of a cervical cancer patient, from which 4500 mg_{Ra} h was contributed by the vaginal applicator and the remainder by the intrauterine tube.

The Paris system was similar to the Stockholm system and consisted of both intravaginal cork cylinders (colpostats) and the intrauterine catheter. The treatment was given in a single implant over five days to deliver 7200 to 8000 mg_{Ra} h. Almost equal amounts of radium were used in the uterine and vaginal portions of the applicator. The intrauterine tube contained three sources, the two most superior ones typically of equal strength and the most inferior one typically half the strength of the others. Each colpostat had one source of almost the same strength as the top intrauterine source. Uterine sources in both systems were arranged in a line extending from the external os to nearly the top of the uterine cavity. Both systems preferred the longest possible intrauterine tube to increase the dose to the paracervical region, pelvic region, and pelvic lymph nodes.

Both the Stockholm and Paris systems specified treatment in units mg_{Ra} h and not exposure or dose. Prescription specified in exposure started with the Manchester system.

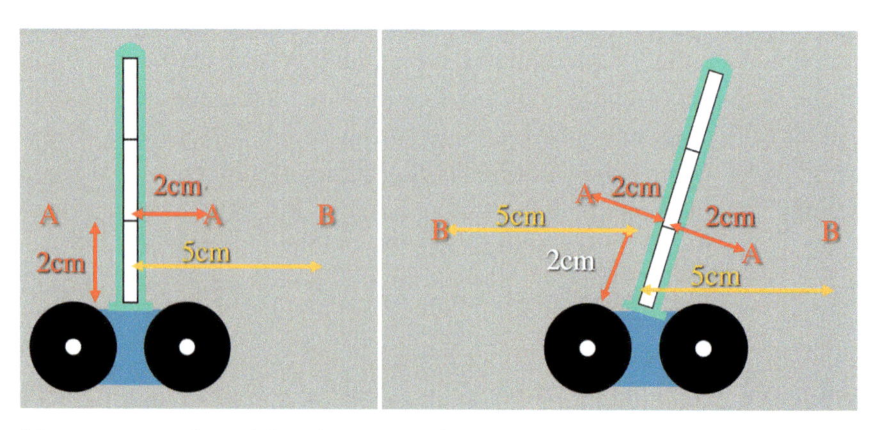

Figure 6–8 Graphic representation of the classical definition of point A and B. If the intrauterine tandem is not aligned with the patient midline, the two points B are not in line with the points A.

The Manchester system, developed in 1938 and modified in 1953 (Tod and Meredith 1953), was most commonly used until the computerized dosimetry systems became available. The prescription point, point A, was chosen as it was thought to be reproducible as well as representative of dose to the target at the lateral margin of a typical uterus, and crossing of the ureter and the uterine artery, thought at the time to be the dose-limiting structures. Point A was defined as 2 cm lateral to the central canal of the uterus and 2 cm cephalad from the superior mucous membrane of the lateral fornix in the axis of the uterus (Figure 6–8) Point A was used to assess absorbed dose in the para-cervical region. Point B was located 5 cm from midline and 2 cm cephalad from the superior mucous membrane of the lateral fornix, and it was thought to approximately correspond to the location of the obturator lymph nodes. To achieve consistent absorbed-dose rates at point A, a set of strict rules dictating the position and activity of radium sources in the uterine and vaginal applicators was devised. The amount of radium varied based on ovoid size and uterine length, such that the same exposure (in roentgens) would be delivered to point A, regardless of the size of the patient or the size and shape of the tumor, uterus, and vagina.

The Fletcher system was developed in 1953 (Haas et al. 1980). As in the Paris system, source-mass-duration was used for prescription under the premise that with any geometric arrangement of specified sources, absorbed dose at any point is proportional to the amount of radioactivity and the implant duration. Though previous systems had used source-mass-duration and clinical experience to determine the radiation tolerance of tissues, Fletcher believed that better results and less morbidity would ensue if knowledge of the absorbed dose at various points in the pelvis—such as the bladder, rectum, and pelvic lymph nodes—could be determined. The system did not include a concept of target dose.

6.3.1.2 Success and Limitation of the Standard Systems

The historical brachytherapy systems have been continuously evolving. Point A has been used as the historic dose specification point for cervical cancer brachytherapy, and it is still used now in clinical practice for prescription and reporting. The exact definition of point A has changed over the years. The current definition has been established by the ABS (Viswanathan et al. 2012) and is shown in Figure 6-9. Dose prescription at point A has numerous weaknesses. Prescription to point A without regard to cervical anatomy can lead to substantial underdosing of a large cervix. Also, the original concept for the dose-limiting paracervical triangle is now considered invalid in light of clinical data, which shows that the tolerance of this area far exceeds that of the bladder or the rectum. The most important shortcoming of the standard systems was the reliance on radiographic imaging, which lacked precise target and normal structures delineation and, consequently, limited the ability to assess the dose delivered to these structures (Nag et al. 2004). The lack of integration of 3D

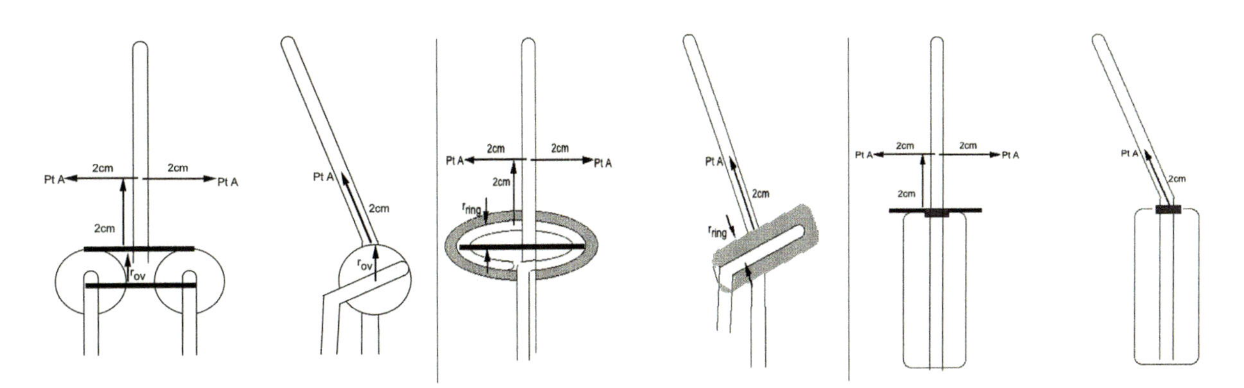

Figure 6–9 American Brachytherapy Society definition of point A, based only on applicator geometry, for tandem and ovoid (left), tandem and ring (center), and tandem and cylinder (right) (Viswanathan et al. 2012).

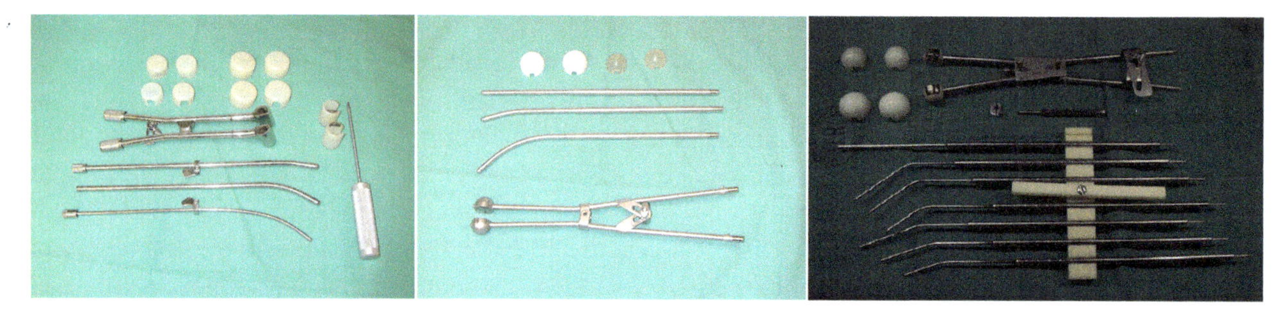

Figure 6–10 Left: Fletcher-Suit-Declos LDR cervix applicator. Middle: Henschke LDR cervix applicator. Right: Henschke HDR cervic applicator.

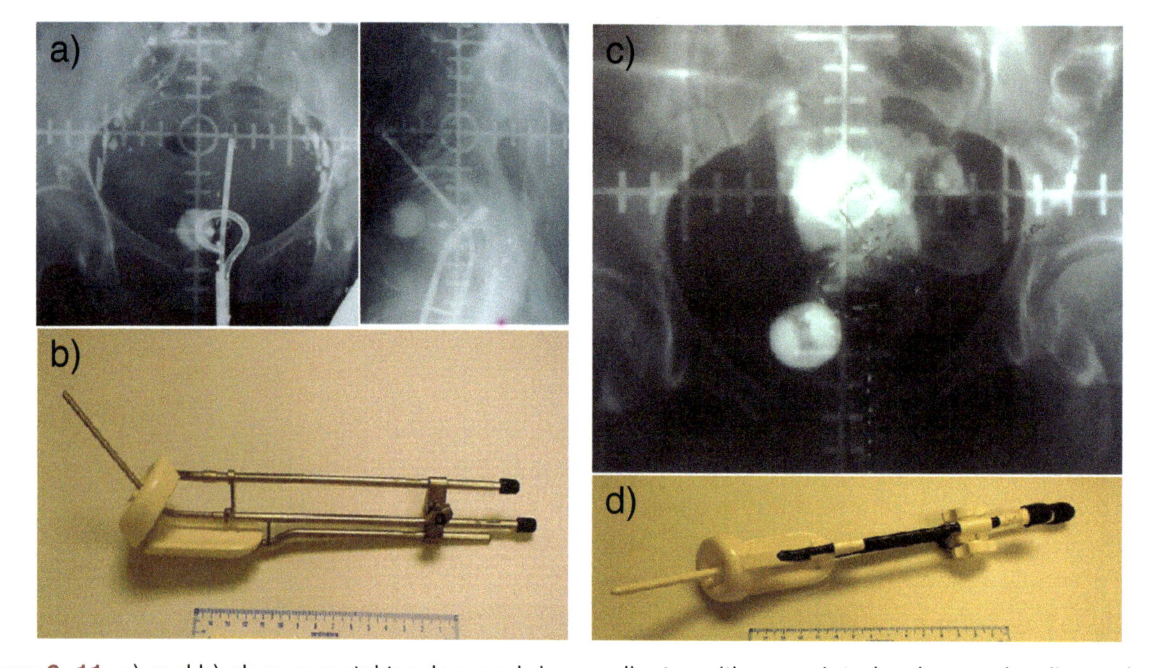

Figure 6–11 a) and b) show a metal tandem and ring applicator with associated orthogonal radiographs. c) and d) show a CT/MR compatible tandem and ring applicator with an associated AP radiograph. In both radiographs images, the exact locations of the entire organs at risk relative to the applicator are difficult to define. This shortcoming is negated by the use of CT or MR imaging.

imaging information in the standard systems was mainly due to inexistent or limited access to that technology at the time the standard systems were popularized. The transition to 3D imaging is discussed later in this chapter.

Intracavitary applicators have also evolved throughout the years. In Figure 6–10, examples are shown of applicators used historically, and in Figure 6–11, examples of modern applicators are depicted.

6.3.1.3 ICRU Reports 38 and 89

In the 1980s, cervix brachytherapy was, in general, performed using either the Stockholm system, the Paris system, the Manchester system, or a derivation of one of those. At the same time, the increase in use of computers for dose calculation allowed the generation of dose distributions. Ideally, the same concepts of dose reporting in EBRT that had been introduced by Report 29 of the International Commission on Radiation

Units and Measurements (ICRU) would be applied to brachytherapy. Yet, in 1985, the ICRU's Report 38 on dose and volume specification for cervix brachytherapy acknowledged that due to the high dose gradients present around brachytherapy sources, a different reporting system had to be developed. ICRU 38 recommended reporting the following parameters:

1. Complete description of the treatment technique and method of dose specification.
2. Total reference air-kerma (TRAK). For a given source geometry, distribution, and loading, the doses delivered at the different tissues or organs was thought to be directly proportional to the total reference air kerma.
3. Reference volume, described in terms of the height, width, and thickness:
   a. For LDR, a volume enclosed in the 60 Gy isodose surface.
   b. For HDR, a dose level that is thought to equate biologically to the LDR 60 Gy isodose surface.

   For intracavitary brachytherapy combined with EBRT, reference dose became 60 Gy less the EBRT dose.
4. Absorbed dose specifications for organs at risk at defined representative points (rectum, bladder) and structures related to the skeleton (lymphatic trapezoid and pelvic lymph nodes).
5. The time-dose schedule including the associated EBRT and the time in-between EBRT and intracavitary brachytherapy.

The practice of cervix brachytherapy since the 1980s has evolved. A transition from LDR to PDR and HDR techniques has permitted increased customization of the treatment plan through optimization. Moreover, access to volumetric imaging technology (CT and MR) allowed the identification of target volumes and the introduction of adaptive brachytherapy, which could account for tumor response during EBRT. Treatment planning based on 3D image sets permits the calculation of dose-volume histograms, which can then be used to evaluate dose-response effects on the tumor and organs at risk. Also, it has become more common to use interstitial needles in addition to the intracavitary applicator to permit escalation of dose to the contoured target volume while sparing surrounding normal structures. At the same time as technological advances have allowed increased complexity in cervix brachytherapy, a majority of the cervical cancer cases treated in the world are performed in developing countries where technologic availability and logistical considerations may limit the transition from the historical systems to volume-based planning. In 2013, the ICRU published Report 89 in conjunction with the GEC-ESTRO. This report aims to provide an updated framework of concepts and terminologies to enable a meaningful sharing of treatment information among practices. Based on the previously published GEC-ESTRO recommendations, the central difference between ICRU Report 89 and 38 is the shift toward the adaptive approach, where reporting of the 60 Gy reference volume covering a target essentially established through clinical examination at presentation is substituted by clinical target volume concepts defined at different temporal points during treatment (Figure 6–12). Aside from the target definitions, ICRU Report 89 also addresses organs-at-risk concepts, radiobiological considerations for the summation of EBRT and brachytherapy treatments, dose metrics descriptions of target and organs-at-risk coverage, and physics aspects of planning and dose calculation. Finally, Report 89 also adopts point A (defined related to the applicator, as in the ABS recommendation) as a reference point, in addition to the newly defined target dose-volume concepts. This report is referenced in the remaining sections of this chapter as it represents the most current reference for the practice of cervix brachytherapy.

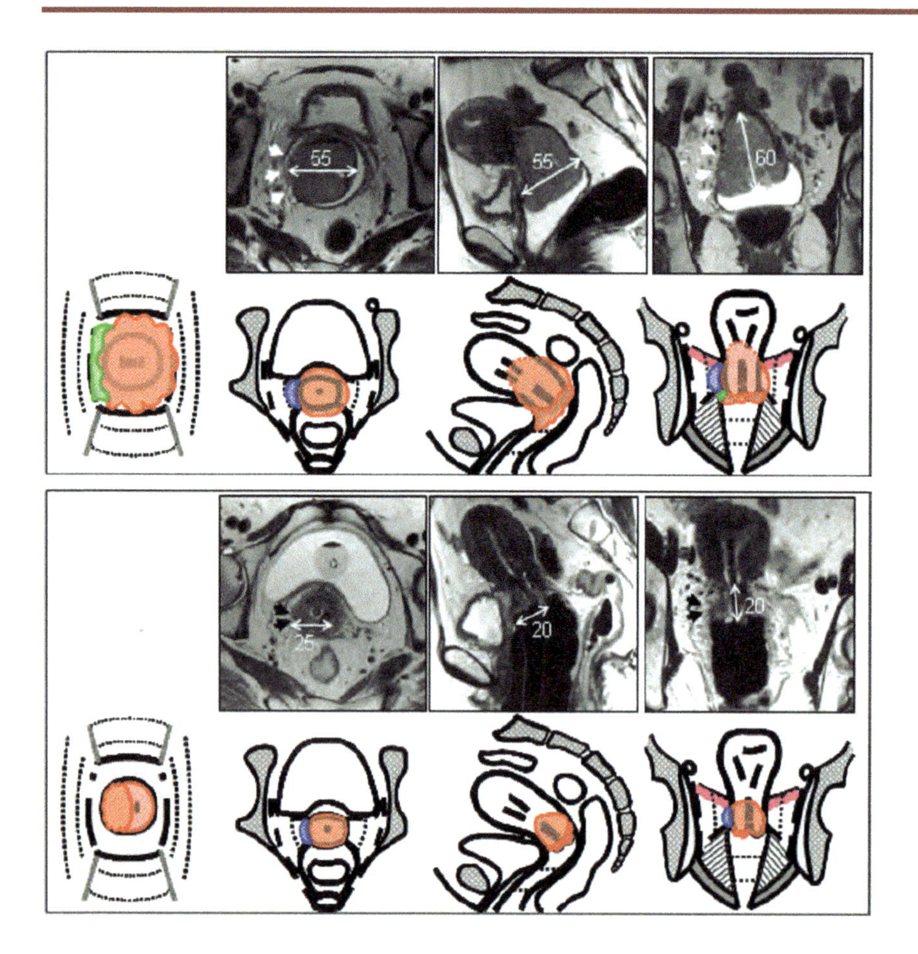

Figure 6–12 From ICRU Report 89 (Figure 5.2): MR and clinical drawings at presentation and at the time of brachytherapy illustrate the adaptive target definition concept. Reprinted with permission of the International Commission on Radiation Units and Measurements, http://ICRU.org.

6.3.2 Image-guided Cervical Cancer Brachytherapy

6.3.2.1 Role of Volumetric Imaging

Volumetric imaging has been shown to increase overall survival while decreasing toxicity (Charra-Brunaud et al. 2012). Volumetric imaging allows defining 3D volumes of target structures or organs at risks. This is different from 2D planning, where dose distributions are plotted in a single plane to estimate dimensions of certain isodose volumes or to predict the dose to points located in these planes. Three-dimensional treatment planning by combining radiographs from two directions has been used in treatment planning for cervical brachytherapy for several decades. This technique allowed the reconstruction of dwell or source positions and reference dose points in 3D. By definition, this kind of radiograph-based planning is already 3D based, with calculation of dose to any given points. In contrast the 3D, volumetric-based approach—based on CT, US, MRI, or other volume-imaging methods—allows volumetric assessment, especially reporting dose-volume parameter relationships, for example with analysis using a dose-volume histogram (DVH). The main limitation of the point-based techniques is the simplification of the highly heterogeneous dose distribution resulting from the brachytherapy part of cervix cancer treatments. In summary, point A would be related to a minimum target dose of a CTV which is symmetrically 4 cm in width at a height of 2 cm above the vaginal applicator surface. This is an appropriate average value. However there is a large variety in symmetry and size. Point A reporting often results in substantial over- or underestimations of the dose delivered to the CTV (Tanderup et al. 2010). The use of DVH allows a much more detailed way of reporting. The main parameter became the $D_{90\%}$, the dose received by at least 90% of the volume. In order to report the substantial heterogeneity, additional parameters—such as $D_{98\%}$ (as a near minimum dose) or $\dot{D}_{50\%}$ (as indicator for high doses)—should be added.

For organs at risk, the use of dose points also resulted in a severe simplification of the analyzed dose distribution. Similar to target dose reporting for OAR, the dose can be reported in terms of minimum dose values received by at least a certain volume (e.g., D_{2cm^3}). A study on intracavitary radiotherapy for cervical cancer (Pelloski et al. 2005) indicated that the estimated dose to the ICRU rectal point may be a reasonable surrogate for the D_{2cm^3} (the minimum dose to the 2-cm^3 volume receiving the highest dose of radiation in a given organ). In contrast, the dose to the ICRU bladder point does not appear to be a reasonable surrogate for the D_{2cm^3}.

The combination of DVH parameters for defined volumes and dose reporting at reference anatomy points allows one to even describe approximately the spatial dose distribution within organs (Nkiwane et al. 2015). Volumetric imaging of the anatomy together with the applicator allows, therefore, a comprehensive dose reporting for the situation close to dose delivery. It can be compared to in-room imaging in EBRT, because the volumetric image data contains the dose delivery device and the target volumes and organs in one coordinated system directly together.

In addition to imaging, the clinical gynecological examination remains an essential part of any target definition. It has to be emphasized that the target definition of ICRU Report 89 includes the findings of the clinical examination. This limits the approaches of automatic target contouring by using only software.

6.3.2.2 CT Imaging

CT has been widely used for 3D treatment planning for EBRT much earlier than brachytherapy, and only relatively recently has CT been implemented in brachytherapy treatment planning for cervical carcinoma (Fellner, Potter et al. 2001). One benefit of CT for cervix and other gynecological disease sites is that it is available in almost any radiation oncology department. Clinical benefits of CT include accurate delineation of the organs at risk (Viswanathan et al. 2007). Image-guided brachytherapy using CT has become the standard of practice for gynecological brachytherapy, in particular in North America (Viswanathan and Erickson 2009). The impact of CT imaging on transition from radiographic to volumetric treatment planning has been demonstrated for gynecological brachytherapy over the last two decades. Traditional orthogonal films taken in the OR after the gynecological brachytherapy applicator system insertion had good results in terms of local control, especially for early-stage disease, but unexplained toxicities and treatment failures remain (Katz and Eifel 2000). Improvements in brachytherapy applicators and planning have more recently paved the way for IGBT, thus the increases in tumor control, reduced toxicity, and improved outcome prediction (Vargo and Beriwal 2014). The long-term results of CT-based planning for HDR brachytherapy of the cervix has been recently reported from the largest cohort of 216 locally advanced cervical patients with good local control and acceptable toxicity (Zolciak-Siwinska et al. 2016). All cases followed the GEC-ESTRO guidelines for HDR CT-based planning (Haie-Meder et al. 2005; Potter et al. 2006). In comparison with the historical series, a substantial benefit of CT-based planning was shown in terms of severe late effects. Other clinical studies have reported similar findings (Hallock et al. 2011; Simpson et al. 2015).

The major limitation of CT for cervical brachytherapy treatment planning is the poor soft tissue contrast. MRI, on the other hand, enables tumor visualization and imaging of the local spread of tumor into the vagina, parametria, and uterine body cavity (see below) (Hricak et al. 2007). However, CT of the cervix is equivalent in the assessment of enlarged lymph nodes to MRI. CT is useful for delineating the borders of the intact cervix (and the uterus) but of limited use for image-guided assessment of the boundaries of the tumor. Viswanathan and colleagues (Viswanathan et al. 2007) showed that CT tumor contours can significantly overestimate the tumor width, resulting in significant differences in the $D_{90\%}$, $D_{100\%}$, and volume treated to the prescription dose or greater for the CTV_{HR} compared with that using MRI. Hence, MRI remains the standard for CTV definition. Recently, the ABS cervical cancer task group (Viswanathan et al. 2012) recommended CT imaging for the width of the cervix and any parametrial extension. The superior border of the cervix should extend at least 1 cm above either the uterine vessels, identified by IV contrast, or the location where the uterus begins to enlarge. If these cannot be identified, a 3-cm height should be contoured for the cervix.

6.3.2.3 MR Imaging

MR has become a new standard imaging method for several clinical disease sites, such as cervical cancer, vaginal cancer, or endometrium cancer. The excellent soft-tissue contrast resolution and the multiplanar capability is a major advantage compared to other imaging modalities. The possibility of assessment of parametrial involvement for cervical cancer is essential for advanced cases. MRI is used for staging, but also by many centers to routinely to assess treatment response. However, the use of MRI for treatment planning is less frequent.

One often-mentioned limitation of the MRI technique is its reduced availability for radiotherapy. The situation is very different throughout the world. One common problem is that MRI is not standard equipment in radiotherapy centers and, therefore, only limited time slots are available through the radiology departments; this involves logistical problems and extra costs. In contrast, MRI is increasingly used for many different disease sites in EBRT, and brachytherapy should be careful not to become a second-class radiotherapy methodology. Obtaining an MRI for applicator placement causes the highest problems in terms of logistics because it is often difficult to keep the assigned time slots given brachytherapy intervention involves different clinical aspects that are not always predictable.

Another limitation of MRI is the exclusion of certain materials implanted either as applicators or other implants (e.g., hip implant, pacemakers). Ring or ovoid applicators composed of steel or steel tandems cause severe artifacts, even for low magnetic field strengths. For MRI, the applicator material used is mainly plastic or titanium. However, for high field strengths of 3T and more, titanium also produces increasing artifacts and should be avoided in those cases. Other kinds of artifacts can be taken into account or can even be desirable. For instance, reproducible artifacts, such as the depiction of titanium needle tips, can be used to identify applicators with high accuracy once the location of the artifact has been assigned to the real applicator surface in phantom acceptance tests.

The reconstruction of the source path for intracavitary applicators with MRI has been summarized in Section 9.1 of the ICRU Report 89 and before in dedicated GEC-ESTRO recommendations (Hellebust et al. 2010). The highest accuracy and fastest approach in daily clinical practice is to rely on a predetermined source path in relationship to the applicator. This is further discussed in Section 6.3.4.1.

There are various work flows using MRI for treatment planning of cervix brachytherapy. The use of an MRI preceding a brachytherapy implant has already made a major impact on target definition (Potter et al. 2016). While not used with any fusion methods, the clinician can account for the CTV dimensions and often also estimate the distance from the CTV border to the intrauterine channel in different directions. These measures can then be applied when contouring the CTV (or GTV) on a CT with the applicator in place, although the applicator can alter the anatomic dimensions. The most sophisticated and accurate approach is to perform an MRI with the applicator in place. Target contours, organ contours, and the applicator can be defined on one single MRI dataset, preferable automatically fused (via DICOM) to other MR-image orientations or with 3D sequences.

6.3.2.4 US Imaging

Ultrasound imaging is widely used in medicine and constitutes a cornerstone of obstetrical gynecology care. Yet its use in gynecological brachytherapy is not widespread, and it has often been limited to applicator insertion guidance (van Dyk et al. 2015). The use of US to identify misplaced applicators (Carson et al. 1975) or uterine perforation (Granai et al. 1990), was recognized more than 40 years ago in an era when radiographic-based dosimetry was the norm. The inherent real-time nature of US imaging, of course, makes it a perfect tool for applicator, needle, and catheter insertion guidance. As such, the ABS guidelines for cervix cancer brachytherapy underline that either a trans-abdominal ultrasound (TAUS) or trans-rectal ultrasound (TRUS) probe may be used for applicator insertion in order to minimize perforation (Viswanathan et al. 2012). The demonstration of TRUS as an image-guidance modality in template-based interstitial gynecological brachytherapy followed not long after that of prostate HDR interstitial brachytherapy (Stock et al. 1997) and was used to provide target visualization at the time of implant and better catheter placement (Sharma et al. 2010) when significant disease extension was present. Trans-vaginal ultrasound guidance has also been discussed in the literature (Rose et al. 2014).

The use of US as the sole imaging device for treatment planning in cervical cancer has been studied by van Dyk et al. (2009) by comparing 2D US to orthogonal radiographs using MRI as the assumed true images. That study found significant differences between US and MRI plans relative to traditional film-based plans. Furthermore, 2D US plans were not significantly different than 2D MRI plans. As such, one might envision that US could be a low-cost and efficient solution for brachytherapy of the cervix in parts of the world where access to MRI is already difficult for diagnostic purposes, let alone fully dedicated to brachytherapy.

More recently, Epstein et al. (2013) have compared tumor definition from TAUS and MRI (both 1.5 T and 3 T) as part of a multi-center trial from a cohort of 182 patients. In this study, the pathological examination was used as standard for the presence, size, and extension of the tumors. TAUS was found to agree with pathological finding at detecting the presence of bulky tumors as well as identifying parametrial extension, with TAUS being better than MRI for the latter. Similarly, Mahantshetty et al. (2012) found generally strong correlations between TAUS and MRI measurements of cervix, central disease, and external surface of the uterus relative to the tandem. Van Dyk et al. (2014) conducted a similar study on a larger cohort of 192 patients and 1,668 measurements (see Figure 6–13). The differences between US and MRI were generally larger, up to 3.7 mm for the anterior uterine surface relative to the posterior surface (1.5 mm or less). The authors concluded that these differences were clinically acceptable, but no dosimetry study was performed to demonstrate that point.

Schmid et al. (2013) found strong correlation between TRUS-delineated target width and thickness relative to MRI. A study by the same group was extended and compared MRI and TRSU pre-brachytherapy to

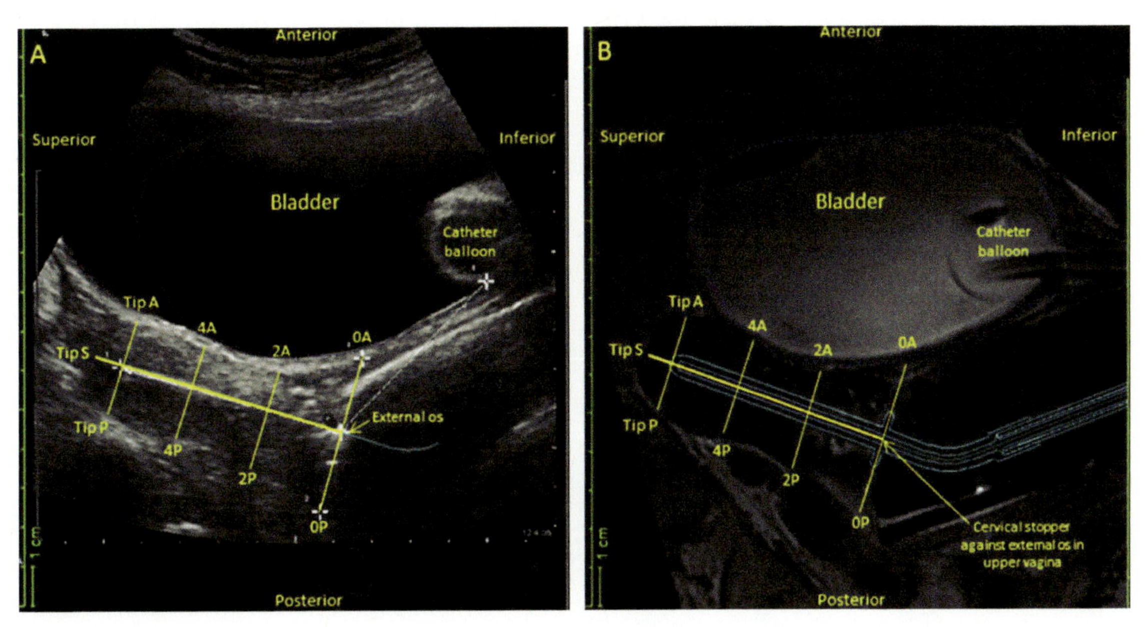

Figure 6–13 From Dyk et al., an ultrasound (left) and MRI (right) view of cervix and uterus, with corresponding measurements points (van Dyk et al. 2014).

MRI, CT, and TRUS after insertion in order to validate TRUS imaging for adaptive brachytherapy in cervical cancer (Schmid et al. 2016). In the case of the CTV_{HR}, TRUS imaging was within the intra-observer variability of the MRI and was found to be superior to CT. The quantitative analysis shows no significant difference between the MRI taken after insertion and TRUS at the time of insertion for maximum target width, target width at the level of point A, as well as target thickness at the level of point A, but with some differences for maximum target thickness. However, the differences between MRI and CT were significant for all parameters reported.

Based on the above-mentioned results, Nesvacil et al. (2016) proposed combining TRUS and CT imaging for image-guided, adaptive brachytherapy for cervical cancer using CT to define key organs at risk and TRUS for real-time guidance and CTV_{HR} delineation. This approach appears promising, and the treatment plan generated as part of this proof-of-concept study fulfilled all clinical acceptance criteria. St-Amant et al. also recently successfully demonstrated a similar approach, but using a 3D TAUS system that is calibrated in the same reference frame as a CT scan (Elekta Clarity, Montreal, Canada), such that both image sets are automatically registered, providing a common reference frame. While more validation studies are necessary (in particular relative to the strength of the generated treatment plans), US and, in particular, modern 3D US could become an interesting and useful tool for cervical brachytherapy.

6.3.2.5 Target Definition

ICRU Report 89 defines a clear target concept. For brachytherapy the most important volumes are the GTV_{res}, the CTV_{HR} and the CTV_{IR} (Figure 6–14). The GTV_{res} is the residual macroscopic disease that is left after EBRT. This target volume should receive the highest dose. In most cases, this usually happens as the GTV is in the center of the implant, often around the intracervical channel with the tandem in place. However, for advanced cases the dose to the GTV needs special attention and also an optimization of the implant procedure with interstitial needles.

The main target volume for those centers used to the point A tradition is the CTV_{HR}. The CTV_{HR} is related to the target volume as it appears at the time of brachytherapy—always including the entire cervix plus any residual tumor tissue. In most cases, this will be the remaining parametrial involvement. Typically, the approximate width of the cervix without or with very small parametrial involvement is 4 cm at the level of point A. Therefore, the doses applied to the CTV_{HR} are close to previous point A prescriptions: around 80 Gy equi-effective dose (EQD2, see Section 6.3.4.3). The CTV_{IR} is an additional target volume that rep-

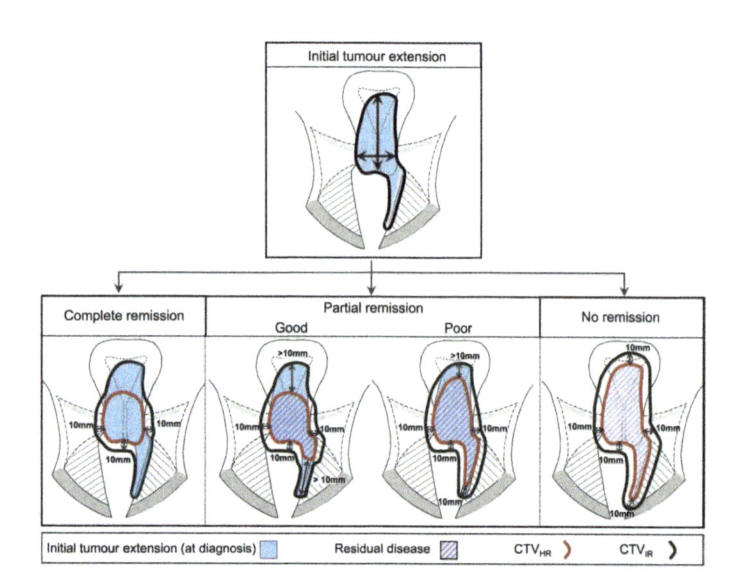

Figure 6–14 From ICRU Report 89 (Figure 5.14): Diagram showing schematically the concepts of CTV_{HR} and CTV_{IR}. Reprinted with permission of the International Commission on Radiation Units and Measurements, http://ICRU.org.

resents the area of the initial GTV superimposed on the topography at the time of brachytherapy plus a margin. This target volume usually gets lower dose (around 60 Gy EQD2). In daily clinical routine, it is often sufficient to focus dose optimization to the CTV_{HR}, as then the GTV and the CTV_{IR} receive appropriate doses without additional optimization. However, in case of very advanced GTV—or if the CTV_{IR} is large due to initial vaginal or endometrial involvement or lack of regression during EBRT—it becomes essential to also follow DVH constraints for optimizing the dose to these target volumes.

6.3.3 Treatment Planning: Applicator Reconstruction

6.3.3.1 Basic Concepts

Modern cervical brachytherapy requires localization for dosimetry using volume imaging techniques with good soft-tissue contrast. The main modalities are CT and MR imaging; ultrasound could be used, and some work has been presented using that, but the lack of registration for image planes with commercial systems proves a major impediment (Van Dyk et al. 2009; ICRU International Commission on Radiation Units and Measurements 2016; Petric and Kirisits 2016). For a discussion of radiographic localization, see the work by Paliwal and colleagues (1997).

MR is the preferred imaging methodology because of its superior soft-tissue contrast for delineating the GTV and normal uterine anatomy. While beneficial for delineating soft tissues, MR imaging adds difficulties and complexities to digitizing the applicator. Much work has gone into clinical localization using MRI for applicator dwell positions. See, for example, Schindel et al. (2013). Markers containing a solution of $CuSO_4$, and even simple saline, show adequate contrast and definition for live localization, yet few facilities moved to using such. The problem is localizing the contrast solution into small packets along a marker to indicate the dwell position. The localization issue did not originate with MR; CT markers frequently only delineate the source pathway starting from the first position without demarcating individual or selected positions. Given the starting position and the source track, the TPS can determine the location of the dwell positions. Making a linear marker by filling a catheter with contrast, while challenging for geometric reasons, should not preclude development, construction, and use. However, the small diameter needed to pass through many applicators results in reduced contrast when used in a patient. Another approach has become more common, as described below.

Hellebust et al. discuss in great detail the challenges of applicator reconstruction with MR localization, and anyone starting an MR-guided cervical brachytherapy program should study this work (Hellebust et al. 2007; Hellebust et al. 2010; Hellebust et al. 2012). The first step is commissioning the applicators for the procedure. The commissioning entails imaging the applicators in MR and through more conventional CT or radiography and using the latter to establish the true dwell positions in an applicator and, using fiducial markers, to transfer that information to the MR images. Particularly for highly curved applicators, such as a ring applicator, each dwell position must be imaged separately with the non-active source simulator at location according to the distance setting. Except for the first dwell position, the location of the source when sent to a position is different from that indicated by inserting the x-ray marker in the applicator. The difference comes from the support for markers indicating middle dwell position by the wire connecting the dwell markers, compared with the source, which at the end of the cable tends to drag along the outer edge of the source path.

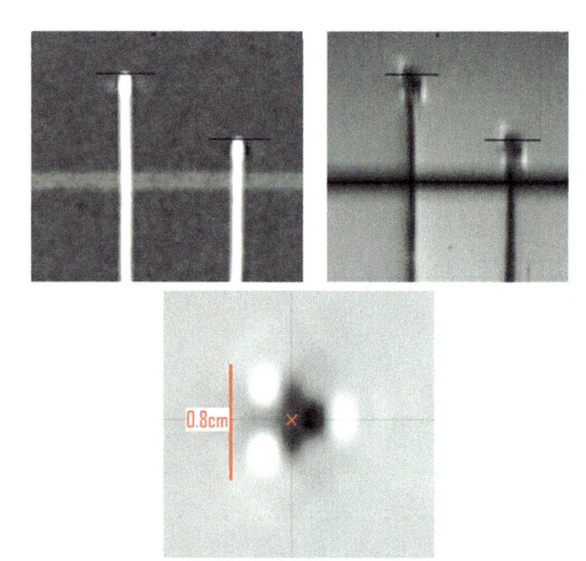

Figure 6–15 a) Coronal reconstruction of transverse CT scan and (b) T2-weighted MR scan of a water phantom with titanium needles. MRI susceptibility artifacts are seen in relation to the needle tip. c) Para-transaxial alignment of the tandem applicator on MRI. Bright susceptibility artifacts are seen around the tandem. The red X marks the position of the center of the source channel. (From Hellebust, Kirisits et al. 2010.)

The information obtained as described above can serve as the library version of the applicator for use in treatment planning. Figure 6–15 compares images of an applicator between CT and MR (Hellebust et al. 2010). A major problem with MR images is that the image of an applicator or needle frequently projects beyond the physical ends, as shown in Figure 7–15.

6.3.3.2 Model-based Applicator Reconstructions

Manual or software-based templates from libraries relate the visualized outer dimensions of the applicator with the possible dwell positions (Figure 6–16). This geometry has to be verified at commissioning and can then be used for all subsequent cases. With an MRI available, the template can then be positioned into the MR dataset by using the outer visible applicator surface or landmarks (e.g., needle holes, screws, connecting parts). This approach does not require MRI markers within the applicator during imaging. The approach also does not require fusion with other imaging modalities, such as CT. Especially dangerous are fusion methods based on the bony anatomy, as the fixed target with applicator geometry moves relative to the bones. While useful in EBRT planning, it would be wrong to rely on such methods in pelvic brachytherapy. Aligning the applicator model with the image carries an uncertainty. The hope is that this uncertainty remains less than the uncertainty for identifying and digitizing the applicator *de novo*. An alternative is to perform both MR and CT imaging and fuse the two studies, using the CT to enter the applicator information and the MR for the tar-

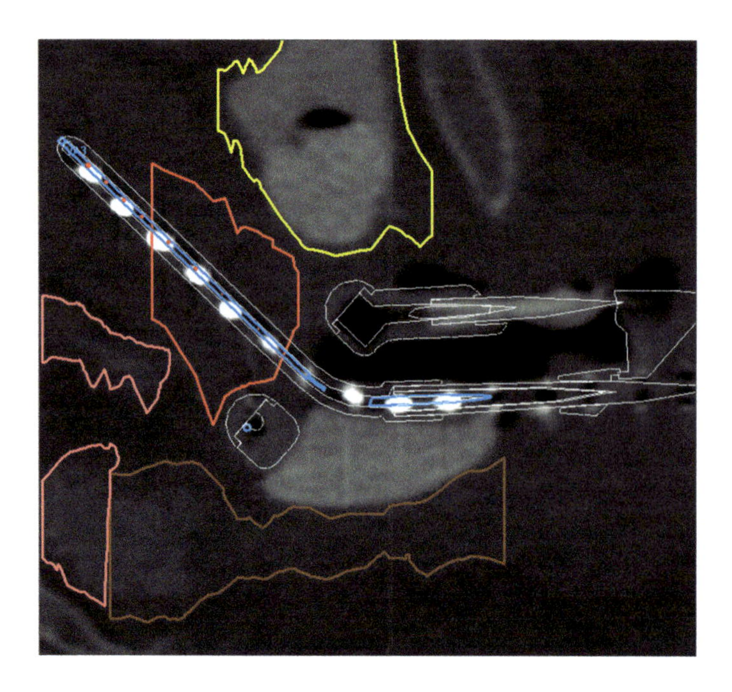

Figure 6–16 Example of a digital model of the applicator (in white) overlaid over a CT scan of a patient.

get contour. The OAR contours could be delineated from either or both image sets. For applicators with arbitrary curvature—such as supplemental interstitial needles that may be bent at the time of implantation either intentionally to direct their path or inadvertently upon hitting a hard part of a tumor—a library graphic will not work due to the different geometry between the stored applicator and the one in the patient. In such cases, the first dwell position and the source track in the needle requires individual delineation, either through fusion with a CT or by using a contrast-enhancing marker in the needle. As noted in Chapter 1, attention must be paid to the direction of digitization to prevent the needle direction from being reversed in the computer.

Several of the commercial planning systems now offer applicator models or libraries of applicators. In most instances, the applicator models were created from original computer-aided designs. Applicator models, or libraries, completely and accurately describe the applicators made by a particular vendor. This means that once an applicator model is inserted into an image set, it can be aligned with the applicator in the image, and the source path is automatically incorporated in the applicator. This eliminates any manual digitization of the source path, which can be time saving and also help mitigate errors associated with incorrect digitization of the source path, which is discussed further in the section below. Once commissioned, and after some training and practice on use, applicator models can be incorporated into daily use during treatment planning.

Just as with starting any new treatment technique or implementing a new technology, it is extremely important to commission the applicator models prior to clinical deployment. One of the key ingredients in commissioning the applicator models is to verify that the model geometry accurately corresponds to the physical applicator used in the clinic. For example, if a tandem and ring applicator model is to be commissioned, then the model dimensions must correspond to the tandem and ring available for use in the clinic. The tandem must be the same length and diameter, and the ring must also be the same length and diameter and be available with the proper cap attachment if there are multiple sizes available. Additionally, as heterogeneity corrections become more prevalent in routine brachytherapy planning, it will become increasingly important to verify that the applicator models correctly represent the true composition and density of the applicator.

6.3.3.3 Uncertainties in Applicator Reconstruction

There are a large number of uncertainties throughout the entire process of brachytherapy treatment, and applicator reconstruction is one component. A report entitled "Review of clinical brachytherapy uncertainties: Analysis guidelines of GEC-ESTRO and the AAPM" published in 2014 in *Radiotherapy and Oncology* provides a thorough analysis and excellent synopsis of total uncertainties in brachytherapy practice (Kirisits et al. 2014). Additionally, the GEC-ESTRO group has also published recommendations strictly dealing with applicator reconstruction in 3D image-based planning (Hellebust et al. 2010). In this paper, the authors state a dose gradient of approximately 5% to 12% per mm in intracavitary applications at distances of 1 to 3 cm from the source. It is easy to recognize that with this steep dose gradient it is possible to have substantial dose deviations due to uncertainty in applicator reconstruction. Therefore, it is extremely important during the commissioning process to properly assess the true source path inside an applicator, especially applicators with large curves, such as a ring or ovoids. The true source path inside these applicators can deviate from the intended path of the geometry of the applicator. Care must also be taken in considering the type of imaging that will be used for treatment planning, such as CT or MRI. Each of these imaging modalities will require its own set of considerations when commissioning applicators, as has been covered in the literature (Haack et al. 2009; Hellebust et al. 2010).

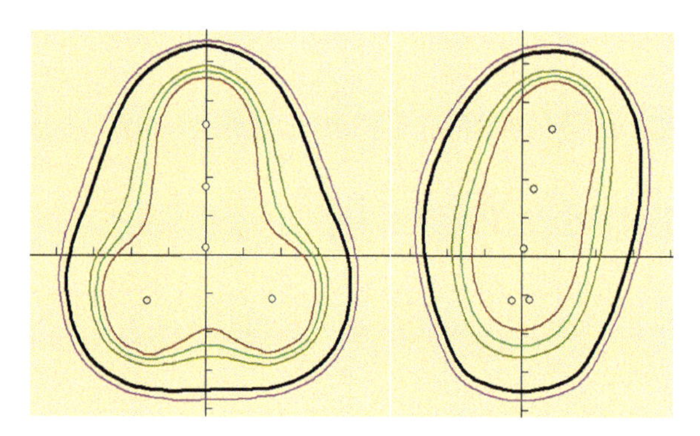

Figure 6–17 LDR Henschke dose distribution with a ^{137}Cs LDR loading. The dose distribution appears pear-shaped in the coronal projection (left panel) but not in the sagittal (right panel).

6.3.4 Treatment Planning: Dose Distribution and Optimization

6.3.4.1 Standard Loading Systems

The typical dose distribution for cervix cancer brachytherapy is commonly referred to as "pear-shaped," and it reflects the dose distribution that would be obtained from a classic LDR application (Figure 6–17). The exact implementation of the pear-shaped distribution varies, but the general principles remain of loading the applicators to obtain the 100% dose line to pass through the A points and bow out inferiorly by the vaginal component of the applicator. With LDR brachytherapy, this was typically achieved by loading cesium tube sources of equal strength in the ovoids and the most inferior two slots of the tandem (depending on the size of the ovoids), and a higher-strength tube at the tip of the tandem. In HDR brachytherapy, this loading scheme has, in general, been converted using two different approaches. In one approach (Kirisits et al. 2005), all dwell times are kept equal, but the dwell positions are unevenly distributed in the tandem, with more closely spaced dwell position activated at the tip. For a ring applicator, dwell positions in the two lateral sides are activated. In a second approach, all dwell positions are uniformly distributed in the tandem, but the dwell times are unevenly allocated to increase the contribution of the tip tandem dwell positions. This can be achieved by optimizing the dwell time to a series of optimization points: a set of points along the tandem 2 cm lateral to the uterine canal, tapered toward the tip, with a dose goal of 100% of the prescription dose; and a set of points at the surface of the ring or ovoid with a dose goal of 160% of the prescription dose. Table 6–3 describes the points for this optimization example, with additional points placed radially from the active ring dwells at 6 mm. These two approaches are equivalent. In both systems, the standard plan ensures scaling the dwell times in the standard loading system to provide doses to point A when averaging the prescription dose.

6.3.4.2 Achieving the Sculpted Pear

The main benefit of a standard loading system is to provide a starting point for further optimization. While standard plans are still used clinically in some institutions, the practice is shifting toward more patient-specific optimization based on the clinical target volume (typically identified on MRI) and organs-at-risk, volume-based metrics (Tanderup et al. 2010). Most commercial TPSs provide various optimization tools, including:

Table 6–3 Example of optimization points to achieve a sculpted pear

Tandem Dwell Position	4 cm Tandem	6 cm Tandem	8 cm Tandem
1	12 mm	12 mm	12 mm
3	16 mm	14 mm	14 mm
5	20 mm	16 mm	14 mm
7	20 mm	18 mm	16 mm
9	20 mm	20 mm	18 mm
11+ to level of point A and not beyond	20 mm	20 mm	20 mm

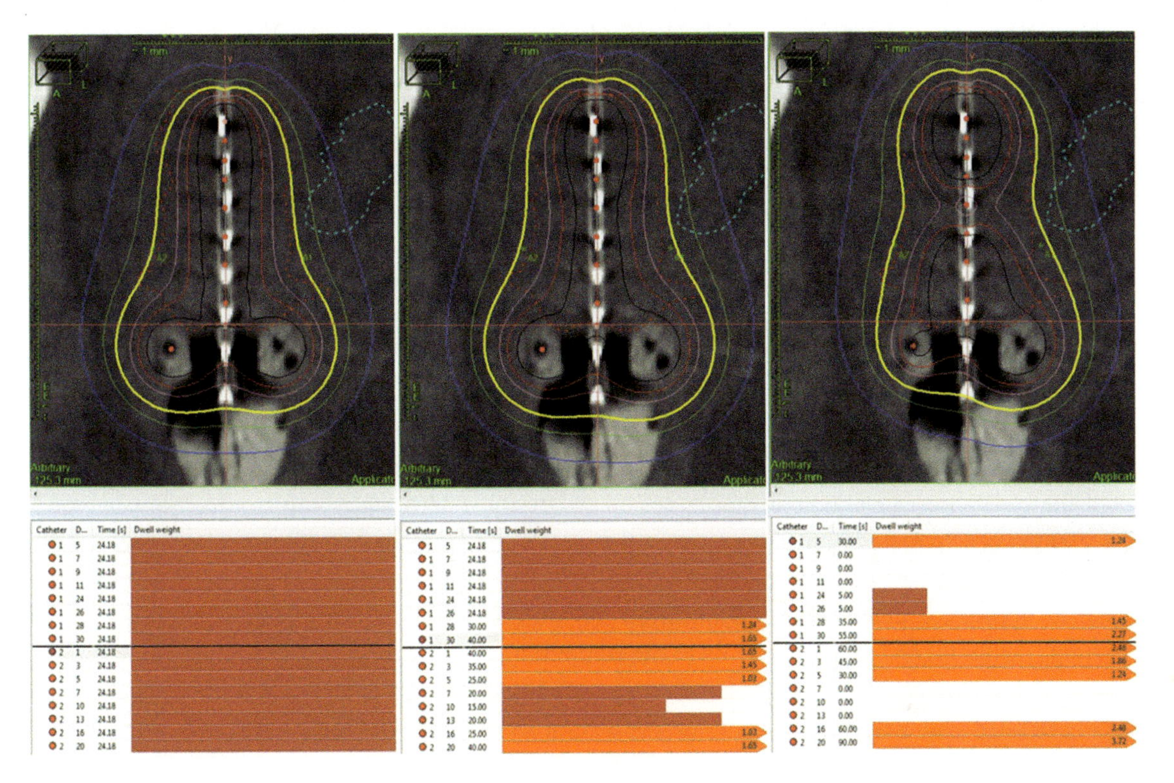

Figure 6–18 Left: A standard loading. Center: sculpting of the pear-shaped dose distribution through small changes to the dwell loading. Right: A greater departure from the standard loading.

- manual optimization, in which the user changes the dwell times directly;
- graphical optimization, in which the user manipulates the isodose lines and the software adjust the dwell times accordingly;
- point-optimization, in which the user modifies the location of a set of optimization points; and
- inverse optimization, in which a set of objectives are entered by the user, and the software automatically allocates all the dwell times to reach those objectives.

The manual, graphic, and point-optimization methods allow for a "sculpting" of the standard plan to adjust it to the needs of each specific patient. The sculpting of the pear-shaped plan can change the dose distribution throughout the implant, depending on the specific shape of the CTV. There are no hard rules on the amount of sculpting that is correct, and different clinics have used different approaches (Tanderup et al. 2010; Nomden et al. 2013). Generally speaking, the CTV shape and volume is very different than the volume encompassed by the 100% isodose line in a standard plan. Therefore, it is theoretically possible to achieve the same coverage of the CTV with dose distributions that do not resemble the standard plan's pear shape. An example of different isodose lines providing equivalent coverage to the CTV is show in Figure 6–18. Some clinics prefer to apply gentle modification to the pear-shaped plan by manually changing the dwell times or by using graphical optimization. The ABS recommends to not optimize exclusively based on volumetric constraints, but to consider the 3D dose distribution in the optimization process (Viswanathan et al. 2012). The use of inverse optimization should then be validated with care as optimization only based on CTV, rectum, bladder, sigmoid, and bowel will likely result in dose distributions significantly different than a pear shape.

6.3.4.3 Dose Objectives and Constraint

Cervix cancer brachytherapy is typically a boost treatment following EBRT. As such, dose metrics analysis without accounting for the prior dose received by the tumor and the normal tissue is not clinically meaningful. The GEC-ESTRO and ABS recommend the use of the EQD2 formalism, which aims at calculating the "equi-effective" dose (Bentzen et al. 2012) from each component of the radiation treatment referenced to a nominal 2 Gy fraction scheme. Formulas to calculate the EQD2 for LDR, PDR, and HDR can be found in ICRU Report 89, and spreadsheets for EQD2 calculation endorsed by the ABS and GEC-ESTRO can be downloaded from the ABS website (https://www.americanbrachytherapy.org/guidelines/index.cfm). Historically, EBRT was delivered with 3D plans, e.g., a four-field box delivering uniform dose to the entire area being re-irradiated with brachytherapy. Under this assumption, the prescription dose from the EBRT course could safely and conservatively be used to estimate the dose received by the tumor and the OARs. As intensity-modulated techniques become more common, sparing of regions of the OARs and hot spots in other regions of the OARs may invalidate this assumption. In theory, a deformable registration between the EBRT anatomy and the brachytherapy anatomy would alleviate the uncertainty in dose summation. In practice, the lack of validation tools for this deformable image registration (DIR) and the very different anatomical conditions (due to applicator placement) has hindered the utilization of DIR for dose summation in cervical brachytherapy. Similar considerations apply to the summation of the dose received from different brachytherapy insertions. Dose objectives for organs at risk are typically expressed in $D_{2\,cm^3}$. A limit of 75 Gy EQD2 has been recommended for rectum and sigmoid and a limit of 90 Gy EQD2 for bladder (Viswanathan et al. 2012). There is no established dose objective for bowel dose, although more than 60% of clinics in North America report routinely contouring the bowel in cervical cancer brachytherapy cases (Grover et al. 2016; Damato et al. 2017). Dose objective for the tumor is, in general, defined as $D_{90\%}$. An objective of $D_{90\%} >$ 85 Gy is recommended for CTV_{HR}, which is defined on MRI. The study protocol for the EMBRACE II recommends for the $D_{90\%}$ a planning aim dose of 90 Gy to 95 Gy ("soft constraint") for the CTV_{HR}, and a dose constraint of 85 Gy ("hard constraint") (www.embracestudy.dk). The same objective is typically used also for CT-based contouring of the CTV (Viswanathan et al. 2007).

6.3.5 Hybrid Applications

A special approach to treatment planning is needed for combined intracavitary and interstitial implants, as shown in Figure 6–19 (Dimopoulos et al. 2006; Kirisits et al. 2006; Petric et al. 2009; Tanderup et al. 2010; Nomden et al. 2012). The dose distribution should remain, in principle, comparable to intracavitary implants,

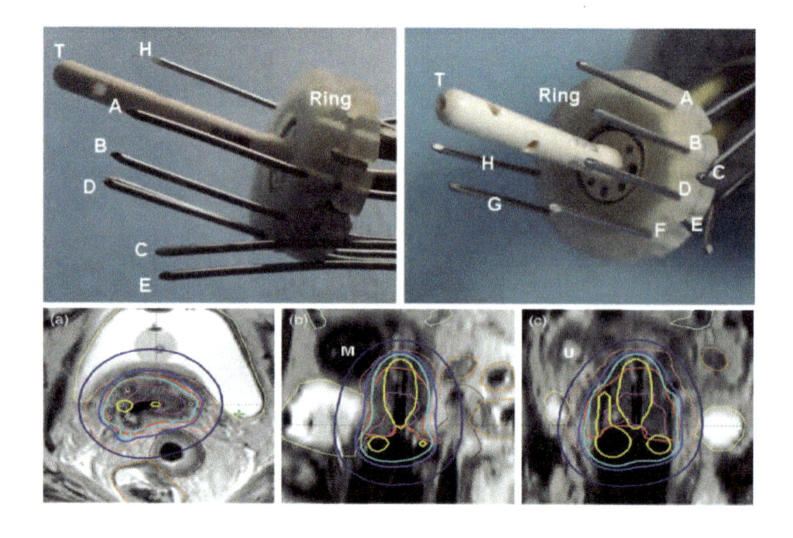

Figure 6–19 From ICRU Report 89 (Figures A.6.5 and A.6.6). Top panel: example of a hybrid applicator with needles inserted through the cap in ring. Bottom panel: Example of isodose distribution in a hybrid applicator plan. Reprinted with permission of the International Commission on Radiation Units and Measurements, http://ICRU.org.

as this was, and is, the state of the art for cervical cancer. The unique dose gradients within the implant ensure very high doses at the GTV, appropriate dose at the CTV_{HR}, and still enough dose for the microscopic diseases within the CTV_{IR}. All dose constraints for target volumes and OARs are set for single-dose parameters, but rely on the fact that the underlying dose profiles with organ walls are comparable. Therefore, a common optimization process for combined implants starts as for intracavitary implants with optimizing tandem and vaginal applicator dwell positions. However, the optimization process should try to cover as much as possible, but already taking into account that the OARs should be spared. In the next step, the available dwell positions inside interstitial needles may be activated according to the target and implant geometry. Usually in a step-wise procedure, the dwell times in these needles are increased by 10% to 20% over the dwell times applied for a non-optimized intracavitary implant normalized to point A. After a step-wise process, the total dwell time of all needle positions is set to 10% to 20% of the overall treatment time. Exceptions are made for very advanced cases with more than one row of interstitial needles (Lindegaard and Tanderup 2012).

6.3.6 Quality Assurance

General brachytherapy HDR quality assurance and quality management concepts apply (Kubo et al. 1998), and specific translation of those concepts to cervix brachytherapy are discussed in the ABS guidelines (Viswanathan et al. 2012). Attention to technical aspects and formal consistency of the treatment plan, prescription, and treatment delivery is paramount. Given the image-guided nature of cervix brachytherapy treatments, the large fraction sizes typically involved, and the invasive nature of the procedure, the QMP should also focus on imaging and contouring consistency (Kirisits et al. 2015), work flow analysis and optimization (Damato et al. 2015), dose metrics evaluation (Damato et al. 2013; Deufel and Furutani 2014), reasonableness tests (Takahashi et al. 2012), and conformity of conditions between planning and delivery (Schindel et al. 2013). Availability of imaging technology in the treatment room varies, with some clinics having no access to in-room imaging, others relying on radiographs or CBCT (in particular if the treatment room is in a linac vault), and others using a CT (Lee et al. 2013) or, in some cases, MRI. While guidelines for image-based treatment verification of cervix cancer brachytherapy are not available, clinics should evaluate their work flows to identify when image verification is warranted. In particular, patient transfer from the CT or MRI scanner to the treatment room is common to most clinical work flows, and applicator slippage may occur. Moreover, changes in anatomical configurations, e.g., bladder filling, between the planning scan and treatment would introduce a discrepancy between dose received and planned. *In vivo* dosimetry techniques have also been suggested for verification of organ dose (Waldhausl et al. 2005; Hassouna et al. 2011) and for on-line error detection (Andersen et al. 2009), but those techniques have not yet seen broad clinical implementation.

References

Advisory Committee on the Medical Uses of Isotpoes of the US Nuclear Regulatory Commission. (2014). "Meeting Agenda, Advisory Committee on the Medical Uses of Isotpoes, September 29–30, 2014." Retrieved 04-24-2015 from http://pbadupws.nrc.gov/docs/ML1426/ML14260A235.pdf.

Andersen, C. E., S. K. Nielsen, J. C. Lindegaard, and K. Tanderup. (2009). "Time-resolved in vivo luminescence dosimetry for online error detection in pulsed dose-rate brachytherapy." *Med. Phys.* 36(11):5033–43.

Bentzen, S. M., W. Dorr, R. Gahbauer, R. W. Howell, M. C. Joiner, B. Jones, D. T. Jones, A. J. van der Kogel, A. Wambersie, and G. Whitmore. (2012). "Bioeffect modeling and equieffective dose concepts in radiation oncology—terminology, quantities and units." *Radiother. Oncol.* 105(2):266–68.

Beriwal, S., D. J. Demanes, B. Erickson, E. Jones, J. F. De Los Santos, R. A. Cormack, C. Yashar, J. J. Rownd, A. N. Viswanathan, and S. American Brachytherapy. (2012). "American Brachytherapy Society consensus guidelines for interstitial brachytherapy for vaginal cancer." *Brachytherapy* 11(1):68–75.

Brown, L. C., I. A. Petersen, M. G. Haddock, J. N. Bakkum-Gamez, L. J. Lee, N. C. Cimbak, R. S. Berkowitz, and A. N. Viswanathan. (2015). "Vaginal brachytherapy for early-stage carcinosarcoma of the uterus." *Brachytherapy* 14(4):433–39.

Carson, P. L., W. W. Wenzel, P. Avery, and W. R. Hendee. (1975). "Ultrasound imaging as an aid to cancer therapy-I." *Int. J. Radiat. Oncol. Biol. Phys.* 1(1–2):119–32.

Chapman, C. H., J. I. Prisciandaro, K. E. Maturen, Y. Cao, J. M. Balter, K. McLean, and S. Jolly. (2016). "MRI-based evaluation of the vaginal cuff in brachytherapy planning: Are we missing the target?" *Int. J. Radiat. Oncol. Biol. Phys.* 95(2):743–50.
Charra-Brunaud, C., V. Harter, M. Delannes, C. Haie-Meder, P. Quetin, C. Kerr, B. Castelain, L. Thomas, and D. Peiffert. (2012). "Impact of 3D image-based PDR brachytherapy on outcome of patients treated for cervix carcinoma in France: results of the French STIC prospective study." *Radiother. Oncol.* 103(3):305–13.
Creutzberg, C. L., W. L. van Putten, P. C. Koper, M. L. Lybeert, J. J. Jobsen, C. C. Warlam-Rodenhuis, K. A. De Winter, L. C. Lutgens, A. C. van den Bergh, E. van de Steen-Banasik, H. Beerman, and M. van Lent. (2000). "Surgery and postoperative radiotherapy versus surgery alone for patients with stage-1 endometrial carcinoma: multicentre randomised trial. PORTEC Study Group. Post Operative Radiation Therapy in Endometrial Carcinoma." *Lancet* 355(9213):1404–11.
Damato, A. L., I. Buzurovic, M. S. Bhagwat, R. A. Cormack, P. M. Devlin, S. Friesen, J. Hansen, L. J. Lee, M. M. Manuel, L. P. Cho, D. O'Farrell, and A. N. Viswanathan. (2017). "The value of systematic contouring of the bowel for treatment plan optimization in image-guided cervical cancer high-dose-rate brachytherapy." *Brachytherapy* pii: S1538-4721(17)30014-4. doi: 10.1016/j.brachy.2017.01.008. [Epub ahead of print].
Damato, A. L., R. A. Cormack, and A. N. Viswanathan. (2015). "A novel intracavitary applicator design for the treatment of deep vaginal fornices: preliminary dose metrics and geometric analysis." *J. Contemp. Brachytherapy* 7(1):48–54.
Damato, A. L., L. J. Lee, M. S. Bhagwat, I. Buzurovic, R. A. Cormack, S. Finucane, J. L. Hansen, D. A. O'Farrell, A. Offiong, U. Randall, S. Friesen, and A. N. Viswanathan. (2015). "Redesign of process map to increase efficiency: Reducing procedure time in cervical cancer brachytherapy." *Brachytherapy* 14(4):471–80.
Damato, A. L., A. N. Viswanathan, and R. A. Cormack. (2013). "Validation of mathematical models for the prediction of organs-at-risk dosimetric metrics in high-dose-rate gynecologic interstitial brachytherapy." *Med. Phys.* 40(10):101711.
Deufel, C. L. and K. M. Furutani. (2014). "Quality assurance for high dose rate brachytherapy treatment planning optimization: using a simple optimization to verify a complex optimization." *Phys. Med. Biol.* 59(3):525–40.
Dimopoulos, J. C., C. Kirisits, P. Petric, P. Georg, S. Lang, D. Berger, and R. Potter. (2006). "The Vienna applicator for combined intracavitary and interstitial brachytherapy of cervical cancer: clinical feasibility and preliminary results." *Int. J. Radiat. Oncol. Biol. Phys.* 66(1):83–90.
Epstein, E., A. Testa, A. Gaurilcikas, A. Di Legge, L. Ameye, V. Atstupenaite, A. L. Valentini, B. Gui, N. O. Wallengren, S. Pudaric, A. Cizauskas, A. Masback, G. F. Zannoni, P. Kannisto, M. Zikan, I. Pinkavova, A. Burgetova, P. Dundr, K. Nemejcova, D. Cibula, and D. Fischerova. (2013). "Early-stage cervical cancer: tumor delineation by magnetic resonance imaging and ultrasound—a European multicenter trial." *Gynecol. Oncol.* 128(3):449–53.
Fallon, J., S. J. Park, L. Yang, D. Veruttipong, M. Zhang, T. Van, P. C. Wang, A. M. Fekete, M. Cambeiro, M. Kamrava, M. L. Steinberg, and D. J. Demanes. (2016). "Long term results from a prospective database on high dose rate (HDR) interstitial brachytherapy for primary cervical carcinoma." *Gynecol. Oncol.* pii: S0090-8258(16)31496-2. doi: 10.1016/j.ygyno.2016.10.020. [Epub ahead of print].
Fellner, C., R. Potter, T. H. Knocke, and A. Wambersie. (2001). "Comparison of radiography- and computed tomography-based treatment planning in cervix cancer in brachytherapy with specific attention to some quality assurance aspects." *Radiother. Oncol.* 58(1):53–62.
Granai, C. O., F. Doherty, P. Allee, H. G. Ball, H. Madoc-Jones, and S. L. Curry. (1990). "Ultrasound for diagnosing and preventing malplacement of intrauterine tandems." *Obstet. Gynecol.* 75(1):110–13.
Grover, S., M. M. Harkenrider, L. P. Cho, B. Erickson, C. Small, W. Small, Jr., and A. N. Viswanathan. (2016). "Image Guided Cervical Brachytherapy: 2014 Survey of the American Brachytherapy Society." *Int. J. Radiat. Oncol. Biol. Phys.* 94(3):598–604.
Haack, S., S. K. Nielsen, J. C. Lindegaard, J. Gelineck, and K. Tanderup. (2009). "Applicator reconstruction in MRI 3D image-based dose planning of brachytherapy for cervical cancer." *Radiother. Oncol.* 91(2):187–93.
Haas, J. S., R. D. Dean, and C. M. Mansfield. (1980). "Evaluation of a new Fletcher Applicator using cesium-137." *Int. J. Radiat. Oncol. Biol. Phys.* 6(11):1589–95.
Haie-Meder, C., R. Potter, E. Van Limbergen, E. Briot, M. De Brabandere, J. Dimopoulos, I. Dumas, T. P. Hellebust, C. Kirisits, S. Lang, S. Muschitz, J. Nevinson, A. Nulens, P. Petrow, N. Wachter-Gerstner, and G.E.C.E.W.G. Gynaecological. (2005). "Recommendations from Gynaecological (GYN) GEC-ESTRO Working Group (I): concepts and terms in 3D image based 3D treatment planning in cervix cancer brachytherapy with emphasis on MRI assessment of GTV and CTV." *Radiother. Oncol.* 74(3):235–45.
Hallock, A., K. Surry, D. Batchelar, L. Vanderspek, J. Yuen, A. Hammond, J. Radwan, B. Yaremko, G. Rodrigues, and D. D'Souza. (2011). "An early report on outcomes from computed tomographic-based high-dose-rate brachytherapy for locally advanced cervix cancer: A single institution experience." *Pract. Radiat. Oncol.* 1(3):173–81.
Harkenrider, M. M., A. M. Block, K. M. Alektiar, D. K. Gaffney, E. Jones, A. Klopp, A. N. Viswanathan, and W. Small, Jr. (2017). "American Brachytherapy Task Group Report: Adjuvant vaginal brachytherapy for early-stage endometrial cancer: A comprehensive review." *Brachytherapy* 16(1):95–108.
Harkenrider, M. M., S. Grover, B. A. Erickson, A. N. Viswanathan, C. Small, S. Kliethermes, and W. Small, Jr. (2016). "Vaginal brachytherapy for postoperative endometrial cancer: 2014 Survey of the American Brachytherapy Society." *Brachytherapy* 15(1):23–29.
Hassouna, A. H., Y. A. Bahadur, C. Constantinescu, M. E. El Sayed, H. Naseem, and A. F. Naga. (2011). "In vivo diode dosimetry vs. computerized tomography and digitally reconstructed radiographs for critical organ dose calculation in high-dose-rate brachytherapy of cervical cancer." *Brachytherapy* 10(6):498–502.
Hellebust, T. P., C. Kirisits, D. Berger, J. Perez-Calatayud, M. De Brabandere, A. De Leeuw, I. Dumas, R. Hudej, G. Lowe, R. Wills, and K. Tanderup. (2010). "Recommendations from Gynaecological (GYN) GEC-ESTRO Working Group (III): considerations and pitfalls in commissioning and applicator reconstruction in 3D image-based treatment planning of cervix cancer brachytherapy." *Radiother. Oncol.* 96(2):153–60.
Hellebust, T. P., C. Kirisits, D. Berger, J. Perez-Calatayud, M. De Brabandere, A. De Leeuw, I. Dumas, R. Hudej, G. Lowe, R. Wills, K. Tanderup, and G.E.C.E.W.G. Gynaecological. (2010). "Recommendations from Gynaecological (GYN)

GEC-ESTRO Working Group: considerations and pitfalls in commissioning and applicator reconstruction in 3D image-based treatment planning of cervix cancer brachytherapy." *Radiother. Oncol.* 96(2):153–60.

Hellebust, T. P., P. Petric, K. Tanderup, D. Berger, E. Fidarova, R. Pötter, C. Haie-Meder, and E. Malinen. (2012). "Spatial dosimetric sensitivity analysis of contouring uncertainties in GYN 3D based brachytherapy." *Radiother. Oncol.* 103, Suppl. 2:34.

Hellebust, T. P., K. Tanderup, E. S. Bergstrand, B. H. Knutsen, J. Roislien, and D. R. Olsen. (2007). "Reconstruction of a ring applicator using CT imaging: impact of the reconstruction method and applicator orientation." *Phys. Med. Biol.* 52(16):4893–904.

Holloway, C. L., E. A. Macklin, R. A. Cormack, and A. N. Viswanathan. (2011). "Should the organs at risk be contoured in vaginal cuff brachytherapy?" *Brachytherapy* 10(4):313–17.

Hricak, H., C. Gatsonis, F. V. Coakley, B. Snyder, C. Reinhold, L. H. Schwartz, P. J. Woodward, H. K. Pannu, M. Amendola, and D. G. Mitchell. (2007). "Early invasive cervical cancer: CT and MR imaging in preoperative evaluation—ACRIN/GOG comparative study of diagnostic performance and interobserver variability." *Radiology* 245(2):491–98.

Huq, M. S., B. A. Fraass, P. B. Dunscombe, J. P. Gibbons, Jr., G. S. Ibbott, A. J. Mundt, S. Mutic, J. R. Palta, F. Rath, B. R. Thomadsen, J. F. Williamson, and E. D. Yorke. (2016). "The report of Task Group 100 of the AAPM: Application of risk analysis methods to radiation therapy quality management." *Med. Phys.* 43(7):4209.

ICRU. International Commission on Radiation Units and Measurements. (2016). *Prescribing, Recording, and Reporting Brachytherapy for Cancer of the Cervix.* Report 89. *Journal of the ICRU* 13.

Katz, A. and P. J. Eifel. (2000). "Quantification of intracavitary brachytherapy parameters and correlation with outcome in patients with carcinoma of the cervix." *Int. J. Radiat. Oncol. Biol. Phys.* 48(5):1417–25.

Keys, H. M., J. A. Roberts, V. L. Brunetto, R. J. Zaino, N. M. Spirtos, J. D. Bloss, A. Pearlman, M. A. Maiman, J. G. Bell, and G. Gynecologic Oncology. (2004). "A phase III trial of surgery with or without adjunctive external pelvic radiation therapy in intermediate risk endometrial adenocarcinoma: a Gynecologic Oncology Group study." *Gynecol. Oncol.* 92(3):744–51.

Kim, R. Y., P. Pareek, J. Duan, H. Murshed, and I. Brezovich. (2002). "Postoperative intravaginal brachytherapy for endometrial cancer; dosimetric analysis of vaginal colpostats and cylinder applicators." *Brachytherapy* 1(3):138–44.

Kirisits, C., M. Federico, K. Nkiwane, E. Fidarova, I. Jurgenliemk-Schulz, A. de Leeuw, J. Lindegaard, R. Potter, and K. Tanderup. (2015). "Quality assurance in MR image guided adaptive brachytherapy for cervical cancer: Final results of the EMBRACE study dummy run." *Radiother. Oncol.* 117(3):548–54.

Kirisits, C., S. Lang, J. Dimopoulos, D. Berger, D. Georg, and R. Potter. (2006). "The Vienna applicator for combined intracavitary and interstitial brachytherapy of cervical cancer: design, application, treatment planning, and dosimetric results." *Int. J. Radiat. Oncol. Biol. Phys.* 65(2):624–30.

Kirisits, C., R. Potter, S. Lang, J. Dimopoulos, N. Wachter-Gerstner, and D. Georg. (2005). "Dose and volume parameters for MRI-based treatment planning in intracavitary brachytherapy for cervical cancer." *Int. J. Radiat. Oncol. Biol. Phys.* 62(3):901–11.

Kirisits, C., M. J. Rivard, D. Baltas, F. Ballester, M. De Brabandere, R. van der Laarse, Y. Niatsetski, P. Papagiannis, T. P. Hellebust, J. Perez-Calatayud, K. Tanderup, J. L. Venselaar, and F. A. Siebert. (2014). "Review of clinical brachytherapy uncertainties: analysis guidelines of GEC-ESTRO and the AAPM." *Radiother. Oncol.* 110(1):199–212.

Klopp, A., B. D. Smith, K. Alektiar, A. Cabrera, A. L. Damato, B. Erickson, G. Fleming, D. Gaffney, K. Greven, K. Lu, D. Miller, D. Moore, D. Petereit, T. Schefter, W. Small, Jr., C. Yashar, A. N. Viswanathan, and O. American Society for Radiation. (2014). "The role of postoperative radiation therapy for endometrial cancer: Executive summary of an American Society for Radiation Oncology evidence-based guideline." *Pract. Radiat. Oncol.* 4(3):137–44.

Kubo, H. D., G. P. Glasgow, T. D. Pethel, B. R. Thomadsen, and J. F. Williamson. (1998). "High dose-rate brachytherapy treatment delivery: report of the AAPM Radiation Therapy Committee Task Group No. 59." *Med. Phys.* 25(4):375–403.

Lee, L. J., A. L. Damato, and A. N. Viswanathan. (2013). "Clinical outcomes following 3D image-guided brachytherapy for vaginal recurrence of endometrial cancer." *Gynecol. Oncol.* 131(3):586–92.

Lee, L. J., A. L. Damato, and A. N. Viswanathan. (2013). "Clinical outcomes of high-dose-rate interstitial gynecologic brachytherapy using real-time CT guidance." *Brachytherapy* 12(4):303–10.

Lee, L. J., I. J. Das, S. A. Higgins, A. Jhingran, W. Small, Jr., B. Thomadsen, A. N. Viswanathan, A. Wolfson, P. Eifel, and S. American Brachytherapy. (2012). "American Brachytherapy Society consensus guidelines for locally advanced carcinoma of the cervix. Part III: low-dose-rate and pulsed-dose-rate brachytherapy." *Brachytherapy* 11(1):53–57.

Lindegaard, J. C. and K. Tanderup. (2012). "Counterpoint: Time to retire the parametrial boost." *Brachytherapy* 11(2):80–83; discussion 84.

Mahantshetty, U., N. Khanna, J. Swamidas, R. Engineer, M. H. Thakur, N. H. Merchant, D. D. Deshpande, and S. Shrivastava. (2012). "Trans-abdominal ultrasound (US) and magnetic resonance imaging (MRI) correlation for conformal intracavitary brachytherapy in carcinoma of the uterine cervix." *Radiother. Oncol.* 102(1):130–34.

Manuel, M. M., L. P. Cho, P. J. Catalano, A. L. Damato, D. T. Miyamoto, C. M. Tempany, E. J. Schmidt, and A. N. Viswanathan. (2016). "Outcomes with image-based interstitial brachytherapy for vaginal cancer." *Radiother. Oncol.* 120(3):486–92.

Nag, S., H. Cardenes, S. Chang, I. J. Das, B. Erickson, G. S. Ibbott, J. Lowenstein, J. Roll, B. Thomadsen, M. Varia, and G. Image-Guided Brachytherapy Working. (2004). "Proposed guidelines for image-based intracavitary brachytherapy for cervical carcinoma: report from Image-Guided Brachytherapy Working Group." *Int. J. Radiat. Oncol. Biol. Phys.* 60(4):1160–72.

Nakano, T., S. Kato, T. Ohno, H. Tsujii, S. Sato, K. Fukuhisa, and T. Arai. (2005). "Long-term results of high-dose rate intracavitary brachytherapy for squamous cell carcinoma of the uterine cervix." *Cancer* 103(1):92–101.

Nesvacil, N., M. P. Schmid, R. Potter, G. Kronreif, and C. Kirisits. (2016). "Combining transrectal ultrasound and CT for image-guided adaptive brachytherapy of cervical cancer: Proof of concept." *Brachytherapy* 15(6):839–44.

Nkiwane, K. S., R. Potter, L. U. Fokdal, P. Hoskin, R. Pearcey, B. Segedin, U. Mahantshetty, and C. Kirisits. (2015). "Use of bladder dose points for assessment of the spatial dose distribution in the posterior bladder wall in cervical cancer brachytherapy and the impact of applicator position." *Brachytherapy* 14(2):252–59.

Nomden, C. N., A. A. de Leeuw, M. A. Moerland, J. M. Roesink, R. J. Tersteeg, and I. M. Jurgenliemk-Schulz. (2012). "Clinical use of the Utrecht applicator for combined intracavitary/interstitial brachytherapy treatment in locally advanced cervical cancer." *Int. J. Radiat. Oncol. Biol. Phys.* 82(4):1424–30.

Nomden, C. N., A. A. de Leeuw, E. Van Limbergen, M. de Brabandere, A. Nulens, R. A. Nout, M. Laman, M. Ketelaars, L. Lutgens, B. Reniers, and I. M. Jurgenliemk-Schulz. (2013). "Multicentre treatment planning study of MRI-guided brachytherapy for cervical cancer: comparison between tandem-ovoid applicator users." *Radiother. Oncol.* 107(1):82–87.

Nout, R. A., V. T. Smit, H. Putter, I. M. Jurgenliemk-Schulz, J. J. Jobsen, L. C. Lutgens, E. M. van der Steen-Banasik, J. W. Mens, A. Slot, M. C. Kroese, B. N. van Bunningen, A. C. Ansink, W. L. van Putten, C. L. Creutzberg, and P. S. Group. (2010). "Vaginal brachytherapy versus pelvic external beam radiotherapy for patients with endometrial cancer of high-intermediate risk (PORTEC-2): an open-label, non-inferiority, randomised trial." *Lancet* 375(9717):816–23.

Ostrom, L. T., P. Rathbun, R. Cumberlin, J. Horton, R. Gastorf, and T. J. Leahy. (1996). "Lessons learned from investigations of therapy misadministration events." *Int. J. Radiat. Oncol. Biol. Phys.* 34(1):22734.

Paliwal, B., B. Thomadsen, and D. Petereit. "Imaging Applications in Brachytherapy." In *Principles and Practice of Brachytherapy.* S. Nag, Ed. Armonk NY: Futura Publishing Co., 1997.

Park, S. J., M. Chung, D. J. Demanes, R. Banerjee, M. Steinberg, and M. Kamrava. (2013). "Dosimetric comparison of 3-dimensional planning techniques using an intravaginal multichannel balloon applicator for high-dose-rate gynecologic brachytherapy." *Int. J. Radiat. Oncol. Biol. Phys.* 87(4):840–46.

Pelloski, C. E., et al. (2005). "Comparison between CT-based volumetric calculations and ICRU reference-point estimates of radiation doses delivered to bladder and rectum during intracavitary radiotherapy for cervical cancer." *Int. J. Radiat. Oncol. Biol. Phys.* 62.1:131–37.

Petric, P., R. Hudej, and M. Music. (2009). "MRI assisted cervix cancer brachytherapy pre-planning, based on insertion of the applicator in para-cervical anaesthesia: preliminary results of a prospective study." *J. Contemp. Brachytherapy* 1(3):163–69.

Petric, P. and C. Kirisits. (2016). "Potential role of TRAns Cervical Endosonography (TRACE) in brachytherapy of cervical cancer: proof of concept." *J. Contemp. Brachytherapy* 8(3):215–20.

Potter, R., J. Dimopoulos, P. Georg, S. Lang, C. Waldhausl, N. Wachter-Gerstner, H. Weitmann, A. Reinthaller, T. H. Knocke, S. Wachter, and C. Kirisits. (2007). "Clinical impact of MRI assisted dose volume adaptation and dose escalation in brachytherapy of locally advanced cervix cancer." *Radiother. Oncol.* 83(2):148–55.

Potter, R., M. Federico, A. Sturdza, I. Fotina, N. Hegazy, M. Schmid, C. Kirisits, and N. Nesvacil. (2016). "Value of magnetic resonance imaging without or with applicator in place for target definition in cervix cancer brachytherapy." *Int. J. Radiat. Oncol. Biol. Phys.* 94(3):588–97.

Potter, R., P. Georg, J. C. Dimopoulos, M. Grimm, D. Berger, N. Nesvacil, D. Georg, M. P. Schmid, A. Reinthaller, A. Sturdza, and C. Kirisits. (2011). "Clinical outcome of protocol based image (MRI) guided adaptive brachytherapy combined with 3D conformal radiotherapy with or without chemotherapy in patients with locally advanced cervical cancer." *Radiother. Oncol.* 100(1):116–23.

Potter, R., C. Haie-Meder, E. Van Limbergen, I. Barillot, M. De Brabandere, J. Dimopoulos, I. Dumas, B. Erickson, S. Lang, A. Nulens, P. Petrow, J. Rownd, and C. Kirisits. (2006). "Recommendations from gynaecological (GYN) GEC ESTRO working group (II): concepts and terms in 3D image-based treatment planning in cervix cancer brachytherapy-3D dose volume parameters and aspects of 3D image-based anatomy, radiation physics, radiobiology." *Radiother. Oncol.* 78(1):67–77.

Potter, R., C. Haie-Meder, E. Van Limbergen, I. Barillot, M. De Brabandere, J. Dimopoulos, I. Dumas, B. Erickson, S. Lang, A. Nulens, P. Petrow, J. Rownd, C. Kirisits, and G. E. W. Group. (2006). "Recommendations from gynaecological (GYN) GEC ESTRO working group (II): concepts and terms in 3D image-based treatment planning in cervix cancer brachytherapy-3D dose volume parameters and aspects of 3D image-based anatomy, radiation physics, radiobiology." *Radiother. Oncol.* 78(1):67–77.

Rose, J., F. Bachand, D. Petrik, D. Batchelar, M. Schmid, and J. Crook. (2014). "Transvaginal Ultrasound-Guided Interstitial Brachytherapy for. Vaginal Tumors: A Fixed Template Technique." *Int. J. Radiat. Oncol. Biol. Phys.* 90:S502–S502.

Schindel, J., M. Muruganandham, F. C. Pigge, J. Anderson, and Y. Kim. (2013). "Magnetic resonance imaging (MRI) markers for MRI-guided high-dose-rate brachytherapy: novel marker-flange for cervical cancer and marker catheters for prostate cancer." *Int. J. Radiat. Oncol. Biol. Phys.* 86(2):387–93.

Schindel, J., W. Zhang, S. K. Bhatia, W. Sun, and Y. Kim. (2013). "Dosimetric impacts of applicator displacements and applicator reconstruction-uncertainties on 3D image-guided brachytherapy for cervical cancer." *J. Contemp. Brachytherapy* 5(4):250–57.

Schmid, M. P., N. Nesvacil, R. Potter, G. Kronreif, and C. Kirisits. (2016). "Transrectal ultrasound for image-guided adaptive brachytherapy in cervix cancer - An alternative to MRI for target definition?" *Radiother. Oncol.* 120(3):467–72.

Schmid, M. P., R. Potter, P. Brader, A. Kratochwil, G. Goldner, K. Kirchheiner, A. Sturdza, and C. Kirisits. (2013). "Feasibility of transrectal ultrasonography for assessment of cervical cancer." *Strahlenther. Onkol.* 189(2):123–28.

Schwarz, J. K., S. Beriwal, J. Esthappan, B. Erickson, C. Feltmate, A. Fyles, D. Gaffney, E. Jones, A. Klopp, W. Small, Jr., B. Thomadsen, C. Yashar, and A. Viswanathan. (2015). "Consensus statement for brachytherapy for the treatment of medically inoperable endometrial cancer." *Brachytherapy* 14(5):587–99.

Sharma, D. N., G. K. Rath, S. Thulkar, S. Kumar, V. Subramani, and P. K. Julka. (2010). "Use of transrectal ultrasound for high dose rate interstitial brachytherapy for patients of carcinoma of uterine cervix." *J. Gynecol. Oncol.* 21(1):12-17.

Simpson, D. R., D. J. Scanderbeg, R. Carmona, R. M. McMurtrie, J. Einck, L. K. Mell, M. T. McHale, C. C. Saenz, S. C. Plaxe, T. Harrison, A. J. Mundt, and C. M. Yashar. (2015). "Clinical outcomes of computed tomography-based volumetric brachytherapy planning for cervical cancer." *Int. J. Radiat. Oncol. Biol. Phys.* 93(1):150–57.

Small, W., Jr., S. Beriwal, D. J. Demanes, K. E. Dusenbery, P. Eifel, B. Erickson, E. Jones, J. J. Rownd, J. F. De Los Santos, A. N. Viswanathan, D. Gaffney, and S. American Brachytherapy. (2012). "American Brachytherapy Society consensus guidelines for adjuvant vaginal cuff brachytherapy after hysterectomy." *Brachytherapy* 11(1):58–67.

Small, W., Jr., B. Erickson, and F. Kwakwa. (2005). "American Brachytherapy Society survey regarding practice patterns of postoperative irradiation for endometrial cancer: current status of vaginal brachytherapy." *Int. J. Radiat. Oncol. Biol. Phys.* 63(5):1502-07.

Sorbe, B., G. Horvath, H. Andersson, K. Boman, C. Lundgren, and B. Pettersson. (2012). "External pelvic and vaginal irradiation versus vaginal irradiation alone as postoperative therapy in medium-risk endometrial carcinoma—a prospective randomized study." *Int. J. Radiat. Oncol. Biol. Phys.* 82(3):1249–55.

St-Amant, P., W. Foster, M.-A. Forment, S. Aubin, and L. Beaulieu. (2016). "Use of 3D-Ultrasound for Cervical Cancer Brachytherapy: An Imaging Technique to Improve Treatment Planning." *Brachytherapy* 15: S92–S93.

Stitt, J. A., B. R. Thomadsen, and J. F. Fowler. (1992). "High-dose-rate brachytherapy for cervical carcinoma." *Int. J. Radiat. Oncol. Biol. Phys.* 24(3):574.

Stock, R. G., K. Chan, M. Terk, J. K. Dewyngaert, N. N. Stone, and P. Dottino. (1997). "A new technique for performing Syed-Neblett template interstitial implants for gynecologic malignancies using transrectal-ultrasound guidance." *Int. J. Radiat. Oncol. Biol. Phys.* 37(4):819–25.

Syed, A. M., A. A. Puthawala, N. N. Abdelaziz, M. el-Naggar, P. Disaia, M. Berman, K. S. Tewari, A. Sharma, A. Londrc, S. Juwadi, J. M. Cherlow, S. Damore, and Y. J. Chen. (2002). "Long-term results of low-dose-rate interstitial-intracavitary brachytherapy in the treatment of carcinoma of the cervix." *Int. J. Radiat. Oncol. Biol. Phys.* 54(1):67–78.

Takahashi, Y., M. Koizumi, I. Sumida, F. Isohashi, T. Ogata, Y. Akino, Y. Yoshioka, S. Maruoka, S. Inoue, K. Konishi, and K. Ogawa. (2012). "The usefulness of an independent patient-specific treatment planning verification method using a benchmark plan in high-dose-rate intracavitary brachytherapy for carcinoma of the uterine cervix." *J. Radiat. Res.* 53(6):936–44.

Tanderup, K., S. K. Nielsen, G. B. Nyvang, E. M. Pedersen, L. Rohl, T. Aagaard, L. Fokdal, and J. C. Lindegaard. (2010). "From point A to the sculpted pear: MR image guidance significantly improves tumour dose and sparing of organs at risk in brachytherapy of cervical cancer." *Radiother. Oncol.* 94(2):173–80.

Tod, M. and W. J. Meredith. (1953). "Treatment of cancer of the cervix uteri, a revised Manchester method." *Br. J. Radiol.* 26(305):252–57.

Townamchai, K., R. Berkowitz, M. Bhagwat, A. L. Damato, S. Friesen, L. J. Lee, U. Matulonis, D. O'Farrell, and A. N. Viswanathan. (2013). "Vaginal brachytherapy for early stage uterine papillary serous and clear cell endometrial cancer." *Gynecol. Oncol.* 129(1):18–21.

Townamchai, K., L. Lee, and A. N. Viswanathan. (2012). "A novel low dose fractionation regimen for adjuvant vaginal brachytherapy in early stage endometrioid endometrial cancer." *Gynecol. Oncol.* 127(2):351–55.

van Dyk, S., S. Kondalsamy-Chennakesavan, M. Schneider, D. Bernshaw, and K. Narayan. (2014). "Comparison of measurements of the uterus and cervix obtained by magnetic resonance and transabdominal ultrasound imaging to identify the brachytherapy target in patients with cervix cancer." *Int. J. Radiat. Oncol. Biol. Phys.* 88(4):860–65.

van Dyk, S., K. Narayan, R. Fisher, and D. Bernshaw. (2009). "Conformal brachytherapy planning for cervical cancer using transabdominal ultrasound." *Int. J. Radiat. Oncol. Biol. Phys.* 75(1):64–70.

van Dyk, S., M. Schneider, S. Kondalsamy-Chennakesavan, D. Bernshaw, and K. Narayan. (2015). "Ultrasound use in gynecologic brachytherapy: Time to focus the beam." *Brachytherapy* 14(3):390–400.

Vargo, J. A. and S. Beriwal. (2014). "Image-based brachytherapy for cervical cancer." *World J. Clin. Oncol.* 5(5):921–30.

Viswanathan, A. N., S. Beriwal, J. F. De Los Santos, D. J. Demanes, D. Gaffney, J. Hansen, E. Jones, C. Kirisits, B. Thomadsen, and B. Erickson. (2012). "American Brachytherapy Society consensus guidelines for locally advanced carcinoma of the cervix. Part I: general principles." *Brachytherapy* 11(1):33–46.

Viswanathan, A. N., S. Beriwal, J. F. De Los Santos, D. J. Demanes, D. Gaffney, J. Hansen, E. Jones, C. Kirisits, B. Thomadsen, B. Erickson, and S. American Brachytherapy. (2012). "American Brachytherapy Society consensus guidelines for locally advanced carcinoma of the cervix. Part II: high-dose-rate brachytherapy." *Brachytherapy* 11(1):47–52.

Viswanathan, A. N., J. Dimopoulos, C. Kirisits, D. Berger, and R. Potter. (2007). "Computed tomography versus magnetic resonance imaging-based contouring in cervical cancer brachytherapy: results of a prospective trial and preliminary guidelines for standardized contours." *Int. J. Radiat. Oncol. Biol. Phys.* 68(2):491–98.

Viswanathan, A. N. and B. A. Erickson. (2009). "Three-dimensional imaging in gynecologic brachytherapy: a survey of the American Brachytherapy Society." *Int. J. Radiat. Oncol. Biol. Phys.* 76(1):104–09.

Viswanathan, A. N., B. Thomadsen, C. American Brachytherapy Society Cervical Cancer Recommendations and S. American Brachytherapy. (2012). "American Brachytherapy Society consensus guidelines for locally advanced carcinoma of the cervix. Part I: general principles." *Brachytherapy* 11(1):33–46.

Waldhausl, C., A. Wambersie, R. Potter, and D. Georg. (2005). "In-vivo dosimetry for gynaecological brachytherapy: physical and clinical considerations." *Radiother. Oncol.* 77(3):310–17.

Wooten, C. E., M. Randall, J. Edwards, P. Aryal, W. Luo, and J. Feddock. (2014). "Implementation and early clinical results utilizing Cs-131 permanent interstitial implants for gynecologic malignancies." *Gynecol. Oncol.* 133(2):268–73.

Zolciak-Siwinska, A., E. Gruszczynska, M. Bijok, J. Jonska-Gmyrek, M. Dabkowski, J. Staniaszek, W. Michalski, A. Kowalczyk, and K. Milanowska. (2016). "Computed tomography-planned high-dose-rate brachytherapy for treating uterine cervical cancer." *Int. J. Radiat. Oncol. Biol. Phys.* 96(1):87–92.

Example Problems

(Answers are found at the end of the book.)

1. What choice of optimization points is recommended by the American Brachytherapy Society for vaginal cuff brachytherapy with a vaginal cylinder?
    a. ABS recommends points at a fixed distance from the cylinder central channel.
    b. The ABS does not recommend the use of optimization points but volume-based optimization.
    c. ABS recommends points following the curved dome of the cylinder and an additional point at the apex.
    d. ABS recommends points at least 5 mm away from the vaginal surface.
2. Dose prescription for vaginal cuff brachytherapy is specified:
    a. to the D_{90} of the clinical target volume
    b. to points at least 5 mm depth from the vaginal surface
    c. to the apex
    d. to points at either the vaginal surface or at a specified depth, typically 5 mm
3. Regarding the role of imaging in vaginal cylinder planning, which of the following statements is true?
    a. 3D planning of vaginal cylinders for the first fraction is clinically necessary.
    b. 3D planning of vaginal cylinders for the first fraction is commonly performed in North America.
    c. Because of inter-fraction variations of organs-at-risk dose, replanning at each insertion is recommended.
    d. Use D_{2cm^3} values for rectum, bladder, sigmoid colon, and bowel, and D_{90} and V_{100} values for the clinical target volume.
4. Regarding quality assurance needs for vaginal cuff brachytherapy, which of the following statements is true?
    a. Due to the simple nature of this treatment, no reports of errors in vaginal cuff brachytherapy exist in modern times.
    b. 2D or 3D imaging, while useful for treatment verification, are not considered mandatory by the ABS guidelines.
    c. *In vivo* dosimetry at each fraction is required by the Nuclear Regulatory Commission.
    d. In-room imaging capability, either with films, CT, or MRI, is necessary.

5. According to the current American Brachytherapy Society definition, Point A should be defined as ______.

 a. 2 cm superiorly along the tandem from the apices of the vaginal fornices and 2 cm perpendicularly lateral to the tandem

 b. 2 cm superiorly along the tandem from the flange, and 2 cm perpendicularly lateral to the tandem

 c. 2 cm superiorly along the tandem from the top of the ovoids/ring/cylinder, and 2 cm perpendicularly lateral to the tandem

 d. Points A are currently not defined by the American Brachytherapy Society, in favor of utilizing only volumetric dose metrics such CTV.

 e. D_{90}.

6. According to the ICRU Report 89, dose from EBRT and brachytherapy should be reported:

 a. summing the total dose in Gy from prior radiation to the total dose in Gy from the brachytherapy

 b. calculating the total equi-effective dose (EQD2) according to the linear-quadratic model

 c. only if inhomogeneity correction was applied to both EBRT and brachytherapy dose calculations, or neither of them

 d. using deformable registration and summation of the deformed dose matrices

7. According to the American Brachytherapy Society, optimization of cervical cancer brachytherapy should be performed:

 a. with the only goal of minimizing D_{2cm^3} doses for the organs-at-risk, and maximizing D_{90} dose for the target

 b. with the only goal of meeting

 c. with D_{2cm^3} constraints (typically 70–75 Gy EQD2 for rectum and bladder and 90 Gy EQD2 for bladder) and D_{90} constraints (typically D_{90} > prescription dose)

 d. with the goal of meeting dose metrics constraints, in unison with a slice-by-slice analysis of the isodose lines

 e. with the goal to have the dose point A greater than the prescription dose

8. Regarding applicator reconstruction, which of the following statements apply?

 a. When using CT images, a slice thickness $\leq$3 mm should be used.

 b. The internal lumen of the applicator is not distinguishable from the applicator material in T2 MRI images, making manual applicator reconstruction challenging.

 c. The use of applicator libraries is fast and simple, and less prone to reconstruction errors.

 d. Whenever possible, contouring and applicator reconstruction should be performed on the same image set.

 e. All of the above are true.

9. The high-risk clinical target volume (CTV_{HR}) is defined as:
   a. the residual GTV at the time of brachytherapy, the whole cervix, and adjacent residual pathologic tissue
   b. the whole parametria, the whole uterus, and the upper part of the vagina
   c. the GTV at presentation, the whole cervix, and adjacent residual pathologic tissue
   d. The GTV and cervix as observed only on T2 MRI.
10. Which is the most commonly used imaging techniques to assist with applicator placement during the brachytherapy insertion in North America?
   a. ultrasound
   b. CT
   c. fluoroscopy
   d. MRI

Chapter 7

Skin Brachytherapy

Regina K. Fulkerson[1], Ivan Buzurovic[2], Antonio Damato[3], and Zoubir Ouhib[4]

[1]RKF Consultants, LLC
Dundee, New York

[2]Dana-Farber/Brigham and Women's Cancer Center
Boston, Massachusetts

[3]Memorial Sloan Kettering Cancer Center
New York, New York

[4]Lynn Regional Cancer Center
Delray Beach, Florida

7.1 Introduction

Skin cancer is the most common form of cancer. Basal and squamous cell cancers are the most prevalent types of cancer, accounting for approximately 3.3 million cases per year. Several treatment modalities exist for these types of lesions, including surgery, cryotherapy, topical drugs, photodynamic therapy, and radiation therapy (American Cancer Society 2016). Historically, radiation therapy treatments of these lesions have been achieved with linear accelerator-based electron beams and superficial x-ray beams (Fulkerson et al. 2014). Recently, there has been a renewed interest in the use of brachytherapy for surface lesions. This chapter discusses the types of skin cancers most often treated with brachytherapy, reviews the dosimetric aspects of the various sources and applicators available, and addresses the main items of interest for the treatment process (focusing on the physicist responsibilities).

7.2 Types of Lesions Treated with Brachytherapy

As expected, basal cell and squamous cell carcinomas are the most common skin cancers treated with brachytherapy. Basal cell lesions are most commonly characterized as papulonodular lesions with a pearly transparent rim, and often have telangiectasias. Central ulcerations can be present if the lesion is large enough, and may appear cyst-like due to their clear color and soft consistency. Basal cell carcinomas have low metastatic potential and consist of cells that look similar to the basal epidermis layer, and they occur most frequently on the scalp, face, and ear.

The second most common skin cancer (squamous cell carcinoma) is characterized by scaly, ulcerated lesions with sharply demarcated borders. Squamous cell carcinomas typically present in regions of the body lined with epithelium, such as the oral cavity, lips, genital regions, and hands (Fulkerson et al. 2014). Basal cell and squamous cell carcinomas are often found while still small and do not typically require staging, blood work, or imaging tests for diagnosis. If high-risk features are present—such as a tumor thickness greater than 4 mm; diameter greater than 20 mm; tumor invasion into the lower dermis, perineural, or lymphovascular invasion; moderate or poor differentiation; or if the tumor involves high-risk sites such as the scalp, lip, ear, eyelids, or nose—staging may be done according to the American Joint Committee on Cancer and TNM scale. It should be noted that radiation therapy is not typically used for primary melanomas due to poor clinical response of the lesions. Brachytherapy is effective for both small (≤20 mm in diameter) and large (>20 mm in diameter) lesions, and it is well suited for lesions of the nose, lips, and ears, especially if it is not clear how extensive any surgical defects would be or how complicated the reconstruction may be. Brachytherapy is also a good alternative for high-risk surgical patients, such as those with significant comorbidities. Finally, brachytherapy has been successful in the treatment of non-melanoma lesions such as primary cutaneous lymphomas and Kaposi sarcomas (Ouhib et al. 2015).

7.3 Sources and Systems

The treatment of skin cancer with radiation therapy has a long history and has been carried out using many methodologies. External beam radiation therapy is still a popular treatment option, either with electron beams in linac rooms using orthovoltage machines or with dedicated x-ray machines designed for surface treatments. Electron therapy patient setup and dosimetry can be challenging, especially for large-area skin tumors, tumors that are difficult to reach, or tumors that are very curvaceous. Dedicated machines for surface therapy are not widely utilized, but some new machines have recently become available: Esteya (Elekta (Stockholm, Sweden), Sensus (Boca Raton, FL), and XStrahl (Camberley, England). Brachytherapy is, therefore, an attractive option, as it offers the advantage of simpler and more precise clinical setups as the applicators are positioned directly on the area of interest. In a recent survey of practice patterns by Likhacheva et al. (2017), 41% of respondents indicated shorter treatment courses as a reason to choose brachytherapy over external beam), while 34% indicated the higher conformality to the target for irregular and curvaceous lesions. A majority (69%) reported using HDR, radionuclide-based brachytherapy, with 31% of the respondents indicating the use of electronic brachytherapy (eBT).

Skin brachytherapy has been practiced since 1899 (Alam et al. 2011), originally using radium applications, and subsequently switching to LDR (Berridge and Morgan 1997) and HDR treatments. The most common radiation source to be used in HDR skin brachytherapy is ^{192}Ir (Guix et al. 2000), given its wide adoption for brachytherapy in general. Multiple vendors offer ^{192}Ir sources. Typically these sources are 2–4 mm in length, <1 mm in diameter, and are encapsulated by claddings, resulting in external dimensions of 3–5 mm in length and 1 mm in diameter. The sources are mounted on a steel cable controlled by a robotic afterloader that permits the precise positioning of the source in various dwell locations within the applicators for precisely defined dwell times.

HDR ^{60}Co sources have also been used (Strohmaier et al. 2011) for skin brachytherapy treatments. With dimensions similar to ^{192}Ir sources, ^{60}Co sources have a form factor that is consistent with skin brachytherapy treatments. Given the lower specific activity of ^{60}Co compared to ^{192}Ir, ^{60}Co may be on average at a lower source strength, resulting in longer treatment times. This may be a consideration for clinics that expect treatments of large areas requiring large flaps or custom applicators. Diagrams representing various ^{192}Ir and

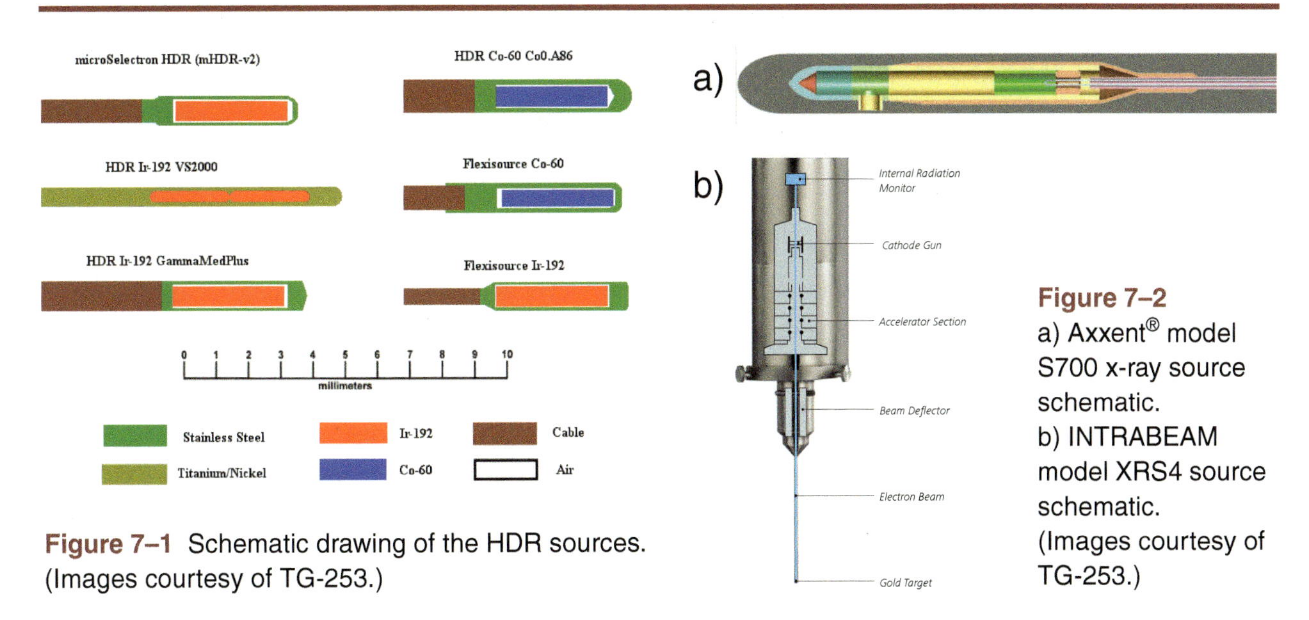

Figure 7–1 Schematic drawing of the HDR sources. (Images courtesy of TG-253.)

Figure 7–2 a) Axxent® model S700 x-ray source schematic. b) INTRABEAM model XRS4 source schematic. (Images courtesy of TG-253.)

^{60}Co sources are shown in Figure 7–1 [from the TG-253 report, under review]. Despite their different energy spectra, ^{192}Ir (average energy of 0.4 MeV) and ^{60}Co (1.2 MeV) are dosimetrically equivalent in treatment conditions (Richter et al. 2008), and they both provide penetrations in tissue that are greater than a typical surface therapy machine. While this may at times be advantageous in the reduction of dose hot spots to the skin for treatments prescribed at deeper-than-usual depths, typically this results in increased dose to deep tissue compared to other modalities, and it is not desirable.

Reports exist on the use of other radioactive isotopes for skin brachytherapy. For instance, early reports exist of using beta-emitting ^{90}Sr for the treatment of skin lesions (Stewart et al. 1977; Anger 1958). These techniques do not appear to be in clinical use anymore, although some interest in these applications still exists (Ferreira et al. 2016). More recently, the use of ^{125}I has been proposed, which would combine the advantage of using a source that is readily available and with an energy spectrum better suited to skin applications compared to ^{192}Ir or ^{60}Co (Ferreira et al. 2017).

HDR brachytherapy can also be performed without the use of radionuclides (Doggett et al. 2015; Rong and Welsh 2010; Bhatnagar and Loper 2010; Paravati et al. 2015). The eBT sources consist of a miniature x-ray tube typically operated at 50 kVp. Compared to radionuclides, these sources are physically larger, and have lower energies and comparable dose rate. The energy spectrum is equivalent to a 20–30 keV source, thus much less penetrating than radionuclide-based HDR sources. Typically, eBT sources are used for the treatment of relatively small lesions, while HDR sources are used for both small and large fields. Examples of eBT sources include the Axxent and INTRABEAM (iCAD, Sunnyvale CA) shown in Figure 7–2 [from the TG-253 report, under review]. The eBT sources were quickly adopted by private clinics and by dermatology offices, with a slower adoption by bigger centers with an already-established practice in radionuclide-based brachytherapy (Grant-Kels et al. 2014). This is due in part to an initial lower regulatory burden associated with eBT compared to isotope-based brachytherapy. The regulatory landscape is evolving, as is the reimbursement pattern for eBT. However, eBT maintains a logistical advantage over HDR isotope-based brachytherapy in that it requires less shielding. This may have a greater impact on small clinics than on larger clinics where HDR brachytherapy may already be in use for other indications.

7.4 Description of Applicators

7.4.1 Solid Conical Applicators

In the past few years, several applicators with a fixed circular geometry ranging from 10 to 50 mm in diameter have been made available for skin brachytherapy. These applicators are designed to be in full contact with the skin when treating lesions are ideal for flat surfaces and use a single dwell position in the center of the applicator diameter. The main applicators available are the Valencia (Figure 7–3) (Granero et al. 2008) and Leipzig from Elekta (Figure 7–4) (Evans et al. 1997; Hwang and Leung 2003, Niu et al. 2004; Perez-Calatayud et al. 2005), the Leipzig-style from Varian (Figure 7–5) (Fulkerson et al. 2014a; Fulkerson et al. 2014b), and surface applicators from iCAD, Inc. and INTRABEAM (Figure 7–6 and Figure 7–7), respectively (Rivard et al. 2006; Dickler et al. 2007; Dickler et al. 2008; Hiatt et al. 2008; Fowler et al. 2004; Eaton 2015). The Elekta Leipzig applicators and Varian Leipzig-style applicators are available in two types: hori-

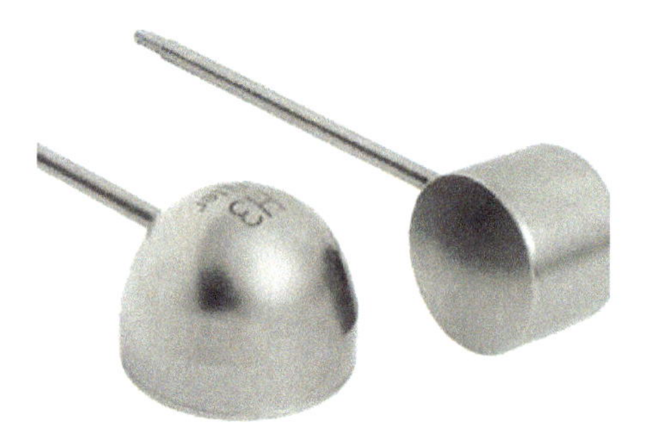

Figure 7–3 Valencia applicators. (Image courtesy of Perez-Calatayud.)

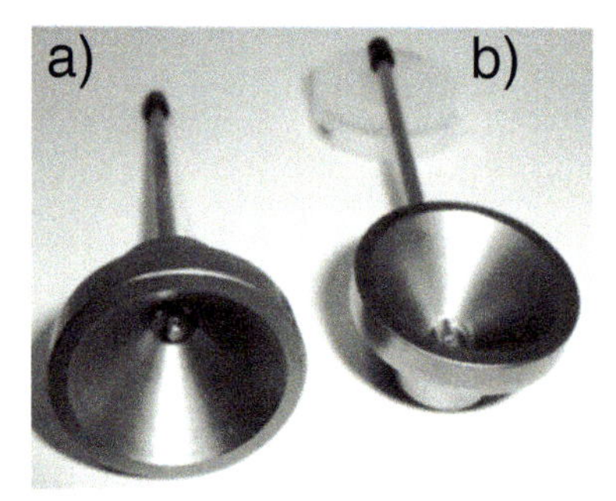

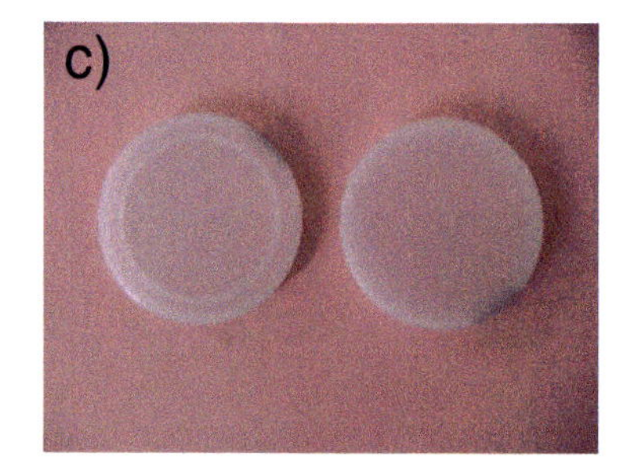

Figure 7–4 Leipzig applicators: a) vertical b) horizontal c) plastic caps (simulation cap, left, and treatment cap, right. (Images courtesy of Perez-Calatayud.)

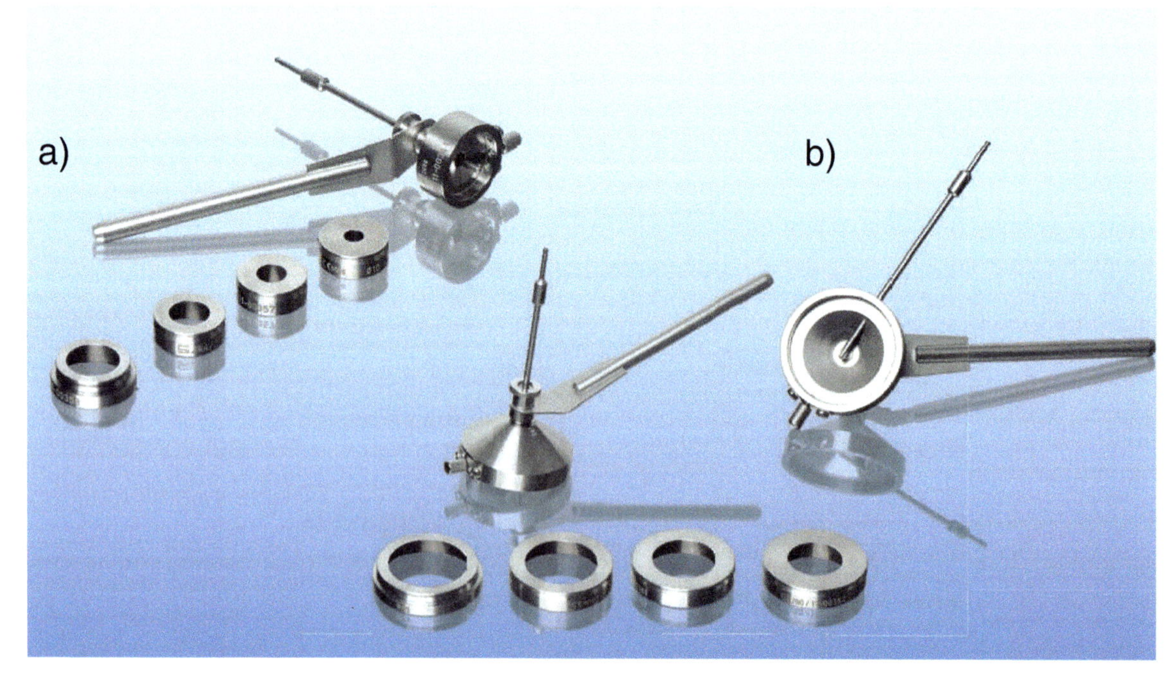

Figure 7–5 Leipzig applicators from Varian: a) vertical, b) horizontal. (Image courtesy of Varian Medical Systems).

zontal with the source long-axis parallel to the skin surface and vertical with the source being perpendicular. The vertical source position has been indicated for use with lesions that may be deeper seated as the dose profile is more penetrating (Perez Calatayud et al. 2005). These applicators have focus-to-skin distances of less than 30 mm. A summary of relevant applicator parameters is presented in Table 7–1.

The Xoft device has four circular applicators (Figure 7–6) with diameters of 10, 20, 35, and 50 mm. It also has an integrated flattening filter to improve the dose profile (Figure 7–10). Each of the treatment cones has a unique SSD ranging from 20.7 to 30.0 mm for the 10 mm and 30 mm cones, respectively, along with a flattening filter positioned at the apex of the cone to ensure a flat treatment area at the applicator exit window. Custom shields (Axxent FlexiShield™) made from tungsten-loaded silicone can be used to shape the useful beam. There are two types of applicators available for the treatment of skin lesions with the INTRABEAM system: a "flat" applicator and a "surface" applicator. The surface applicator was designed to create a homogeneous flat radiation field at the exit surface of the applicator, and the flat applicator was designed to create a homogeneous flat radiation field 5 mm (depth) from the applicator exit surface. The flat applicator has a polyetherimide flattening filter within it to create the flattened field at depth. The surface applicator has cone sizes ranging from 10–40 mm in diameter and 10–60 mm in diameter for the flat applicator. The SSD and the

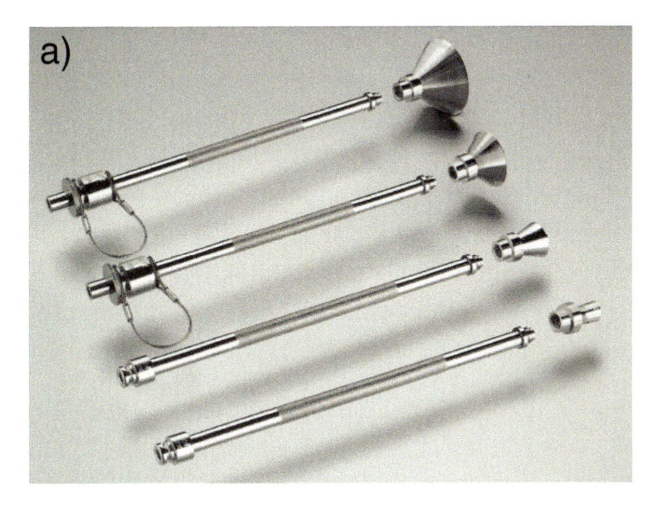

Figure 7–6 Applicators for the Xoft system: a) applicators and b) caps for Xoft applicators. (Images courtesy of iCAD, Inc.)

Figure 7–7 INTRABEAM a) flat and b) surface applicators. Lower left: flat 50 mm applicator with position marker. Lower right: surface 40 mm applicator with position marker. (Images courtesy of TG-253.)

Table 7–1 Applicator characteristics

Applicator Type	Source Orientation	Focus-to-Skin Distance	Applicator Diameter (mm)	Effective Energy (bare source)
Elekta Leipzig	horizontal	16.1 mm	10, 20, 30	380 keV (^{192}Ir)
Elekta Leipzig	vertical	16.1 mm	10, 20, 30	380 keV (^{192}Ir)
Elekta Valencia	horizontal	16.1 mm	20, 30	380 keV (^{192}Ir)
Varian Leipzig	horizontal	12.5 mm	30, 40, 45, 50	380 keV (^{192}Ir)
Varian Leipzig	vertical	12.5 mm	10, 15, 20, 25	380 keV (^{192}Ir)
Xoft	vertical	varies (~21 mm)	10, 20, 35, 50	50 kVp (Miniature x-ray source)
INTRABEAM	vertical	10–26 mm	10–60	50 kVp

filter thickness (if appropriate) vary with applicator size. The flat and the surface applicators can be used for various treatment sites during intraoperative radiotherapy, in a tumor bed, or on the body surface. The flat applicators produce fields with a maximum flatness at a depth of 5 mm from the most distal end of the applicator, while the surface applicators produce fields with a maximum flatness directly at the most distal end of the applicator (Schneider et al. 2014; Goubert et al. 2015). While similar in design, these applicators have varying dosimetric properties. Dose profiles are shown in figures 7–8, 7–9, and 7–10. The presence of filters in the Valencia and Xoft applicators provides a flat profile (Figure 7–8 and Figure 7–10). As expected, the presence of this filter increases the treatment time significantly (42% and 92% respectively for the 20 and 30 mm Valencia applicators) when compared to the Leipzig applicator for the same treatment depth (~3 mm). One of the challenges when using these applicators is proper immobilization of the treatment area and applicator to ensure accurate treatment. These applicators are used with plastic caps (Figure 7–4) mounted at the end of the applicator to absorb electron contamination originating from the tungsten/lead used for shielding and to provide a flat geometry when in contact with the patient's skin. No treatment should be delivered without use of the plastic cap to prevent a significantly higher skin dose (Evans et al. 1997) and a higher dose to target with tissue protruding inside the applicator, which could result in a medical event. Because of contact with the treatment surface, these caps should be cleaned after each fraction per the manu-

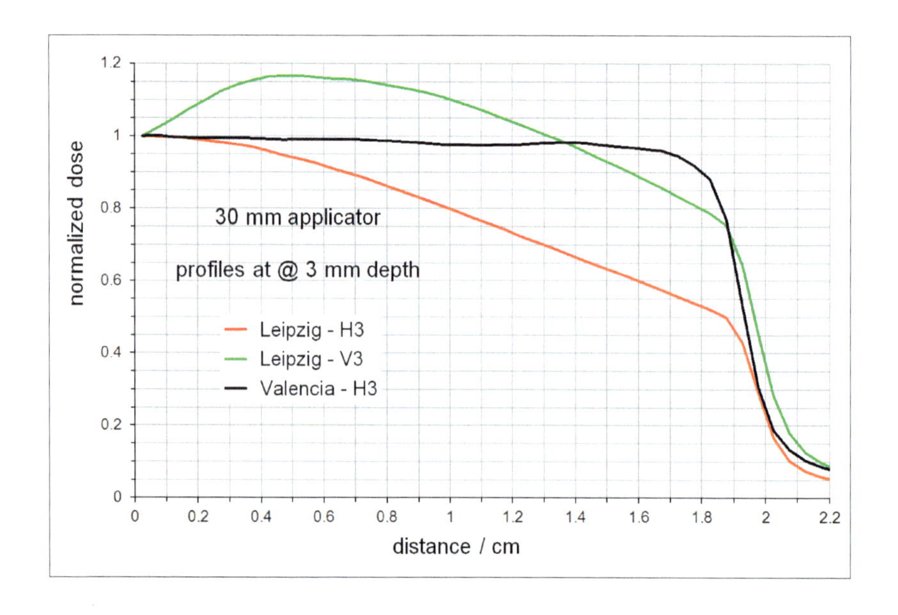

Figure 7–8 Normalized dose profiles for the Elekta Leipzig and Valencia applicators.

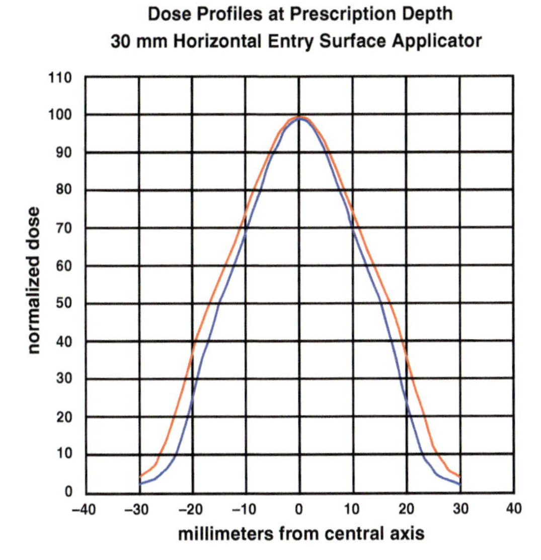

Figure 7–9 Profiles of Varian horizontal applicators. Blue = 3 mm depth; red = 5 mm

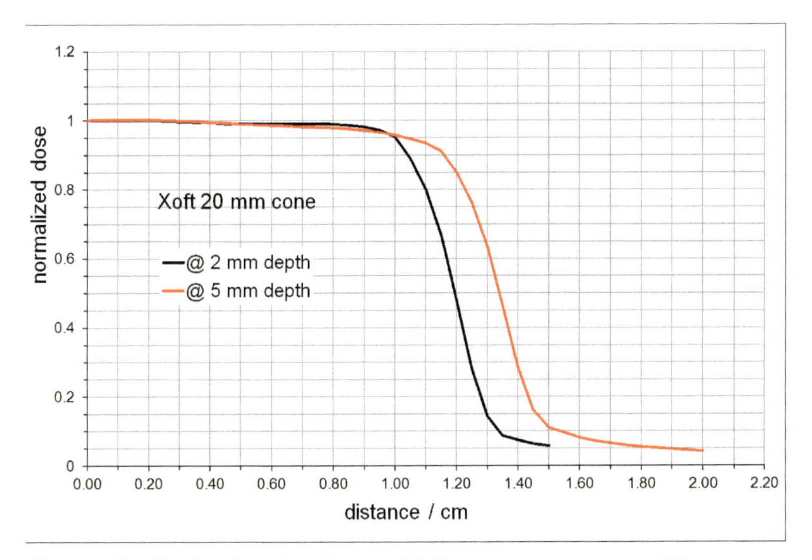

Figure 7–10 Profiles for 20 mm Xoft applicator at two different depths (2 mm and 5 mm).

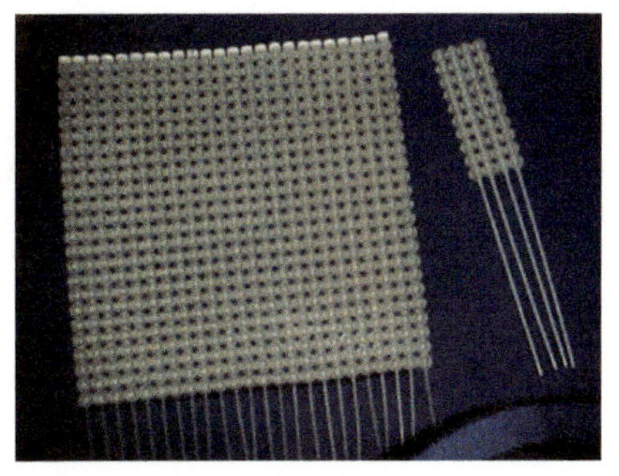

Figure 7–11 Elekta Flap applicator.

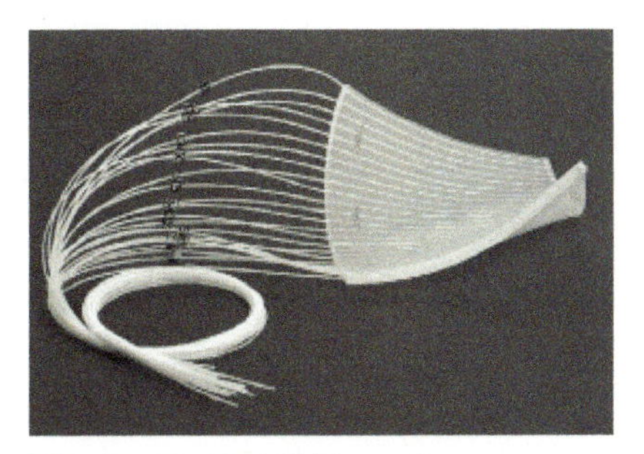

Figure 7–12 HAM applicator available from Bebig. (Image courtesy of Bebig.)

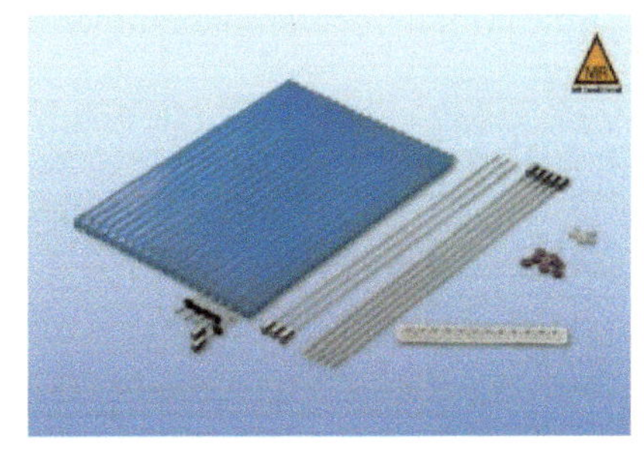

Figure 7–13 Catheter flap available from Varian. (Image courtesy of Varian Medical Systems.)

facturer's recommendations. To minimize the surface skin dose, the recommended prescribed depth for these applicators is 3 mm or less (Ouhib et al. 2015).

7.4.2 Flap-style Applicators

There are currently three varieties of flaps (figures 7–11, 7–12, and 7–13) available: the Freiburg from Elekta, the Flap from Varian Medical Systems, and the Harrison–Anderson–Mick applicator (also known as the HAM from E&Z Bebig). These flaps are made from silicon rubber spheres or bolus material with channels where catheters are inserted at a depth of 5 mm. These flaps should not be used for treatments at depths greater than 5 mm beyond the surface of the skin because the dose to the surface becomes clinically unacceptable. For treatment depths >5 mm, interstitial implants (figures 7–14 and 7–15) or electron beam treat-

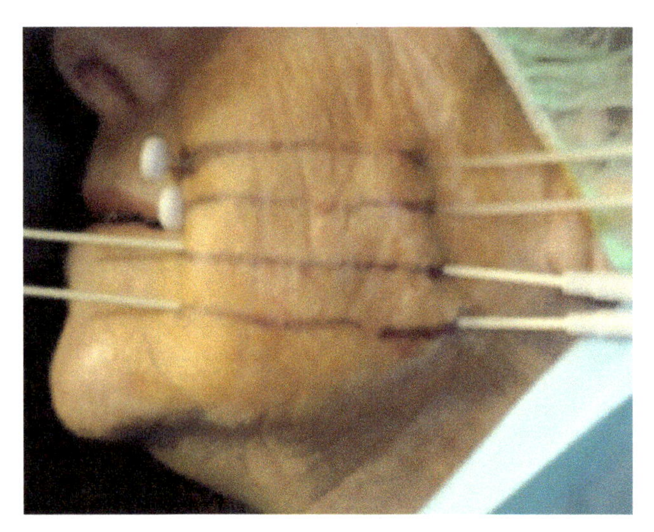

Figure 7–14 Clinical case of interstitial surface brachytherapy. (Image courtesy of Perez Calatayud.)

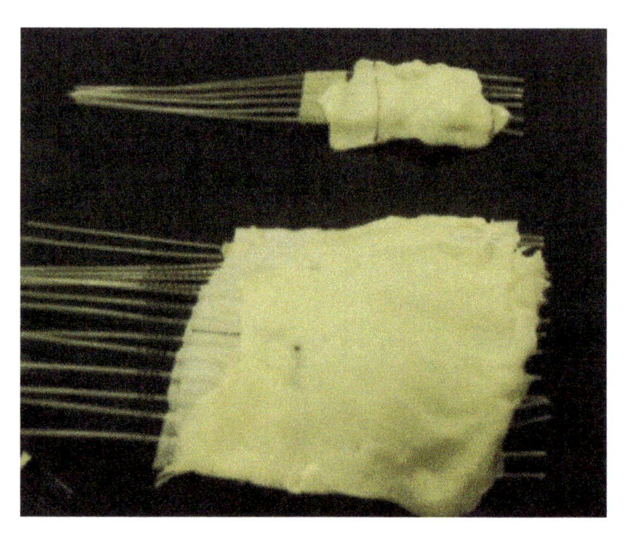

Figure 7–15 Example of mold. (Image courtesy of Perez Calatayud.)

ments are advised. These flaps provide the convenience of a fixed geometry and separation between catheters. They also have other advantages, including their flexibility to conform to the patient's skin, although they may cause gaps between the flap and the skin if the surface has concavities. The flap may be reusable depending on the manufacturer. To ensure treatment reproducibility, markers, tattoos, or other means should be used for accurate applicator placement. As can be seen in Figure 7–6, the three-point technique (when possible) has shown to be a reliable method.

7.4.3 Mold-style Applicators

Custom flaps or molds are also used to treat skin lesions, particularly those with challenging locations and geometry (e.g., tip of the nose, the ear). When preparing these molds, one must make sure to maintain homogeneous thickness and an adequate number of catheters and separation for dose optimization. If heterogeneity corrections are not applied in treatment planning, the material used for mold fabrication should have an electron density equivalent to water to ensure accurate dose calculation since all treatment planning systems using the TG-43 formalism assume water-equivalent material in their calculations. Like the use of flaps, one must ensure treatment reproducibility when using these applicators.

7.5 Dosimetry

7.5.1 Methods for Source and System Output Validation for Conical Applicators

Although the effects of the applicator must be considered, determination of the bare source strength for quality assurance and treatment planning is necessary. For radionuclide HDR brachytherapy sources, the NIST traceable quantity for source strength is air kerma strength (AKS [$U = \mu Gy\ m^2\ h^{-1}$]), and is typically measured with a well-type ionization chamber. Additional safety checks of the treatment delivery system should be performed as indicated by the manufacturer (e.g., power requirements, emergency stops, and monitor display accuracy).

7.5.2 Applicator Acceptance and Quality Assurance

Upon delivery of surface applicators, the integrity of each applicator should be assessed, and it is good practice to label applicator diameters, inserts, and any other associated accessories. It is common to use a color system to help keep applicators and inserts organized. After measuring the base source strength and entering that value into the treatment console or computer, the output and dose distribution of each solid conical applicator must be characterized. Necessary measurements to characterize these applicators include percent depth dose, profiles at the skin surface and relevant treatment depths (1, 3, and 5 mm are common), and reference output. Depth dose measurements can be performed in a water tank with a waterproofed parallel-plate ionization chamber, or in a solid phantom (if the chamber geometry accommodates this). Care must be taken to ensure the ionization chamber is small enough as to prevent volume averaging, yet sensitive enough to provide accurate measurements. Several studies have utilized the PTW 34013 and the Exradin A20 chambers for output verification of solid conical applicators (Evans 1997; Perez-Calatayud 2006; Fulkerson 2014). The quantity of interest for output measurements is air kerma as there is no absorbed dose-to-water standard in the United States for eBT or ^{192}Ir sources. Ionization chambers used for these measurements are calibrated for air kerma at the appropriate beam qualities (^{192}Ir or relevant x-ray beams). Measurements of air kerma can be converted to a dose to water at the surface or at depth via an appropriate dosimetry protocol. Studies have been published describing methods for a dose-to-water determination for the Varian Leipzig-style applicators and Xoft surface applicators (Fulkerson 2014; Rong 2010) based on Monte Carlo-determined correction factors and a modified version of the TG-61 dosimetry protocol, respectively. The output of the Elekta Leipzig applicators can be determined via a consistency check measurement with the applicator positioned on top of a well chamber (Perez-Calatayud 2006). Other types of passive dosimeters may be used for determination of applicator output, such as TLDs, MOSFETS, or radiochromic film. Radiochromic film can also be used to measure dose profiles and dose distributions, as well as depth dose information if the film dosimetry system is well characterized and calibrated. Additional details regarding output verification measurements are discussed in the TG-253 report on surface brachytherapy (Fulkerson et al., in press). Typically, output and dose distribution measurements are not performed for flaps or molds as they vary between patients and are determined in the treatment planning systems (as discussed in Section 7.2.3). However, it is still essential to check the flap or mold integrity during manufacture and prior to each fraction.

7.5.3 Methods for Dose Calculations

As discussed in Chapter 1, the updated AAPM Task Group No. 43 report (TG-43 U1) was introduced in 2004 so that the development of the clinical sources could be followed (Rivard et al. 2004). The intent of this update was: a) to provide a revised definition for the air kerma strength, b) to eliminate any apparent activity for the specification of source strengths, c) to eliminate the anisotropy constant in favor of the distance-dependent one-dimensional anisotropy function, d) to provide guidance on extrapolating tabulated TG-43 parameters to both longer and shorter distances, and e) to eliminate minor inconsistencies and omissions in the original protocol and its implementation (Rivard et al. 2004).

Recently, advanced brachytherapy dose calculation algorithms were introduced (see Chapter 1). They are: a) the Monte Carlo simulation (MC), b) the collapsed-cone (CC) superposition/convolution method (S/C), and c) deterministic solutions to the linear Boltzmann transport equation (BTE). These algorithms are known as model-based dose calculation algorithms (MBDCA). The principal reason for introducing these MBDCAs was the fact that the TG-43-based dose calculation was prone to inaccurate dose calculation when air, tissue, or another medium was approximated by water.

The current patient dose calculation is different for various sources and surface applicators. For instance, the dose calculation for HDR skin brachytherapy using ^{192}Ir sources depends on the applicators used. Flap-style applicators and customized molds can use the TG-43 U1 formalism. For the treatment time and dose calculations for the Leipzig- and Valencia-style solid conical applicators, Varian initially recommended the use of tabulated dose rates normalized to a depth of 5 mm. In this case, the TG-43 formalism is not sufficient due to various factors, such as the material of the applicator, increased scatter from the applicator's inner wall, etc. Recently, Varian has implemented the model of the solid conical applicators into the BrachyVision™ TPS that allows for the model-based dose calculation. The Oncentra Brachy TPS from Elekta has recently introduced an advanced collapsed-cone engine for dose calculation implemented according to the TG-186 standards. This calculation method can be used in plan analysis, but not for the treatment plan delivery. Skin treatments using ^{60}Co sources and electronic brachytherapy sources mostly utilize the TG-43 U1 formalism.

7.6 Treatment Procedures

In general, the treatment process for skin brachytherapy includes the initial consult, pretreatment imaging (ultrasound or computed tomography), contouring, prescription definition and applicator selection/manufacture, immobilization, dose calculation, and treatment delivery. There are, of course, many sub-steps associated with each of these general steps, and institutions should apply proper risk analysis in the development of operating procedures for skin brachytherapy. A detailed discussion of the processes associated with skin brachytherapy can be found in the AAPM TG-253 report.

7.6.1 Special Considerations for Radionuclide-based Treatments

7.6.1.1 Mold Design and Manufacture

When designing a mold for skin treatment, there are few parameters for the user to keep in mind:

a. Treatment area and geometry to provide adequate dose coverage (20 mm beyond the edge of the target). Activated dwell positions are usually needed beyond the edge of the target volume to provide adequate dose coverage.

c. Reproducibility: the applicator should maintain its integrity, size, and geometry throughout the entire treatment.

c. Conformal to the target. Targets with curvature can be a challenge with rigid molds and, therefore, one must ensure that once the mold is dry it remains conformal to the skin surface.

d. Source pathway patency. If curvatures are severe, source obstruction or jamming should be evaluated prior to patient treatment. Users should consult with manufacturers for maximum curvature angles.

7.6.1.2 Catheter Placement, Fixation, and Identification

Catheter orientation should be chosen to minimize unnecessary radiation to areas other than the intended target. In addition, there should be easy patient access in case emergency response is needed (e.g., patient condition or source malfunction requires applicator removal). Catheter movement along its axis for molds should be carefully monitored to ensure correct delivery of the planned dose for each fraction. Ink marks or other means should be used to identify the proper position of these catheters within the mold. Radio-lucent buttons designed for this purpose can also be used to ensure no movement of catheters within the mold will occur (Figure 7–16). Catheter numbering (or labeling) should be easily visible on the mask. Figure 7–17 is an example where the radiation therapist can easily identify the catheter number while connecting it without any difficulties. An in-house standard policy is recommended to label catheters, for example, from left-to-right or superior to inferior. It is good practice to standardize as much of the procedure as possible to prevent errors due to incorrect catheter connections.

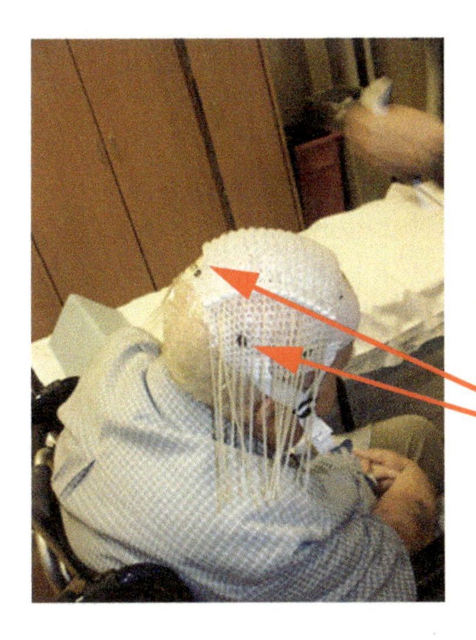

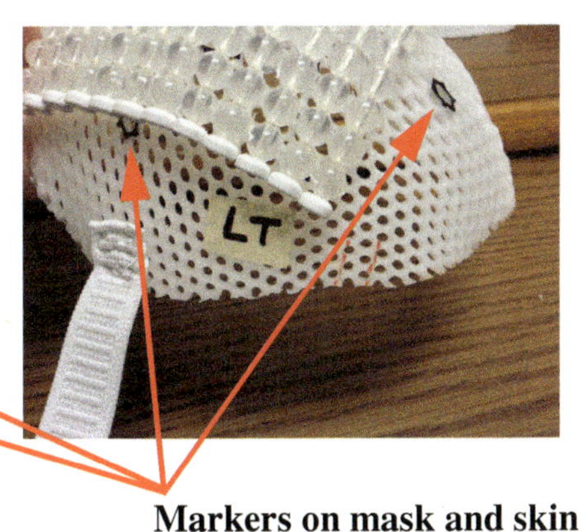

Figure 7–16 Superficial brachy-therapy mask with CT radio-opaque markers.

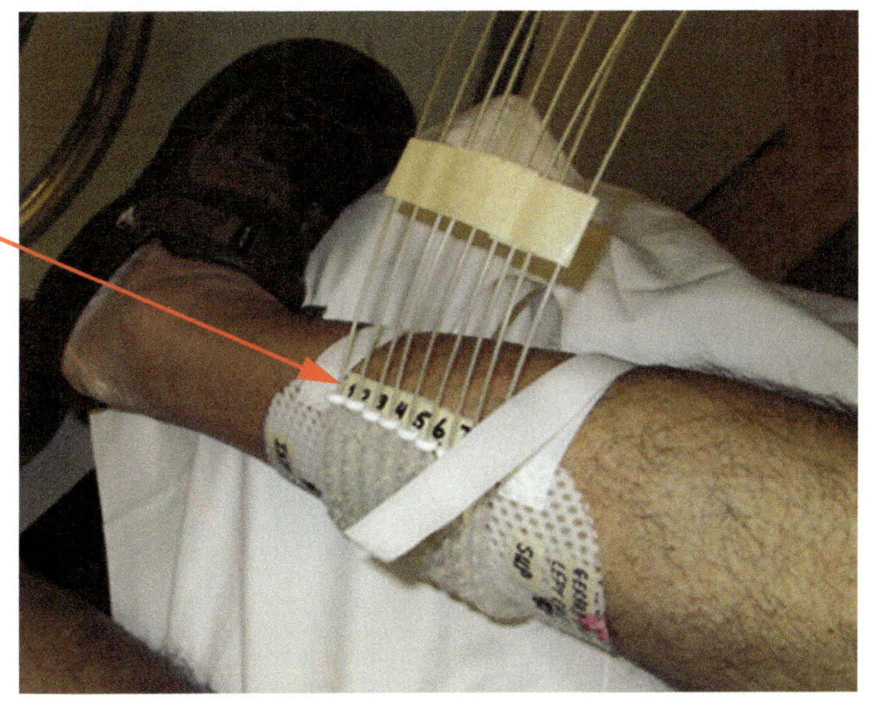

Figure 7–17 Catheter labeling for guiding staff for treatment delivery.

7.6.1.3 Flap Creation (Mask Example), Modification, Fixation, and Repositioning

While there are several ways for making a mask and flap combination, a common method is described here. The process requires separate patient appointments before the start of treatment. In the first one, the patient is brought in to make the mask (e.g., Aquaplast™). During this session, the target with margin is drawn on the patient's skin. The mask is prepared and mounted on the patient's skin. The mask should remain on the patient's skin until it is completely dry. Premature removal could result in a mask that will be less conformal and, therefore, might have to be redone. Once the mask is dry, the target is reproduced on the surface of the mask using a marker pen. This target delineation will be used to select the proper size and orientation of the flap. The patient is then sent home and returns for the second appointment for simulation. The flap can be secured using dental floss or other means. While sewing the flap, the source path should remain clear of any obstruction. Once the fixation has been completed, the patient returns (second appointment) for simulation and CT image acquisition. At the end of imaging, three skin marks (tattoos) that can also be identified on the mask should be placed for a reliable mask placement during each treatment (Figure 7–16). The patient name, mask orientation, catheter number, and any other useful information should be documented on the mask. Any catheter replacement should be brought to the attention of the planning physics staff since the length of the catheter should be taken in consideration and, if different, will require a treatment plan modification.

As it is the case for circular applicators, treatment reproducibility for flap and mold will depend, in part, on the patient's comfort. The patient has to be comfortable with the applicator in place prior to each treatment and should never be asked to hold it during treatment.

7.6.2 Special Considerations for Electronic Brachytherapy-based Treatments

While the treatment process with applicators utilizing electronic brachytherapy sources is similar to that of radionuclide-based conical applicators, there are a few differences. If a shielding cutout is to be used between the skin and the applicator exit window, this is typically completed after target delineation and applicator selection. Further, the electronic brachytherapy systems discussed in this chapter have an associated mobile treatment computer which must be preprogrammed to deliver the appropriate dose based on the source strength as measured that day. This source strength measurement is semi-automated, and the treatment dose should be compared to a hand calculation prior to treatment delivery.

7.7 Conclusions

Treatment of skin lesions is one of the oldest clinical uses for brachytherapy applications. The treatment options available today provide the ability to conform the dose distributions and make clinical decisions based on RBE (eBT versus radionuclide-based therapies, see Chapter 3). As with all forms of radiation therapy, patient safety is a primary concern, and all steps of the process should be analyzed with risk assessments from all members of the radiation therapy team. Several publications have addressed this specifically for skin brachytherapy and provide an excellent comparison for reference in developing a skin brachytherapy program (Ibanz-Rosello et al. 2016).

References

Alam, M., S. Nanda, B. B. Mittal, N. A. Kim, and S. Yoo. (2011). "The use of brachytherapy in the treatment of nonmelanoma skin cancer: A review." *J. Am. Acad. Dermatol.* 65(2):377–88 .doi:10.1016/j.jaad.2010.03.027.

American Cancer Society. *Cancer Facts & Figures 2016*. Atlanta: American Cancer Society, 2016.

Anger, H. (1958). "Radiotherapy of skin diseases with strontium 90 and yttrium 90." *Dermatologische Wochenschrift* 138(33):897–99.

Beaulieu, L., Å. Carlsson Tedgren, J. F. Carrier, et al. (2012) "Report of the Task Group 186 on model-based dose calculation methods in brachytherapy beyond the TG-43 formalism: Current status and recommendations for clinical implementation." *Med. Phys.* 39(10):6208–36.

Bhatnagar, A. and A. Loper. (2010). "The initial experience of electronic brachytherapy for the treatment of non-melanoma skin cancer." *Radiat. Oncol.* 5:87. doi:10.1186/1748-717X-5-87.

Berridge, J. K. and D. A. Morgan. (1997). "A comparison of late cosmetic results following two different radiotherapy techniques for treating basal cell carcinoma." *Clin. Oncol.* 9(6):400–02.

Dickler, A., M. C. Kirk, N. Seif, et al. (2007). "A dosimetric comparison of MammoSite high dose-rate brachytherapy and Xoft Axxent electronic brachytherapy." *Brachytherapy* 6:164–68.

Dickler, A., M. C. Kirk, A. Coon, et al. (2008). "A dosimetric comparison of Xoft Axxent electronic brachytherapy and iridium-192 high-dose rate brachytherapy in the treatment of endometrial cancer." *Brachytherapy* 7:351-54. doi: 10.1016/j.brachy.2008.05.003.

Doggett, S., M. Willoughby, C. Willoughby, E. Mafong, and A. Han. (2015). "Incorporation of electronic brachytherapy for skin cancer into a community dermatology practice." *J. Clin. Aesthet. Dermatol.* 8(11):28–32.

Eaton, D. J. (2015). "Electronic brachytherapy—current status and future directions." *Br. J. Radiol.* 88:20150002. doi: 10.1259/bjr.20150002.

Evans, M. D. C., M. Yassa, E. B. Podgorsak, T. N. Roman, L. J. Schreiner, and L. Souhami. (1997). "Surface applicators for high dose rate brachytherapy in AIDs-related Kaposi's sarcoma." *IJROBP* 39(3):769–74.

Ferreira, C., D. Johnson, K. Rasmussen, S. Ahmad, and J. Jung. (2016). "Investigation of beta-emitter ^{90}Sr-^{90}Y dose distribution using gafchromic EBT3 film for application on conformal skin brachytherapy device." *Med. Phys.* 43(6):3625. doi:10.1118/1.4956885.

Ferreira, C., D. Johnson, K. Rasmussen, C. Leinweber, S. Ahmad, and J. W. Jung. (2017). "A novel conformal superficial high-dose-rate brachytherapy device for the treatment of nonmelanoma skin cancer and keloids." *Brachytherapy* 16(1):215–22. doi:10.1016/j.brachy.2016.09.002.

Fowler, J. F., R. G. Dale, and T. W. Rusch. (2004). "Variation of RBE with dose and dose rate for a miniature electronic brachytherapy source." *Med. Phys* 31:1927. doi:10.1118/1.1776415.

Fulkerson, R. K., J. A. Micka, and L. A. DeWerd. (2014). "Dosimetric characterization and output verification for conical brachytherapy surface applicators. Part I: Electronic brachytherapy source dosimetric characterization and output verification for conical brachytherapy surface applicators. Part I." *Med. Phys.* 41:22103. doi:10.1118/1.4862505.

Goubert, M. and L. Parent. (2015). "Dosimetric characterization of INTRABEAM® miniature accelerator flat and surface applicators for dermatologic applications." *Phys. Medica.* 31(3):224. doi:10.1016/j.emp.2015.01.009.

Granero, D., J. Perez-Calatayud, J. Gimeno-Olmos, et al. (2008). "Design and evaluation of a HDR skin applicator with flattening filter." *Med. Phys.* 35(2):495. doi:10.1118/1.2825622.

Grant-Kels, J. M. and M. J. VanBeek. (2014) "The ethical implications of 'more than one way to skin a cat': increasing use of radiation therapy to treat nonmelanoma skin cancers by dermatologists." *J. Am. Acad. Dermatol.* 70(5):945–47. doi:10.1016/j.jaad.2014.01.849.

Guix, B., F. Finestres, J. Tello, C. Palma, A. Martinez, J. Guix, et al. (2000). "Treatment of skin carcinomas of the face by high-dose-rate brachytherapy and custom-made surface molds." *Int. J. Radiat. Oncol. Biol. Phys.* 47(1):95–102.

Hiatt, J.,G. Cardarelli, J. Hepel, et al. (2009) "A commissioning procedure for breast intracavitary electronic brachytherapy systems." *J. Appl. Clin. Med. Phys.* 9:58–68.

Hwang, I. M. and H. Leung. "Dosimetry characteristics of Leipzig applicators." In *Proceedings of the 1st Far East Radiotherapy Treatment Planning Workshop*. Veenendaal, The Netherlands: Nucletron-Oldelft, 1996.

Ibanez-Rosello, B., J. A. Bautista-Ballesteros, J. Bonaque, et al. (2016). "Failure mode and effects analysis of skin electronic brachytherapy using Esteya® unit." *J. Contemp. Brachytherapy* 8(6):518–24. doi: 10.5114/jcb.2016.64745.

Likhacheva, A. O., P. M. Devlin, S. M. Shirvani, C. A. Barker, P. Beron, A. Bhatnagar, et al. (2017). "Skin surface brachytherapy: A survey of contemporary practice patterns." *Brachytherapy* 16(1):223–29. http://dx.doi.org/10.1016/j.brachy.2016.10.006.

Ouhib Z., M. Kasper, J. Perez-Calatayud, et al. (2015). "Aspects of dosimetry and clinical practice of skin brachytherapy: The American Brachytherapy Society working group report." *Brachytherapy* 14(6):840–58. doi:10.1016/j.brachy.2015.06.005.

Nath, R, L. Anderson, G. Luxton, K. A. Weaver, J. Williamson, and A. S. Meigooni. (1995). "Dosimetry of interstitial brachytherapy sources: Recommendations of the AAPM Radiation Therapy Committee Task Group No. 43." *Med. Phys.* 22:209–34.

Niu, H., W. C. Hsi, J. C. H. Chu, M. C. Kirk, and E. Kouwenhoven. (2004). "Dosimetric characteristics of the Leipzig surface applicators used in the high dose rate brachy radiotherapy." *Med. Phys.* 31(12):3372–77. doi:10.1118/1.1812609.

Paravati, A. J., P. G. Hawkins, A. N. Martin, G. Mansy, D. A. Rahn, S. J. Advani, et al. (2015). "Clinical and cosmetic outcomes in patients treated with high-dose-rate electronic brachytherapy for nonmelanoma skin cancer." Pract. Radiat. Oncol. 5(6):e659-664. doi: 10.1016/j.prro.2015.07.002.

Perez-Calatayud, J., D. Granero, F. Ballester, et al. (2005). "A dosimetric study of the Leipzig applicators." *Int. J. Radiat. Oncol. Biol. Phys.* 62:579–84. doi:http://dx.doi.org/10.1016/j.ijrobp.2005.02.028.

Pérez-Calatayud, J., D. Granero, F. Ballester, V. Crispin, and R. van der Laarse. (2006). "Technique for routine output verification of Leipzig applicators with a well chamber." *Med. Phys.* 33(7):16–20. doi:10.1118/1.2207237.

Richter, J., K. Baier, and M. Flentje. (2008). "Comparison of ^{60}cobalt and ^{192}iridium sources in high dose rate afterloading brachytherapy." *Strahlenther. Onkol.*184(4):187–92. doi:10.1007/s00066-008-1684-y.

Rivard, M. J., B. M. Coursey, L. A. DeWerd, et al. (2004). "Update of AAPM Task Group No. 43 Report: A revised AAPM protocol for brachytherapy dose calculations." *Med. Phys.* 31(3):633–74.

Rivard, M. J., S. D. Davis, L. A. DeWerd, et al. (2006). "Calculated and measured brachytherapy dosimetry parameters in water for the Xoft Axxent xray source: an electronic brachytherapy source." *Med. Phys.* 33:4020–32.

Rong, Y. and J. S. Welsh. (2010). "Surface applicator calibration and commissioning of an electronic brachytherapy system for nonmelanoma skin cancer treatment." *Med. Phys.* 37(10):5509–17. doi:10.1118/1.3489379.

Schneider F., S. Clausen, J. Thölking, F. Wenz, and Y. Abo-madyan. (2014). "A novel approach for superficial intraoperative radiotherapy (IORT) using a 50 kV X-ray source: a technical case report." *J. Appl. Clin. Med. Phys.* 15(1):167–76.

Stewart, W. M., J. Nouel-Midoux, P. Lauret, E. Thomine, M. C. Boullie, and D. Amice. (1977). "Strontium 90 in the treatment of pre-cancerous lesions and of some superficial skin cancers (author's transl)." *Annales de dermatologie et de venereologie* 104(12):855–59.

Strohmaier, S. and G. Zwierzchowski. (2011). "Comparison of ^{60}Co and ^{192}Ir sources in HDR brachytherapy." *J. Cont. Brachy.* 3(4):199–208. doi:10.5114/jcb.2011.26471.

Example Problems

(Answers are found at the end of the book.)

1. In general, high-risk skin lesions are:

   a. greater than 20 mm in diameter and greater than 4 mm thickness

   b. greater than 10 mm in diameter and greater than 4 mm thickness

   c. less than 30 mm in diameter and greater than 5 mm thickness

   d. greater than 30 mm in diameter and greater than 5 mm thickness

2. End caps are needed for treatments with solid conical applicators.

   true or false

3. Which imaging modality is commonly used for pretreatment or daily imaging of skin lesions?

   a. ultrasound

   b. MRI

   c. angiography

   d. x-ray

4. Which is not a main reason that MBDCAs were introduced?

   a. TG-43 dose calculations assume a water medium.

   b. TG-43 dose calculations cannot be easily checked with a hand calculation.

   c. TG-43 dose calculations can be inaccurate when air is approximated by water.

   d. TG-43 dose calculations can be inaccurate when tissue is approximated by water.

5. What is *not* an important factor to consider when constructing a surface mold?

   a. reproducibility

   b. conformality

   c. degree of curvature

   d. source strength

6. Per general practice, what is the maximum lesion depth that can successfully be treated with surface molds or applicators?

   a. 3 mm

   b. 4 mm

   c. 5 mm

   d. 10 mm

7. Flap-based treatments are advantageous for larger lesions because they are:

   a. flexible

   b. water equivalent

   c. have a fixed geometry

   d. can be easily visualized on x-ray

8. The traceable quantity for source strength for radionuclide sources in the United States is:
   a. apparent activity
   b. activity
   c. air kerma rate
   d. air kerma strength
9. Radiochromic film is an acceptable tool for measuring dose distribution information from solid conical applicators.
   true or false
10. Custom molds are well suited for the following clinical sites:
    a. nose
    b. ear
    c. cheek
    d. all of the above

Chapter 8

Breast Brachytherapy

Deidre Batchelar[1], Regina Fulkerson[2], Zoubir Ouhib[3], Sujatha Pai[4], Susan Richardson[5], Mark J. Rivard[6], Daniel Scanderbeg[7], Bruce Thomadsen[8], and Dorin Todor[9]

[1]British Columbia Cancer Agency
Kelowna, British Columbia Canada

[2]RKF Consultants, LLC
Dundee, New York

[3]Lynn Regional Cancer Center
Delray Beach, Florida

[4]Memorial Hermann Texas Medical Center, LMP
Houston, Texas

[5]Swedish Medical Center-Tumor Institute
Seattle, Washington

[6]Tufts University School of Medicine
Boston, Massachusetts

[7]University of California–San Diego
La Jolla, California

[8]University of Wisconsin
Madison, Wisconsin

[9]Virginia Commonwealth University
Richmond, Virginia

8.1 Introduction and Historic Perspective

It is 1896. Antoine Henri Becquerel serendipitously discovers radioactivity. Two years later, Marie and Pierre Curie announce the discovery of radium. Three more years pass by, and in 1901, Pierre Curie suggests to a French physician, Henry Danlos, that he insert a radioactive source into a tumor, thus making 1901 the birth year of brachytherapy.

In 1894, Halsted devised a radical procedure for breast cancer treatment involving removal of the breast, the pectoral muscle, and the regional lymph nodes. It took a few decades more until Geoffrey Keynes at St. Bartholomew's Hospital in London proposed a breast-conserving treatment in 1932 involving implanting ^{226}Ra needles in the breast. With randomized trials starting in the 1970s, it was 2002 when the results showed that radical mastectomy might be unnecessary because breast-conserving surgery followed by radiotherapy was equally effective in most patients. In 1991, a visit to New Orleans by a Venezuelan woman in seeking treatment for her breast cancer became a turning point in breast brachytherapy. Dr. Robert Kuske, Chairman of Radiation Oncology at the Ochsner Clinic in New Orleans at the time, offered her a novel treatment, emulating a technique developed at Memorial Sloan Kettering Cancer Center, to treat sarcomas. The idea was placing empty plastic catheters in and around the surgery site immediately after the mass was removed and then loading the catheters with strands of low-dose-rate (LDR) sources. Radiation was delivered in the next few days and the catheters removed, thus completing the whole course of treatment—an otherwise lengthy procedure—in just a few days.

Brachytherapy for breast cancer is intimately related to the concept of partial breast irradiation (PBI)—the belief that the tissue at risk for recurrence and target of radiation treatment is mostly proximal to the lumpectomy cavity. Interstitial brachytherapy using LDR ^{192}Ir seeds was historically the first modality used for PBI and, consequently, has the longest follow-up evaluation. More recently, external beam radiation therapy (EBRT) has been used, either as photon beams or highly collimated electron beams, the latter being more often associated with intra-operative techniques relatively popular in Europe. New brachytherapy sources have been developed, one example being the miniature low-energy (~50 kV) source known as electronic brachytherapy. Existent sources (e.g., ^{103}Pd) commonly associated with treatment of prostate cancer are also used for permanent breast seed implant (PBSI) (Rusch et al. 2004; Rivard et al. 2009; Pignol et al. 2006). However, most commonly (at least in the United States), APBI is associated with high-dose-rate (HDR) treatment using ^{192}Ir. Since the radioactive source is placed in the body but must be kept from contact with tissues, one has to use an "applicator" to both contain and guide the source in the desired positions. The simplest applicator is a plastic catheter inserted in the tissue, and a multitude (15–25) of well-spaced such catheters is common for an interstitial breast implant. Each catheter is connected to a channel in the afterloader using a transfer (or guide) tube.

Pierquin et al. (1977) described a multi-catheter implant approach to irradiate the lumpectomy cavity as a boost in addition to external-beam radiotherapy for breast-conserving radiotherapy. The first technique developed for APBI was the multi-catheter interstitial brachytherapy (King et al. 2000). Even though from a technical perspective the interstitial implant is the most flexible technique, it requires specialized training and considerable experience. In an effort to make APBI more accessible, a number of devices were created with the goal of reducing the level of invasiveness, making the technique simpler and more reproducible (Edmundson et al. 2002; Rusch et al. 2004; Cuttino et al. 2005). There are two main categories of devices: balloon-based devices (MammoSite®, Contura MLB®, MammoSite ML®, and the Double-Balloon Best Applicator), and strut-based devices (SAVI®). MammoSite, historically the first balloon device for breast brachytherapy, was, in fact, following in the footsteps of another balloon device for brain brachytherapy: GliaSite RTS®, a balloon filled with liquid ^{125}I (Edmundson et al. 2002; Monroe et al. 2001). A "noninvasive" version of PBI called AccuBoost® has also been developed. It is a combination of imaging and treatment, where applicators are placed outside the skin on a breast compressed by a mammography machine (Rivard et al. 2006).

8.2 Fundamentals of Physics with Relevance for Breast Brachytherapy

Dose calculation for breast brachytherapy is typically based on the TG-43 formalism which calculates dose to water and does not account for any material inhomogeneities or for the influence of radiation scatter conditions specific to patient dimensions (Nath et al. 1995). As is described later in this chapter, each breast brachytherapy device or applicator comes with its own challenges and complexities when considering the actual dose delivered, and few generalities can be made. An APBI applicator may be filled with air or with a mixture of water and contrast material; therefore, a heterogeneous dose calculation may show that one applicator increases the delivered dose to the PTV and that another decreases the dose to the PTV over what was originally calculated with the TG-43 formalism. Shallow implants near the skin surface may not have the full radiation equilibrium conditions inherent to the TG-43 formalism and can exhibit decreased surface dose. Chapter 1 discusses these radiological issues and model-based dose calculation algorithms that are used in brachytherapy; Section 1.4.2 describes breast brachytherapy dose calculations for a range of applications.

8.3 Description of Devices and Techniques Available

8.3.1 Interstitial Multi-catheter Implants

Multi-catheter interstitial brachytherapy has the longest history of use for delivery of APBI (King et al. 2000). Two different approaches can be used for catheter insertion: template-based or free-hand. Templates for breast brachytherapy consist of two matched plates mounted at either end of a breast bridge (Figure 8–1). The plates are positioned on the medial and lateral sides of the breast in the region of the lumpectomy cavity, as identified on mammography, ultrasound (US), or, most commonly, CT (Vicini et al. 1998). The separation between the plates is adjusted to provide breast compression and lift the breast tissue away from the chest

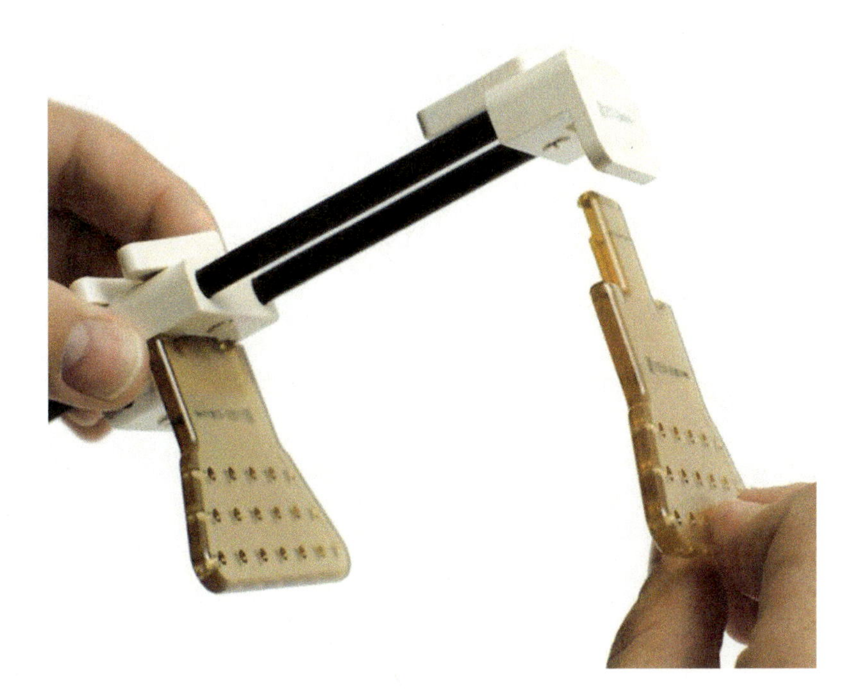

Figure 8–1 A multi-catheter breast brachytherapy template system consisting of A) two identical templates attached to B) a bridge. (Image courtesy of Elekta.)

wall. Each plate is perforated with the same pattern of holes, generally permitting implantation of parallel catheters in triangular or square arrangements. Several plate sizes are available to accommodate a variety of patient anatomies. This basic template set can be augmented with an additional guide template and stabilization rails (Figure 8–2). As the names suggest, these components add to stability and geometrical fidelity of the implant. The stabilization rails keep the medial and lateral templates aligned rotationally and aid in keep-

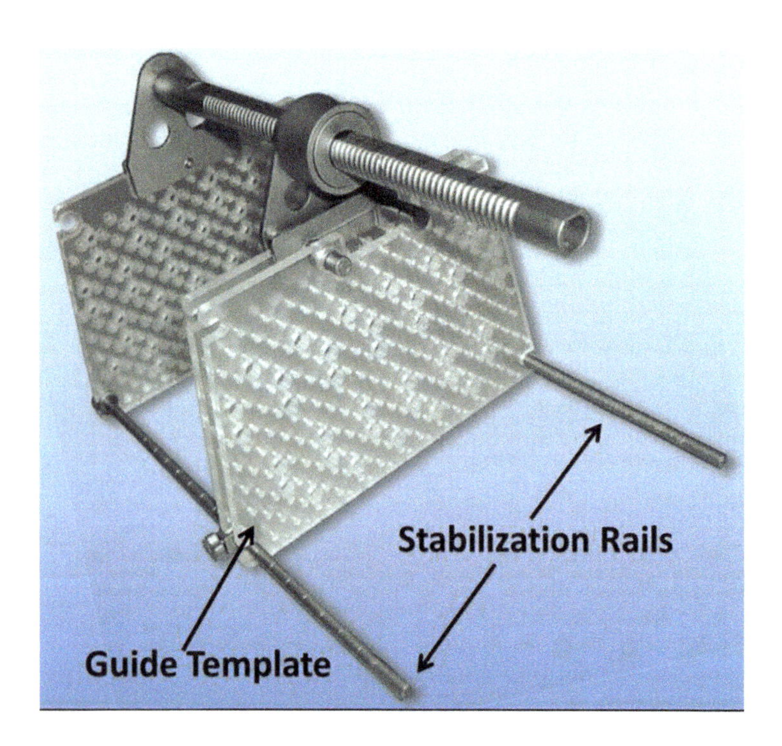

Figure 8–2 A multi-catheter breast template system including stabilization rails and a guide template. (Image courtesy of Varian.)

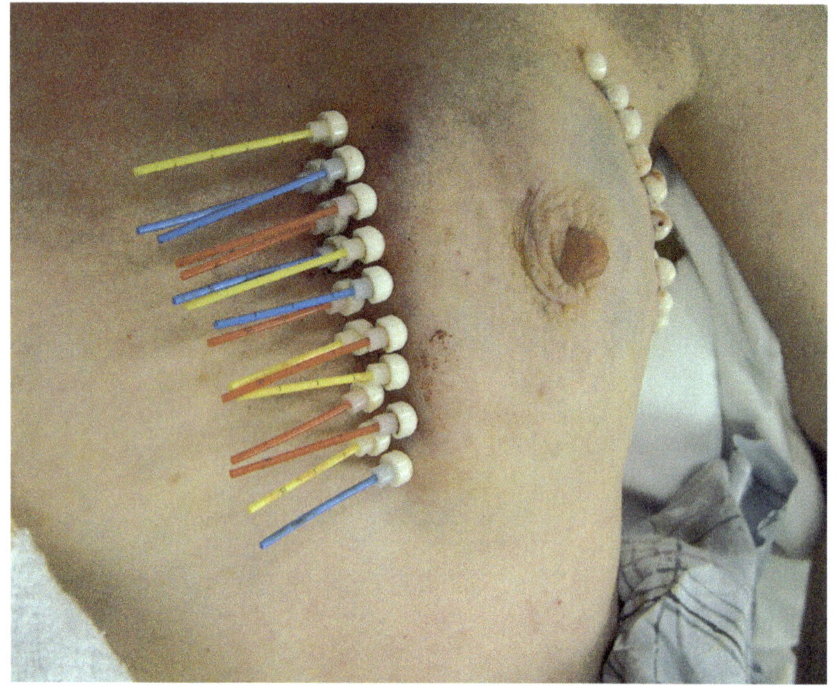

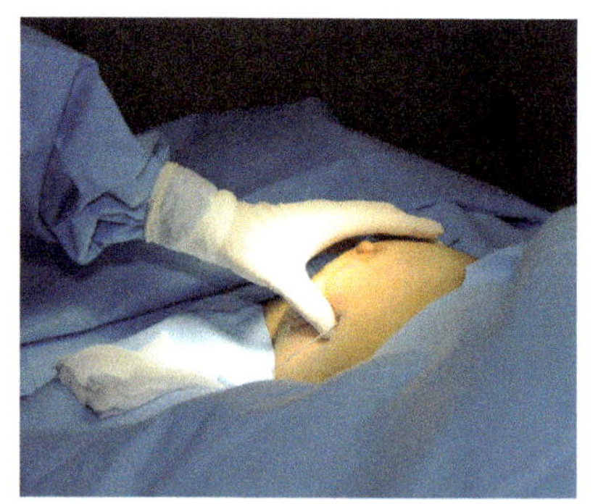

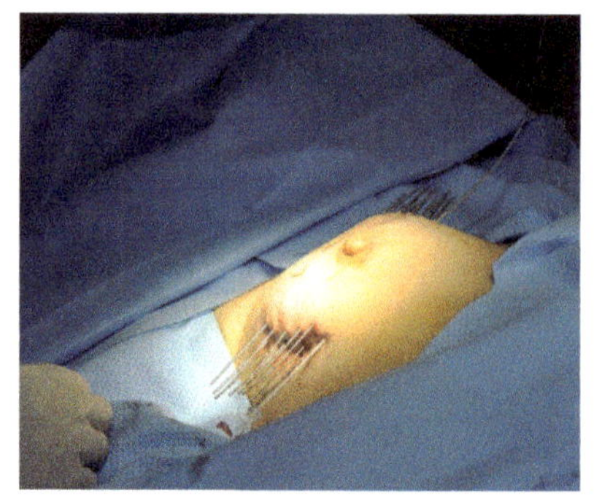

Figure 8–3 Free-handed interstitial HDR ^{192}Ir breast brachytherapy implant. (Image courtesy of Virginia Commonwealth University.)

ing the two templates parallel as the breast is compressed. The guide template is affixed on the implant side of the bridge and consists of either a third identical plate or a thicker plate with the same arrangement of holes. The purpose of this plate is to ensure that the implant needles are inserted straight toward the exit holes and, thus, to increase the parallelism of the implant. Once the template is positioned, hollow metal needles are inserted throughout the region of the planning target volume (PTV). The templates are then removed and the needles replaced with flexible plastic catheters fixed with buttons at both the entrance and exit points.

The free-hand approach (Figure 8–3), which is more challenging and operator-dependent, is used when a template is unsuitable, such as in cases of small breast size or extremely medial or lateral lumpectomy cavities. For this approach, the cavity is usually identified with US, with the extent of the cavity outlined on the skin (Tang et al. 2014). Needle exit and entrance points are identified on the skin, hollow guide needles are inserted, and then replaced with flexible plastic catheters as in the template-guided case.

8.3.2 Balloon-based Devices

Balloon-based devices for the treatment of breast cancer first arrived to the market in 2002 (Edmundson et al. 2002; Raffi et al. 2010). The first balloon-based devices had single lumens (Mammosite; Hologic, Inc., Marlborough, MA). Several multi-lumen devices followed a few years after the MammoSite, including Contura, MammoSiteML, and Best Double-Balloon® (Best Medical International, Inc, Springfield, VA). These single-use devices are comprised of a silicone catheter with lumens for the source to traverse and dwell inside of a polyurethane or silicone balloon. During use, the balloons are typically filled with saline, though inclusion of a contrast agent is often used for better visualization. The effect of contrast on the dose calculation has been studied extensively and should be accounted for during dose calculations (Kirk et al. 2004; Kassas et al. 2004; Oh et al. 2009; Slessinger et al. 2011). The dose perturbation was quantified as a heterogeneity correction factor (HCF) for various balloon radii and contrast concentration levels. The dose perturbation is larger for larger balloon radii and higher-contrast concentrations. Based on a validated Monte Carlo simulation, the calculated HCF values were 0.99 for a 2-cm radius balloon and 0.98 for a 3-cm radius balloon at 6% contrast concentration levels. The effects of air bubbles within the balloons have also been studied and can have an impact on the dose distribution up to 6% (Cheng 2005). The single-lumen MammoSite device is available in two spherical sizes (4–5 cm and 5–6 cm), and one ellipsoidal size (4 × 6 cm) (Figure 8–4). Clinical decisions regarding patient and applicator selection are discussed in Section 8.3. In general, for the single-lumen MammoSite device, a spherical balloon is chosen if the lumpectomy cavity length and width are approximately equal. When selecting a balloon size, the size of the cavity is considered, allowing for inflation of the balloon to a volume greater than that of the lumpectomy cavity to ensure maximum tissue-to-balloon conformance. The minimum cavity diameter recommended for use with the 4–5 cm device is 3 cm, and 4 cm for the 5–6 cm device. Given that the length along the shaft is fixed, one can produce an ellipsoidal balloon or a spherical one by simply varying the inflation volume. In addition to determining the optimal balloon size, the volume of fluid used to fill the selected balloon should fall within the range shown in Table 8–1.

Table 8–1 Suggested balloon fill volumes (Hologic 2014)

Balloon Shape and Size	Balloon Fill Volume (cm^3)
Ellipsoidal (4 × 6 cm)	60–65
Spherical (4–5 cm)	35–70
Spherical (5–6 cm)	70–125

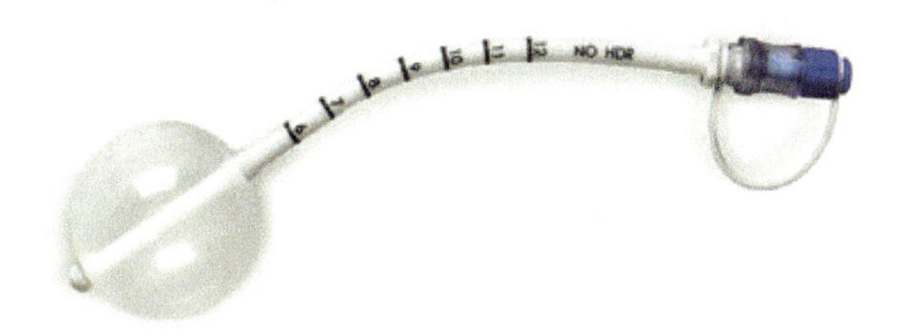

Figure 8–4 MammoSite spherical balloon device. (Image courtesy of Hologic, Inc.)

Table 8–2 Suggested balloon fill volumes for Contura applicators (Hologic 2014)

Balloon Shape	Balloon Diameter (cm)	Balloon Fill Volume (cm^3)
Variable ellipsoidal	3.5–4.0	23–32
Nominally spherical	4.0–5.0	32–59

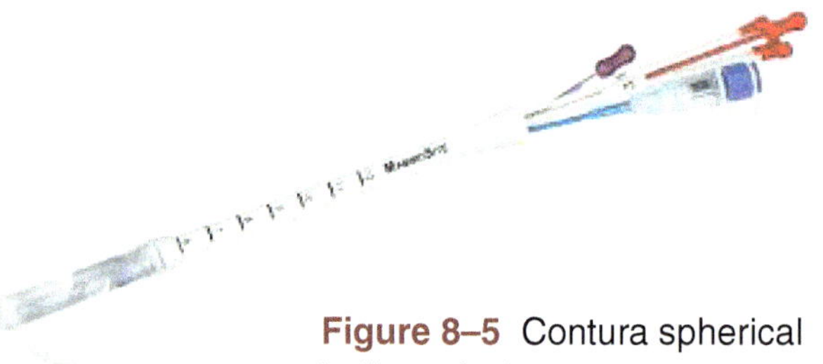

Figure 8–5 Contura spherical balloon device.

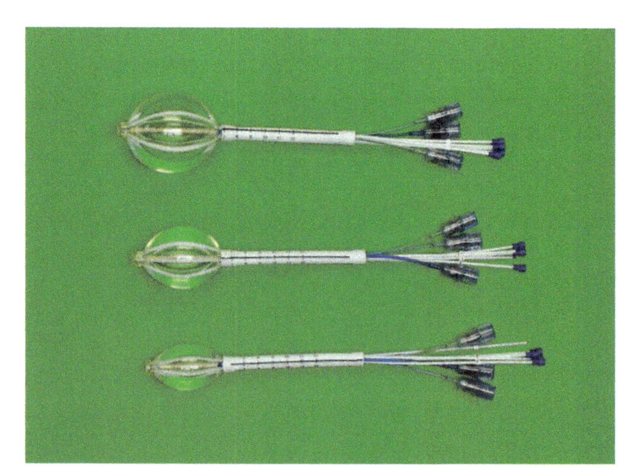

Figure 8–6 Best Double-Balloon Applicator. Top: Model 1995, 6.5 cm diameter balloon. Middle: Model 1922, 5.5 cm diameter balloon. Bottom: Model 1994, 4.5 cm diameter balloon. (Image courtesy of Best Medical International, Inc.)

The Contura device is like the single-lumen MammoSite device with respect to the balloon sizes and shapes available. However, there are four lumens positioned equidistant from and around the central lumen for the source to traverse (Figure 8–5). These additional lumens allow for flexibility in creating more conformal dose distributions. The recommended volumes for filling the Contura devices are shown in Table 8–2.

The newest commercially available balloon-based device is the Best Double-Balloon Applicator. This device has four lumens, one central and four peripheral. Like other devices, however, there are two balloons that inflate independently from one another. The outer balloon expands within the resection cavity, and the inner balloon allows for treatment lumens to expand within the outer balloon, optimizing dose conformity and homogeneity. The inner balloon expands to 50% of the outer balloon diameter, and the device comes in three different sizes (Figure 8–6).

8.3.3 Strut-based Devices

With increased utilization of APBI techniques, a few companies saw potential drawbacks to the previous two techniques described, interstitial multi-catheter implants and balloon-based devices. Multi-catheter implants are technically challenging, and not all physicians are well trained and comfortable with the technique. The balloon-based devices sought to overcome this obstacle with the single-entry aspect; however, the ability to alter the dose distribution with the internal catheters was limited. Innovators sought a hybrid solution that included the dosimetric flexibility of multi-catheter implants with the ease of use of the single-entry balloon devices. Out of this goal, strut-based devices were born. These devices include the strut-adjusted volume implant (SAVI) (Cianna Medical, Aliso Viejo, CA) and ClearPath (North American Scientific, Chatsworth, CA).

8.3.3.1 Strut-adjusted Volume Implant—SAVI

The strut-adjusted volume implant, commonly known as the SAVI, was introduced in 2006, with early clinical implementation and results reported by Gurdalli, Scanderbeg, and Yashar (Gurdalli et al. 2011; Scanderbeg et al. 2009; Yashar et al. 2009). The SAVI comes in four different sizes: the 6-1 mini, 6-1, 8-1, and 10-1, shown in Figure 8–7. The first digit corresponds to the number of peripheral channels in the device. The dimensions of each of the devices along with placement guidelines are shown in Figure 8–8. The central channel of the device is longer than the peripheral channels as it is recessed further inside the "nose-cone" of the device. The peripheral channels terminate just inside the nose-cone. The device is inserted in a collapsed state and then deployed by rotating the expansion key clockwise by approximately three revolutions to achieve maximum expansion. Once expanded, the peripheral channels rest against the lumpectomy cavity walls. Each of the peripheral channels can be used for source positions, in addition to the central channel. With the peripheral channels against the lumpectomy walls and target area, local dose sculpting can be achieved similar to interstitial brachytherapy; however, this can also lead to hot spots, and care must be taken to ensure dosimetric criteria are met prior to treatment. Figure 8–9 shows an extremely small skin bridge that is the thickness of tissue between the skin and the cavity, and subsequent dose distributions with the SAVI device.

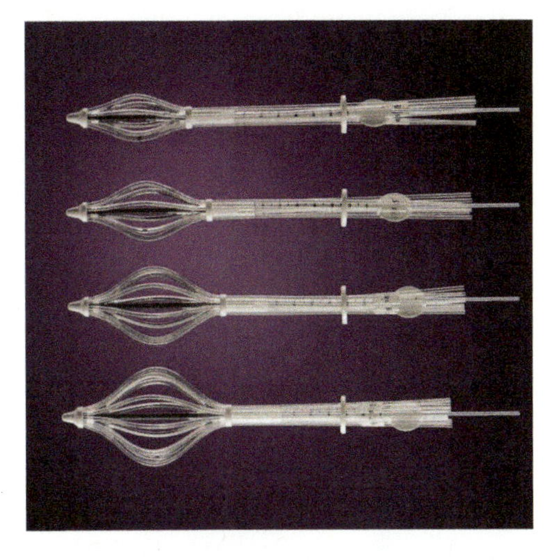

Figure 8–7 SAVI devices. 6-1 mini, 6-1, 8-1, and 10-1, from top to bottom. (Image courtesy of Cianna Medical.)

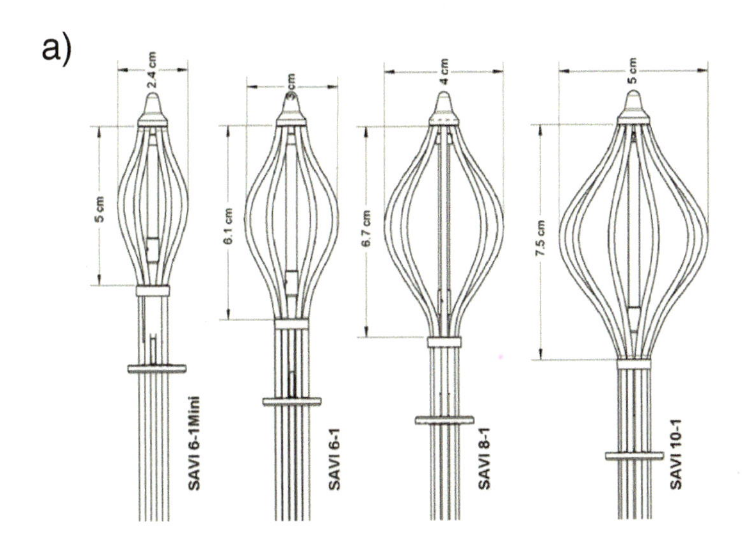

b)

Long Axis \ Diameter	2-3cm	3-4cm	4-5cm
2-3cm	SAVI Prep (20cc) 6-1Mini	-	-
3-4cm	SAVI Prep (20cc) 6-1Mini	SAVI Prep (20cc) 6-1Mini	-
4-5cm	SAVI Prep (20cc) 6-1Mini	SAVI Prep (20cc) 6-1Mini	SAVI Prep (30cc) 6-1
5-6cm	SAVI Prep (30cc) 6-1	SAVI Prep (30cc) 6-1	SAVI Prep (40cc) 8-1
6-7cm	SAVI Prep (30cc) 6-1	SAVI Prep (40cc) 8-1	SAVI Prep (40cc) 8-1
7-8cm	SAVI Prep (40cc) 8-1	SAVI Prep (60cc) 10-1	SAVI Prep (60cc) 10-1

Figure 8–8 a) SAVI device dimensions and b) placement suggestions. (Image courtesy of Cianna Medical.)

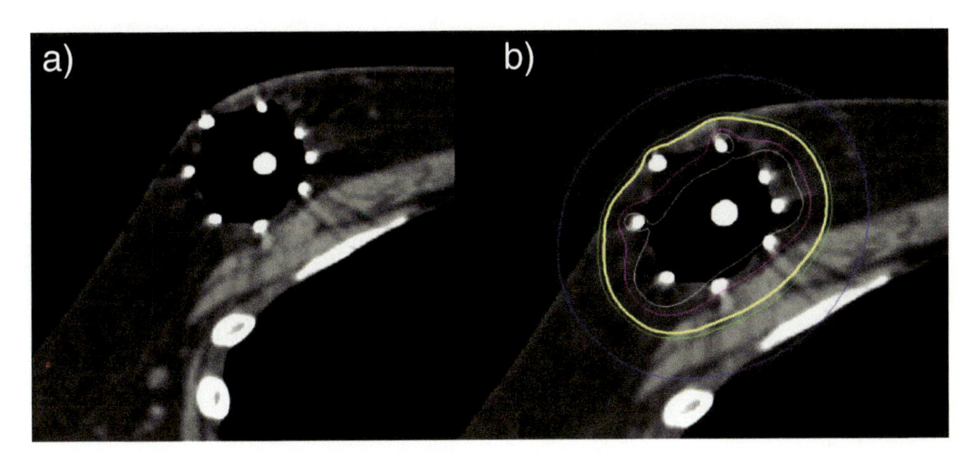

Figure 8–9 a) SAVI device in patient with minimal skin bridge and b) dose distribution. The levels for the isodose lines are (from inside outward) 6.8 Gy, 5.1 Gy, 3.4 Gy, 3.06 Gy, and 1.7 Gy.

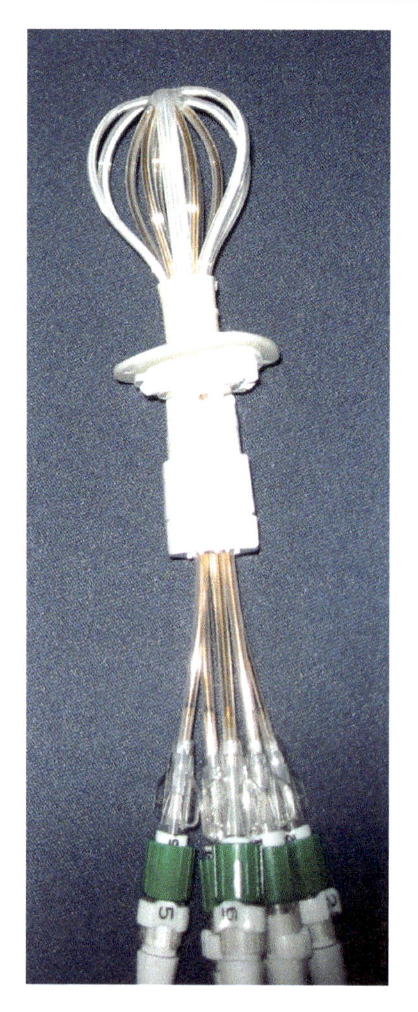

Figure 8–10 ClearPath device.

Figure 8–11 ClearPath device with handle removed and cover cap in place.

The SAVI channels are most easily contoured by inserting dummy wires into the channels prior to CT scanning. However, every device has radiopaque markers on channels 2, 4, and 6 to facilitate localization and identification. Additionally, the channels have a nitinol wire (nickel-titanium shape memory alloy) under them that allows them to expand and collapse to fit the lumpectomy cavity. These wires can be seen easily on x-ray. When using this wire to assist in contouring the channels, it is important to note that the wire is physically located beneath the channel so that the channel contour must lie outside this wire. There is also a metal band near the device's distal tip that is identified on CT for aiding localization.

8.3.3.2 ClearPath

The ClearPath was developed by North American Scientific and introduced in 2006. The ClearPath was a double-cage device in that it has six outer struts and six inner struts, along with a central channel, shown in Figure 8–10. The outer struts are strictly for opening the lumpectomy cavity and supporting the device; the HDR source cannot travel through these channels. The six inner channels and central channel are for the HDR source to travel and dwell to treat the target. One of the unique features of this device was how the device stem would lay against the patient's skin after deployment. A rubber gasket-like piece would sit flush with the patient's skin after the placement and be sutured into place. Then the handle of the device could be removed, leaving a small device profile emerging from the patient, shown in Figure 8–11. Early studies by Beriwal et al. and Dickler et al. compared dosimetry of the ClearPath to that of the MammoSite by using retrospective data of previously treated MammoSite patients and superimposing the ClearPath device over the balloon applicator (Beriwal et al. 2008; Dickler et al. 2008). ClearPath is not currently being marketed.

8.3.4 Non-invasive Devices

Breast brachytherapy may be delivered using less invasive methods than the aforementioned interstitial and balloon-based approaches (Hepel and Wazer 2012). Marketed as AccuBoost® (Advanced Radiation Therapy, LLC, Tyngsboro, MA), this noninvasive, image-guided breast brachytherapy immobilizes the breast with collimated applicators positioned through mammographic image guidance. It has been used as a boost to whole-breast, external-beam treatments since 2007 (Hamid et al. 2012). The system includes a mammography unit that is used to locate the target with nominal 30 kVp x-rays with respect to a radiopaque grid which offers tissue discernment with high contrast. With the breast compressed and immobilized and the target identified, tungsten-alloy applicators are rigidly positioned on the mammography breast compression paddles in a parallel-opposed configuration (Iftimia et al. 2015) at the designated coordinates. The breast is then decompressed and another parallel-opposed set of beams is applied orthogonally to the first set. Unlike with electrons, the beams are not directed into the

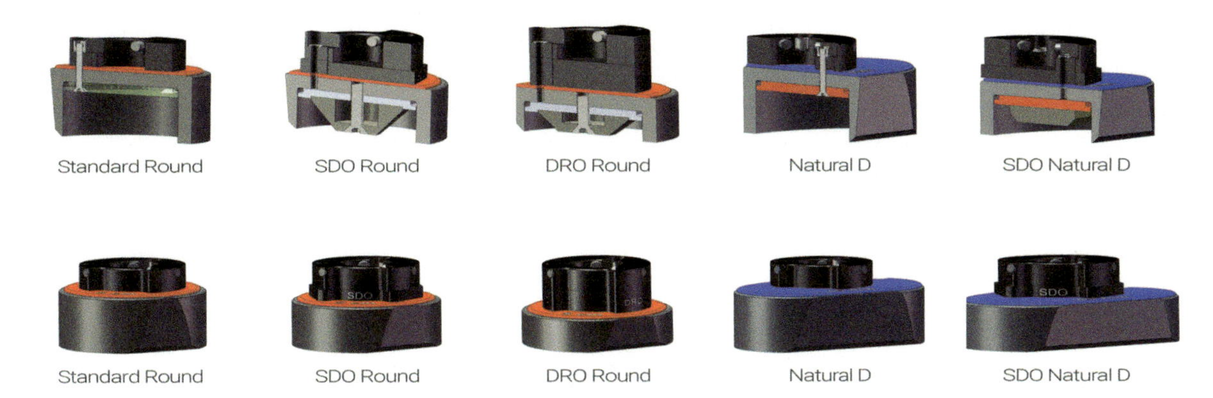

Figure 8–12 AccuBoost applicator cross sections (top) and their profiles (bottom). Applicators having internal conical shields are depicted in the 2nd, 3rd, and 5th columns for minimizing skin dose or increasing the dose rate (when compared to the standard one). (Image courtesy of Advanced Radiation Therapy, LLC.)

patient thoracic cavity where healthy tissues, such as the heart and lungs, reside. The combination of field orientation and tungsten shielding reduces dose nominally by a factor of 10 (Sioshansi et al. 2011). Radiation is administered using HDR ^{192}Ir with the brachytherapy source moving along the interior edge of each applicator. In this way, collimation is maximized without perturbing the unobstructed field. This technique of using parallel-opposed radiation beams with ^{192}Ir photons is possible only with anatomy having a fairly small thickness, as is possible with mild mammographic compression of the breast.

Unlike for megavoltage x-rays emitted from linacs, the energy of ^{192}Ir photons does not offer skin sparing from a single set of AccuBoost parallel-opposed beams. However, skin dose is minimized through use of the orthogonal beam set (4-field box approach) to distribute the radiation dose over a larger solid angle. The maximum skin dose using these applicators does not exceed the target dose for mammographic compression producing a breast separation of less than 10 cm.

The AccuBoost applicators come in sets of round and D-shaped devices (Rivard et al. 2009; Yang and Rivard 2009). The round applicators deliver circular radiation beams that sum when adding the orthogonal beam set to make a cylindrical dose distribution for target coverage (Rivard et al. 2015). The D-shaped applicators have a flat edge and are often used when the target is located close to the chest wall. This permits irradiation of targets that would be located outside the field if round applicators were used. Additional AccuBoost applicators have been developed that contain internal high-Z cones (Figure 8–12) to block radiation near the skin (Yang and Rivard 2010; Yang et al. 2011), which permit treatment of thicker breast separations. Because the maximum dose tolerance of skin is a function of both radiation dose and coverage area, development of the skin-sparing applicators expands the range of use for APBI (Hepel et al. 2014). Unlike with the AAPM TG-43 formalism, dose calculations to breast tissue are derived using Monte Carlo calculations validated by dose measurement in phantoms.

8.3.5 Permanent ^{103}Pd Seed Implants

Permanent breast seed implants (PBSI) using ^{103}Pd seeds apply the same principles of permanent seed implants for prostate to APBI for early stage breast cancer treatment. First introduced by Pignol et al. (2006), PBSI is a single-day procedure that relies on template-based needle placement guided by both freehand US and a fiducial needle (Pignol et al. 2006; Batchelar et al. 2014).

8.3.5.1 Patient Selection

Patients eligible for PBSI are aged ≥50 years with ductal carcinoma *in situ* or low-risk invasive ductal carcinoma (unifocal, negative margins, estrogen receptor positive, no lymphovascular invasion, and no evidence of metastasis) (Pignol et al. 2015). Technical eligibility criteria include a seroma that is <3 cm in diameter (equivalent sphere) that can be clearly delineated on both US and CT, with a final PTV volume not >125 cm^3. Additionally, the seroma location and breast size must be feasible for implant (Pignol et al. 2015; Morton et al. 2016).

8.3.5.2 Description of Equipment

PBSI requires a template, stabilizer arm, fiducial needles, and needles loaded with ^{103}Pd sources. Ancillary equipment includes a breast board, an US unit with breast transducer, a digital protractor, and some means of accurately measuring displacements from setup marks in the reference system of the treatment room. Currently, there is one commercially available kit for PBSI: the Breast Microseed Treatment^{TM} set (Concure Oncology, Mercer Island, WA). This kit consists of two templates and the stabilizer arm. Fiducial needles are supplied with each seed order. As this technique is still novel, some academic groups are pursuing technical improvements that may become commercially available.

Template. PBSI templates (Figure 8–13) have a rectilinear grid of holes at 0.5 cm spacing. Unlike prostate templates, they are mounted on their horizontal axis to permit rotation of the template to match the planned implant angle. In addition to the needle insertion holes, there is a set of holes running horizontally across the mid-point of the template (C in Figure 8–13). Fiducial needle can be inserted in any of these holes and then locked in place using a central locking barrel in the template. This fixes the fiducial needle depth while allowing all other needles to be inserted freely. In usual practice, a single fiducial needle is inserted through the hole at the center of the template in both the vertical and horizontal directions.

Stabilizer Arm. The commercial stabilizer arm is a device that mounts to the side rail of an operating room bed (Figure 8–14). The arm can be adjusted with nine degrees of freedom to position the template in the planned position relative to the PTV.

Fiducial Needles. The fiducial needles are 20-cm long, solid steel needles with a complex pattern of scoring, which causes a hyperechoic signal in US (Figure 8–15). Unlike in prostate seed brachytherapy, the template is not inherently coupled to the localization imaging system, and since the plane imaged is not parallel with the template, it is not possible to physically visualize the deepest plane of the implant; there is no

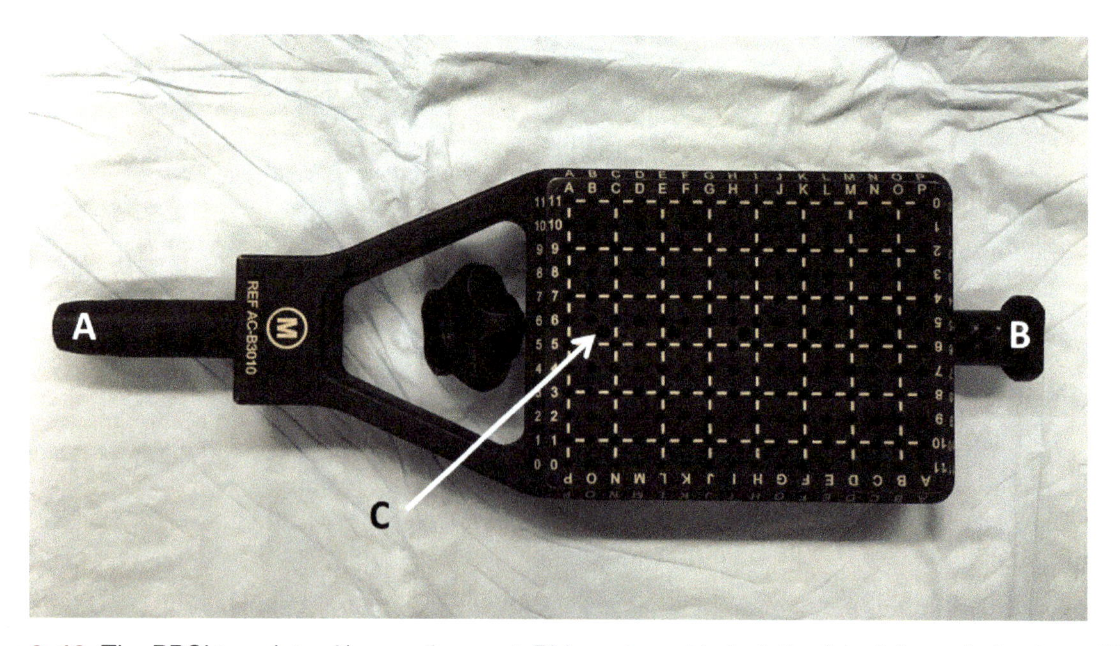

Figure 8–13 The PBSI template: A) mounting post, B) barrel used to lock the fiducial needle in place, C) row of five fiducial needle holes across the template center.

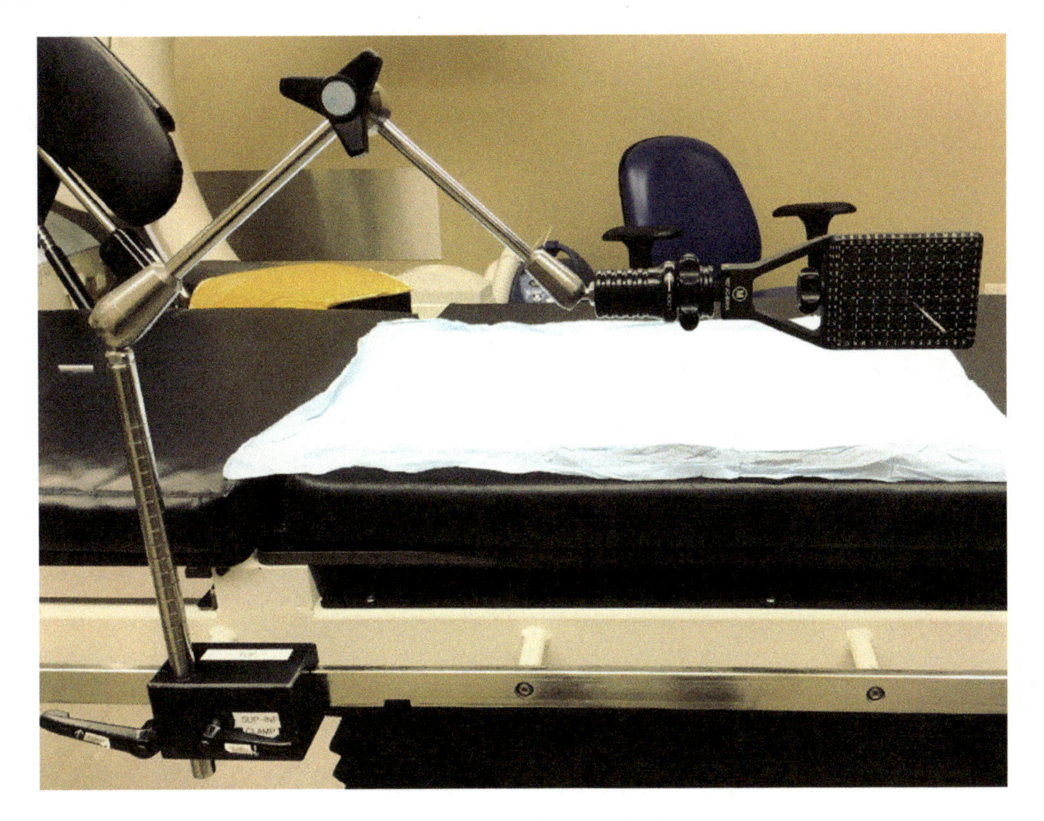

Figure 8–14 The stabilizer arm is mounted to the OR table rails.

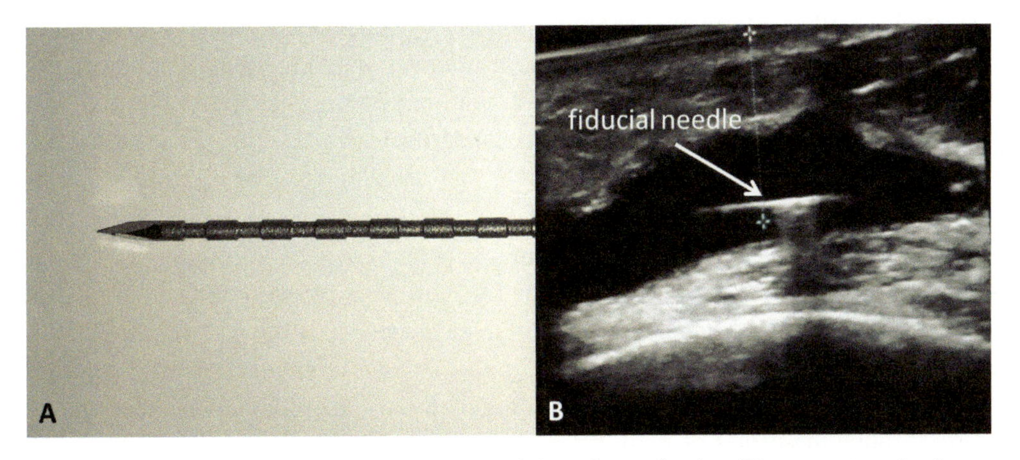

Figure 8–15 A) Detail of the scoring on the tip of the fiducial needle that B) generates the hyperechoic signal seen in US.

base to aim for, only normal breast tissue (Batchelar et al. 2014). The fiducial needle determines the angle of the implant, the height of the implant with respect to the chest wall, as well as the overall depth of the implant.

Pre-procedure loaded ^{103}Pd needles. ^{103}Pd was chosen over ^{125}I for PBSI to reduce the effective dose to patients' partners to be consistently below 5 mSv per year, the effective dose limit for caregivers of radionuclide patients, following recommendations of the National Council on Radiation Protection and Measurements (Keller et al. 2005; Keller et al. 2008). As with prostate seed brachytherapy, needles loaded before the procedure with ^{103}Pd are ordered to match a pre-plan. Stranded seeds are required to minimize seed displacement within a fluid cavity. Typical seed strength is 2.5 U.

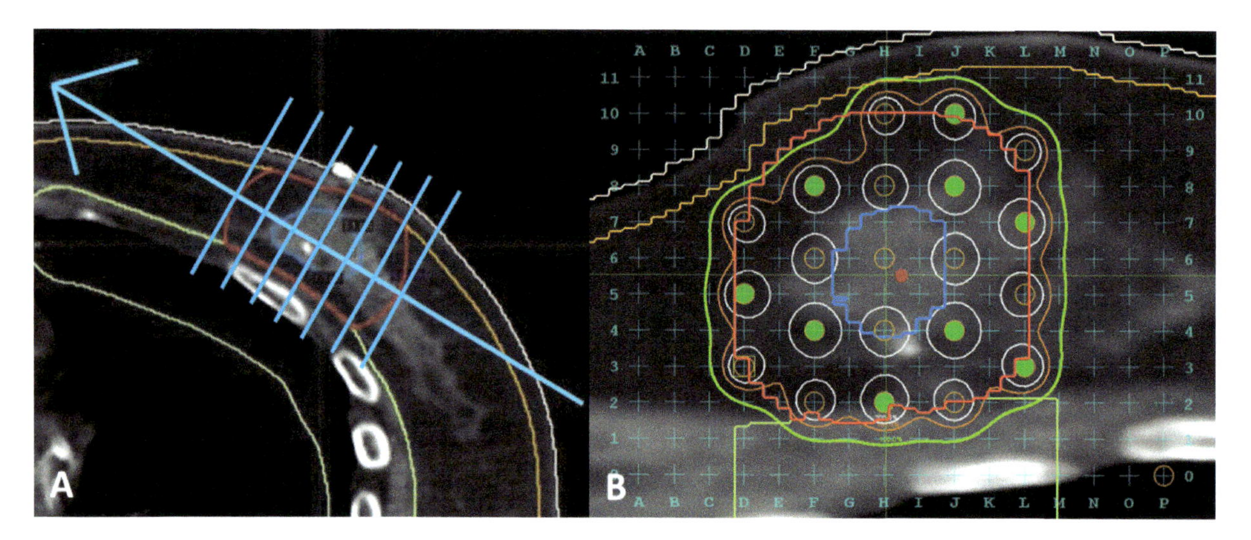

Figure 8–16 A) The planning CT images (parallel blue lines) are reformatted and resliced at 0.5 cm intervals perpendicular to the direction of the fiducial needle (blue arrow). B) The template is overlaid on the resulting images and a plan created.

8.3.5.3 Pre-implant Imaging and Planning

Final eligibility for PBSI is determined at the time of treatment simulation, which includes both a breast US and a CT imaging. The breast US is performed first to evaluate seroma visibility, size, and distance from the skin. A CT will provide further information about the 3D aspect of the cavity, its largest cross section, and its position within the breast—all relevant for the choice of orientation and number of catheters to be used. To minimize treatment delays if brachytherapy is not feasible, patients are positioned on a breast board and undergo CT simulation as for EBRT. In addition to the standard tattoos, a CT marker is placed on the inferior medial quadrant of the ipsilateral nipple.

The seroma and surrounding fibrosis are contoured as a clinical target volume (CTV) on the CT images, and the PTV is defined as seroma plus a 1.0–1.5 cm margin (Pignol et al. 2006; Keller et al. 2012; Watt et al. 2015; Hilts et al. 2015a), cropped to the chest wall and 0.5 cm from the skin surface. The needle insertion angle is chosen to be tangential to the chest wall to minimize the risk of lung perforation. The CT images are then reformatted to generate a series of images at 5-mm intervals orientated perpendicular to the needle insertion angle that encompasses the implant volume (Figure 8–16). The fiducial needle is planned for insertion at this angle, through the center of the CTV, and this defines the template position and implant depth (Keller et al. 2012; Morton et al. 2015).

In the plan, needle positions are selected at 1-cm intervals throughout the PTV, in either square or triangular arrangements. Peripheral needles are loaded with ^{103}Pd seeds spaced 1 cm apart, with neighboring needles staggered in depth to ensure inter-digitation of seeds within adjacent needles. Central needles are loaded with a seed-spacer-spacer-seed configuration to reduce hot-spots. Final needle and seed patterns are adjusted to achieve 100% coverage of the CTV with the minimum peripheral dose of 90 Gy and PTV coverage of $V_{90\%} > 98\%$, $V_{100\%} > 95\%$, $V_{150\%} < 70\%$, and $V_{200\%} < 25\%$ (all dosimetric indices are defined in Section 1.4.4). The dose to the maximally irradiated 1 cm^2 of the skin is limited to 90% of prescription.

8.3.5.4 Pre-operative Simulation

On the day of implant, the patient is situated on the CT simulator and is positioned using room lasers, EBRT tattoos, and alignment marks on the breast board. The seroma is re-assessed using US. The fiducial needle

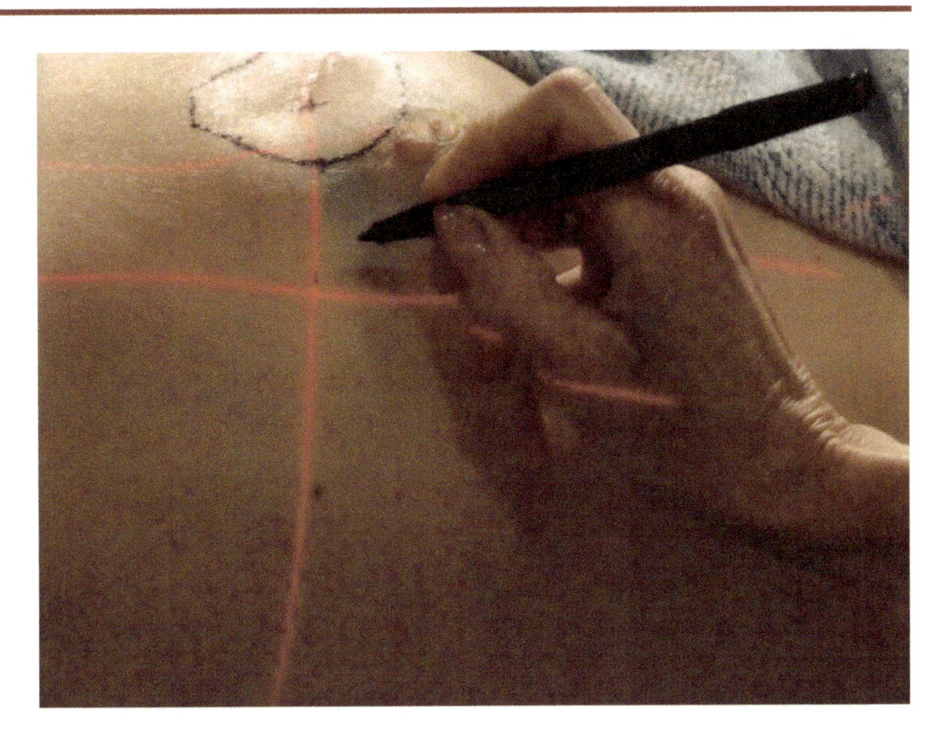

Figure 8–17 Marking the fiducial needle entry point and PTV projection on skin.

entry point and implant center projected perpendicular to the needle insertion direction are marked on the skin using lasers to guide room-coordinate moves (generated from the treatment plan) from the nipple marker or lateral tattoo. The PTV skin projection is also marked about the implant center to guide US imaging during the implant (Figure 8–17) (Batchelar et al. 2014).

8.3.5.5 Implantation

At implant, the patient is placed in treatment position under general anesthetic. Following preparation and draping of the patient, the template is placed so that the fiducial needle will enter at the correct position and angle. Free-hand US is then used to guide the insertion of the fiducial needles to obtain proper positioning and depth relative to the seroma. Once the fiducial needle position is satisfactory, the loaded needles are inserted through the template according to the plan using freehand US guidance to judge position of each needle relative to the seroma, chest wall, skin, and fiducial needle. As there is no stepper for PBSI, the depth of insertion for each needle is guided by the depth of the fiducial needle. The duration for surgical implantation is 1–2 hours.

8.3.5.6 Post-operative Imaging and Evaluation

Immediately following recovery from anesthesia, post-implant CT images are acquired with the patient in the same position as when she was implanted (Day 0). These images provide immediate feedback regarding implant quality. Images are also obtained at Day 30 to monitor seed migration. Post-implant delineation of the seroma is challenging due to the presence of the seeds and the artifacts they create. Image registration, particularly deformable image registration, can be used to transfer the pre-operative CTV to the post-operative scans (Hilts et al. 2015a). This is followed by careful evaluation and adaptation of the contour by the implanting radiation oncologist. Implant quality is assessed on an evaluative PTV (PTVeval) defined as the 0.5–1.0 cm beyond the CTV, with a goal of covering this with 90% of the prescription dose (Hilts et al. 2015b). Maximum skin dose, evaluated as the maximally irradiated 1 cm^2 (D_{1cm^2}), approximated as $D_{0.2cm^2}$ of a 0.2-cm thick skin structure, should be less than 90 Gy.

8.4 Process and Work Flow Description

8.4.1 Pre-implant Imaging and Planning

8.4.1.1 Interstitial Brachytherapy

Interstitial applicators are placed either during breast conservation surgery (BCS) or 3–4 weeks after BCS. In the perioperative brachytherapy, flexible brachytherapy applicators are placed during surgery in the tumor bed. In the latter case, the interstitial implantation is performed after the surgical scar heals and after receiving the final histo-pathological diagnosis.

For interstitial implants after BCS, a pre-plan can help design an optimal implant. In this case, a pre-implant CT is obtained for evaluating the tumor bed size, shape, and location with respect to the skin. Surgical clips left during BCS are used to determine the tumor bed and shape. A plan and placement of the applicators is designed in collaboration between the radiation oncologist and the medical physicist. The radiation oncologist inserts catheters under image guidance. Many forms of imaging have been used. Surgical clips are visualized, which assist in placement of catheters in the tumor bed in accordance with a plan. Figure 8–18 shows a photograph of a patient with a multi-catheter implant.

8.4.1.2 Single-entry Intracavitary Balloon Applicators

Implantation processes for the balloon or strut applicators are similar to one another. A pre-implant CT is performed after the patient is selected for APBI treatment. The pre-implant CT of the breast is acquired with ≤0.3 cm slices with no gaps between slices. Scans are acquired with breath hold (if the patient can tolerate it) in order to obtain an accurate volume measurement. The entire breast is scanned to ensure that the tumor bed plus a minimum of 2 cm margin superiorly and inferiorly is included in the scans. The tumor bed and margins are delineated on the axial scans to determine the cavity volume. The radiation oncologist and the medical physicist determine the cavity size, its placement with respect to the skin, long and short axis dimensions, and assess the best insertion site and angle. Using this data, the most appropriate applicator type and size is planned. If the surgeon is implanting the applicator, the cavity parameters are communicated to the surgeon by the radiation oncology team. The primary goal of the pre-implant CT is to determine the type and the size of the device that will be used for the treatment, as well as the orientation. As the surgeon and the radiation oncologist gain experience in this modality, the additional step of pre-implant CT may be replaced by the clinical judgment of the breast surgeon. The cavity parameters needed for the applicator implantation are taken at the time of BCS.

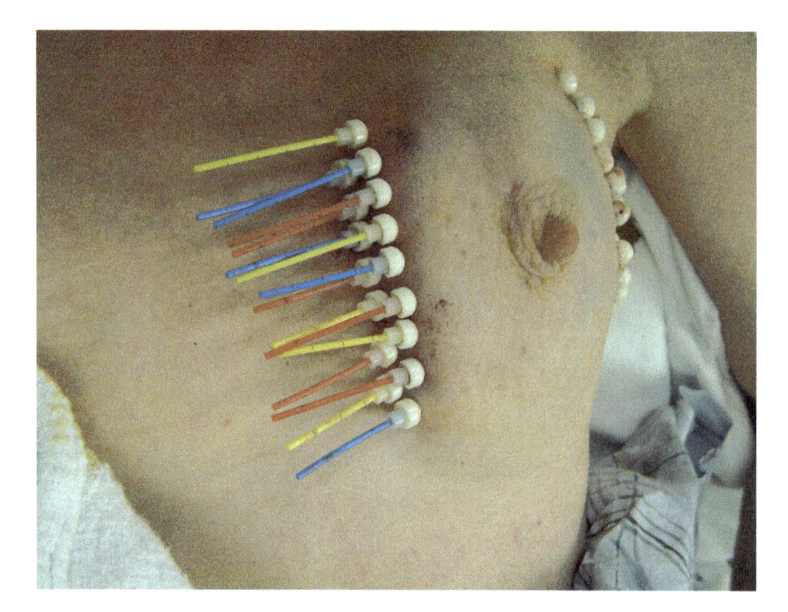

Figure 8–18 Interstitial implant.

8.4.2 Device Implantation

In the case of interstitial brachytherapy, sterilized breast templates, trocars, and breast catheters are prepared for the implant. The radiation oncologist implants the appropriate number of catheters as per a pre-implant treatment plan under image guidance in the procedure room.

Implantation process of balloon and strut based intracavitary applicators have evolved over the past decade based on institutional resources and physician experience. Single-entry intracavitary applicators may be placed during the lumpectomy or nodal dissection, but they may need to be removed prior to the start of treatment if pathology reports an unfavorable nodal status. Surgeon preference, patient convenience, and cosmetic considerations decide this process. Placement of the intracavitary applicator in a separate procedure after the pathology is complete is the more common practice because it does not waste an expensive device and lessens patient disappointment and discomfort if the procedure must be aborted. However, insertion delays of several weeks can cause difficulty in finding the tumor bed if too much time has elapsed. With US localization, the lumpectomy site is normally visible after a couple of weeks.

The device must be thoroughly examined before implantation to make sure that the device is intact with no visible damage, while preserving the sterility of the device. After insertion, the patient is brought to the radiation oncology department for a planning CT (within 48 hours after implantation). Immobilization devices such as breast boards or Vac-Loks™ (CIVCO Radiotherapy, Coralville, IA) can be used to keep the patient comfortable and its position reproducible, as patients will be treated in the same position as they were for planning. While immobilization is not necessary, it definitely makes the treatment QA much easier. CT laser alignment marks are placed on patient. AP and lateral scout images are acquired to encompass the entire breast. Axial scans of the breast are acquired with ≤0.3 cm slices with no gaps between slices. Scans are acquired with modest breath-hold if tolerated by the patient. The expansion of the device is checked by the radiation oncologist and adjusted if necessary. The radiation oncologist may also manipulate the applicator to vacuum out air inside the balloon. Next, the lengths of each catheter and transfer guide tubes are measured and recorded for use in treatment planning. Skin markings are made next to the orientation line of the applicator, and a photograph at the time of CT is taken. For each fraction, one ensures that the relative position of skin markings and orientation line are the same, thus ensuring that the applicator did not rotate. There are subtle differences between the breast intracavitary APBI applicator markings due to their design, and they are explained below.

8.4.2.1 MammoSite

MammoSite balloon is usually filled with saline and radiographic contrast agent, which helps in better imaging of the balloon and ensures that there are no leaks to the tumor bed. Three-dimensional images are evaluated to assess the balloon cavity conformance, skin distance from surface of balloon (minimum 0.7 cm), balloon-rib distance, and balloon symmetry. At the time of CT for planning, final appropriateness of the implant is decided based on the parameters listed above. The balloon rotational orientation (as indicated by the shaft orientation line) is documented via a picture of catheter and orientation line (Figure 8–19). While the rotational position of the central lumen would be irrelevant for a symmetrical balloon, most balloons exhibit some level of asymmetry that is taken into account in planning, thus rendering some usefulness to preserving the rotational position.

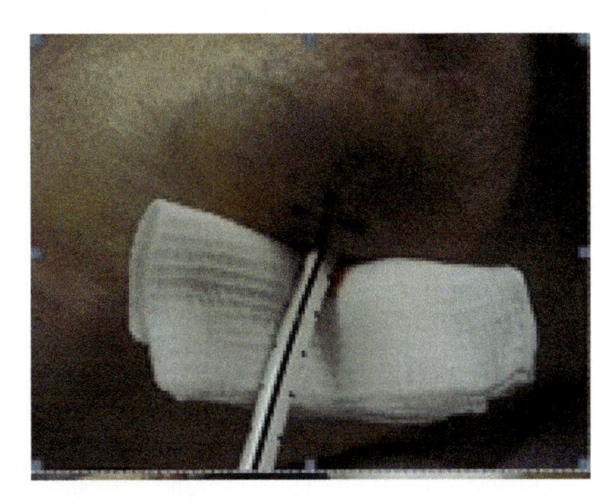

Figure 8–19 MammoSite applicator and patient markings.

8.4.2.2 Contura Multi-lumen Balloon

Contura MLB has a central catheter surrounded by four fixed catheters in a configuration within an inflatable balloon that allows them to spread from the central catheter slightly. Before the CT scan, the radiation oncologist makes sure to check the inflation of the device. Two vacuum ports on the proximal and distal end of the balloon are used to remove possible air or seroma (blood serum buildup between the applicator and the breast tissue) after the MLB inflation through a sixth catheter. A

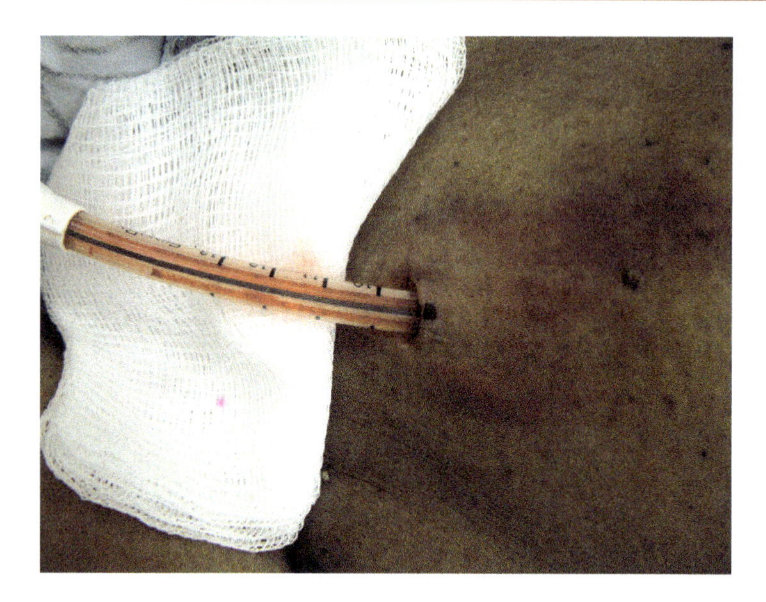

Figure 8–20 Contura MLB applicator and patient markings.

black mark along the shaft of the catheter running parallel to lumen #1 is used to confirm the device did not rotate between imaging and treatment. Alignment of an external skin mark to this black line is a reliable way to assure no significant rotation of the balloon in the patient (Figure 8–20). Configuration of the orientation line on the applicator in relation to the line drawn on the skin of the patient should be checked before each treatment.

8.4.2.3 Strut-adjusted Volume Implant

Before the CT scans, applicator expansion is checked by the radiation oncologist. The device is collapsed and inserted into the lumpectomy cavity, and then the peripheral struts are expanded using the knurled knob (expansion tool). The struts are secured in place by the pressure exerted by the cavity walls. The expansion tool is reinserted prior to each treatment in order to allow the operator to collapse and remove the device in the case of an emergency event during treatment. The CT should show minimal invagination or air gaps. After CT, the distance from skin surface to the catheter handle is measured and documented for QA before each treatment (used as an axial assessment tool). The applicator and skin are marked with a continuous line as a rotational assessment tool (quality check) before each treatment (Figure 8–21).

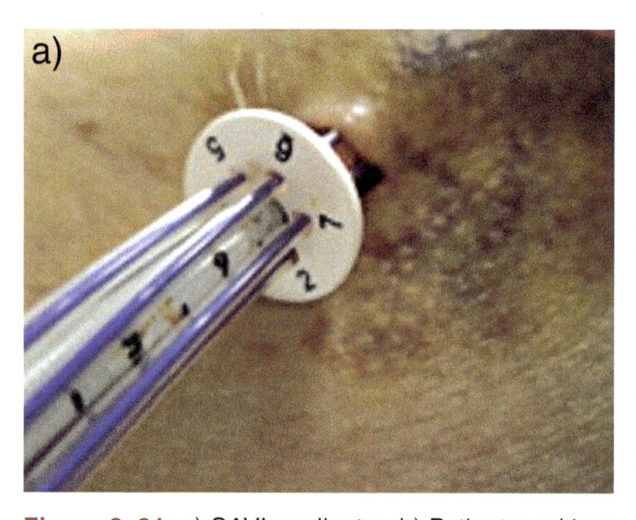

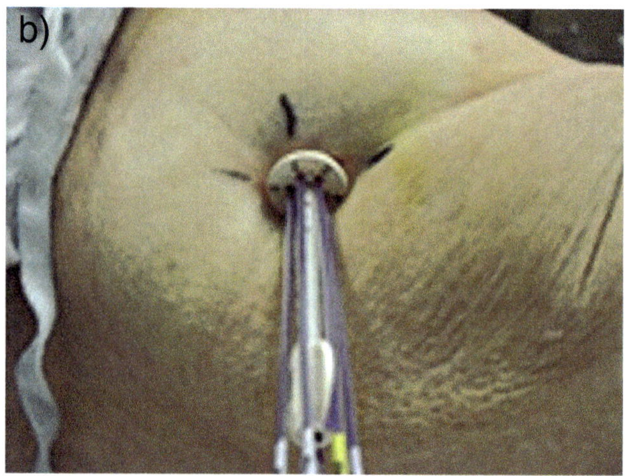

Figure 8–21 a) SAVI applicator. b) Patient markings.

8.4.3 Imaging for Planning and Device Geometry Optimization

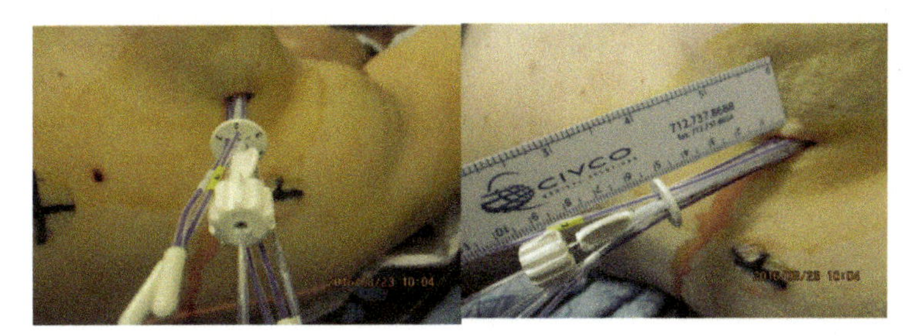

Figure 8–22 Digital photograph of implanted APBI device and QA measurements.

There are a variety of imaging modalities available in radiation oncology that can be used for treatment planning, including US, orthogonal radiographs or mammograms, CT or CBCT, or MRI. Often US is used for cavity evaluation pre-implant or during the implant, but it is largely not relied upon during the treatment planning process. Orthogonal imaging was typical for early implants, but it fails to show the soft tissues. CT has become the most common imaging used for treatment planning for breast brachytherapy. The role of MRI in treatment planning for APBI has yet to be determined. Digital photography of the implant can also be useful during reconstruction and for identifying and labeling the catheters for pre-treatment QA (Figure 8–22).

For CT scanning, thin slice thickness, such as 0.125 cm, can be useful in reconstructing the device. Additionally, breath hold during a CT scan can help aid in reducing motion artifact during the scan. Adjusting slice thickness depending on the directionality of the implant can also be useful and, generally, smaller slice thickness can aid in the reconstruction, especially if the implant is laterally implanted. For example, if the implant is placed in the superior-inferior direction (same direction as CT couch travel), digital reconstruction of the catheters is typically easier than if the implant is placed in the lateral direction (perpendicular to the CT couch direction of travel). Dummy wires can also aid in the reconstruction of the catheters, although a number of the APBI devices have radiopaque markers to assist in visualization and identification.

8.4.4 Planning

8.4.4.1 Structure Delineation

Once the implant or device is in place and a CT dataset has been acquired, the next steps are the delineation of structures (anatomical and for planning), delineation of applicators, plan optimization, and evaluation. The conventional construct for the target of a PBI treatment starts with delineation of the lumpectomy cavity for the interstitial implants, or with surrogate structures for it (balloon surface or for SAVI, the surface obtained by connecting the struts). The margin needed in order to create CTV varied over time and is also device dependent. The vast majority of authors describing early experience with interstitial implants report a clinical target volume (CTV) defined as 2.0 cm beyond the lumpectomy cavity for interstitial approaches (Arthur et al. 2003; Kuske et al. 1998; Nag et al. 2001). It is not clear when the accepted interstitial margin became 1.5 cm, but most likely it is related to the CT-based planning when careful, systematic, reproducible, and true 3D volumes were created. It is also likely that it occurred at the time of writing the National Surgical Breast and Bowel Project (NSABP) protocol B-39 (NSABP B-39/RTOG 0413 2006). The 1.5 cm expansion could have been chosen as a compromise between the 2.0 cm value for interstitial implants and the 1.0 cm value for the "new" balloon applicator that was to be included in the brachytherapy arm of the trial.

When the balloon brachytherapy method using MammoSite was introduced, Edmunson and colleagues demonstrated (using the concept of "effective thickness") that the target irradiated by the MammoSite was in fact nearly 2.0 cm (Edmundson et al. 2002). This demonstrates that at the time (2002) the 2.0 cm margin was still the accepted margin for interstitial implants. Dickler and colleagues also calculated the effective thickness for 13 patients and produced a mean result of 1.6 cm (Dickler et al. 2004). These results were consistent with the CTV margins used for multi-catheter interstitial brachytherapy, supporting the hypothesis that the two APBI modalities treat effectively similar target volumes. Submitted in 2003, the work by Dickler and colleagues indicated that the margin for interstitial implants was transformed from 2.0 cm to 1.5 cm sometime between 2002 and 2003.

A more recent paper, acknowledging that the effective thickness might not be uniform due to the fact that the tissue adjacent to the balloon may compress or stretch, used deformable registration to propagate contours from one CT set to another (Shaitelman et al. 2012). They concluded, "The effective CTV treated by the MammoSite was on average 7% ±10% larger and 38% ±4% smaller than 3D–CRT (Conformal Radio-Therapy). CTVs created using uniform expansions of 1.0 and 1.5 cm, respectively. The average effective CTV margin was 1.0 cm, the same as the actual MammoSite CTV margin. However, the effective CTV margin was nonuniform and could range from 0.5 to 1.5 cm in any given direction."

Contouring of the lumpectomy cavity is known to be associated with low inter-observer concordance due to a number of factors: dense breast parenchyma, benign calcifications in the breast, and tissue stranding from the surgical cavity. Surgically placed clips after lumpectomy as radiographic surrogates of the cavity have been shown to help with the delineation of the lumpectomy cavity. In balloon-based treatment methods, the uncertainty of delineating the lumpectomy cavity is removed and replaced by the much more reliable delineation of the balloon surface. Of course, the implicit assumption is that the balloon is correctly placed in the center of lumpectomy cavity and that the inflated balloon stretches the tissue isotropically. For increased visibility on CT images and automatic delineation, a small quantity of iodine-based contrast is used in the water. A planning target volume for evaluation (PTV_EVAL) is created using a 1.0 cm expansion from the balloon surface of the CTV but limited by the first 0.5 cm from the skin surface and abutting the chest wall. A 1.0 cm expansion is used in the case of balloon-based brachytherapy devices due to stretching of the tissue surrounding the lumpectomy cavity. If air or fluid is retained within the lumpectomy cavity after inflation of the balloon, the displacement of the tissue away from the balloon surface must be accounted for. The air-fluid trapped should be contoured separately on each CT slice, and a total volume should be calculated to determine the percentage of the PTV_EVAL that is displaced. When determining the PTV_EVAL dose coverage, this displaced percentage must be subtracted. For example, if the percentage of PTV_EVAL displaced by trapped air-fluid is calculated to be 5%, then to comply with minimum dose coverage criteria, the dose coverage must be at least 95% of the PTV_EVAL receiving 90% of the prescribed dose.

For planning purposes, the use of an "avoidance" structure can be highly efficient in order to "contain" the prescription dose to PTV (Figure 8–23). An example is shown here for a balloon-based treatment and an interstitial implant. For the interstitial implants, the avoidance ring structure should be used in conjunction

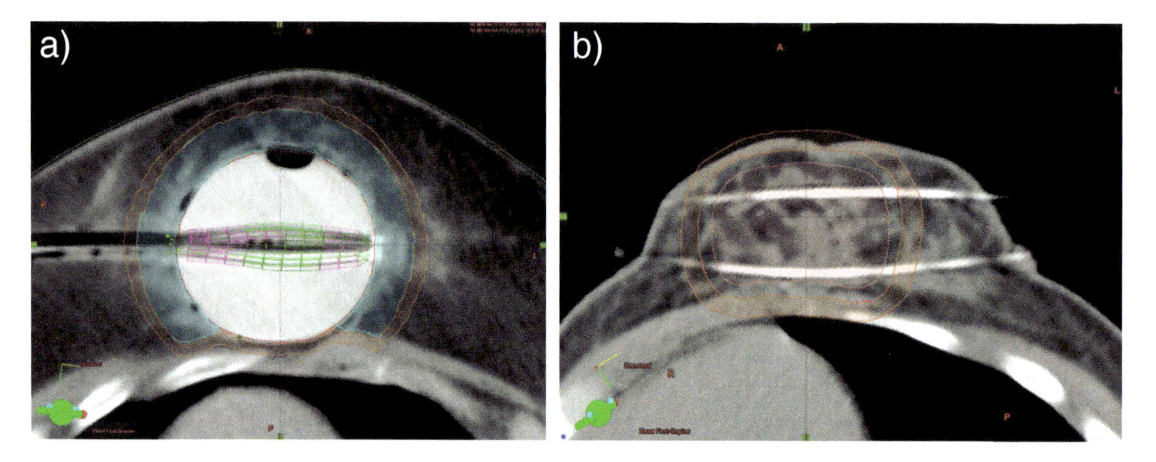

Figure 8–23 Avoidance structure used for planning is shown for a) a balloon plan and b) an interstitial implant. Due to different placement of dwell positions relative to CTV, the avoidance structure is created next CTV for balloons and at some distance (3–5 mm) from CTV for interstitial implants and SAVI.

with another planning structure, essentially the inner surface of the ring, used to activate dwell positions strictly inside the ring. Skin, chest wall, pectoralis muscle, breast tissue, heart, and lung are also segmented as part of the plan.

Treatment catheters are delineated one by one. All current TPS have the ability to assist the user with an intensity threshold-based detection of applicators. Thin wire markers can facilitate outlining the catheters, as in Figure 8–24.

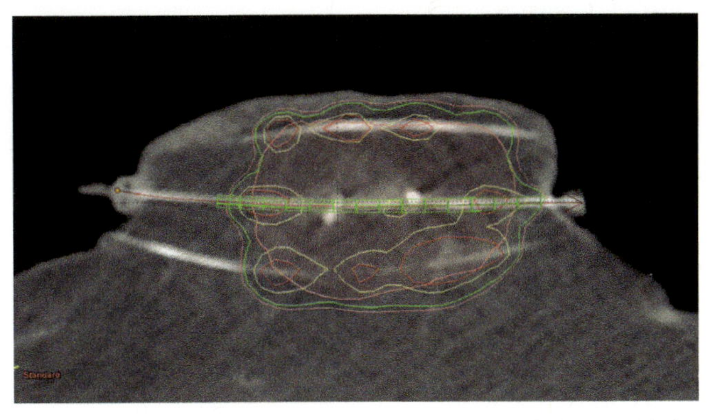

Figure 8–24 Sagittal cross section through interstitial implant showing an outlined catheter between the two buttons on the skin. Thin, continuous wire markers were used in each catheter.

The Contura balloon has the advantage of fixed relative geometry catheters, permitting creation of an applicator library device. The device is then imported over a new set of CT images. Since the position of the applicator relative to CT images is not the same for all patients, the device can then be adjusted to "fit" in the images using translations and rotations. Moving all the catheters of the device as an ensemble preserves their physical relationship and hastens the process, while eliminating potential for errors (e.g., mislabeling catheters). Such a method is not possible with SAVI, the Best Double balloon, and interstitial implants since every applicator is different, and a library template could not be generalized.

The SAVI applicator presents additional difficulties in outlining catheters because each catheter is sustained by a metal strut which, while visible in CT, is not the path of the source (Figure 8–25).

Identification (labeling) of catheters is extremely important, as incorrect labeling has the potential to transform a correct plan into an incorrect delivery. In interstitial implants, photos and quality check (QC) forms that rely on CT-based length measurements combined with geometrical measurements can and should be used to check the correlation of catheters with HDR unit channels. In Contura balloons, the most distal point in each lumen is visualized by a radiopaque plug and catheter #1 is marked by a radiopaque wire under-

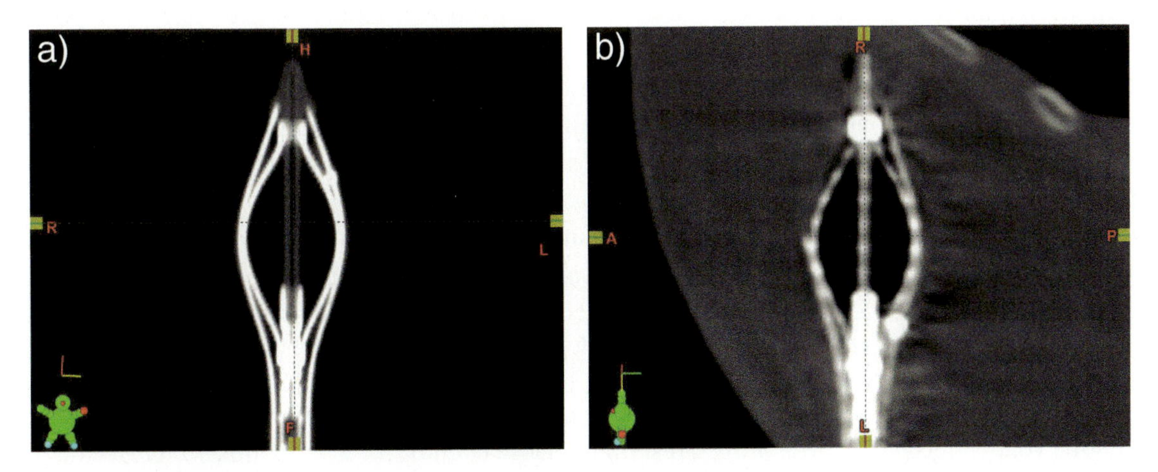

Figure 8–25 a) Wire markers are inserted in all catheters except the central one. One can distinguish between catheters and supporting struts. b) Patient image showing a wire marker partially inserted. Outlining the actual catheter and its tip is left to imagination, unless a thin wire marker is used. Depending on CT image quality, window and level settings, and device orientation relative to scanner axis, catheter delineation can, in some cases, be affected to produce large errors.

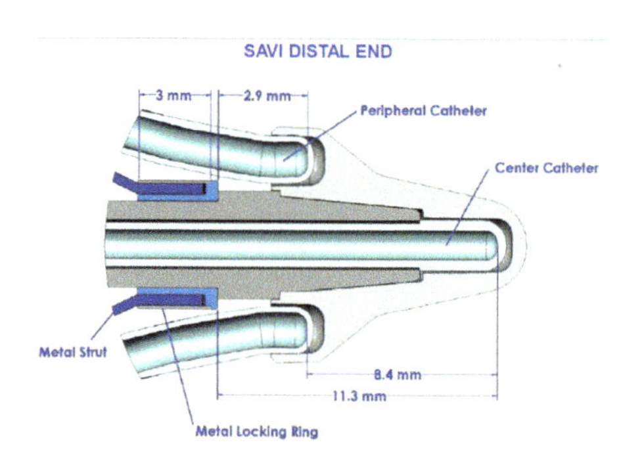

Figure 8–26 Distal end blueprint for the SAVI device (image courtesy of Cianna Medical, Aliso Viejo, CA).

neath. The order of catheters is clockwise when looking along the balloon shaft with the tip of the catheters pointing away from the viewer. In SAVI, catheters #2, #4, and #6 are identified by small metal bands in the distal, medial, and proximal aspects of the catheters. Unfortunately, there are no features helping to establish the catheter tips. Instead, the user is trained to measure distances from a metal locking ring, the only device feature in that area visible with CT imaging (Figure 8–26).

8.4.4.2 Dose Optimization

Before discussing dose optimization, one needs to stress the importance of optimal device placement. While there are many constraints in how a device is being implanted—many of them related to patient's anatomy and patient convenience, lumpectomy placement within breast, and position of OARs relative to the cavity—optimal placement should also account for specific device features and the way dose distributions are created.

Similar considerations apply for interstitial catheter implants and for SAVI. An optimal placement is the foundation for an optimal plan. While the degrees of freedom in HDR planning are sufficient most of the time, there is a fine line between suboptimal and mediocre for achieving an optimal plan with a less-than-optimal applicator placement. Small deviations from geometrical optimality of an implant can and should be compensated for by optimal planning. A bad implant (one with large distances between catheters or target tissue away from the struts or the surface of a balloon) *cannot* be compensated for by planning.

For balloon and SAVI devices, multiple scans may be taken in order to find the device's best orientation, rotation, and inflation. For an interstitial implant, we recommend imaging (US, CT, or CBCT) during the insertion phase to ensure adequate catheter positioning relative to lumpectomy cavity, adequate spacing between catheters, etc.

While there are many ways of creating a dose distribution—among them, manual (or graphical) and geometrical optimization—the only "true" optimization is the so-called "volume" or DVH-based optimization. Each of the structures that will be subjected to dose-volume constraints is sampled in a number of points, typically hundreds to a few thousand, depending on the volume. The reason for doing this is that in order to speed up the optimization process, one can choose to calculate dose in a much smaller number of points than the total number of dose voxels that the structure inherently has, based on the resolution of the dose matrix. While upper and lower constraints are placed on the dose volume histogram for the target, the avoidance structure placed outside the target is the one limiting higher-than-prescription dose "spilling" into the normal tissue and, thus, driving the dose conformance. Good structures and good constraints will produce desirable treatment plans. If one needs to manually alter an already optimized plan, it is very likely that either the optimizer is not performing adequately (might be too sensitive and easily "captured" by local minima) or the set of constraints does not adequately, completely, or reasonably reflect the treatment planning goals.

During "volume optimization," dose-volume constraints are placed on the histograms for the PTV_EVAL and the avoidance structure (an example with typical constraints is shown in Figure 8–27). Depending on the shape of the PTV_EVAL, one can limit the volume of the avoidance structure receiving prescription dose to a few percent (1% to 10%), thus increasing dose conformality to the PTV_EVAL. The PTV_EVAL volume receiving 150% of the Prescription Dose (PD) is, for a balloon-based device, dependent on its diameter (around 25% to 30% of the total volume). For an interstitial implant, $V_{150\%}$ should be less than 15% to 20% of $V_{100\%}$ (~CTV volume), for the Dose Homogeneity Index (DHI = 1 – $V_{150\%} / V_{100\%}$) to be greater than 0.75.

SAVI planning is very much similar to an interstitial implant, but given the typically smaller target volumes and significantly higher (percentage-wise) $V_{150\%}$ and $V_{200\%}$, these plans tend to have DHI <0.6.%

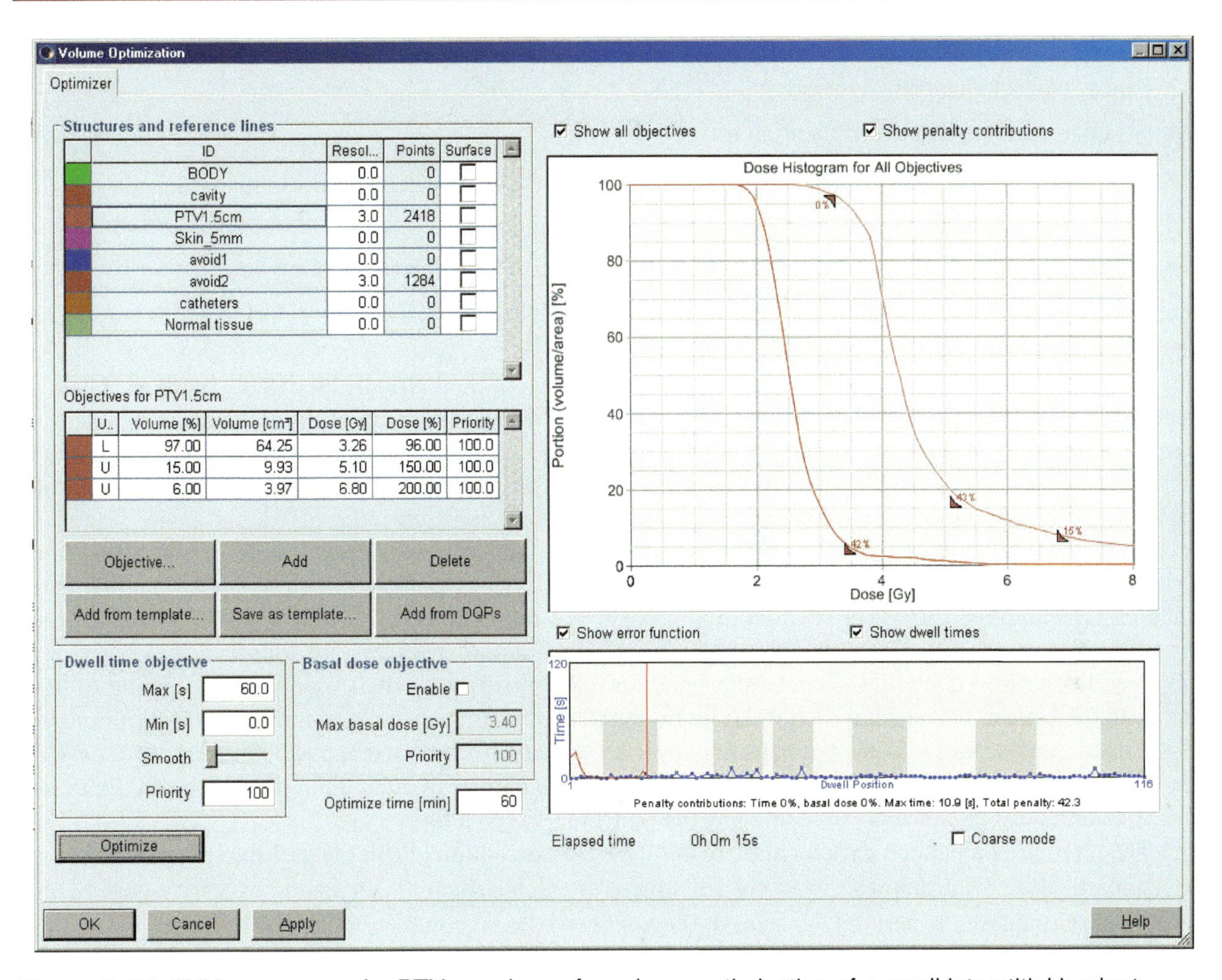

Figure 8–27 DVH parameters for PTV are shown for volume optimization of a small interstitial implant.

8.4.4.3 Plan Evaluation

Plans are typically evaluated by $V_{90\%}$, $V_{95\%}$, $V_{150\%}$, and $V_{200\%}$. DHI, used mostly for interstitial implants, is a homogeneity index using the ratio $V_{150\%} / V_{100\%}$. Initially, the only available document that set forth constraints for these parameters was the NSABP B-39/RTOG 0413 protocol (NSABP 2005). The ABS recently revised its recommendations regarding dosimetric constraints for interstitial implants and essentially lowered the acceptable limits for $V_{150\%}$ and $V_{200\%}$ (Shah et al. 2013). Maximum skin dose was recommended to be ≤100%PD for interstitial implants and SAVI, but balloons can only achieve that limit if implanted with sufficient balloon-to-skin distance (>1.0 cm).

8.4.5 Imaging for Treatment Delivery, QA, and QC

Many events involving breast brachytherapy occur due to failures related to preparing the patient for treatment delivery. The major items to check before intracavitary treatment include:

- geometry of the treatment appliance: diameter of a balloon, if present; the positions of struts and inner catheters
- position of the appliance in the patient: the applicator conforms to the cavity as in the dosimetry images; the rotation of the applicator is the same as in the dosimetry images

- length of the catheters has been correctly measured
- patient has not changed significantly

Interstitial treatments require verification that:

- length of the catheters has been correctly measured
- patient has not changed significantly
- position of the catheters has not shifted from that during the dosimetric imaging

Before all treatments, attention should be paid to the position of the transfer tubes to avoid draping over the contralateral breast.

Of these items, the axial position of the applicator can be checked visually, as can the rotation of balloon-type intracavitary applicators. The importance of checking the axial position of the catheters comes from the need to keep them a little loose at first because of the edema that follows the implant and typically takes a little over one day to subside. Tightening the catheters after the first day can cause considerable discomfort later in the course (about the third day of treatment) when radiation-induced edema again requires an increased spacing between the buttons that immobilize the catheters. A common approach for consistency entails always having the distal buttons flush with the skin. While that gives more consistency than letting each catheter assume a random position, this process positions the dose distribution at the same distance from the skin on the far side of the catheter connection while the target may be closer to the center of the implanted volume. Attempting to center the catheters daily proves very challenging, as the amount of tissue between the buttons can change each fraction. Fortunately, the method of maintaining contact with one side shifts the dose distribution in the target usually by <0.5 cm, an uncertainty that can be included in the PTV. If the space between the skin and the button on the loose side changes by more than the expected 0.5 cm, serious consideration should be given to imaging the patient and replanning.

The rotation of a balloon catheter also rotates the dose distribution. With the multiple-path catheters used to match the dose distribution to avoid normal structures, such a rotation can lead to excessive dose to those structures. Comparing a mark on the stem of the applicator (most come with one) with a mark made on the patient at the time of localization imaging adequately positions the catheters.

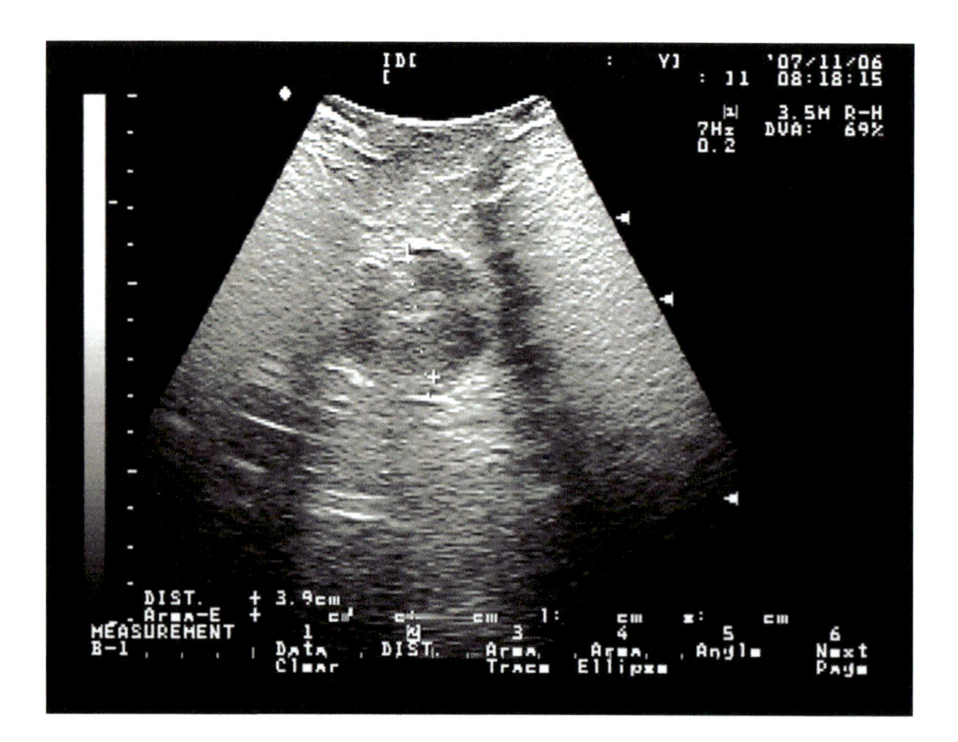

Figure 8–28 An US image of a Contura applicator to verify balloon diameter before treatment.

The other items of the list require imaging at the time of treatment to verify. Ideally, treatment could be delivered in a suite that also housed a CT scanner. Few facilities have co-located HDR treatment units and CT scanners. Moving the patient significantly between imaging adds time and complexity to the procedure, particularly if done with heavily used CT scanners. Aside from the practicality issues, performing a CT before each fraction delivers unnecessary radiation dose to the contralateral breast and other organs in the field of view. Facilities considering CT verification should establish the lowest dose technique compatible with visualizing the applicator adequately to distinguish the detail necessary for verification. Modern protocols can allow fairly low doses for this procedure.

While there may be an advantage to MRI for contouring and subsequent dose calculations, pretreatment MRI verification only adds complexity. MRI is particularly useful with interstitial implants, where there may be no cavity and the target becomes difficult to visualize.

US is a simple, fast, and inexpensive method for verification for balloon-based applicators. Delineation of the balloon takes little instruction, and the images can detect the most dangerous failure mode, that is, loss of volume in the balloon. All modern US units provide on-screen display of the distance between two points marked on the image, which provides a verification that the diameter of the balloon has not changed. Finding a balloon diameter takes experience. Figure 8–28 shows such an image. Due to the lack of three-dimensionality of the images (i.e., no registration between image planes), US is not good at checking the patient for physical changes. US is also not possible with SAVI applicators because the air that often occupies the implanted cavity will reflect the US signal.

Radiography or fluoroscopy can provide verification for several of the items on the list. The biggest challenge to radiographic verification is adequate control of the imaging geometry. Using a conventional simulator simplifies the tasks, such as finding the balloon diameter because of the control of the isocenter and projection of a radiopaque ruler at the isocenter. Using the scout-image facility of a CT scanner also simplifies the procedure as long as the unit has software to prove measurement between points identified on the AP and lateral pair. Figure 8–29 shows such an image pair with the distances between opposite struts on a SAVI indicated. Comparison of these distances with those taken at the time of the dosimetry CT gives an idea of the consistency of the applicator geometry. The use of a stan-

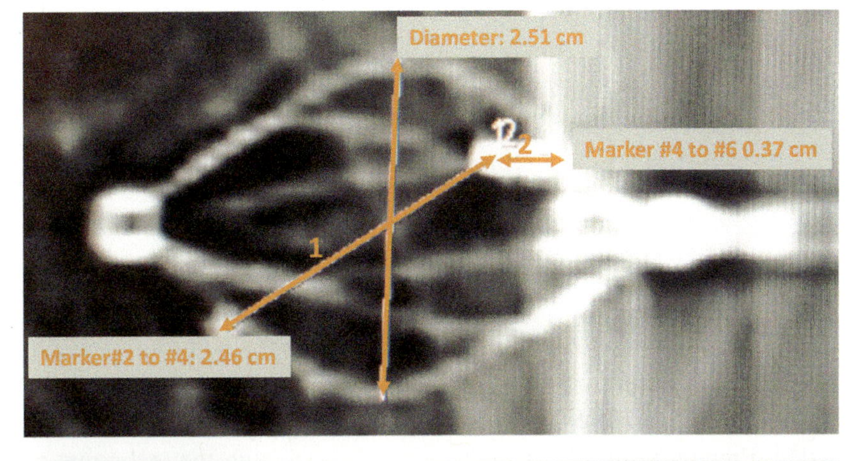

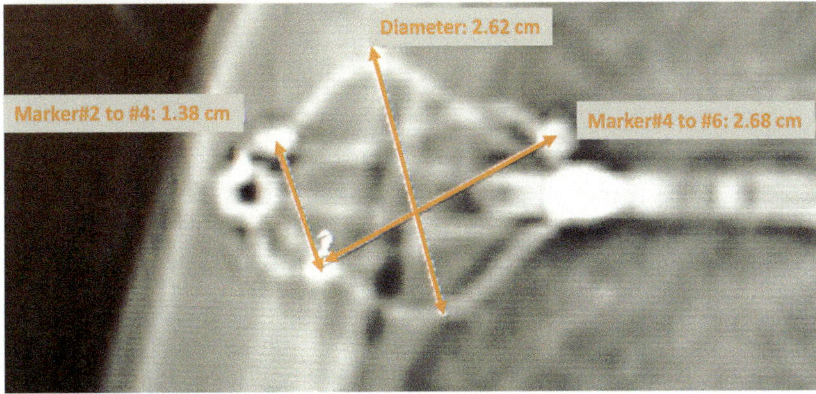

Figure 8–29 Orthogonal scout images of SAVI applicators from a CT scanner. The arrows indicate the distance between markers on opposing struts. (Images courtesy of Firas Mourtada.)

dard radiographic unit, or worse, a portable unit, causes difficulties that increase the uncertainties greatly, mostly due to trying to place a radiopaque ruler at the same focal-target distance as the balloon center. Radiography can be a practical alternative to full CT for checking SAVI applicators. With three markers on the breast and a device geometry that matches an image taken at the time of pre-planning, comparison with an image taken just before treatment highlights any changes in the applicator strut pattern. This approach also works for balloon-based applicators. The radiographic technique provides the ability to validate the catheter length by sending the HDR device check cable to dwell position #1 and visualizing the source in close proximity to the tip of the catheter. If the HDR unit cannot be in the same room as the radiographic unit, the source simulator can be used. It is important to use the same transfer tubes that will be used for the treatment to intercept an error from the use of the wrong transfer tubes.

8.4.6 Intra-op (Single Fraction vs Multiple Fractions)

Breast brachytherapy can be delivered in a variety of fractionation schemes and is dependent on the device and method used. After catheter insertion, patients receiving HDR MIB return on an outpatient basis for a typical dose of 34 Gy in 10 fractions (twice daily). For patients receiving LDR MIB, sources are implanted for 2–5 days, while the patient is admitted as an inpatient. Balloon-based brachytherapy can also be delivered in a single fraction or multiple fractions. For multiple fractions, applicator placement and radiation delivery does not differ greatly from MIB (Njeh et al. 2010). Intraoperative radiation therapy delivered as a single fraction is a more intensive process, as the dosimetry information must either be pre-loaded or planned at the time of applicator placement. Following applicator placement, radiation is delivered directly in the operating room, ensuring appropriate radiation safety measures are taken (rolling shields/walls, external rubberized lead shielding, and minimizing staff in the room). Often, electronic brachytherapy sources are used for this type of treatment (see Chapter 2), and treatment parameters about the source characteristics are preloaded onto the treatment computer. Planning information (time of irradiation) can be done at the time of device insertion or using a pre-determined treatment plan (Pinnaro et al. 2011; Francescatti et al, 2010).

8.5 Target Delineation in APBI

8.5.1 Differences between Brachytherapy and External Beam

The relevant target volumes for APBI are the lumpectomy cavity, or seroma, the clinical target volume (CTV), and a structure for evaluating the final dose distribution (PTV_EVAL) (NSABP 2005; Vicini et al. 2016). Precise definition of the structures varies with the technique used to deliver APBI. In all cases, the lumpectomy cavity is defined as the combination of fluid accumulation, tissue distortion, and fibrosis that can be visualized on CT as an increase in Hounsfield units over the background breast tissue (NSABP 2005; Wong et al. 2006; Smith et al. 2009; Major et al. 2015). If present, surgical clips are used to aid in the delineation of the lumpectomy cavity.

For APBI using EBRT, the CTV is defined as the lumpectomy cavity plus a uniform expansion of 1–2 cm (Smith et al. 2009; NSABP 2005; Wong et al. 2006), limited to 0.5 cm from the skin surface and excluding the chest wall and pectoralis muscle. The PTV is at least a 1 cm uniform expansion of the CTV. This provides a margin around the CTV to compensate for setup variability and breathing motion. As such, the exact magnitude of the PTV expansion should be influenced by treatment technique and institutional experience. While the PTV is the appropriate volume for defining the treatment beams, it is not appropriate for evaluating the dose coverage, as it will often extend outside the patient or into the lungs. Instead, dose is assessed using PTV_EVAL, defined as the PTV clipped to exclude the extent beyond the ipsilateral breast and within 5 mm of the skin, as well as to exclude the pectoralis muscle and chest wall (Figure 8–30).

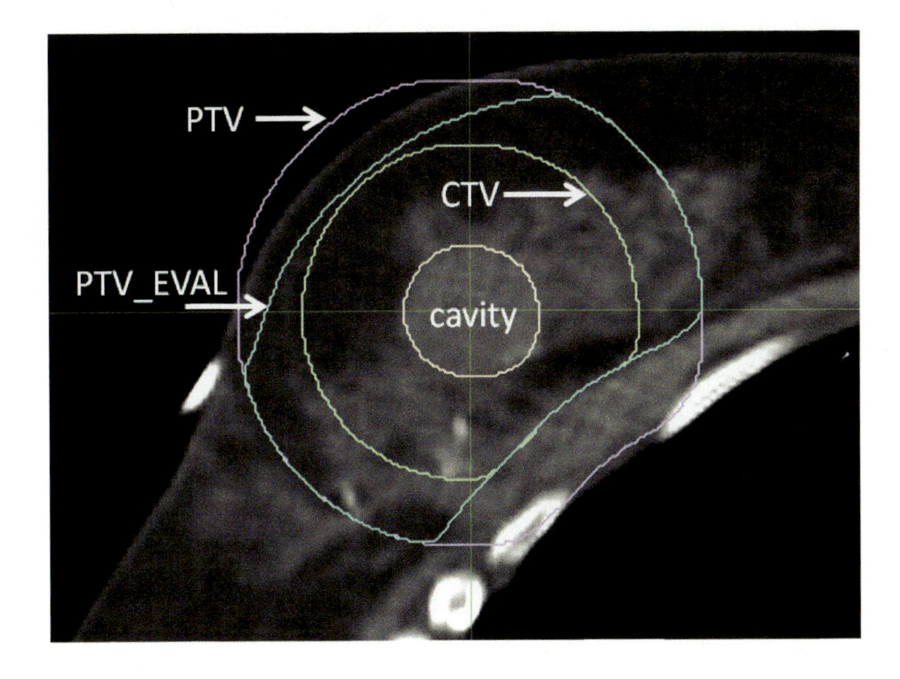

Figure 8–30 External beam radiotherapy APBI targets. Cavity = CT-defined lumpectomy cavity; CTV = cavity + 1.5 cm, excluding pectoralis muscles, chest wall, and 0.5 cm from skin; PTV = CTV + 1.0 cm; PTV_EVAL = PTV clipped to pectoralis muscle, chest wall, and 0.5 cm from skin.

In brachytherapy the definition of the CTV depends on the technique used. For balloon-based procedures, the CTV is defined as the breast tissue volume bounded by a uniform expansion of 1 cm from the balloon surface, less the balloon volume (Shah et al. 2013), while for the multi-catheter interstitial technique, the CTV is the lumpectomy cavity plus a uniform expansion of 1.0–1.5 cm (NSABP 2005; Vicini et al. 2016). In both cases, the CTV is clipped to exclude 0.5 cm from skin and the pectoralis muscle and chest wall. As the brachytherapy applicators move with the target, there is no need to compensate for breathing motion or setup variability. Therefore, for all brachytherapy techniques, CTV equals PTV (figures 8–31 and 8–32).

Regarding surrounding normal tissues, all APBI techniques should have the skin and ipsilateral breast delineated for planning and evaluation purposes. For EBRT, the contralateral breast, thyroid, lungs, and heart

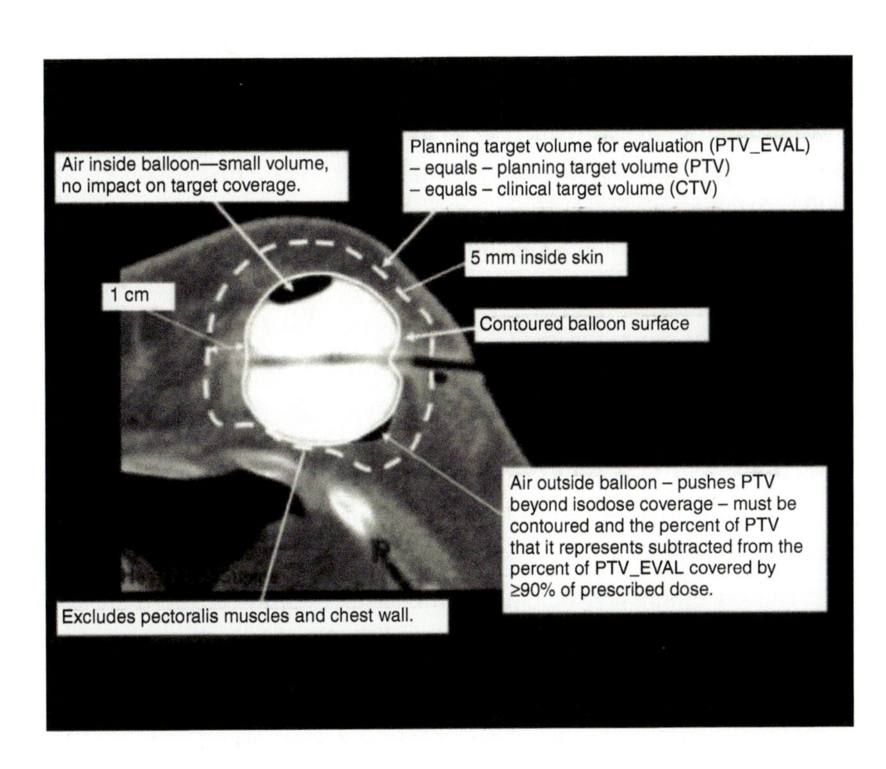

Figure 8–31 Balloon brachytherapy targets. Cavity = contoured balloon surface; CTV = 1 cm of tissue surrounding balloon, excluding the balloon itself, the pectoralis muscles, chest wall, and 0.5 cm from skin; PTV = CTV.

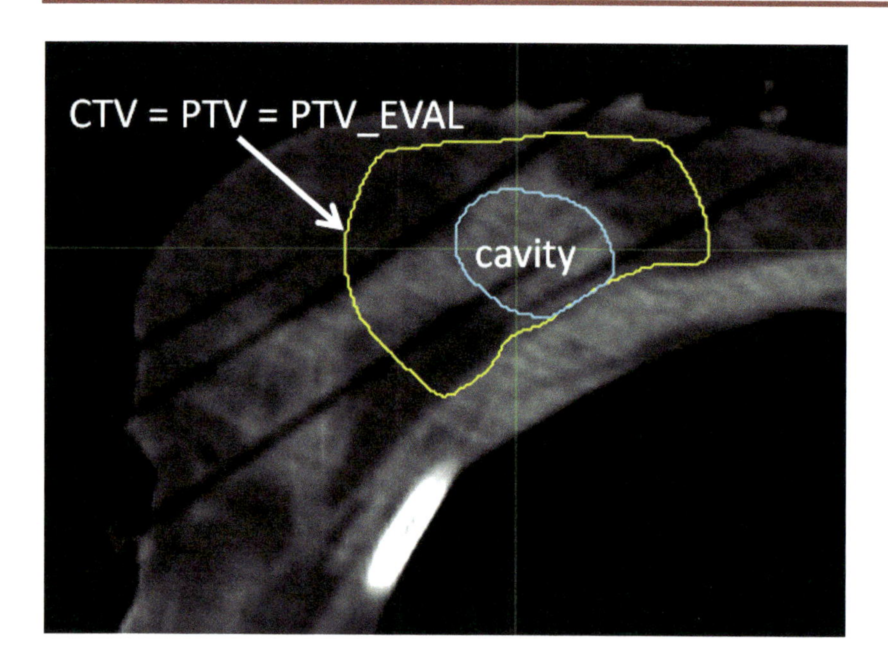

Figure 8–32 Multi-catheter interstitial brachytherapy targets. Cavity = CT defined lumpectomy cavity; CTV = cavity + 1.5 cm, excluding pectoralis muscles, chest wall, and 0.5 cm from skin; PTV = CTV = PTV_EVAL.

should also be contoured due to the wide variety of possible beam angles as well as the increased penetration of the treating x-rays relative to brachytherapy radionuclides.

8.5.2 NSABP B39/RTOG 0413 Recommendations vs GEC-ESTRO

A large amount of data published from several landmark randomized trials (National Institute of Oncology-Hungary; GEC-ESTRO; NABP-39) supports APBI as a viable, safe, and attractive treatment option in appropriately selected patients (Shah et al. 2011; Strnad et al. 2016; Polgár et al. 2010; Khan et al. 2012; Wazer et al. 2002; Arthur et al. 2003). Despite the potential benefits for patients, APBI adoption has been very slow (Wazer et al. 2002; Arthur et al. 2003). This could be because APBI using the interstitial brachytherapy technique used in several randomized trials (National Institute of Oncology; GEC-ESTRO) is a complex procedure (Lovey et al. 2007). In addition, early European APBI studies had less-than-satisfactory results because of inappropriate patient selection or suboptimal treatment techniques (Lovey et al. 2007; Polgár et al. 2007). However, after the introduction of intracavitary breast applicators, many centers in United States have adopted applicator-based brachytherapy to deliver APBI (Shah et. al. Mammosite Trial 2013; Wazer et al. 2002; Arthur et al. 2003; Shah et al. 2013; Zauls et al. 2012; Abbott et al. 2013). After the initial experience evaluating single-lumen applicators, multi-lumen and strut applicators have been developed which have been shown to improve target coverage and reduce dose to organs at risk, further improving outcomes (Khan et al. 2013).

As interest and utilization of APBI grew, several societies in addition to ASTRO have released consensus guidelines for APBI treatment off-protocol (Smith et al. 2009; Zauls et al. 2012; Haviland et al. 2013; Shah et al. ABS consensus Statement for APBI 2013; ASBS consensus Statement for APBI 2012). However, most of these studies have failed to show a correlation between local failure and consensus groupings (Husain et al. 2015). Nevertheless, since the publications of the ASTRO consensus guidelines, data has demonstrated an increase in the use of APBI for suitable candidates (Husain et al. 2015).

The key components of successful partial breast irradiation are patient selection, target delineation, technique, dosimetry, and quality assurance. Keeping this as the primary goal, two large, multi-institutional phase III trials—one in Europe (Phase III GEC-ESTRO trial) and another one in United States (Phase III NSABP B-39/RTOG 0413 trial)—were opened (GEC-ESTRO 2010; NSABP B-39/RTOG 0413 2011). They were both designed to definitively compare APBI with whole breast irradiation in a prospective randomized fashion and to define the role of APBI in the management of early stage breast cancer. There are many similarities between the two trials and a few subtle differences. The largest distinction between the NSABP B-39/RTOG 0413 and the GEC-ESTRO multicenter Phase III trial is the technique of delivering partial breast irradiation (figures 8–33 and 8–34) (GEC-ESTRO 2010; NSABP B-39/RTOG 0413 2011).

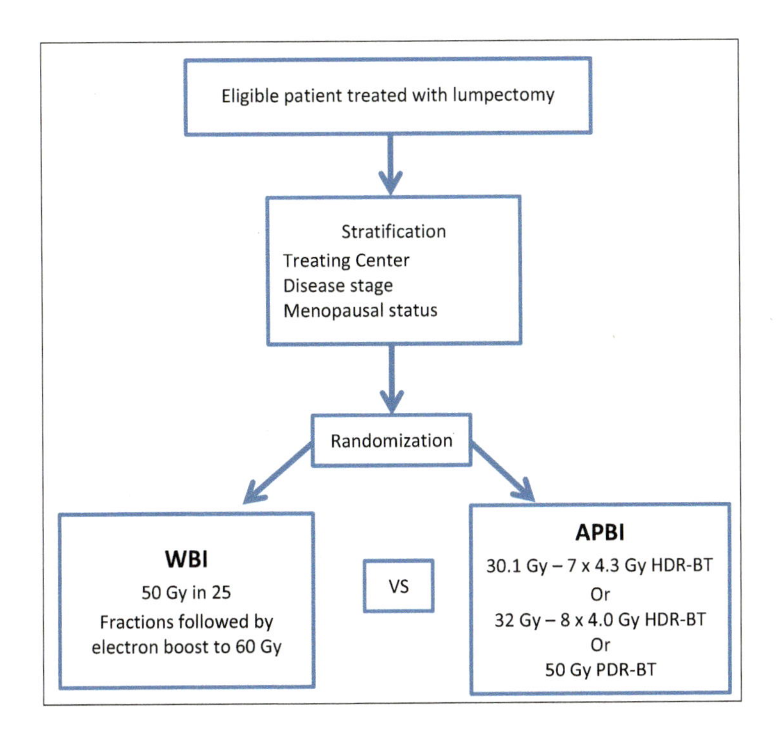

Figure 8–33 Groupe European de Curietherapie European Society for Therapeutic Radiology and Oncology GEC-ESTRO Multicenter Phase III Trial.

WBI = Whole Breast Irradiation
APBI = Accelerated Partial Breast Irradiation
HDR-BT = High Dose Rate Brachytherapy
PDR-BT = Pulsed Dose Rate Brachytherapy

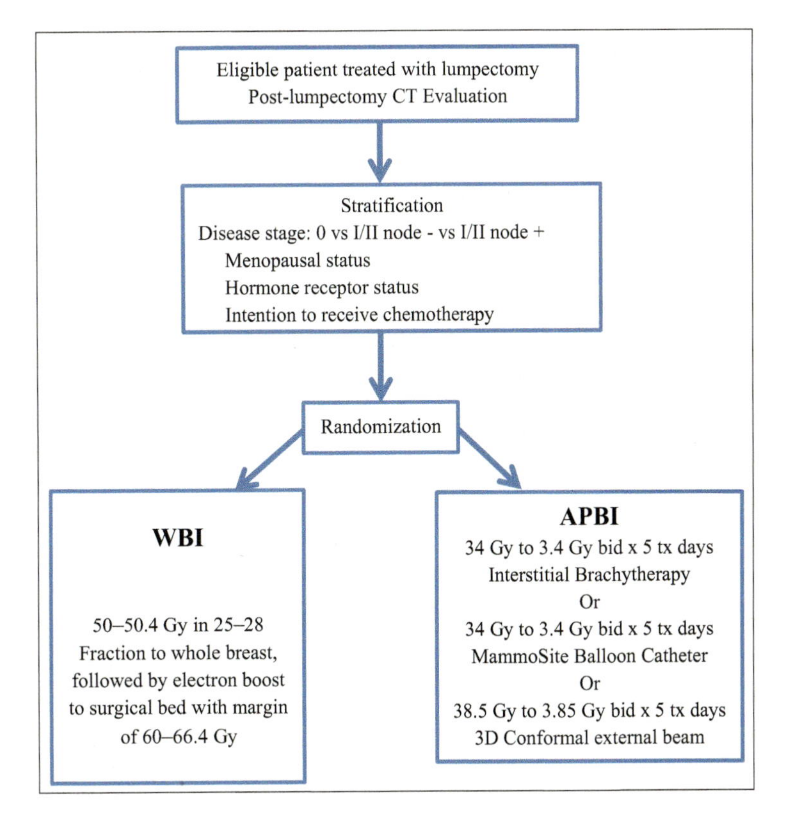

Figure 8–34 National Surgical Adjuvant Breast and Bowel Project B39 and Radiation Therapy Oncology Group 0413 (NSABP B39/RTOG 0413) protocol schema.

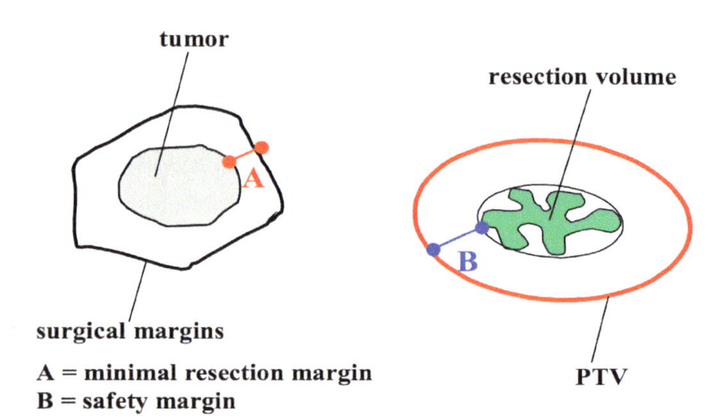

Figure 8–35 Schematic representation of the target volume. (GEC-ESTRO Multicenter Phase III Trial).

Both trials follow a strict QA in target delineation and treatment management. Since GEC-ESTRO (the Groupe Européen de Curiethérapie and the European SocieTy for Radiotherapy & Oncology) multicenter Phase III trial only allows interstitial APBI technique, it includes pre-implant PTV definition by surgical clips or pre-implant CT image-based pre-planning of the implant geometry (Figure 8–35) (GEC-ESTRO Multicenter Phase III Trial).

The NSABP B-39/RTOG 0413 trial includes intracavitary applicator brachytherapy, which is not as adaptable as multi-catheter brachytherapy. Applicator APBI can only treat safely to a nominal 1.0 cm distance from the balloon/applicator surface. Therefore, CTV is defined by identifying the limits of the lumpectomy cavity and then by expanding it to include a 1.0 cm margin. Due to the effect of applicator inflation in the lumpectomy cavity and subsequent compression of the breast tissue, investigators have reported the treatment distances to be between 1.0–1.5 cm from the balloon surface (Gregory et al. 2002).

More data is expected in the next decade from both the NSABP B-39/RTOG 0413 and the GEC-ESTRO multicenter Phase III trials. As data from these and other trials mature, they will hopefully support the implementation of APBI into routine clinical practice.

The CTV is defined as the excision cavity plus 2.0 cm margin minus the minimum clear pathological margin (Figure 8–36).

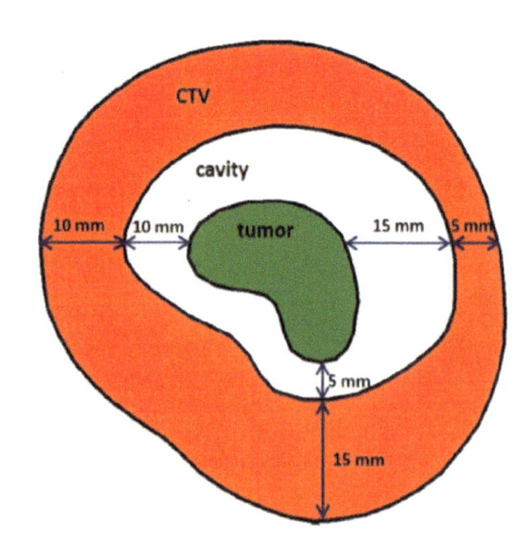

Tumour margins defined by preresection CT and post resection clips

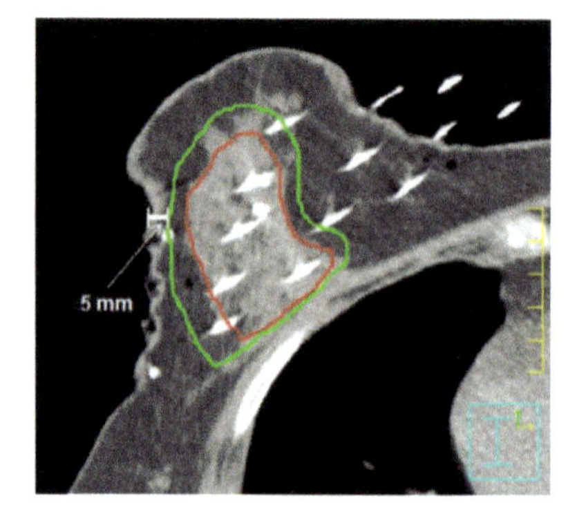

Limited to chest wall and 5mm from skin

Figure 8–36 Tumor margins defined by preresection CT and postresection clips. (GEC-ESTRO Multicenter Phase III Trial.)

8.6 Differences and Comparisons between Devices

As discussed, the important structures for contouring in APBI are the CTV, PTV, and PTV_EVAL, and that the definition of the CTV depends on the technique used. Based on the NSABP B-39/RTOG 0413 trial, CTV for balloon-based devices was defined as the breast tissue volume within a uniform expansion of 1.0 cm from the balloon surface (less the balloon volume) (Shah et al. 2013; Gregory 2002). Later, when the strut device SAVI was included, the same margin was used, but the surface of the balloon was replaced by the pseudo-surface defined by the struts. For multi-catheter interstitial implants and permanent interstitial implants using ^{103}Pd, the CTV is typically defined as the lumpectomy cavity plus a uniform expansion of 1.0–1.5 cm (RTOG 0413; Vinci et al. 2016; Pignol et al. 2006; Keller et al. 2012; Watt et al. 2015; Hilts et al. 2015a). For all brachytherapy methods discussed, the CTV is typically clipped to exclude 0.5 cm from the skin and chest wall, and there is no need to compensate for breathing motion or setup variability.

Several studies have compared the overall effectiveness and characteristics of the various breast brachytherapy techniques (Lu et al. 2012; Njeh et al. 2010; Hepel and Wazer 2012). Most often, these studies also include a comparison with photon- and electron-based EBRT techniques. When comparing or selecting a treatment modality for breast radiotherapy, important factors to consider are the dose coverage, quality of life, and cost effectiveness. According to a literature review by Njeh and colleagues, there was no difference between 3D-CRT and MammoSite (which were both better than interstitial), at the 90% coverage level, yet 3D-CRT resulted in better PTV coverage compared with MammoSite or MIB. However, this improved coverage with 3D-CRT came at a cost of a higher integral dose to the remaining normal breast tissue. Furthermore, quality-of-life studies indicated that patients receiving brachytherapy-based APBI were happy with the cosmetic outcome, compared with whole-breast radiation therapy. When selecting a treatment method, brachytherapy techniques can be preferred over EBRT as they directly target the tumor bed and do not require large PTV margins (compared with EBRT) to account for uncertainties in targeting due to respiration (Hepel and Wazer 2012). In addition, small-breasted patients may be better suited for IORT, while larger breasted patients may be best treated with balloon-based brachytherapy techniques (Njeh et al. 2010). Hepel and Wazer compared the various breast brachytherapy techniques and offered a table comparing the advantages of each (Table 8–3). The largest body of clinical knowledge exists for MIB techniques regarding local control and normal tissue toxicities. MIB is the most flexible treatment option to conform to the sometimes-complex treatment geometry. However, MIB demands a specialized expertise, and since it requires the placement of multiple catheters, it may not be acceptable for some patients. Balloon-based brachytherapy is considered to be simpler, both technically and dosimetrically. Using a single-catheter entry, balloon placement is often easier for the patient and clinician. However, it has been noted that an increased risk of infection is present with balloon-based techniques, and this is countered with increased antibiotics. Balloon-based methods perform well when the tumor cavity conforms well to the geometry, but lacks flexibility in shaping the dose if this is not the case or if the skin or chest wall are in close proximity to the balloon surface. Based on these drawbacks, multi-lumen balloons were developed to improve dose conformality and offer the same placement advantages as single-channel devices. Recent publications have focused on the differences and potential advantages of the newer multi-lumen devices, as the inherent dose distributions, and thus indication for use, differ based on their design (Figure 8–37). As described by Lu et al., the tissue coverage capabilities of the Mammosite and SAVI were similar based on the stretching of the surrounding tissue when expanded, indicating that a milder degree of tissue distortion may be produced with the smaller intracavitary devices. As a result, the feasibility of achieving 1.0 cm tissue margins using a free-margin technique with the SAVI and Contura was investigated by Lu and colleagues for patients considered to be high-risk, such as those with close or positive margins and patients fitted with smaller intracavitary brachytherapy devices (Lu et al. 2012). In addition, the ability of the SAVI and Contura to conform to asymmetric margins was assessed.

Table 8–3 Comparison of the current available APBI techniques (MIB = multicatheter interstitial brachytherapy, IORT = intraoperative radiation therapy, RCT = randomized clinical trials, OAR = organ at risk) From Hepel and Wazer (2012)

	MIB	Balloon-based Brachytherapy			Hybrid-based Brachytherapy		External Beam			IORT	
		Mammosite	Axxent Electronic	Contura	SAVI	ClearPath	Photons	Electrons	Protons	Electrons	Photons
Prescription Point	1.5–2 cm	1 cm	1 cm	1 cm	1 cm	1 cm	1.5–2 cm	1.5–2 cm	1.5–2 cm	10–30 mm	2 mm
Coverage of Target Volume	Variable	Good	Good	Good	Good	Good	Best	Good	Best	Good	Good
Dose Homogeneity	Fair	Fair	Fair	Fair	Fair	Fair	Best	Fair	Best	Fair	Fair
Sparing of OAR	Good	Good	Better	Better	Better	Better	Least	Varies	Good	Good	Best
Skin Dose	Least	Variable	Variable	Variable	Variable	Variable	Least	Maximum	Least	Least	Least
Expertise Required	High	Average	Average	Average	Average	Average	Average	Least	High	Very High	High
Suitability for Various Tumor Size, Location, and Shape	Not suitable if inadequate tissue or near axilla	Not suitable for large/ irregular cavities or at the periphery	Not suitable large cavities	Not suitable large cavities	Not suitable large cavities	Not suitable large cavities	May not be suitable for small breast	Not suited for deep-seated cavities in large breast	Superficial tumor	Not suitable for tumors near brachial plexus/axilla or skin	Not suitable for large irregular cavities or at the periphery of breast
Potential for Widespread Use	Fair	Very good	Very good	Very good	Very good	Very good	Very good	Very good	Limited	Limited	Fair
Clinical Outcome Data	11 years case studies	5 years case studies	None	Limited	Limited	None	4.5 years case studies	8 years case studies	Limited	4 years case studies	4 years RCT
Main Drawback	High expertise required and QA	Stringent QA is required. Cavity shape and size	Cavity shape and size	Cavity shape and size	Treatment planning complex	Treatment planning complex	Setup and breathing errors	High skin dose	Expensive and 2nd neutrons	Pathology not available	Pathology not available

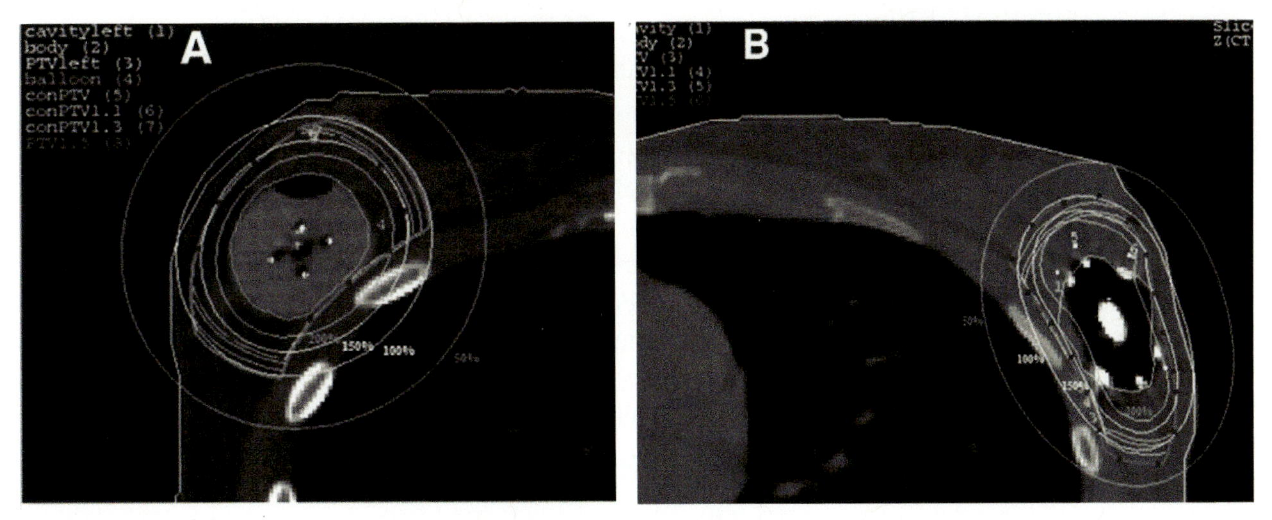

Figure 8–37 Example dose distribution for A) Contura and B) SAVI breast brachytherapy applicators. (Image courtesy of Lu et al. 2012 with permission from Elsevier.)

One of the main conclusions from this study demonstrated that the SAVI and Contura could treat more than a 1.0 cm tissue margin, but that the Contura could not maintain the published dosimetry guidelines for V_{200} when covering margins >1.1 cm. Regarding the ability to treat asymmetrically, the SAVI and Contura both possess this capability, but the SAVI demonstrated more dose flexibility than the Contura.

8.7 Dose, Fractionation, and Homogeneity

8.7.1 HDR Brachytherapy

The most commonly used dose prescription in HDR APBI is 3.4 Gy per fraction for 10 fractions, given twice a day separated by at least 6 hours. The treatment should be given over a period of 5 to 10 days. This treatment regimen was used in the NSABP B-39 protocol (RTOG 2005). Other HDR protocols utilized include 4 Gy twice daily for 8 fractions and 4.3 Gy twice daily for 7 fractions (Strauss and Dickler 2009). The prescribed dose is defined as minimum dose to the periphery of the CTV. Most often, volumetric constraints are used, initially with $V_{90\%} > 90\%$PD, later $V_{95\%} > 95\%$PD. Unlike EBRT, the goal of a brachytherapy dose distribution is not necessarily uniform dose throughout the target volume. Due to the nature of the dose falloff from the source itself, brachytherapy dose distributions are usually quite heterogeneous, and most dose prescriptions are, in fact, prescriptions of a minimum peripheral dose (MPD). For a balloon-based treatment, Dickler and colleagues (2004) used a 6-point dose technique, but the points are technically on the periphery of CTV.

The dose homogeneity requirements depend on the applicator type used. For multi-catheter or multi-channel brachytherapy, more optimization can be used to modulate the dwell times near the skin, so tighter skin dose constraints can be met; consequently, the surface dose should typically not exceed the prescribed dose. The DHI should be determined and volumes of regions receiving more than the prescribed dose calculated: particularly $V_{150\%}$ and $V_{200\%}$. The volume of tissue receiving more than 150% of the prescribed dose ($V_{150\%}$) should be $\leq$70 cm^3, and the volume of tissue receiving more than 200% of the prescribed dose ($V_{200\%}$) should be $\leq$20 cm^3. Recently, these limits were updated by an ABS panel of experts to 45 cm^3 and 14 cm^3, respectively (ABS Consensus Report for Interstitial partial breast irradiation, in press). The DHI should be $\geq$0.75 according to the protocol standard, but a value greater than 0.85 is preferred (NSABP B39/RTOG 0413). These constraints reduce the hot spots within the irradiated volume and strive to achieve limited fat necrosis, fibrosis, and telangiectasia, which are related to higher doses. Cuttino and colleagues (2005) reported an average DHI of 0.82 with a CT-guided, multi-catheter technique.

For the single-lumen MammoSite device, the protocol constraints require $V_{150\%}$ ≤50 cm^3 and $V_{200\%}$ ≤10 cm^3. Tighter constraints in the normal tissue are required for balloon-based devices because a large portion of the high dose should be within the balloon itself and excluded during the creation of the CTV. The skin dose will be much higher, however, since the use of a single lumen prevents much optimization. The desired skin spacing for MammoSite delivery is greater than 0.7 cm, although 0.5 cm is acceptable if the maximum skin dose is ≤145% of the prescribed dose. For a multi-lumen, single-entry balloon device such as the Contura, the dosimetric goals can be slightly tighter on the skin due to the ability to off-load catheters near the skin surface. In a Phase IV trial, Arthur and colleagues used a maximum skin dose of 125% of the prescribed dose, with the same $V_{150\%}$ and $V_{200\%}$ constraints (Arthur et al. 2011). The SAVI applicator can also be used with minimal skin doses and $V_{150\%}$ and $V_{200\%}$ with volumes less than 30 cm^3 and 15 cm^3, respectively (Yashar et al. 2011).

Other treatment regimens that have been successful with interstitial brachytherapy include the William Beaumont protocol of 32 Gy delivered in 8 fractions of 4 Gy. The implant treated the lumpectomy cavity plus a 1 or 2 cm margin (Baglan et al. 2001). The Hungarian experience delivered a total dose of 30.3 Gy or 36.4 Gy in 7 fractions over 4 days (Polgár et al. 2007).

Future directions are leaning toward shorter and shorter treatment regimens. Four-year results were recently presented with successful treatment and toxicity levels of HDR APBI in 28 Gy given in 4 fractions over 2 days (Wilkinson et al. 2012). Currently, the phase 2 TRIUMPH-T trial is accruing patients for treatment with 3 fractions of 7.5 Gy each (NCI et al. 2015).

8.7.2 Other Types of Brachytherapy

Other techniques of delivery have a widely variable dose prescription depending on the mechanism of dose delivery. Permanent low-dose-rate implants with ^{103}Pd give 90 Gy in a single implant (Pignol et al. 2006). The AccuBoost system provides either conventional APBI (34 Gy in 10 fractions) or a tumor bed boost to patients after completion of whole breast brachytherapy (Hepel et al. 2014). The prescriptions for the boost technique are typically 2 Gy per fraction for 5 to 8 fractions (Kuruvilla et al. 2016). LDR interstitial brachytherapy techniques have widely variable prescriptions, but range from 45 to 60 Gy given in 4 to 5 days (Lawenda et al. 2003; Fentiman et al. 1991). HDR single-fraction intraoperative breast brachytherapy has been given in a single fraction of 12.5 Gy (Trifiletti et al. 2015). Finally, electronic brachytherapy has been used for 20 Gy in a single fraction to the surface of the balloon in the intraoperative setting (Hanna et al. 2015).

8.8 Outcome Measures and Toxicity Analysis

8.8.1 HDR Brachytherapy

HDR breast brachytherapy has been shown to be a technique that is comparable to EBRT in terms of loco-regional control with comparable toxicity profiles. Initially, the APBI technique was criticized due to concerns about increased mastectomy rates and potential infections. However, prospective clinical trials have consistently found the clinical outcome and toxicity results of APBI are similar or superior to WBI (Polgár et al. 2007; Presley et al. 2012). Since APBI is a relatively new technique, long-term follow-up data is limited. However, some trials and registries are now reporting 5-year data, and the most mature data has 10 years of follow-up. Subsequently, while APBI loco-regional control and acute toxicity data can be compared to other techniques, results are sparse for distant metastatic rates. However, in a small matched-pair analysis, APBI was found to have a distant metastatic rate of approximately half that of WBI (4.5% vs 10.1%, respectively) (Shah et al. 2011).

Shah and colleagues (2013) recently made a final report from the MammoSite Breast Brachytherapy Registry (Phase IV post-marketing approval trial) in which 1,449 patients were enrolled. This trial represents the largest series with the longest follow-up in balloon applicator-based APBI with the registry collecting data from May 2002 through July 2004. Their reported outcomes indicated a 5-year actuarial rate of 3.8% for ipsilateral breast recurrence (IBTR) and a 5.8% 7-year rate. The rate of distant metastases was 3.1% at 6 years, and disease-free survival, cause-specific survival, and overall survival were consistent with previous reports of 83.0%, 98.2%, and 90.4%, respectively. Physician-assessed cosmetic results at 5 years was good or excellent for over 90% of patients and reported fat necrosis rates of 2.5%. Telangiectasia was reported in 13% of cases. Dragun and colleagues (2006) reported on the cosmesis in a small study of 100 patients implanted with the MammoSite applicator, and they also found good or excellent rates in 90%. They found the skin-to-balloon distance was a statistically significant indicator of acute skin toxicity and recommended >0.8 cm spacing.

Multi-lumen APBI applicators have also successfully been used for APBI. The multi-institutional Phase 4 trial results reported by Cuttino and colleagues (2014) using the Contura balloon showed 10 patient failures out of a 342-patient cohort with a 3-year medial follow-up. Additionally, they achieved an 88% good or excellent cosmetic result. Finally, they noted that at high-volume centers, 95% of patients had a good or excellent cosmetic result. Yashar and colleagues (2016) recently published a multi-institutional 5-year study result for 250 patients treated with a strut-based device. Following the established trend, their five-year IBTR rate was 3.6% and cosmetic result was 85.9%. As noted in their publication, 10 out of 11 institutions had cosmetic results higher than 96%, with one institution reporting 57.9%. While this cosmetic result is lower than other publications, it may be considered that multi-lumen devices are used with smaller skin spacing and potentially more challenging treatment areas.

Multi-catheter interstitial breast brachytherapy studies report similar data. The Pooled Registry of Multicatheter Interstitial Sites (PROMIS) reports a 10-year actuarial rate of regional failure at 2.3% and distant metastasis at 3.8% (Kamrava et al. 2015). Cosmesis was reported as good or excellent in 84% of patients with more than 5 years of follow-up. Polgár, et al. (2007) reported significantly better cosmetic outcome in a randomized trial with HDR interstitial multi-catheter treatment compared to WBI with a good or excellent result at 81.2% vs 62.9%, respectively. In a matched-pair analysis with 10 years of follow-up, Antonucci and

Table 8–4 Summary findings of IBTR, distant metastasis, and cosmetic results (G = good and E = excellent)

Brachytherapy Device	Authors	Metric % (Median Follow-up in Years)		
		IBTR	Distant Mets	Cosmesis G/E
MammoSite	Shah et al. (2013)	3.8% (5)	3.1% (6)	90.6% (7)
	Dragun et al. (2006)			90 (2)
Interstitial	Kamrava et al. (2015)	7.6% (10)	3.8% (10)	84 (5+)
	Polgar et al. (2007)	4.7%		81.2 (5)
	Antonucci et al. (2009)	5% (10)		
	Arthur et al. (2011)	3% (5)		
Multi-lumen	Cuttino et al. (2014)	3% (3)		88 (3)
	Yashar et al. (2016)	3.6% (5)		85.9 (5)

colleagues (2009) found no statistical difference between the rates of Ipsilateral Breast Tumor Recurrence (IBTR) rates for APBI vs WBI.

A summary of IBTR in patients treated with HDR brachytherapy, distant metastasis, and cosmetic results from the references cited here are shown below. An excellent summary describing the location of and patterns of failure is given by Skowronek and colleagues (2012). Finally, a complete summary of trials, techniques, and findings was published in a 2013 consensus statement by the ABS in the support of APBI (Table 8–4) (Shah et al 2013).

8.8.2 LDR and Other Brachytherapy Techniques

Five-year data from permanent LDR ^{103}Pd implants has been reported with excellent results. The overall survival rate was 97.4% and the disease-free survival rate was 96.4%. Some skin toxicity and telangiectasia was reported at levels comparable to WBI (Pignol et al. 2015). This remains a promising alternative for patients who are ineligible for other APBI protocols or who desire more compact treatment schedules.

The University of Virginia group has pioneered an intraoperative HDR radiotherapy technique delivering 12.5 Gy in a single fraction with a multi-lumen balloon. In their Phase 1 trial, they reported no Grade 3 toxicity, and 21% had Grade 2 events. A reported 93% had good or excellent cosmetic result at their last follow-up, and the study is being evaluated for Phase 2 (Showalter et al. 2016). Electronic brachytherapy has also been used in the intraoperative setting. One-year follow-up results from Syed and colleagues 2016) showed Grade 2 or greater side effects in approximately 15% of patients, but only a single serious side effect (1/100).

8.9 Uncertainties in Planning and Delivery, QA for Delivery

Brachytherapy processes carry an uncertainty that limits the accuracy with which dose is delivered to the target and normal structures are protected. The limit of accuracy for the dose calculation was assessed by DeWerd and colleagues in the AAPM Task Group 138 report (DeWerd et al. 2011). They found that likely uncertainty for an HDR ^{192}Ir source using the dosimetry protocol of TG-43 was about 6.8% for two standard deviations (Nath et al. 1995; Rivard et al. 2004). It is not well established how the use of model-based dose calculation algorithms changes that uncertainty (Beaulieu et al. 2012). In addition to the dose-calculation uncertainties, all the rest of the actions in the treatment process add to this uncertainty. Kirisits and colleagues performed a thorough review of the sources of uncertainty (discussed below) and typical resultant values (Kirisits et al. 2014). For a typical partial-breast treatment using a balloon-based catheter, they estimated the dose uncertainty at 26% at two standard deviations. The uncertainties for an interstitial implant approach would likely be slightly less. The interested reader is directed to these publications for the deeper details of their analyses (DeWerd et al. 2011; Kirisits et al. 2014). The calculations performed in both works assumed execution of all the steps involved in the treatment and dose calculation as the state of the art allows. The use of older equipment, factors, or techniques—or the application of state-of-the-art treatments but performed poorly—increases the uncertainties. At some point, poor performance moves from increasing uncertainty to qualify as error. Many factors affect the switch and, in fact, there is a large gray zone over which that happens. Often, however, an error, rather than just poor performance, creates large deviations. Below are some ways that the uncertainty could increase.

1. Causes of increased uncertainties in treatment planning

   Uncertainties in planning can increase as the result of:

   - Inadequate imaging. This can be associated with the choice of imaging (modality, technique, speed etc.).
   - Inappropriate slice thickness for accurate target and normal-structures contouring as their geometry vary from one slice to another.

- Poor patient setup. A patient asked to remain still during imaging acquisition in an uncomfortable position will most likely result in poor images due to motion.
- Inappropriate motion management. The patient respiratory phase used for planning might not also coincide with the one used for treatment, as the patient becomes more comfortable and relaxed.
- Suboptimal image display. The choice of window and level selected for planning can introduce some uncertainties in the volumes of target and other structures.
- Poor contouring. Variation in contouring between individuals, or for an individual at different sittings, results in treatment uncertainties (Landis et al. 2007). Catheter reconstruction depends on parameters such as the slice thickness, catheter curvatures, and the nature of the markers used for reconstruction. A moving source within a catheter might not travel along the same path as a fixed catheter marker. In similar fashion, when markers are not used, choosing the center of the catheter as the path of the source within the catheter (air in this case) can be erroneous and will introduce similar uncertainties. See Section 2.7.1 in this text.

These uncertainties in catheter reconstruction will result in uncertainties on the activation points or dwell points, which will result in an inaccurate dose distribution.

2. Uncertainties in treatment delivery

    The uncertainties in treatment delivery can be associated with three factors: patient, treatment device, and image verification. Unless a reference image was created at the time of image planning with positioning of the patient and applicator as per the treatment, there can be no verification that the application will follow the plan and the calculated dose (except in simple surface applications). Applicator shift or rotation, changes within the target and its surrounding (breast cavity, increase or reduction of seroma, air, swelling, etc.) could introduce other uncertainties. In breast brachytherapy, these changes could lead to rib and skin dose, as an example, different than the expected planned ones.

3. Recommended QMP for treatment delivery

    The purpose of the QC for treatment delivery is to ensure that the treatment is delivered according to the treatment plan—including treating the correct patient, using the correct device, and setup—as verified with pretreatment imaging. To implement a quality management program to include these items, the user should follow the TG-100 approach to identify the key areas where intervention and execution is recommended to prevent deviation from the planning dose at every stage during their treatment delivery process. While there are items that are common for breast brachytherapy, there are some specifics associated with each particular device or technique being used. It is assumed in this discussion that the user performs the required quality checks on the treatment unit and follows routine good practices, such as identifying the patient in the correct manner, for example in Thomadsen et al. (2013).

    The events reports in the database of the U.S. Nuclear Regulatory Commission involving breast brachytherapy give some insight on the types of errors that can happen (as analyzed by the Advisory Committee on the Medical Uses of Isotopes of the US Nuclear Regulatory Commission in 2013). Most of the events could occur in any type of brachytherapy treatment, such as treating the wrong patient or errors in calibration of the source strength, and are not listed here. The remaining events also mostly include problems that occur in many types of brachytherapy, but are common for breast brachytherapy:

- position errors due to erroneous measurement of the length of applicators
- using the wrong version of the plan
- loss of fluid from balloon applicators through several different errors
- digitization of the applicator in the opposite direction from that expected
- per-fraction dose divided by the number of fractions instead of the total dose
- applicator failure due to the use of a damaged applicator

Most of these failures could have been intercepted by QC. While they can provide information on possible risks, quality management activities should not be particularly directed to address the failures that happened. The number of events are few compared with the number of treatments delivered, so basing a QMP on these few data points would bias resources to target these particular types of failures at the expense of preventing all others.

8.10 Conclusions

Breast brachytherapy, and in particular partial breast brachytherapy and its "accelerated" version, is a mature treatment modality. The final results of the largest clinical trial (NSABP B-39/RTOG 0435) set to establish the equivalence of whole breast irradiation with partial breast irradiation are still pending, but preliminary results show that the two modalities are indeed equivalent, both in terms of control and toxicity. While the interstitial multi-catheter implant has the longest follow-up and is probably the most flexible of all techniques available, the field is still developing, both in terms of emergence of new devices (e.g., AccuBoost, Best Medical double balloon) and new treatment paradigms (e.g., intra-operative irradiation and pre-operative irradiation).

References

Abbott, A. M., P. R. Portschy, C. Lee, C. T. Le, L. K. Han, T. Washington, M. Kinney, M. Bretzke, and T. M. Tuttle. (2013). "Prospective multicenter trial evaluating balloon-catheter partial-breast irradiation for ductal carcinoma in situ." *Int. J. Radiat. Oncol. Biol. Phys.* 87(3):494–98. doi: 10.1016/j.ijrobp.2013.06.2056.

American Society of Breast Surgeons. (2012). "Consensus Statement for Accelerated Partial Breast Irradiation." https://www.breastsurgeons.org/statements/PDF_Statements/APBI.pdf.

Antonucci, J. V., M. Wallace, N. S. Goldstein, L. Kestin, P. Chen, P. Benitez, N. Dekhne, A. Martinez, and F. Vicini. (2009). "Differences in patterns of failure in patients treated with accelerated partial breast irradiation versus whole-breast irradiation: a matched-pair analysis with 10-year follow-up." *Int. J. Radiat. Oncol. Biol. Phys.* 74(2):447–52. doi: 10.1016/j.ijrobp.2008.08.025.

Arthur, D. W., F. A. Vicini, D. A. Todor, T. B. Julian, and M. R. Lyden. (2011). "Improvements in critical dosimetric endpoints using the Contura multilumen balloon breast brachytherapy catheter to deliver accelerated partial breast irradiation: preliminary dosimetric findings of a phase iv trial." *Int. J. Radiat. Oncol. Biol. Phys.* 79(1):26–33. doi: 10.1016/j.ijrobp.2009.10.025.

Arthur, D. W., D. Koo, R. D. Zwicker, S. Tong, H. D. Bear, B. J. Kaplan, B. D. Kavanagh, L. A. Warwicke, D. Holdford, C. Amir, K. J. Archer, and R. K. Schmidt-Ullrich. (2003). "Partial breast brachytherapy after lumpectomy: Low-dose-rate and high-dose-rate experience." *Int. J. Radiat. Oncol. Biol. Phys.* 56(3):681–89. doi: 10.1016/S0360-3016(03)00120-2.

Baglan, K. L., A. A. Martinez, R. C. Frazier, V. R. Kini, L. L. Kestin, P. Y. Chen, G. Edmundson, E. Mele, D. Jaffray, and F. A. Vicini. (2001). "The use of high-dose-rate brachytherapy alone after lumpectomy in patients with early-stage breast cancer treated with breast-conserving therapy." *Int. J. Radiat. Oncol. Biol. Phys.* 50(4):1003–11.

Batchelar, D., M. Hilts, T. Rose, M. Brandel, E. Garcia, J. Yanchuk, and J. M. Crook. (2014). "Simulation and intraoperative checks for improved standardization and reproducibility of partial breast seed implant techniques." *Brachytherapy* 13(S1):S84.

Beriwal, S., D. Coon, H. Kim, M. Haley, R. Patel, and R. Das. (2008). "Multicatheter hybrid breast brachytherapy: a potential alternative for patients with inadequate skin distance." *Brachytherapy* 7(4):301–04. doi: 10.1016/j.brachy.2008.07.003.

Cheng, C. W., R. Mitra, X. A. Li, and I. J. Das. (2005). "Dose perturbations due to contrast medium and air in mammosite treatment: an experimental and Monte Carlo study." *Med. Phys.* 32(7):2279–87. doi: 10.1118/1.1943827.

Cuttino, L. W., D. W. Arthur, F. Vicini, D. Todor, T. Julian, and N. Mukhopadhyay. (2014). "Long-term results from the Contura multilumen balloon breast brachytherapy catheter phase 4 registry trial." *Int. J. Radiat. Oncol. Biol. Phys.* 90(5):1025–29. doi: 10.1016/j.ijrobp.2014.08.341.

Cuttino, L. W., D. Todor, and D. W. Arthur. (2005). "CT-guided multi-catheter insertion technique for partial breast brachytherapy: reliable target coverage and dose homogeneity." *Brachytherapy* 4(1):10–17.

Dickler, A., M. Kirk, J. Choo, W. C. Hsi, J. Chu, K. Dowlatshahi, D. Francescatti, and C. Nguyen. (2004). "Treatment volume and dose optimization of MammoSite breast brachytherapy applicator." *Int. J. Radiat. Oncol. Biol. Phys.* 59(2):469–74. doi: 10.1016/j.ijrobp.2003.10.046.

Dickler, A., N. Seif, M. C. Kirk, M. B. Patel, D. Bernard, A. Coon, K. Dowlatshahi, R. K. Das, and R. R. Patel. (2009). "A dosimetric comparison of MammoSite and ClearPath high-dose-rate breast brachytherapy devices." *Brachytherapy* 8(1):14–18. doi: 10.1016/j.brachy.2008.07.006.

Dragun, A. E., J. L. Harper, J. M. Jenrette, D. Sinha, and D. J. Cole. (2007). "Predictors of cosmetic outcome following MammoSite breast brachytherapy: a single-institution experience of 100 patients with two years of follow-up." *Int. J. Radiat. Oncol. Biol. Phys.* 68(2):354–58. doi: 10.1016/j.ijrobp.2006.12.014.

Edmundson, G. K., F. A. Vicini, P. Y. Chen, C. Mitchell, and A. A. Martinez. (2002). "Dosimetric characteristics of the MammoSite RTS, a new breast brachytherapy applicator." *Int. J. Radiat. Oncol. Biol. Phys.* 52(4):1132–39.

Fentiman, I. S., C. Poole, D. Tong, P. J. Winter, H. M. Mayles, P. Turner, M. A. Chaudary, and R. D. Rubens. (1991). "Iridium implant treatment without external radiotherapy for operable breast cancer: a pilot study." *Eur. J. Cancer* 27(4):447–50.

Francescatti, D., O. Ivanov, and A. Dickler. "Single Fraction Breast IORT using Xoft Electronic Brachytherapy." 6th International Conference of the ISIORT, October 14–16, 2012.

Gurdalli, S., R. R. Kuske, C. A. Quiet, and M. Ozer. (2011). "Dosimetric performance of Strut-Adjusted Volume Implant: a new single-entry multicatheter breast brachytherapy applicator." *Brachytherapy* 10(2):128–35. doi: 10.1016/j.brachy.2010.03.002.

Hamid, S., K. Rocchio, D. Arthur, R. Vera, S. Sha, M. Jolly, S. Cavanaugh, E. Wooten, R. Benda, B. Greenfield, B. Prestidge, S. Ackerman, R. Kuske, C. Quiet, M. Snyder, and D. E. Wazer. (2012). "A multi-institutional study of feasibility, implementation, and early clinical results with noninvasive breast brachytherapy for tumor bed boost." *Int. J. Radiat. Oncol. Biol. Phys.* 83(5):1374–80. doi: 10.1016/j.ijrobp.2011.10.016.

Hanna, N. M., A. N. Syed, H. Chang, B. Schwartzberg, A. K. Bremner, A. Bhatnagar, C. Lopez-Penalver, C. A. Vito, O. Ivanov, S. Rahman, and S. L. Golder. (2015). "Feasibility and early outcomes of a multi-center trial of intra-operative radiation therapy using electronic brachytherapy at the time of breast conservation surgery for early stage breast cancer." *Brachytherapy* 14:S41–S42.

Haviland, J. S., J. R. Owen, J. A. Dewar, R. K. Agrawal, J. Barrett, P. J. Barrett-Lee, H. J. Dobbs, P. Hopwood, P. A. Lawton, B. J. Magee, J. Mills, S. Simmons, M. A. Sydenham, K. Venables, J. M. Bliss, J. R. Yarnold, and Start Trialists' Group. (2013). "The UK Standardisation of Breast Radiotherapy (START) trials of radiotherapy hypofractionation for treatment of early breast cancer: 10-year follow-up results of two randomised controlled trials." *Lancet Oncol.* 14(11):1086–94. doi: 10.1016/S1470-2045(13)70386-3.

Hepel, J. T., J. R. Hiatt, S. Sha, K. L. Leonard, T. A. Graves, D. L. Wiggins, D. Mastras, A. Pittier, Group Brown University Oncology Research, and D. E. Wazer. (2014). "The rationale, technique, and feasibility of partial breast irradiation using noninvasive image-guided breast brachytherapy." *Brachytherapy* 13(5):493–501. doi: 10.1016/j.brachy.2014.05.014.

Hepel, J. T. and D. E. Wazer. (2012). "A comparison of brachytherapy techniques for partial breast irradiation." *Brachytherapy* 11(3):163–75. doi: 10.1016/j.brachy.2011.06.001.

Hilts, M., D. Batchelar, and J. Crook. (2015). "Post-implant dosimetry for recent implementation of permanent breast seed implants in british columbia shows a rapid learning curve." *Radiother. Oncol.* 116(1):S3.

Hilts, M., D. Batchelar, J. Rose, and J. Crook. (2015). "Deformable image registration for defining the postimplant seroma in permanent breast seed implant brachytherapy." *Brachytherapy* 14(3):409–18. doi: 10.1016/j.brachy.2014.11.003.

Hologic. 2014a. *MammoSite Radiation Therapy System, Instruction Manual.* MAN-01641-001 Rev. 005.

Hologic. 2014b. "MammoSiteML Radiation Therapy System, Instruction Manual." MAN-02212-001 Rev. 003.

Husain, Z. A., S. Lloyd, C. Shah, L. D. Wilson, M. Koshy, and U. Mahmood. (2015). "Changes in brachytherapy-based APBI patient selection immediately before and after publication of the ASTRO consensus statement." *Brachytherapy* 14(4):490–95. doi: 10.1016/j.brachy.2015.03.003.

Iftimia, I., M. Talmadge, R. Ladd, and P. Halvorsen. (2015). "Commissioning and quality assurance for the treatment delivery components of the AccuBoost system." *J. Appl. Clin. Med. Phys.* 16(2):5156. doi: 10.1120/jacmp.v16i2.5156.

Kamrava, M., R. R. Kuske, B. Anderson, P. Chen, J. Hayes, C. Quiet, P. C. Wang, D. Veruttipong, M. Snyder, and D. J. Demanes. (2015). "Outcomes of breast cancer patients treated with accelerated partial breast irradiation via multicatheter interstitial brachytherapy: The Pooled Registry of Multicatheter Interstitial Sites (PROMIS) experience." *Ann. Surg. Oncol.* 22 Suppl 3:S404–11. doi: 10.1245/s10434-015-4563-7.

Kassas, B., F. Mourtada, J. L. Horton, and R. G. Lane. (2004). "Contrast effects on dosimetry of a partial breast irradiation system." *Med. Phys.* 31(7):1976–79. doi: 10.1118/1.1763006.

Keller, B. M., J. P. Pignol, E. Rakovitch, R. Sankreacha, and P. O'Brien. (2008). "A radiation badge survey for family members living with patients treated with a (103)Pd permanent breast seed implant." *Int. J. Radiat. Oncol. Biol. Phys.* 70(1):267–71. doi: 10.1016/j.ijrobp.2007.08.006.

Keller, B. M., A. Ravi, R. Sankreacha, and J. P. Pignol. (2012). "Permanent breast seed implant dosimetry quality assurance." *Int. J. Radiat. Oncol. Biol. Phys.* 83(1):84–92. doi: 10.1016/j.ijrobp.2011.05.030.

Keller, B., R. Sankreacha, E. Rakovitch, P. O'Brien, and J. P. Pignol. (2005). "A permanent breast seed implant as partial breast radiation therapy for early-stage patients: a comparison of palladium-103 and iodine-125 isotopes based on radiation safety considerations." *Int. J. Radiat. Oncol. Biol. Phys.* 62(2):358–65. doi: 10.1016/j.ijrobp.2004.10.014.

Khan, A. J., F. A. Vicini, and D. Arthur. 2012. "Brachytherapy vs whole-breast irradiation for breast cancer." *JAMA* 308 (6):567; author reply 567-8. doi: 10.1001/jama.2012.8486.

Khan, A. J., F. A. Vicini, S. Brown, B. G. Haffty, T. Kearney, R. Dale, M. Lyden, and D. Arthur. (2013). "Dosimetric feasibility and acute toxicity in a prospective trial of ultrashort-course accelerated partial breast irradiation (APBI) using a multi-lumen balloon brachytherapy device." *Ann. Surg. Oncol.* 20(4):1295–301. doi: 10.1245/s10434-012-2671-1.

King, T. A., J. S. Bolton, R. R. Kuske, G. M. Fuhrman, T. G. Scroggins, and X. Z. Jiang. (2000). LLong-term results of wide-field brachytherapy as the sole method of radiation therapy after segmental mastectomy for T(is,1,2) breast cancer." *Am. J. Surg.* 180(4):299–304.

Kirk, M. C., W. C. Hsi, J. C. Chu, H. Niu, Z. Hu, D. Bernard, A. Dickler, and C. Nguyen. (2004). "Dose perturbation induced by radiographic contrast inside brachytherapy balloon applicators." *Med. Phys.* 31(5):1219–24. doi: 10.1118/1.1705445.

Kuruvilla, A., S. Paryani, N. Paryani, D. Simmons, N. Shah, and J. Caudill. (2016). "Partial Breast Irradiation for Boost Using Image-Guided Brachytherapy: Florida Community Experience and Review of the Literature." *Austin J. Radiat. Oncol. & Cancer* 2(1):1016.

Kuske, R. R., J. S. Bolton, and W. Hanson. "RTOG 95-17: A phase I/II trial to evaluate brachytherapy as the sole method of radiation therapy for stage I and II breast carcinoma." In *Radiation Therapy Oncology Group*, 1-34. Philadelphia: RTOG, 1998.

Landis, D. M., W. Luo, J. Song, J. R. Bellon, R. S. Punglia, J. S. Wong, J. H. Killoran, R. Gelman, and J. R. Harris. (2007). "Variability among breast radiation oncologists in delineation of the postsurgical lumpectomy cavity." *Int. J. Radiat. Oncol. Biol. Phys.* 67(5):1299–308. doi: 10.1016/j.ijrobp.2006.11.026.

Lawenda, B. D., A. G. Taghian, L. A. Kachnic, H. Hamdi, B. L. Smith, M. A. Gadd, T. Mauceri, and S. N. Powell. (2003). "Dose-volume analysis of radiotherapy for T1N0 invasive breast cancer treated by local excision and partial breast irradiation by low-dose-rate interstitial implant." *Int. J. Radiat. Oncol. Biol. Phys.* 56(3):671–80.

Lovey, K., J. Fodor, T. Major, E. Szabo, Z. Orosz, Z. Sulyok, L. Janvary, G. Frohlich, M. Kasler, and C. Polgar. (2007). "Fat necrosis after partial-breast irradiation with brachytherapy or electron irradiation versus standard whole-breast radiotherapy—4-year results of a randomized trial." *Int. J. Radiat. Oncol. Biol. Phys.* 69(3):724–31. doi: 10.1016/j.ijrobp.2007.03.055.

Lu, S. M., D. J. Scanderberg, P. Barna, W. Yashar, and C. Yashar. (2012). "Evaluation of two intracavitary high-dose-rate brachytherapy devices for irradiating additional and irregularly shaped volumes of breast tissue." *Med. Dosim.* 37(1):9–14. doi: 10.1016/j.meddos.2010.12.005.

Major, T., C. Gutierrez, B. Guix, E. van Limbergen, V. Strnad, C. Polgar, and Gec-Estro Breast Cancer Working Group. (2016). "Recommendations from GEC ESTRO Breast Cancer Working Group (II): Target definition and target delineation for accelerated or boost partial breast irradiation using multicatheter interstitial brachytherapy after breast conserving open cavity surgery." *Radiother. Oncol.* 118(1):199–204. doi: 10.1016/j.radonc.2015.12.006.

Monroe, J. I., J. F. Dempsey, J. A. Dorton, S. Mutic, J. B. Stubbs, J. Markman, and J. F. Williamson. (2001). Experimental validation of dose calculation algorithms for the GliaSite™ RTS, a Novel [sup 125]I liquid-filled balloon brachytherapy applicator." *Med. Phys.*28(1)73–85. doi: 10.1118/1.1334608.

Morton, D., M. Hilts, D. Batchelar, and J. Crook. (2016). "Seed Placement in Permanent Breast Seed Implant Brachytherapy: Are Concerns Over Accuracy Valid?" *Int. J. Radiat. Oncol. Biol. Phys.* 95(3):1050–57. doi: 10.1016/j.ijrobp.2016.01.049.

Nag, S., R. R. Kuske, F. A. Vicini, D. W. Arthur, and R. D. Zwicker. (2001). "Brachytherapy in the treatment of breast cancer." *Oncology (Williston Park)* 15(2):195–202, 205; discussion 205–7.

National Cancer Institute. (2015). "Accelerated Partial Breast Radiation Therapy Using High-Dose Rate Brachytherapy in Treating Patients with Early Stage Breast Cancer After Surgery (TRIUMPH-T)." https://clinicaltrials.gov/ct2/show/NCT02526498?term=NCT02526498&rank=1.

Njeh, C. F., M. W. Saunders, and C. M. Langton. (2010). "Accelerated Partial Breast Irradiation (APBI): A review of available techniques." *Radiat. Oncol.* 5:90. doi: 10.1186/1748-717X-5-90.

NSABP National Surgical Adjuvant Breast and Bowel Project. (2006). "B-39, RTOG 0413: A Randomized Phase III Study of conventional whole breast irradiation versus partial breast irradiation for women with stage 0, I, or II breast cancer." *Clin. Adv. Hematol. Oncol.* 4:719–721.

Oh, S., J. Scott, D. H. Shin, T. S. Suh, and S. Kim. (2009). MMeasurements of dose discrepancies due to inhomogeneities and radiographic contrast in balloon catheter brachytherapy." *Med. Phys.* 36(9):3945–54. doi: 10.1118/1.3183497.

Ott, O. J. et al. (2016). "GEC-ESTRO multicenter phase 3-trial: Accelerated partial breast irradiation with interstitial multi-catheter brachytherapy versus external beam whole breast irradiation: Early toxicity and patient compliance." *Radiother. Oncol.* 120(1):119–23,

Pignol, J. P., J. M. Caudrelier, J. Crook, C. McCann, P. Truong, and H. A. Verkooijen. (2015). "Report on the Clinical Outcomes of Permanent Breast Seed Implant for Early-Stage Breast Cancers." *Int. J. Radiat. Oncol. Biol. Phys.* 93(3):614–21. doi: 10.1016/j.ijrobp.2015.07.2266.

Pignol, J. P., B. Keller, E. Rakovitch, R. Sankreacha, H. Easton, and W. Que. (2006). "First report of a permanent breast ^{103}Pd seed implant as adjuvant radiation treatment for early-stage breast cancer." *Int. J. Radiat. Oncol. Biol. Phys.* 64(1):176–81. doi: 10.1016/j.ijrobp.2005.06.031.

Pinnaro, P., S. Arcangeli, C. Giordano, G. Arcangeli, F. A. Impiombato, V. Pinzi, G. Iaccarino, A. Soriani, V. Landoni, and L. Strigari. (2011). "Toxicity and cosmesis outcomes after single fraction partial breast irradiation in early stage breast cancer." *Radiat. Oncol.* 6:155. doi: 10.1186/1748-717X-6-155.

Polgar, C., J. Fodor, T. Major, G. Nemeth, K. Lovey, Z. Orosz, Z. Sulyok, Z. Takacsi-Nagy, and M. Kasler. (2007). "Breast-conserving treatment with partial or whole breast irradiation for low-risk invasive breast carcinoma—5-year results of a randomized trial." *Int. J. Radiat. Oncol. Biol. Phys.* 69(3):694–702. doi: 10.1016/j.ijrobp.2007.04.022.

Polgar, C., T. Major, J. Fodor, Z. Sulyok, A. Somogyi, K. Lovey, G. Nemeth, and M. Kasler. (2010). "Accelerated partial-breast irradiation using high-dose-rate interstitial brachytherapy: 12-year update of a prospective clinical study." *Radiother. Oncol.* 94(3):274–79. doi: 10.1016/j.radonc.2010.01.019.

Polgar, C., E. Van Limbergen, R. Potter, G. Kovacs, A. Polo, J. Lyczek, G. Hildebrandt, P. Niehoff, J. L. Guinot, F. Guedea, B. Johansson, O. J. Ott, T. Major, V. Strnad, and Gec-Estro breast cancer working group. (2010). "Patient selection for accelerated partial-breast irradiation (APBI) after breast-conserving surgery: recommendations of the Groupe Europeen de Curietherapie-European Society for Therapeutic Radiology and Oncology (GEC-ESTRO) breast cancer working group based on clinical evidence (2009)." *Radiother. Oncol.* 94(3):264–73. doi: 10.1016/j.radonc.2010.01.014.

Presley, C. J., P. R. Soulos, J. Herrin, K. B. Roberts, J. B. Yu, B. Killelea, B. A. Lesnikoski, J. B. Long, and C. P. Gross. (2012). "Patterns of use and short-term complications of breast brachytherapy in the national medicare population from 2008–2009." *J. Clin. Oncol.* 30(35):4302–07. doi: 10.1200/JCO.2012.43.5297.

Raffi, J. A., S. D. Davis, C. G. Hammer, J. A. Micka, K. A. Kunugi, J. E. Musgrove, J. W. Winston, Jr., T. J. Ricci-Ott, and L. A. DeWerd. (2010). "Determination of exit skin dose for ^{192}Ir intracavitary accelerated partial breast irradiation with thermoluminescent dosimeters." *Med. Phys.* 37(6):2693–702. doi: 10.1118/1.3429089.

Rivard, M. J., R. J. Bricault, C. S. Melhus, and P. Sioshansi. (2006). "SU-FF-T-396: Stereotactic peripheral brachytherapy and image guidance for the breast." *Med. Phys.* 33(6):2137. doi: 10.1118/1.2241315.

Rivard, M. J., H. R. Ghadyani, A. D. Bastien, N. N. Lutz, and J. T. Hepel. (2015). "Multi-axis dose accumulation of noninvasive image-guided breast brachytherapy through biomechanical modeling of tissue deformation using the finite element method." *J. Contemp. Brachytherapy* 7(1):55–71. doi: 10.5114/jcb.2015.49355.

Rivard, M. J., C. S. Melhus, D. E. Wazer, and R. J. Bricault, Jr. (2009). "Dosimetric characterization of round HDR ^{192}Ir accuboost applicators for breast brachytherapy." *Med. Phys.* 36(11):5027–32. doi: 10.1118/1.3232001.

Rivard, M. J., J. L. Venselaar, and L. Beaulieu. (2009). "The evolution of brachytherapy treatment planning." *Med. Phys.* 36(6):2136–53. doi: 10.1118/1.3125136.

Rusch, T. W., S. D. Davis, L. A. DeWerd, R. R. Burnside, S. Axelrod, and M. J. Rivard. (2004). "Characterization of a new miniature x-ray source for electronic brachytherapy." *Med. Phys.* 31:1807.

Scanderbeg, D. J., C. Yashar, R. Rice, and T. Pawlicki. (2009). "Clinical implementation of a new HDR brachytherapy device for partial breast irradiation." *Radiother. Oncol.* 90(1):36–42. doi: 10.1016/j.radonc.2008.09.024.

Shah, C., J. V. Antonucci, J. B. Wilkinson, M. Wallace, M. Ghilezan, P. Chen, K. Lewis, C. Mitchell, and F. Vicini. (2011). "Twelve-year clinical outcomes and patterns of failure with accelerated partial breast irradiation versus whole-breast irradiation: results of a matched-pair analysis." *Radiother. Oncol.* 100(2):210–14. doi: 10.1016/j.radonc.2011.03.011.

Shah, C., S. Badiyan, J. Ben Wilkinson, F. Vicini, P. Beitsch, M. Keisch, D. Arthur, and M. Lyden. (2013). "Treatment efficacy with accelerated partial breast irradiation (APBI): final analysis of the American Society of Breast Surgeons MammoSite((R)) breast brachytherapy registry trial." *Ann. Surg. Oncol.* 20(10):3279–85. doi: 10.1245/s10434-013-3158-4.

Shah, C., S. Khwaja, S. Badiyan, J. B. Wilkinson, F. A. Vicini, P. Beitsch, M. Keisch, D. Arthur, and M. Lyden. (2013). "Brachytherapy-based partial breast irradiation is associated with low rates of complications and excellent cosmesis." *Brachytherapy* 12(4):278–84. doi: 10.1016/j.brachy.2013.04.005.

Shah, C., F. Vicini, D. E. Wazer, D. Arthur, and R. R. Patel. (2013). "The American Brachytherapy Society consensus statement for accelerated partial breast irradiation." *Brachytherapy* 12(4):267–77. doi: 10.1016/j.brachy.2013.02.001.

Shaitelman, S. F., F. A. Vicini, I. S. Grills, A. A. Martinez, D. Yan, and L. H. Kim. (2012). "Differences in effective target volume between various techniques of accelerated partial breast irradiation." *Int. J. Radiat. Oncol. Biol. Phys.* 82(1):30–36. doi: 10.1016/j.ijrobp.2010.08.059.

Showalter, S. L., G. Petroni, D. M. Trifiletti, B. Libby, A. T. Schroen, D. R. Brenin, P. Dalal, M. Smolkin, K. A. Reardon, and T. N. Showalter. (2016). "A novel form of breast intraoperative radiation therapy with CT-guided high-dose-rate brachytherapy: Results of a prospective Phase 1 clinical trial." *Int. J. Radiat. Oncol. Biol. Phys.* 96(1):46–54. doi: 10.1016/j.ijrobp.2016.04.035.

Skowronek, J., M. Wawrzyniak-Hojczyk, and K. Ambrochowicz. (2012). "Brachytherapy in accelerated partial breast irradiation (APBI) - review of treatment methods." *J. Contemp. Brachytherapy* 4(3):152–64. doi: 10.5114/jcb.2012.30682.

Smith, B. D., D. W. Arthur, T. A. Buchholz, B. G. Haffty, C. A. Hahn, P. H. Hardenbergh, T. B. Julian, L. B. Marks, D. A. Todor, F. A. Vicini, T. J. Whelan, J. White, J. Y. Wo, and J. R. Harris. (2009). "Accelerated partial breast irradiation consensus statement from the American Society for Radiation Oncology (ASTRO)." *Int. J. Radiat. Oncol. Biol. Phys.* 74(4):987–1001. doi: 10.1016/j.ijrobp.2009.02.031.

Strauss, J. B. and A. Dickler. (2009). "Accelerated partial breast irradiation utilizing balloon brachytherapy techniques." *Radiother. Oncol.* 91(2):157–65. doi: 10.1016/j.radonc.2008.12.014.

Strnad, V., O. J. Ott, G. Hildebrandt, D. Kauer-Dorner, H. Knauerhase, T. Major, J. Lyczek, J. L. Guinot, J. Dunst, C. Gutierrez Miguelez, P. Slampa, M. Allgauer, K. Lossl, B. Polat, G. Kovacs, A. R. Fischedick, T. G. Wendt, R. Fietkau, M. Hindemith, A. Resch, A. Kulik, L. Arribas, P. Niehoff, F. Guedea, A. Schlamann, R. Potter, C. Gall, M. Malzer, W. Uter, C. Polgar, Radiotherapy Groupe Europeen de Curietherapie of European Society for Radiotherapy and Oncology. (2016). "5-year results of accelerated partial breast irradiation using sole interstitial multicatheter brachytherapy versus whole-breast irradiation with boost after breast-conserving surgery for low-risk invasive and in-situ carcinoma of the female breast: a randomised, phase 3, non-inferiority trial." *Lancet* 387(10015):229–38. doi: 10.1016/S0140-6736(15)00471-7.

Syed, A. M. N., H. Chang, B. S. Schwartzberg, A. K. Bremner, C. Lopez-Penalver, C. Coomer, S. Boylan, A. Chakravarthy, C. A. Vito, A. Bhatnagar, and G. M Proulx. (2016). "One-year follow-up results of a multi-center trial of intra-operative radiation therapy using electronic brachytherapy at the time of breast conservation surgery for early stage breast cancer." *Cancer Res.* 76.4 Supplement:3–12.

Tang, J. I, P. W. Tan, V. Y. Koh, and S. A. Buhari. (2014). "Multi-catheter interstitial accelerated partial breast irradiation—tips and tricks for a good insertion." *J. Contemp. Brachytherapy* 6(1):85–90.

Trifiletti, D. M., T. N. Showalter, B. Libby, D. R. Brenin, A. T. Schroen, K. A. Reardon, and S. L. Showalter. (2015). "Intraoperative breast radiation therapy with image guidance: Findings from CT images obtained in a prospective trial of intraoperative high-dose-rate brachytherapy with CT on rails." *Brachytherapy* 14(6):919–24. doi: 10.1016/j.brachy.2015.07.001.

Vicini, F. A., D. A. Jaffray, E. M. Horwitz, G. K. Edmundson, D. A. DeBiose, V. R. Kini, and A. A. Martinez. (1998). "Implementation of 3D-virtual brachytherapy in the management of breast cancer: a description of a new method of interstitial brachytherapy." *Int. J. Radiat. Oncol. Biol. Phys.* 40(3):629–35.

Vicini, F., C. Shah, R. Tendulkar, J. Wobb, D. Arthur, A. Khan, D. Wazer, and M. Keisch. (2016). "Accelerated partial breast irradiation: An update on published Level I evidence." *Brachytherapy* 15(5):607–15. doi: 10.1016/j.brachy.2016.06.007.

Watt, E., S. Husain, M. Sia, D. Brown, K. Long, and T. Meyer. (2015). "Dosimetric variations in permanent breast seed implant due to patient arm position." *Brachytherapy* 14(6):979–85. doi: 10.1016/j.brachy.2015.09.008.

Wazer, D. E., L. Berle, R. Graham, M. Chung, J. Rothschild, T. Graves, B. Cady, K. Ulin, R. Ruthazer, and T. A. DiPetrillo. (2002). "Preliminary results of a phase I/II study of HDR brachytherapy alone for T1/T2 breast cancer." *Int. J. Radiat. Oncol. Biol. Phys.* 53(4):889–97.

Wilkinson, J. B., A. A. Martinez, P. Y. Chen, M. I. Ghilezan, M. F. Wallace, I. S. Grills, C. S. Shah, C. K. Mitchell, E. Sebastian, A. S. Limbacher, P. R. Benitez, E. A. Brown, and F. A. Vicini. (2012). "Four-year results using balloon-based brachytherapy to deliver accelerated partial breast irradiation with a 2-day dose fractionation schedule." *Brachytherapy* 11(2):97–104. doi: 10.1016/j.brachy.2011.05.012.

Wong, E. K., P. T. Truong, H. A. Kader, A. M. Nichol, L. Salter, R. Petersen, E. S. Wai, L. Weir, and I. A. Olivotto. (2006). "Consistency in seroma contouring for partial breast radiotherapy: impact of guidelines." *Int. J. Radiat. Oncol. Biol. Phys.* 66(2):372–76. doi: 10.1016/j.ijrobp.2006.05.066.

Yang, Y., C. S. Melhus, S. Sioshansi, and M. J. Rivard. (2011). "Treatment planning of a skin-sparing conical breast brachytherapy applicator using conventional brachytherapy software." *Med. Phys.* 38(3):1519–25. doi: 10.1118/1.3552921.

Yang, Y. and M. J. Rivard. (2009). "Monte Carlo simulations and radiation dosimetry measurements of peripherally applied HDR ^{192}Ir breast brachytherapy D-shaped applicators." *Med. Phys.* 36(3):809–15. doi: 10.1118/1.3075818.

Yang, Y. and M. J. Rivard. (2010). "Dosimetric optimization of a conical breast brachytherapy applicator for improved skin dose sparing." *Med. Phys.* 37(11):5665–71. doi: 10.1118/1.3495539.

Yashar, C., D. Attai, E. Butler, J. Einck, S. Finkelstein, B. Han, R. Hong, L. Komarnicky, M. Lyden, C. Mantz, S. Morcovescu, S. Nigh, K. Perry, J. Pollock, J. Reiff, D. Scanderbeg, M. Snyder, and R. Kuske. (2016). "Strut-based accelerated partial breast irradiation: Report of treatment results for 250 consecutive patients at 5 years from a multicenter retrospective study." *Brachytherapy* 15(6):780–87. doi: 10.1016/j.brachy.2016.07.002.

Yashar, C. M., S. Blair, A. Wallace, and D. Scanderbeg. (2009). "Initial clinical experience with the Strut-Adjusted Volume Implant brachytherapy applicator for accelerated partial breast irradiation." *Brachytherapy* 8(4):367–72. doi: 10.1016/j.brachy.2009.03.190.

Yashar, C. M., D. Scanderbeg, R. Kuske, A. Wallace, V. Zannis, S. Blair, E. Grade, V. H. Swenson, and C. Quiet. (2011). "Initial clinical experience with the Strut-Adjusted Volume Implant (SAVI) breast brachytherapy device for accelerated partial-breast irradiation (APBI): first 100 patients with more than 1 year of follow-up." *Int. J. Radiat. Oncol. Biol. Phys.* 80(3):765–70. doi: 10.1016/j.ijrobp.2010.02.018.

Zauls, A. J., J. M. Watkins, A. E. Wahlquist, N. C. Brackett, 3rd, E. G. Aguero, M. K. Baker, J. M. Jenrette, E. Garrett-Mayer, and J. L. Harper. (2012). "Outcomes in women treated with MammoSite brachytherapy or whole breast irradiation stratified by ASTRO Accelerated Partial Breast Irradiation Consensus Statement Groups." *Int. J. Radiat. Oncol. Biol. Phys.* 82(1):21–29. doi: 10.1016/j.ijrobp.2010.08.034.

Example Problems

(Answers are found at the end of the book.)

1. Which of the following is an accelerated partial breast irradiation technique?

   a. external beam irradiation

   b. brachytherapy using SAVI

   c. interstitial brachytherapy

   d. All of the above are true.

2. Which of the following imaging modalities is not used in checking the SAVI applicator position/dimension for APBI pre-treatment QA?

   a. scout images

   b. orthogonal kV images

   c. ultrasound

   d. CT scan

3. The purpose of the template for multi-catheter interstitial implants is to:
    a. compress the breast
    b. lift breast tissue off the chest wall
    c. protect the catheters during treatment delivery
    d. improve the parallelism of the implanted needles
    e. all of the above
    f. a and b
    g. b and c
    h. a, b, and d
4. True or False?
    a. The PTV in brachytherapy is a 1 cm expansion of the CTV.
    b. The CTV is defined the same for all brachytherapy techniques.
    c. In brachytherapy, the CTV excludes the pectoralis muscle and a 5 mm skin margin.
    d. All forms of APBI use the same seroma-to-CTV expansion.
    e. Skin and ipsilateral breast tissue are organs at risk for all APBI techniques.
5. For multi-catheter interstitial implants and permanent interstitial implants using Pd-103, the CTV is typically defined as the lumpectomy cavity plus a uniform expansion of:
    a. 0.5–1.0 cm
    b. 1.0–1.5 cm
    c. 2.0 cm
    d. 1.5–2.0 cm
6. For breast brachytherapy with balloon-based devices, contrast is often used and should be accounted for in dose calculations:
    a. True
    b. False
7. What are the main attributes of the AccuBoost breast BT applicators?
    a. continuous low-dose-rate treatments that complement EBRT electrons
    b. surface treatments that can also be used for the skin or gynecological disease
    c. treatment applied noninvasively with mammographic image guidance and collimating applicators
    d. computer-controlled interstitial BT using magnetic guidance for avoiding calcifications and ductal tissue
8. HDR APBI has been given successfully in just a single fraction instead of 10 fractions over 5 days.
    a. True
    b. False
9. Assume a breast treatment with a prescription dose of 3.4 Gy/fx and a CTV of 100 cm^3. Assume that all the energy deposited is transferred as heat. Feeling free to make any simplifying assumptions, calculate the increase in temperature that patient's breast would experience. Should the patient be worried about "burns"?

10. A patient is treated with a balloon device and a single dwell position is used in the center of the balloon. The diameter of the balloon is 4 cm, and the CTV is a 1 cm margin from the surface of the balloon, excluding the balloon. The prescription dose is 3.4 Gy / fx $\times$ 10 fxs. Consider now a CTV created with a 2 cm margin around the same balloon. What would have to be the new prescription dose for the two treatments to be considered biologically equivalent?

Chapter 9

Intensity-modulated Brachytherapy

Firas Mourtada[1], Susan L. Richardson[2], Bruce R. Thomadsen[3], and Daniel J. Scanderbeg[4]

[1]Christiana Care Hospital
Newark, Delaware

[2]Swedish Medical Center–Tumor Institute
Seattle, Washington

[3]University of Wisconsin–School of Medicine and Public Health
Madison, Wisconsin

[4]University of California–San Diego
La Jolla, California

9.1 Introduction

9.1.1 IMBT Rationale

Intensity-modulated brachytherapy (IMBT) utilizes high-*Z* materials as shielding in order to improve dose conformity to target tissue while decreasing dose to normal tissues adjacent to the target area. A variety of methods have been designed and researched to achieve these goals, including shielded applicators, both static and dynamic, as well as shielded radiation sources.

Overall, brachytherapy is a highly conformal and effective treatment modality with no entry dose, as with external beam radiotherapy. Good implant geometries maximize dose to the target area while minimizing dose to surrounding tissue(s). However, many targets in brachytherapy are irregularly shaped, and by modifying the cylindrical symmetry of the radiation from the brachytherapy source with IMBT, it can be used to modify the dose for better target coverage with less collateral damage to normal tissue. Technically complex procedures, such as interstitial brachytherapy treatments, can be simplified with IMBT. In these cases, shielded applicators or sources could potentially lead to fewer interstitial needles that are needed for the case, decreasing the technical challenge of implanting a large number of needles, while also leading to less trauma for the patient.

Although the use of shielded applicators and sources has advantages discussed above, it also has some disadvantages. Two disadvantages are the complexity of the physical devices and of the treatment planning. Much of the technical nature of the implant is shifted to physics. One of the largest obstacles to implementation of IMBT is the challenge to characterize the dose distribution accurately from these applicators or sources using conventional brachytherapy treatment planning systems (Rivard et al. 2009) as well as the imaging of high-*Z* applicators in the patient. Even so, commissioning, planning, plan checking, and verification have the potential to become more complicated and time consuming.

9.2 Static Shielding

9.2.1 Description

Shielding in brachytherapy typically refers to the use of high-*Z* materials in the brachytherapy applicator that modify the dose in the direction of the shielding and also in the forward direction for ^{192}Ir due to alteration of the scattering conditions. Usually this is done to protect normal tissues or organs from receiving unnecessary dose from an isotropic radiation source, such as a brachytherapy source. The simplest form of shielding is static shielding, where the shielding is fixed in the applicator in a known geometry. Static-shielded applicators are usually in the form of intracavitary applicators that are bulky or have a substantial thickness. This allows high-*Z* materials—such as steel, lead, or tungsten—to be placed in the applicator without increasing the overall dimensions of the applicator. Brachytherapy applications that have traditionally facilitated the use of shielded applicators include cervical brachytherapy, with tandem and ovoids, and vaginal or rectal brachytherapy, with the use of cylinders. Using shielded applicators requires skillful users who are adept in the placement of the applicators to ensure the shields are in the correct position relative to the patient anatomy to produce the intended effect.

9.2.2 Design

9.2.2.1 Colpostats/Ovoids

One of the most commonly used shields in brachytherapy is those of shielded colpostats or ovoids. Over 50 years ago, the Fletcher applicator system was expanded from the basic Manchester design to incorporate many improvements, including internal shielding, with Suit adding in the concept of afterloading and Delclos the creation of the miniaturized ovoid (Suit et al. 1963). The shielded Fletcher-Suit-Delclos(FSD) tandem and ovoid system—low dose rate (LDR) or high dose rate (HDR)—was a modification from previous generations of tandem and ovoid applicators and was designed to protect the bladder and rectum from excess radiation dose. The bladder, rectum, and vaginal mucosa are typically the dose-limiting structures in cervical cancer brachytherapy, and by placing tungsten alloy shields at the anterior and posterior aspects of the applicators, the normal tissue doses can be reduced. The rectal shield is classically a half disc at 45°, and the bladder shield is a 150° sector. The shielding typically provides dose reduction of 20% to 30% in the shadow of the shields. An example radiograph of shielded ovoids, with the beam collinear to the axes of the ovoids, is shown in Figure 9–1. The shields are designed to maintain dose to the uterosacral and broad ligaments. Figure 9–2 shows an anterior and lateral view with the applicators inserted in a patient and with a geometry such as to shield the critical organs appropriately.

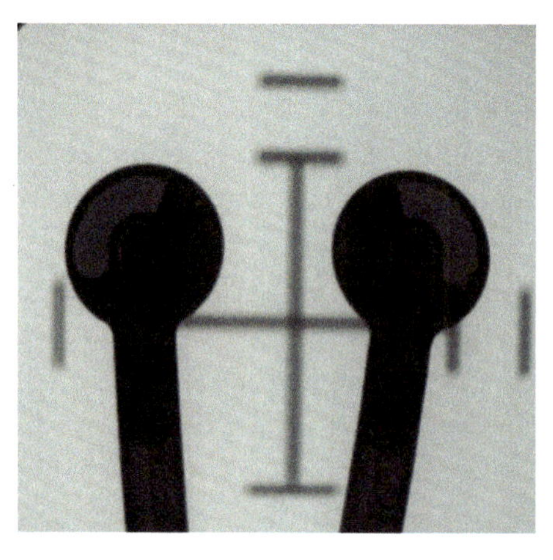

Figure 9–1 Anterior view shielded ovoids, depicting the 45° half-circular rectal shields (black).

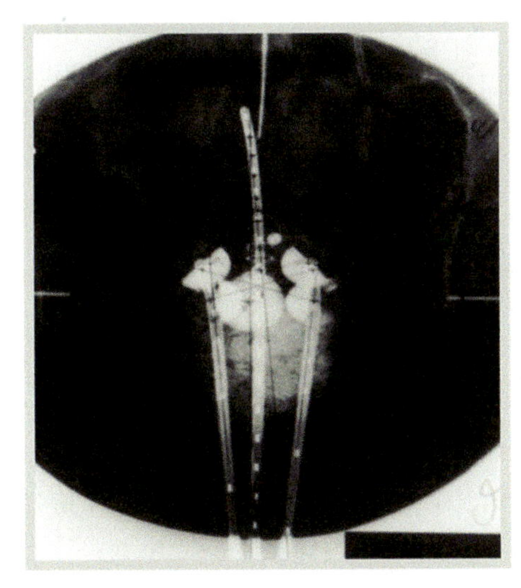

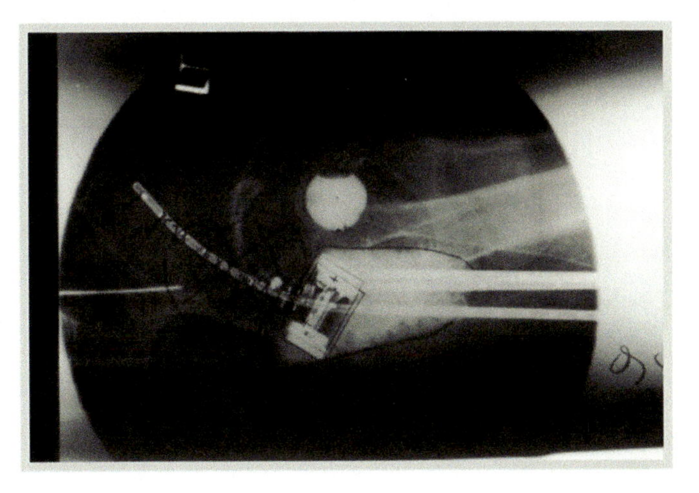

Figure 9–2 Anterior (left) and lateral (right) views of *in vivo* shielded colpostats or ovoids from orthogonal x-ray films. The shields should be oriented to provide maximum protection to bladder and rectal points.

The use of shielded applicators has declined in recent years due to the advent of three-dimensional (3D) imaging and the progression from orthogonal image-based planning to 3D CT and MR-based planning. The soft tissue imaging allows plans to be optimized in a more targeted manner and dose-volume histograms (DVHs) to be calculated more accurately, which can allow clinicians to shape dose distributions and improve organ-at-risk sparing. Imaging shielded applicators with CT and MR can be problematic due to large image artifacts, geometric distortions, and heating (MRI) due to the large pieces of metal present in the applicators (Kalender et al. 1985; Ho 2001).

9.2.2.2 Cylinders

Shielded cylinder applicator sets are offered from a variety of vendors. An example of an applicator set that can be used to treat either cancer of the vagina or cancer of the rectum, depending on the placement of the applicator and the shielding, is shown in Figure 9–3. The inserted shields can offer either 90° or 180° shielding depending on the tungsten alloy shield inserted, while some other vendors also offer an option for 270° of shielding.

The relative dose distributions around a shielded HDR ^{192}Ir vaginal cylinder were described by Waterman and Holcomb (1994) and are shown in Figure 10–4. As can be seen in the figure, significant sparing can be achieved when employing the shields. Transmission factor values of 0.23, 0.17, and 0.16 for the 90°, 180°, and 270° were found, respectively. These values were confirmed via Monte Carlo calculations for clinically relevant regions (Petrokokkinos et al. 2001).

Figure 9–3 Shielded intracavitary applicator set from Varian Medical Systems, Inc. (Image courtesy of Varian Medical Systems, Inc., Palo Alto, CA. © 2017. All rights reserved.)

Poon and colleagues have investigated the use of an eight-channel endorectal applicator with shielding capabilities and analyzed dose distributions calculated using the TG-43 model versus a Monte Carlo simulation (Poon et al. 2008). The silicon applicator,

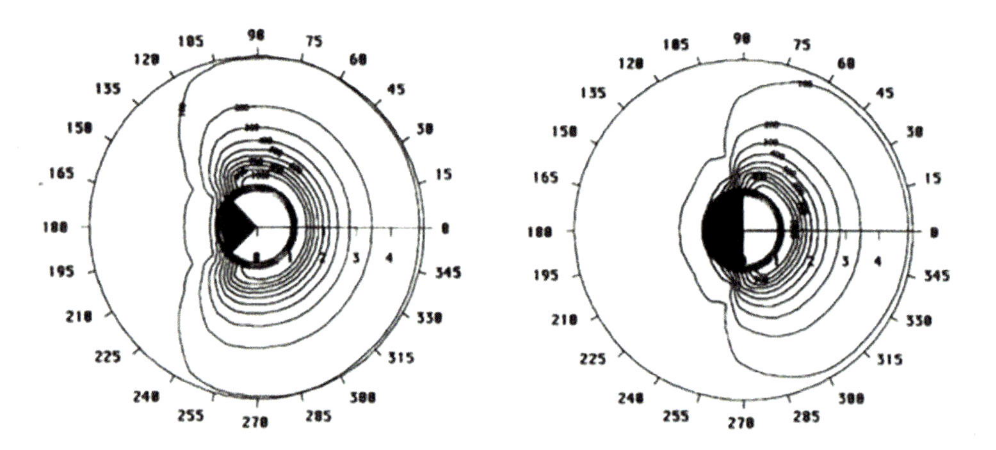

Figure 9–4 Shielded vaginal cylinders with 90° shields (left) and 180° shields (right).

shielding, contrast-filled inflatable balloon, and patient anatomy were all considered. The authors found the shielding decreased normal tissue dose by 24% and reduced the prescribed target dose by 3%. Webster and colleagues also investigated a number of designs for rectal cancer treatments with grooves cut into a tungsten rod (Webster et al. 2013a). They compared designs to an intracavitary mold applicator (ICMA) and calculated dose using Monte Carlo simulation. Results showed their applicators performed better than the ICMA in all dosimetric criteria analyzed; however, treatment times were increased by roughly 30%.

Additionally, Han and colleagues (2016) have proposed a multi-channel tandem (Figure 9–5) made of tungsten in order to provide directional modulation for cervical cancer treatment. Their grooved tandem allows more directional dose distributions as simulated in panel (c) of Figure 9–5.

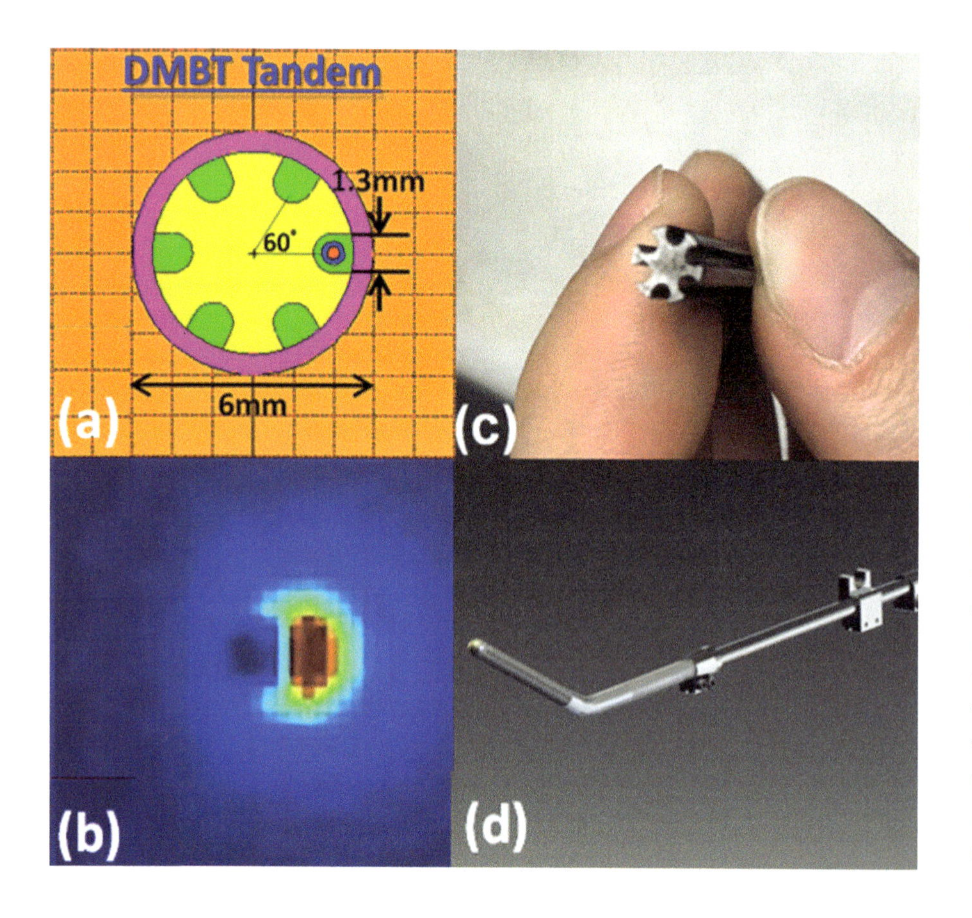

Figure 9–5 The proposed intensity-modulated brachytherapy (IMBT) concept tandem applicator design. (a) IMBT tandem cross section with 6 peripheral holes carved out of a nonmagnetic tungsten alloy rod of 5.4-mm diameter. (b) The Monte Carlo simulated dose distribution of an ^{192}Ir source inside DMBT tandem. (c) A successfully machined-to-specifications tungsten alloy piece to demonstrate the manufacturability of the applicator. (d) An artistic rendering of the concept applicator in full assembly.

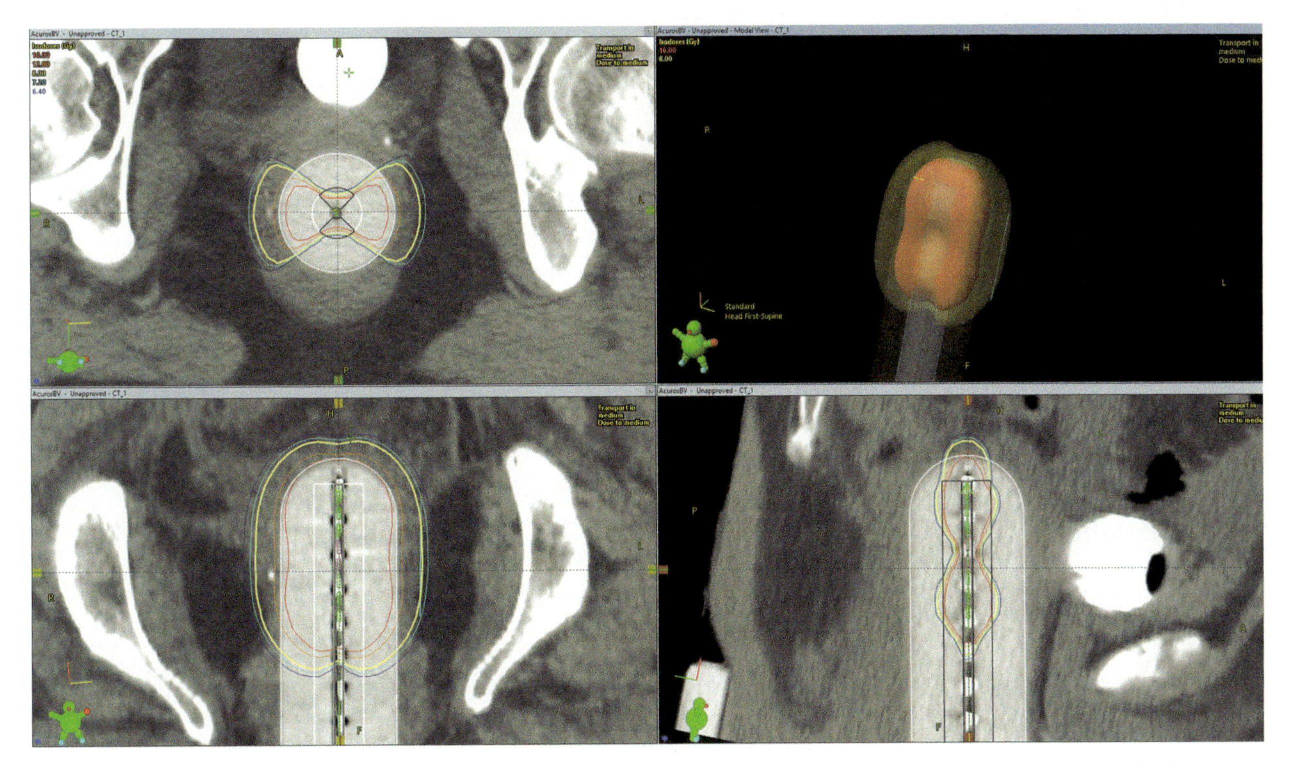

Figure 9–6 Dose distribution from Acuros® BV of a shielded cylinder. Image courtesy of Varian Medical Systems, Inc., Palo Alto, CA. Copyright 2017. All rights reserved.

Treatment planning systems are beginning to incorporate libraries for shielded applicators that can be used to project known anisotropic dose distributions (like the ones from shielded applicators) onto patient anatomy. Most of these types of dose distributions can now be calculated using model-based dose calculation algorithms, such as a grid-based solver, collapsed cone, or Monte Carlo methods. An example dose distribution from Acuros® BV grid-based solver utilizing a shielded cylinder from the applicator library in BrachyVision™ (Varian Medical Systems, Palo Alto, CA) is shown in Figure 9–6. A more thorough discussion of model-based dose calculation can be found in Chapter 2.

9.2.3 Clinical Implementation

As with the incorporation of any new brachytherapy device, good quality assurance and acceptance practices should be followed, as described in the report of the American Association of Physicists in Medicine (AAPM) Task Group 56 (TG-56) (Nath et al. 1997). In terms of quality control, shields should be verified for positional accuracy, ensuring they have not migrated from their original manufactured positions. Historically, with LDR applicators, the ovoids and welds could become brittle and break from repeated sterilization. This is less of an issue in current practice; however, manufacturers still publish guidelines regarding sterilization cycle lifetime or overall age and end-of-life for their applicators and transfer guide tubes. Dimensions and sizes of shields and applicators should be verified initially and annually, as well as initial measurements of the transmission factors or shielding effect that will be used clinically. Implementation should be done carefully to avoid under-dosing gross tumor volumes, and applicator geometry and loading or dwell weighting practices should only be changed after extensive evaluation and comparison of old and new dose distributions (Halperin et al. 2008). Additionally, the shields can be seen positioned well to shield the bladder and rectum, although clinically, positioning is not always reproducible (Figure 9–2). Therefore, implant verification films should be taken to validate appropriate orientation of the shields along with good implant geometry.

9.3 Dynamic Shielding

9.3.1 Description

The previous section described IMBT as the achievement of asymmetric dose distributions through the use of static shielded applicators. In 2002, Ebert proposed the concept of a movable collimated source (Ebert 2002). In their design, the asymmetric radiation source could be rotated about its axis in order to paint dose to irregularly shaped targets with varying distance between source and target edge. The term *dynamic modulated brachytherapy*, or DMBT, was introduced by Webster and colleagues (2013b) to distinguish the difference between IMBT from static shielding and that achieved with dynamic motion of the shielding or collimating aperture. A number of research groups have been studying the use of motion to achieve even more highly conformal treatment plans in brachytherapy, with increased target coverage and decreased dose to organs at risk. Several of these innovations are described below.

9.3.2 Designs

9.3.2.1 Shielded Needle

Adams and colleagues (2014) designed a novel needle, catheter, and radiation source system (^{153}Gd) creating an interstitial rotating shield for prostate brachytherapy as shown in Figure 9–7. This design can theoretically reduce the dose to the urethra by 29% to 44% over conventional high-dose-rate brachytherapy doses and could reduce the rate of stricture. The drawback of this proposed technique is the delivery time, which was significantly longer than with conventional HDR brachytherapy using ^{192}Ir.

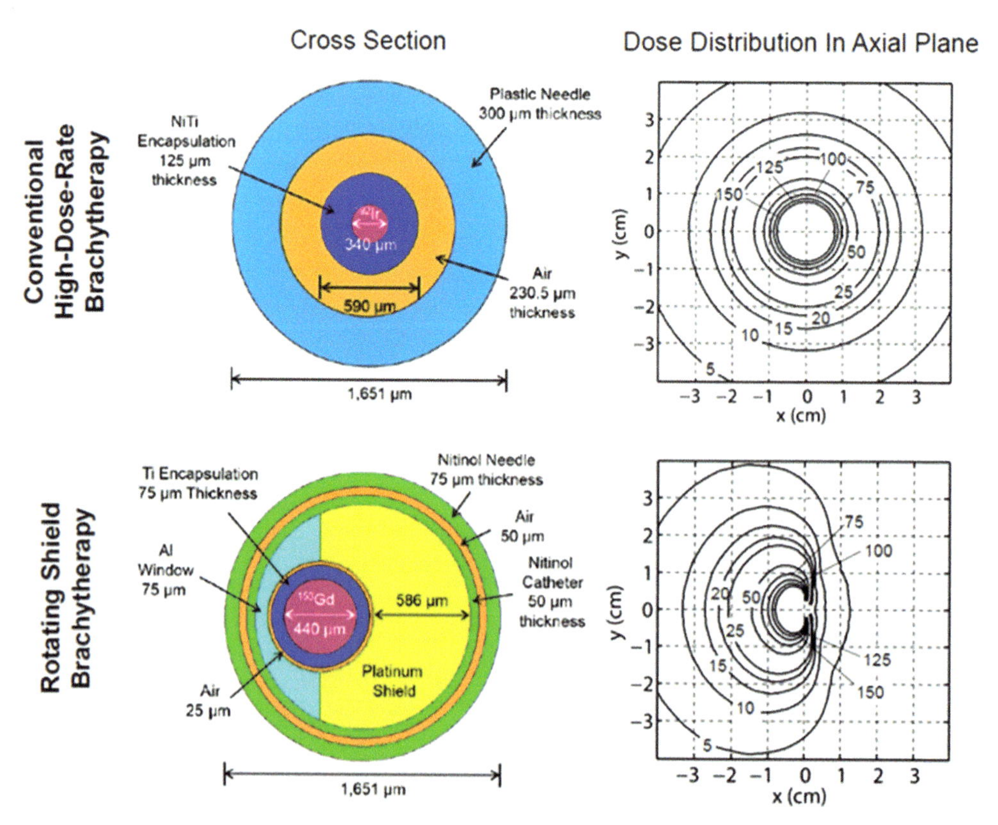

Figure 9–7 Rotating shield brachytherapy for a ^{153}Gd high-dose-rate (HDR) shielded brachytherapy needle (Adams et al. 2014).

9.3.2.2 Paddle-based Rotating Shield Brachytherapy

Another concept under design is the use of paddle-based rotating shield brachytherapy (P-RSBT) to achieve intensity-modulated dose delivery with electronic brachytherapy (Xoft Axxent™, iCAD, Inc., Nashua, NH, USA). The design of a P-RSBT paddle applicator is shown in Figure 9–8. The concept is to use small tungsten paddles that can be inserted and retracted, along with translational and rotational movement of the entire applicator, to create IMBT for electronic brachytherapy. The use of multiple paddles allows for better emission angle selection for target coverage, while keeping treatment duration in a clinically useful range. The theoretical treatment plans of five cervical cancer patients were compared to other methods of shielded brachytherapy, and P-RSBT was found to outperform the others in terms of dosimetry and delivery time (Liu et al. 2015).

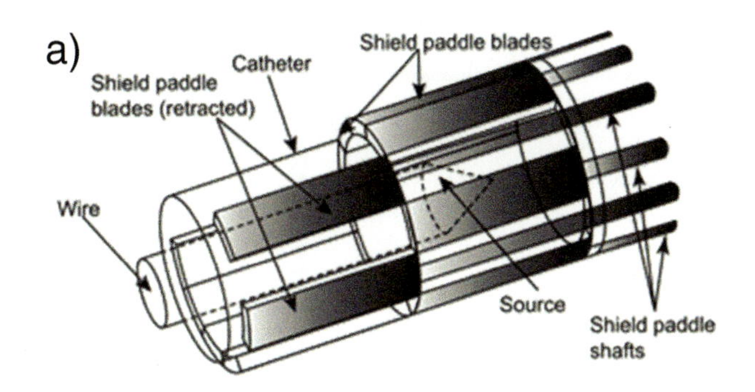

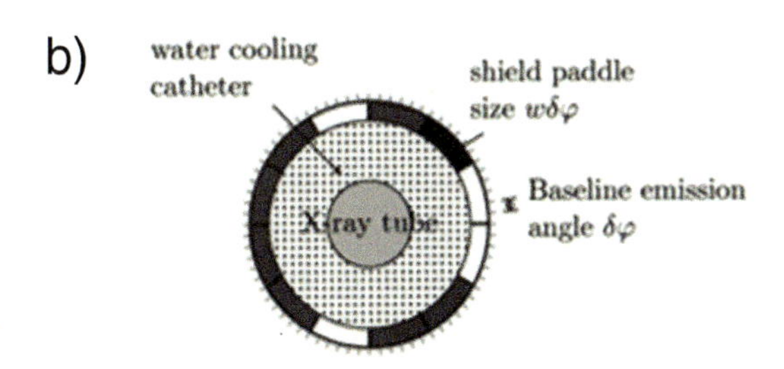

Figure 9–8 A paddle-based rotating shield brachytherapy (P-RSBT) applicator design in a) three-dimensional (3D) view and b) cross section (Liu et al. 2015).

9.3.2.3 Fletcher CT/MR Shielded Applicator with Movable Shields

The current commercially available CT- and MR-compatible brachytherapy applicators for treatment of cervical cancer were designed without the ovoid shields to allow high-quality images for brachytherapy treatment planning, eliminating a key advantage of Fletcher-style applicators (Weeks and Dennett 1990). To address this limitation, Elekta designed a novel CT/MR-compatible applicator with movable shields called the "Fletcher CT/MR Shielded Applicator" (Price et al. 2009; Klopp et al. 2013). This design uses movable shields during CT acquisition, hence the x-ray beam does not pass through the shields and eliminates metal artifact. Furthermore, additional improvements on several aspects of the Fletcher–Williamson applicator were made to make it easier for insertion and placement within the patient anatomy, such as ovoid length optimization to fit more patients, adding graduated marks for locking the tandem instead of preset tandem lengths, and improving the locking mechanism for easier attachment of the tandem to the ovoids. Artifact-free CT or MR scans can be used to generate three-dimensional brachytherapy treatment plans to evaluate dose-volume parameters of normal tissues (Viswanathan et al. 2007; Kim et al. 2011). The scanning procedure is a "step-and-shoot" type method and is described in Figure 9–9. Typical clinical loading of these ovoids has been shown to have an approximately 35% reduction in dose to ROIs (1 cm^3 and 2 cm^3 values) using model-based dose calculations compared to the AAPM TG-43 dose calculation formalism (Price et al. 2009).

9.3.2.4 Dynamic Modulated Brachytherapy Robotics for Rectal Cancer

Work by Webster and colleagues (2013b) describes a tungsten collimator to shield the HDR ^{192}Ir source, along with a robotic applicator that can rotate and translate in the patient, thereby painting a dose distribution much like with IMRT in external beam radiotherapy (Figure 9–10). In the case of rectal applications, the theoretical benefit would decrease the doses to critical structures on the order of 50%, decrease the dose heterogeneity index by 40%, but at the expense of a nearly threefold increase in treatment time.

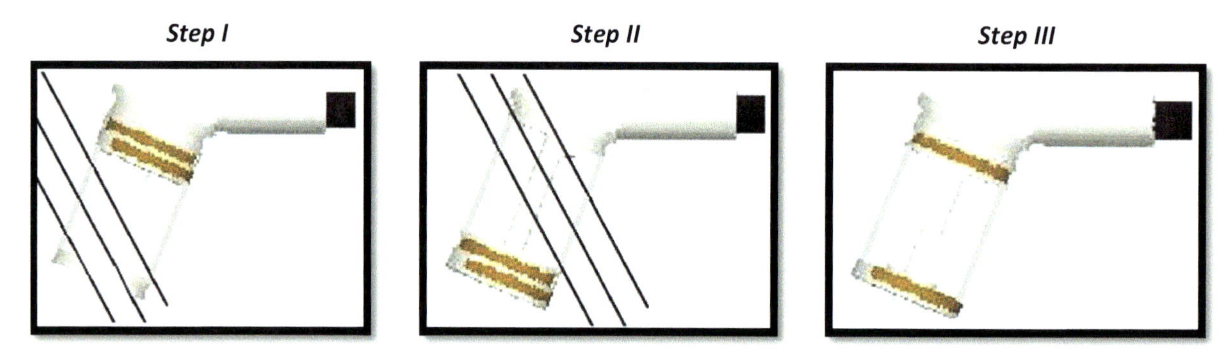

Figure 9–9 Step-and-shoot CT imaging technique. The rectal shield (posterior) is docked at its proximal end of travel, adjacent to the bladder shield (anterior), and scans are taken to the midpoint of the ovoid (Step I). The scanner is then paused, both the rectal and bladder shields are moved to the distal end of travel, and the scan is then resumed and completed (Step II). In Step III, the bladder shield is placed back in its default position (anterior), both shields are locked in position, and a CT scout is taken to verify the final position. (Images courtesy of Dr. Firas Mourtada, Christiana Care Hospital, Newark, DE.)

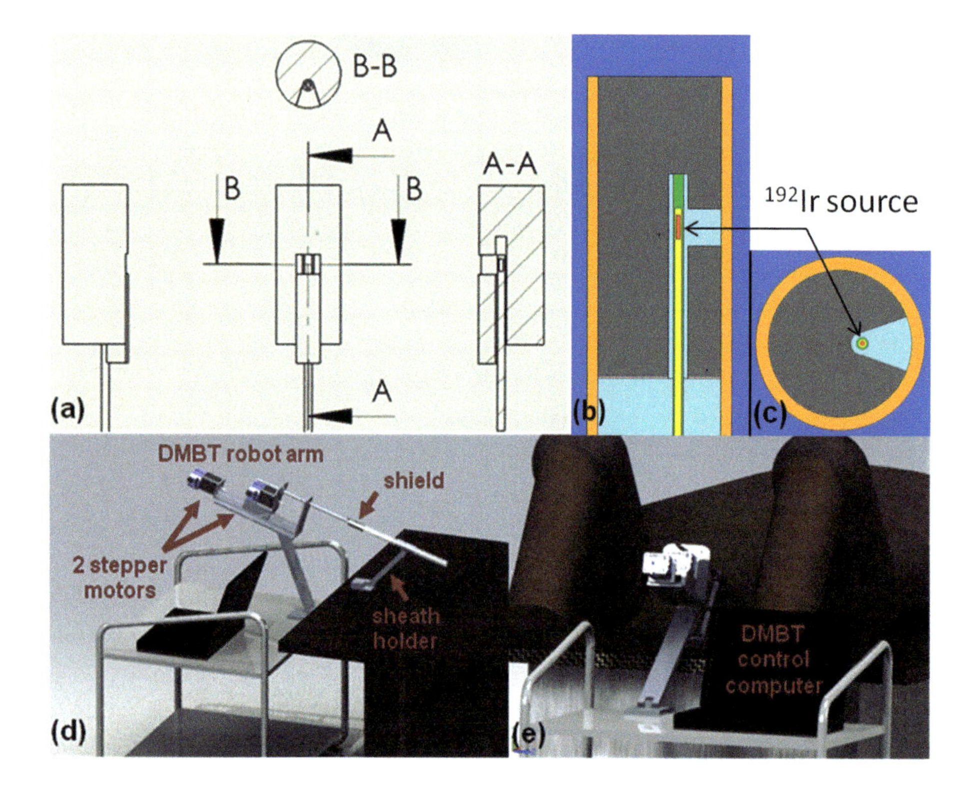

Figure 9–10 Dynamic modulated brachytherapy (DMBT) concept. (a) Drawing of shield design. (b) Sagittal cross section of shield. (c) Axial cross section of shield. Cartoons of (d) imagined setup and (e) imagined treatment. (Image courtesy of Dr. William Song, University of Toronto, Sunnybrook Health Sciences Centre.)

9.3.3 Clinical Implementation

All four of the applicators in the previous sections are unique in that one was designed using its own radiation source (^{153}Gd), one was designed for use in electronic brachytherapy, and although the remaining two use ^{192}Ir, they different in design. All the applicators have the benefit of greater OAR sparing while maintaining dose coverage of the intended target(s). However, for three of the four applicators, this comes at the expense of considerably longer treatment times. This is one obstacle that will need to be addressed prior to clinical implementation of these devices.

Careful quality assurance is necessary for clinical implementation of any new product or technique, as was mentioned in Section 9.2.3. However, clinical implementation of dynamically modulated dose distributions is particularly challenging with the added complexities of additional equipment (robotics) and applicator motion, making dose delivery verification more important to ensure the proper treatment was received by the patient. Additionally, dose calculations are also increasingly complex since moving from TG-43 to TG-186 methodologies to account for material heterogeneities and radiation scatter conditions.

9.4 Directional Sources

9.4.1 Description

As an alternative to shield applicators, non-cylindrical dose distributions can also be produced by directional sources. Directional sources use shielding *within* the source capsule to attenuate radiation emitted in particular directions. Because of the very limited space in the capsule, directional sources pose a greater challenge than a shielded applicator. In addition, much of the construction must be performed with remote handling due to the radiation from the source.

9.4.2 Designs

The dosimetric goal of a directional source is to produce a radiation field that provides a significant dose over a useful solid angle while keeping the dose elsewhere greatly reduced. The words in the previous sentence are intentionally vague because directional sources can be useful in sculpting a dose distribution that satisfies treatment specifications with very different specifications.

Lin and colleagues have designed a directional LDR ^{125}I source, as shown in Figure 9–11 (Lin 2006; Lin et al. 2008). The source demonstrates several important characteristics of most directional sources.

1. *Size.* The ideal directional brachytherapy source would fit inside an interstitial catheter. Making a directional source for intracavitary applications would allow more space and various options, but the

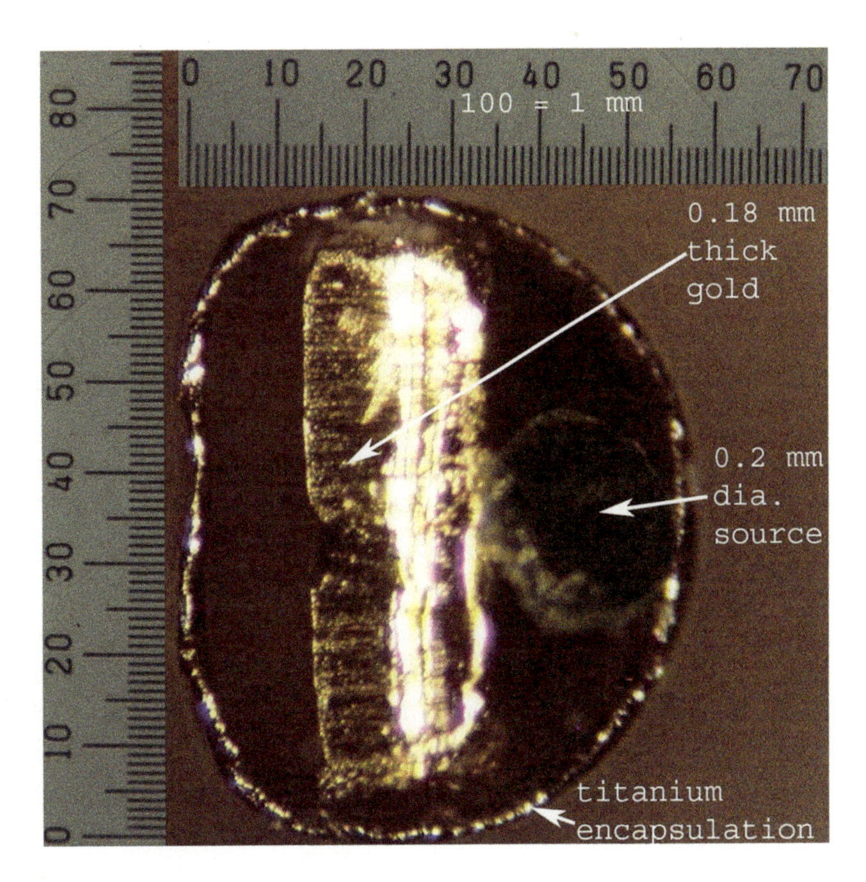

Figure 9–11 An axial cross section of the low-dose-rate (LDR) directional source developed by Lin and colleagues. Figure from (Lin 2006) with permission.

Table 9–1 Some of the more common radionuclides considered for directional brachytherapy sources

Radionuclide	Half-life	Mean Photon Emissions	Notes
^{103}Pd	17.0 d	21 keV	Low penetration
^{125}I	59.4 d	28 keV	Difficult to make highly concentrated
^{153}Gd	240.4 d	60.9 keV	Emits 97.4 keV (29%) and 103.2 keV (21%)
^{169}Yb	32.0 d	93 keV	Relatively high energy
^{270}Ym	128.6 d	165 keV (gold jacket to stop electrons)*	Low photon yield compared to emitted electrons (6:100)*
Electronic brachytherapy	Not applicable	50 kVp x-ray spectrum	Relatively large source

* From Enger, D'Amours, and Beaulieu 2011. Other values from Heredia 2013.

Table 9–2 Materials commonly used as attenuators in a directional source (Heredia 2013)

Element, Symbol	Atomic Number (Z)	Density (g/cm^3)	Tenth Value Layer for ^{103}Pd (10^{-3} cm)
Tantalum, Ta	73	16.69	2.33
Tungsten, W	74	19.25	1.95
Osmium, Os	76	22.59	1.55
Iridium, Ir	77	22.56	1.54
Platinum, Pt	78	21.45	1.52
Gold, Au	79	19.30	1.62
Lead, Pb	82	11.34	2.51

source need not be directional; the modulation of the dose could be accomplished with shielding in the applicator. For a directional LDR source to be compatible with interstitial applications using an 18-gauge needle requires an outer diameter ≤0.83 mm. A source could fit a 17-gauge needle (1 mm diameter), but the larger the needle needed, the less likely the source is to be used. For an HDR ^{192}Ir source, 17 gauge would be similar to those commonly used.

2. *Radionuclide.* Because the next characteristic will be an attenuator to shape the emitted field, the radionuclide energy for the source needs to be low enough for the attenuator to reduce the transmitted radiation to very low levels. Most work has used ^{125}I (as in the source in Figure 10–11), ^{103}Pd, or ^{153}Gd. Table 9–1 lists some of the relevant properties of the most likely radionuclides for directional sources. The conventional brachytherapy workhorse radionuclides, ^{192}Ir or ^{60}Co, have energies far too high to be attenuated by any material that would fit into the source capsule.

3. *Attenuator.* The amount of dose reduction through the attenuator is dependent on the combination of the radionuclide's photon spectrum and the material and thickness of the attenuator. All directional sources use attenuators with high atomic numbers and high densities. As noted above, the radionuclides used in directional sources emit low-energy photons to make use of the enhanced photoelectric-effect-based attenuation in the shields. The high density and mass attenuation coefficient are necessary to fit the attenuator in the source capsule. The most common materials considered for the attenuators are gold and platinum. Lead, while with the highest atomic number of stable elements,

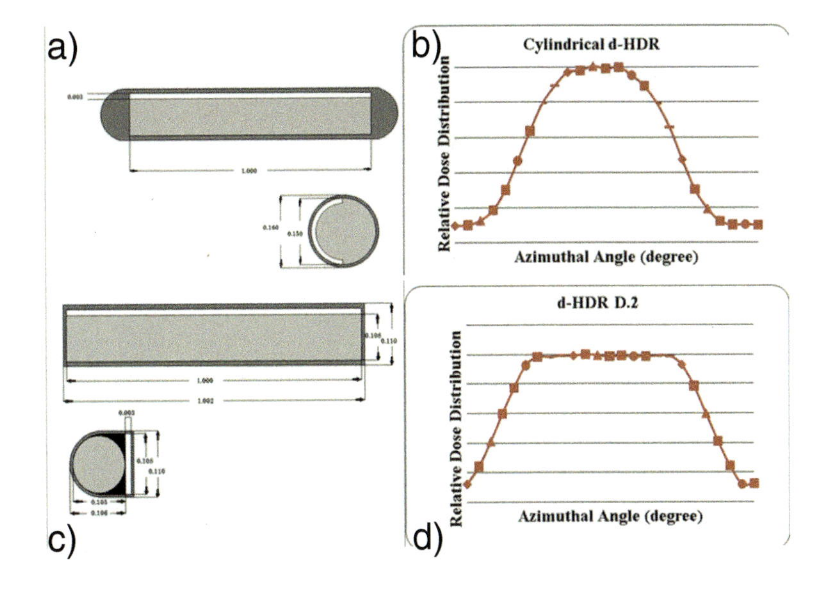

Figure 9–12 The effect of source shape on the dose gradient as a function of azimuthal angle around the source. Panel a) shows a cylindrical radionuclide packet with a semi-cylindrical shied and panel b) the resultant dose function. In panel c), the radionuclide is still cylindrical but the attenuator is flat. Panel d) shows the dose function for the shape in c), with the sharper penumbra and the broader angle for treatment uniformity. Figure from (Heredia 2013) with permission.

has neither the density of the alternatives, nor the resultant attenuation. Table 9–2 gives a list of the materials used for attenuators in directional sources (Heredia 2013).

4. *Geometry.* As Lin and colleagues found, and Heredia verified, the ideal shape for the source is not the usual cylinder with the attenuator forming a partial cylinder around the radionuclide (Lin et al. 2008; Heredia 2013). Assuming that the goal is to provide full dose over some azimuthal angle around the source, while minimizing the dose in others, an additional goal would be to have a sharp gradient dividing the two regions. Figure 9–12 shows how the shape of the source affects the dose gradient. Figure 9–12a shows a cylindrical radionuclide packet partially surrounded by a cylindrical attenuator. Dose as a function of azimuthal angle (Figure 9–12b) shows a relatively small range of angles with the full treatment dose.

Clinically, the concept would be to place the directional sources near locations to be protected. The optimization of source placement poses challenges for most optimization algorithms. There is almost always an advantage of using a directional source in a given position because, oriented correctly, it could deliver a reduced dose to a sensitive structure compared with the dose on the opposite side of the source, delivered to the target. However, that relative benefit may be minuscule if the source position is not near the structure to be spared (Chaswal et al. 2012). A result of excessive use of directional sources is often a greater inhomogeneity of dose through the target. In the midst of the target, and relatively far from the sensitive structures, a non-directional source provides much greater efficiency and uniformity for dose deposition. Properly placed, the directional sources can improve the dose distribution. Figure 9–13 shows two examples of this improvement as simulated with a prostate implant and a breast implant.

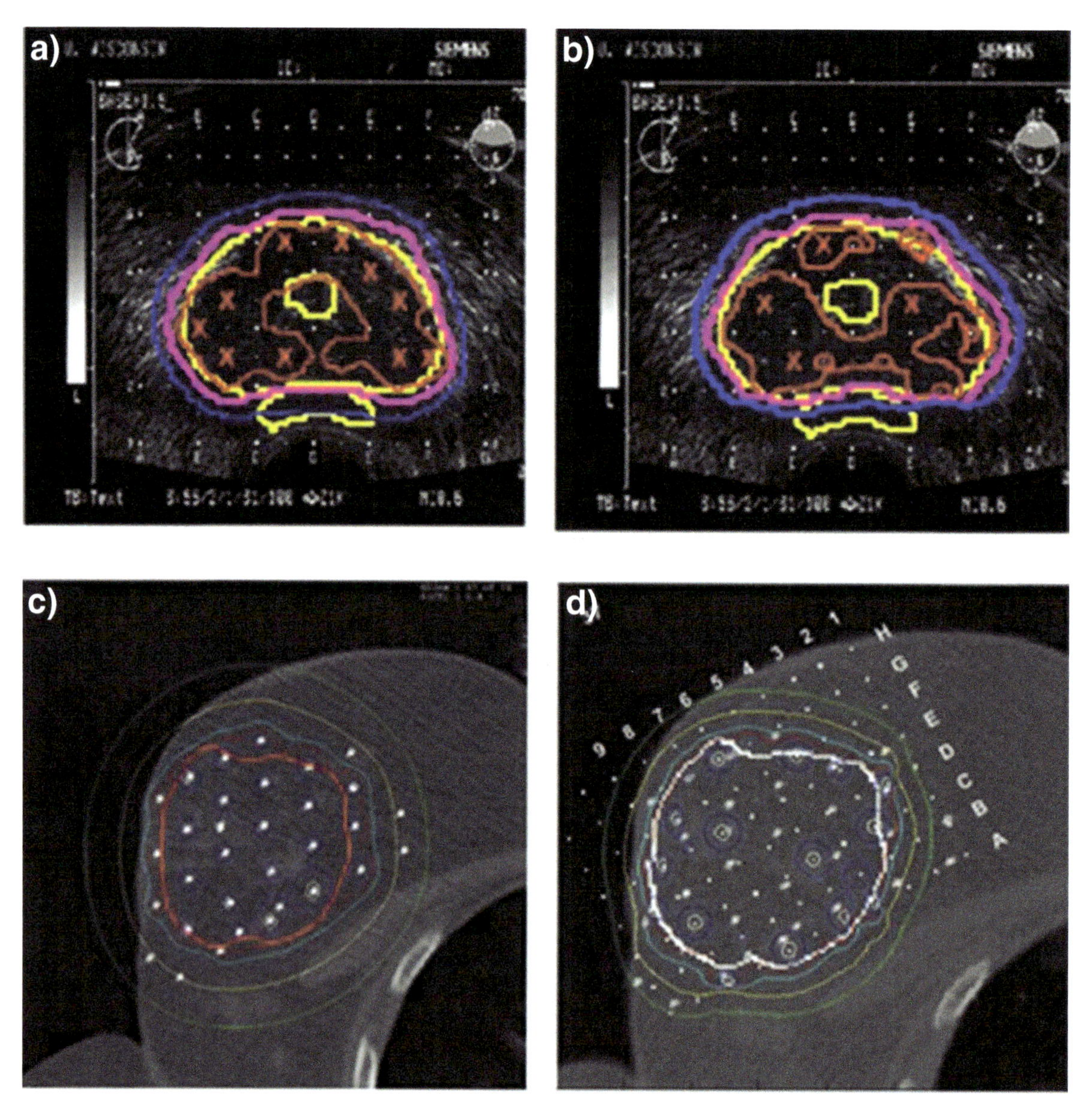

Figure 9–13 Clinical examples of directional sources. Panels a) and b) show a permanent prostate implant using nondirectional sources in a) and replacing some of the sources with directional versions in b). The directional sources are indicated by a "D," with the flat part of the letter in the shielded direction and the non-directional sources indicated by "X." The directional sources reduce the dose to the rectum and the urethra. Panels c) and d) show a similar comparison for a breast implant, where the plan with the directional sources d) reduces the dose to the skin and the pectoralis. Figures from (Lin 2006) with permission.

9.4.3 Clinical Implementation

In general, shielded sources have the same difficulties in implementation as the other methods of IMBT described in this chapter. One of the biggest difficulties to implementation is with treatment verification and ensuring the shielding is placed in the proper direction to get target coverage and spare normal tissues.

The first commercially available directional source was the CivaSheet (CivaTech Oncology, Inc., Durham, NC), shown in Figure 9–14. The design intends that the sheet, a flexible membrane, be sutured to a planar target, such as a post-resection tumor bed. The sheet then irradiates the target, while the gold backing

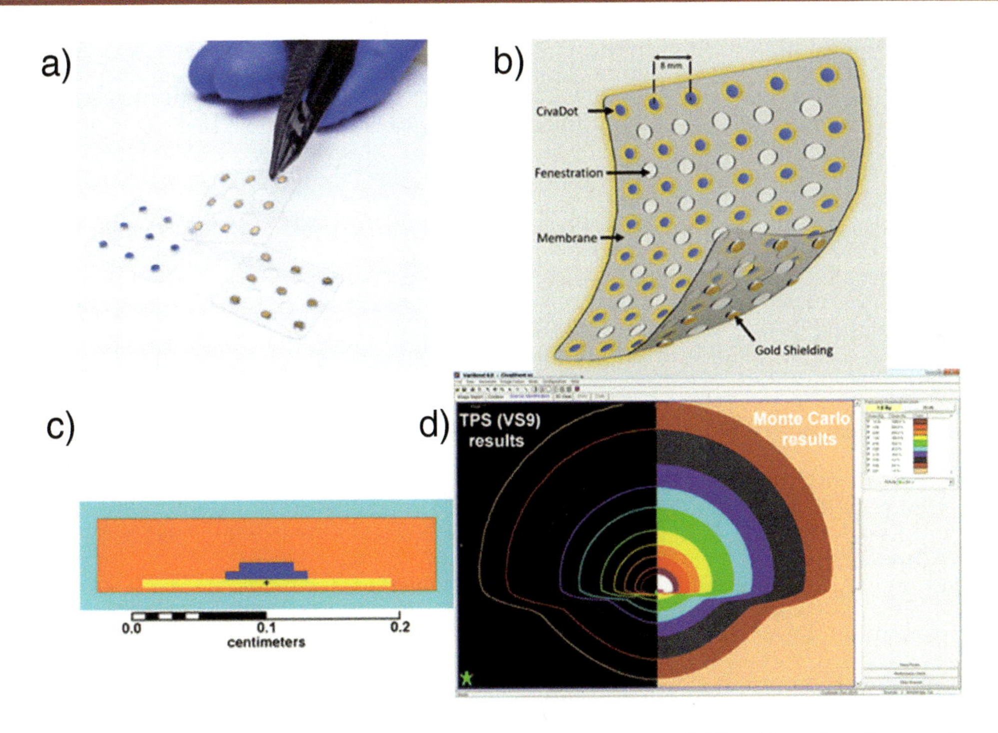

Figure 9–14 The CivaSheet directional brachytherapy membrane. a) The blue dots in the sheet on the left are the ^{103}Pd sources. The gold dots on the other two sheets are the gold attenuator backing that shields the one direction from the treatment radiation. b) A schematic of the sheet with the dimension between the dots. c) A cross-sectional view of the construction of the dots. d) The fluence distribution around a CivaDot. Panels a) and b) courtesy of CivaTech; Panels c) and d courtesy of Dr. Mark Rivard (Rivard 2017).

attenuates much of the radiation emitted in the opposite direction. Each of the ^{103}Pd sources arranged in a gridded pattern in the sheet are called CivaDots. Full dosimetry characterization and parameters for use in treatment planning systems was published by Rivard, and early investigations from some of the first clinical treatments were reported at the 2016 World Congress of Brachytherapy (Rivard 2017; Rivard 2016).

References

Adams, Q. E., J. Xu, E. K. Breitbach, X. Li, S. A. Enger, W. R. Rockey, Y. Kim, X. Wu, and R. T. Flynn. (2014). "Interstitial rotating shield brachytherapy for prostate cancer." *Med. Phys.* 41:051703. doi:10.1118/1.4870441.

Aima, M., J.L. Reed, L.A. DeWerd, and W.S. Culberson. (2015). "Air-kerma strength determination of a new directional ^{103}Pd source." *Med. Phys.* 42: 7144–52. doi: 10.1118/1.4935409.

Beaulieu, L., A. Carlsson Tedgren, J. F. Carrier, S. D. Davis, F. Mourtada, M. J. Rivard, R. M. Thomson, F. Verhaegen, T. A. Wareing, and J. F. Williamson. (2012). "Report of the Task Group 186 on model-based dose calculation methods in brachytherapy beyond the TG-43 formalism: Current status and recommendations for clinical implementation." *Med. Phys.* 39:6208–36. doi: 10.1118/1.4747264.

Chaswal, V., B.R. Thomadsen, and D.L. Henderson. (2012). "Development of an adjoint sensitivity field-based treatment-planning technique for the use of newly designed directional LDR sources in brachytherapy." *Phys. Med. Biol.* 57:963–82. doi: 10.1088/0031-9155/57/4/963.

Ebert, M. A. (2002). "Possibilities for intensity-modulated brachytherapy: technical limitations on the use of non-isotropic sources." *Phys. Med. Biol.* 47:2495. doi: 10.1088/0031-9155/47/14/309.

Enger, S. A., M. D'Amours, and L. Beaulieu. (2011). "Modeling a Hypothetical ^{170}Tm Source for Brachytherapy Applications." *Med. Phys.* 38(10):5307–10. doi:10.1118/1.3626482

Haas, J. S., R. D. Dean, and C. M. Mansfield. (1985). "Dosimetric comparison of the Fletcher family of gynecologic colpostats 1950–1980." *Int. J. Radiat. Oncol. Biol. Phys.* 11:1317–21. doi: 10.1016/0360-3016(85)90247-0.

Halperin, E. C., L. W. Brady, D. E. Wazer, and C. A. Perez. *Perez & Brady's principles and practice of radiation oncology.* Philadelphia: Lippincott Williams & Wilkins, 2013.

Han, D.Y., H. Safigholi, A. Soliman, A. Ravi, E. Leung, D. J. Scanderbeg, Z. Liu, A. Owrangi, and W. Y. Song. (2016). "Direction modulated brachytherapy for treatment of cervical cancer. II: Comparative planning study with intracavitary and intracavitary–interstitial techniques." *Int. J. Radiat. Oncol. Biol. Phys.* 96:440–48. doi: 10.1016/j.ijrobp.2016.06.015.

Heredia, A. Y. "Developing a directional high-dose rate (d-HDR) brachytherapy source." Ph.D. dissertation, University of Wisconsin–Madison, 2013.

Ho, H. S. (2001). "Safety of metallic implants in magnetic resonance imaging." *J. Magn. Reson. Imaging* 14: 472–77. doi: 10.1002/jmri.1209.

Kalender, W. A., R. Hebel, and J. Ebersberger. (1985). "Reduction of CT artifact caused by metallic implants." *Radiology* 164:576–77. doi: 10.1148/radiology.164.2.3602406.

Kim, H., S. Beriwal, C. Houser, and M. Saiful Huq. (2011). "Dosimetric analysis of 3D image-guided HDR brachytherapy planning for the treatment of cervical cancer: Is point A-based dose prescription still valid in image-guided brachytherapy?" *Med. Dosim.* 36:166–70. doi: 10.1016/j.meddos.2010.02.009.

Klopp, A. H., F. Mourtada, Z. H. Yu, B. M. Beadle, M. F. Munsell, A. Jhingran, and P. J. Eifel. (2013). "Pilot study of a computed tomography-compatible shielded intracavitary brachytherapy applicator for treatment of cervical cancer." *Pract. Radiat. Oncol.* 3: 115–23. doi: 10.1016/j.prro.2012.03.014.

Lin, L. "Directional interstitial brachytherapy from simulation to application." Ph.D. dissertation, University of Wisconsin–Madison, 2006.

Lin, L., R. R. Patel, B. R. Thomadsen, and D. L. Henderson. (2008). "The use of directional interstitial sources to improve dosimetry in breast brachytherapy." *Med. Phys.* 35:240–47. doi: 10.1118/1.2815623.

Liu, Y., R. T. Flynn, Y. Kim, H. Dadkhah, S. K. Bhatia, J. M. Buatti, W. Xu, and X. Wu. (2015). "Paddle-based rotating shield brachytherapy." *Med. Phys.* 42:5992–6003. doi: 10.1118/1.4930807.

Nath, R., L. L. Anderson, J. A. Meli, A. J. Olch, J. A. Stitt, and J. F. Williamson. (1997). "Code of practice for brachytherapy physics: Report of the AAPM Radiation Therapy Committee Task Group No. 56." *Med. Phys.* 24:1557–98. doi:10.1118/1.597966.

Petrokokkinos, L., K. Zourari, E. Pantelis, A. Moutsatsos, P. Karaiskos, L. Sakelliou, I. Seimenis, E. Georgiou, and P. Papagiannis. (2011). "Dosimetric accuracy of a deterministic radiation transport based ^{192}Ir brachytherapy treatment planning system. Part II: Monte Carlo and experimental verification of a multiple source dwell position plan employing a shielded applicator." *Med. Phys.* 38:1981–92. doi: 10.1118/1.3567507.

Poon, E., J. F. Williamson, T. Vuong, and F. Verhaegen. (2008). "Patient-specific Monte Carlo dose calculations for high-dose-rate endorectal brachytherapy with shielded intracavitary applicator." *Int. J. Radiat. Oncol. Biol. Phys.* 72:1259–66. doi: 10.1016/j/ijrobp.2008.07.029.

Price, M. J., K. A. Gifford, J. Horton, A. Lawyer, P. Eifel, and F. Mourtada. (2006). "Comparison of dose distributions around the pulsed-dose-rate Fletcher-Williamson and the low-dose-rate Fletcher-Suit-Delclos ovoids: a Monte Carlo study." *Phys. Med. Biol.* 51:4083–94. doi: 10.1088/0031-9155/51/16/014.

Price, M. J., E. F. Jackson, K. A. Gifford, P. J. Eifel, and F. Mourtada. (2009). "Development of prototype shielded cervical intracavitary brachytherapy applicators compatible with CT and MR imaging." *Med. Phys.* 36:5515–24. doi: 10.1118/1.3253967.

Rivard, M. J., C. S. Melhus, G. Granero, J. Perez-Calatayud, and F. Ballester. (2009). "An approach to using conventional brachytherapy software for clinical treatment of planning complex, Monte Carlo-based brachytherapy dos distributions." *Med. Phys.* 36:1968–75. doi: 10.1118/1.3121510.

Rivard, M. J. (2016). "Low-energy brachytherapy sources for pelvic sidewall treatment." *Brachytherapy* 15(S1):S22 (abstract). doi:10.1016/j.brachy.2016.04.011

Rivard, M. J. (2017). "A directional ^{103}Pd brachytherapy device: Dosimetric characterization and practical aspects for clinical use." *Brachytherapy* 16: 421–32. doi:10.1016/j.brachy.2016.11.011.

Suit, H. D., E. B. Moore, G. H. Fletcher, and R. Worsnop. (1963). "Modification of Fletcher ovoid system for afterloading, using standard-sized radium tubes (milligram and microgram)." *Radiology* 81:126–31. doi: 10.1148/81.1.126.

Viswanathan, A. N., J. Dimopoulos, C. Kirisits, D. Berger, and R. Potter. (2007). "Computed tomography versus magnetic resonance imaging-based contouring in cervical cancer brachytherapy: results of a prospective trial and preliminary guidelines for standardized contours." *Int. J. Radiat. Oncol. Biol. Phys.* 68:491–98. doi: 10.1016/j.ijrobp.2006.12.021.

Waterman, F. M. and D. E. Holcomb. (1994). "Dose distributions produced by a shielded vaginal cylinder using a high-activity iridium-192 source." *Med. Phys.* 21:101–6. doi: 10.1118/1.597241.

Webster, M. J., S. Devic, T. Vuong, D. Y. Han, D. Scanderbeg, D. Choi, B.Song, and W. Y. Song. (2013a). "HDR brachytherapy of rectal cancer using a novel grooved-shielding applicator design." *Med. Phys.* 40:091704. doi: 10.1118/1.4816677.

Webster, M. J., S. Devic, T. Vuong, D. Y. Han, J. C. Park, D. Scanderbeg, J. Lawson, B. Song, W. T. Watkins, T. Pawlicki, and W. Y. Song. (2013b). "Dynamic modulated brachytherapy (DMBT) for rectal cancer." *Med. Phys.* 40:011718. doi: 10.1118/1.4769416.

Webster, M. J., D. J. Scanderbeg, C. M. Yashar, D. Y. Han, and W. Y. Song. (2014). "Dynamic modulated brachytherapy for accelerated partial breast irradiation." *Brachytherapy* 13:S53. doi: 10.1016/j.brachy.2014.02.288.

Weeks, K. J. and J. C. Dennett. (1990). "Dose calculation and measurements for a CT compatible version of the Fletcher applicator." *Int. J. Radiat. Oncol. Biol. Phys.* 18:1191–98. doi: 10.1016/0360-3016(90)90457-U.

Example Problems

(Answers are found at the end of the book.)

1. Define IMBT. How is IMBT achieved?
    a. incredibly modern brachytherapy
    b. intensity-modulated brachytherapy
    c. image-modeling brachytherapy
    d. intensely moving brachytherapy
2. In a classical Fletcher Suit applicator, how much dose reduction is expected in the shadow of the shielded rectal or bladder shields?
    a. 15% to 25%
    b. 20% to 30%
    c. 10% to 30%
    d. 20% to 45%
3. Define DMBT. How is DMBT achieved?
    a. direction-modulated brachytherapy
    b. desperate measures brachytherapy
    c. dynamic-modulated brachytherapy
    d. definitely moving brachytherapy
4. In the Fletcher CT/MR shielded applicators that are movable, how much dose reduction can be achieved to ROIs?
    a. approximately 25%
    b. approximately 15%
    c. approximately 35%
    d. approximately 45%
5. With the shielded needle system for prostate implants designed by Adams and colleagues, how much dose reduction to the urethra is theoretically achievable?
    a. 29% to 44%
    b. 19% to 36%
    c. 24% to 40%
    d. 35% to 49%
6. Name a type of dynamic shielding that can be used with electronic brachytherapy.
    a. P-RSBT—Poodle-based rotating shield brachytherapy
    b. P-RSBT—Pollution-resistant rotating shield brachytherapy
    c. P-RSBT—Paddle-based rotating shield brachytherapy
    d. P-RSBT—Pollinating-bee rotating shield brachytherapy

7. Which of the following present challenges using ^{103}Pd as the engine for a directional source?
   a. photon yield
   b. half-life
   c. low penetration in tissue
   e. attenuation in shielding
8. Based on tenth-value layer, what are the three best attenuators for a directional source?
   a. silver, gold, and platinum
   b. tungsten, lead, and mercury
   c. mercury, nickel, uranium
   d. platinum, osmium, and iridium
9. What are four important characteristics of a directional source?
   a. energy, size, attenuator, looks good
   b. size, radionuclide, attenuator, geometry
   c. size, radionuclide, geometry, cost
   d. geometry, radionuclide, attenuator, cool name
10. The first commercially available directional source was the CivaSheet. What metal was used to attenuate the source in the shielded direction?
   a. gold
   b. gadolinium
   c. neutronium
   d. membranous iridium

Chapter 10

Brachytherapy Technologies in Early Clinical Translation

J. Adam M. Cunha[1], Wayne M. Butler[2], Antonio L. Damato[3], and Luc Beaulieu[4]

[1]UC San Francisco
San Francisco, California

[2]Schiffler Cancer Center
Wheeling, West Virginia

[3]Memorial Sloan Kettering Cancer Center
New York, New York

[4]Université Laval
Québec, Canada

10.1 Introduction

This chapter presents a vision of the future for brachytherapy. Detailed in the sections below are five new technologies and processes poised to impact brachytherapy practice in the near future: focal therapy (Section 10.2), 3D printing (Section 10.3), live needle tracking (Section 10.4), *in vivo* dosimetry (Section 10.5), and robotics (Section 10.6). These are presented in order of their level of clinical integration, with focal therapy already seeing considerable use in the clinic to *in vivo* dosimetry and robotics just starting to see clinical trials.

Brachytherapy tends to be a slow-moving field. The years of initial training, continual practice, and highly technical hands-on skill required to practice brachytherapy creates an environment where the experts are extremely competent and the middling practitioners quickly discontinue practice. In this steady state, new technologies can be overlooked and allowed to wallow, since the experts might not imagine much improvement over their current practice, while new practitioners are more inclined to cut their teeth on tried-and-true techniques.

Even as established and vetted as current brachytherapy practice is, there are developments arriving every year that have allowed us to continue to reduce trauma, increase patient eligibility, and improve dose planning and delivery. Brachytherapy practice has advanced over the first part of the 21st century, and a number of technological advances are primed for wide-scale clinical implementation.

The five technologies presented here are all at different stages of implementation, ranging from already implemented in the clinic in some fashion (focal therapy) to still-on-the-bench but ready-for-translation to the clinic (robotics). Each of these technologies individually presents a potential significant improvement to the practice of brachytherapy, yet all could also be incorporated into a single new paradigm. The goal of this chapter is to give the reader a perspective of what brachytherapy practice might look like a decade from now.

10.2 Focal Therapy

10.2.1 Introduction

The goal of focal therapy is to control disease while minimizing morbidity. Accelerated partial breast irradiation via interstitial needles or a variety of commercial intracavitary devices are well-developed examples of focal therapy that is widely used in clinical practice. Focal therapy is also commonly used in a salvage setting after the failure of the primary treatment, where additional therapy carries a heightened risk of morbidity and complications. This section will concentrate on the challenges in applying focal therapy as the primary treatment of prostate cancer.

Various technologies—such as low-dose-rate (LDR) and high-dose-rate (HDR) prostate brachytherapy, high-intensity focused ultrasound (HIFU), cryoablation, photothermal therapy, and radiofrequency ablation— were critically appraised in a consensus report by an international task force (Eggener et al. 2007). A systematic review of disease control and quality-of-life outcomes in retrospective studies and initial clinical trials demonstrated reasonable short-term outcomes and morbidity for various technologies (Valerio et al. 2014; Perera et al. 2016). Many of the studies were proof-of-principle; there have been no randomized comparative studies either between modalities or against active surveillance.

First, the rationale for introducing focal therapy is introduced based on both physics and cost-effectiveness (Section 10.2.2). This is followed in Section 10.2.3 by a discussion of patient selection, including risk stratification, focality of disease, and life expectancy. Methods of lesion localization, including MRI and ultrasound, are covered in Section 10.2.4. The incorporation of the data presented in sections 10.2.2–10.2.4 into brachytherapy practice is tackled in Section 10.2.5.

10.2.2 Rationale

10.2.2.1 Quality of Life

Every curative cancer treatment has a detrimental effect on quality of life: so, too, with prostate brachytherapy. Most men undergoing whole gland brachytherapy suffer an increase in irritative and obstructive urinary symptoms for weeks or months after the implant (Gutman et al. 2006; Bittner et al. 2007). Nearly half of the men who were sexually potent before brachytherapy lose their potency, although most of those men respond well to PDE5 inhibitors to regain erectile function (Taira et al. 2009). In theory, focal brachytherapy in many patients should reduce the radiation dose to the prostatic and bulbomembranous urethra, and it should also reduce the dose to the neurovascular bundles along the prostate and in the penile bulb. High doses and needle puncture trauma to these structures have been implicated in urinary morbidity and erectile dysfunction. However, robust evidence for such benefits of focal therapy is still lacking.

10.2.2.2 Cost Effectiveness

A focal implant will cost less than a whole-gland prostate implant because it will use fewer materials, such as seeds and needles, and it will require less time in the operating room. A similar question arises when considering active surveillance versus radical treatment in men with low-risk and very-low-risk prostate cancer as shown in Figure 10–1. LDR brachytherapy provides the lowest cost for radical treatment. Given the same number of follow-up visits at a fixed cost each, follow-up will increment the cost of each modality equally, assuming no failures. Active surveillance is initially cheaper than LDR brachytherapy, but even with no failures and no subsequent treatments, surveillance has a higher annual incremental cost due to periodic image fusion-guided biopsies. At seven years post-biopsy, surveillance exceeds the cost of LDR brachytherapy and eventually exceeds the cost of HDR and stereotactic body radiotherapy (SBRT). Intensity-modulated radiation therapy (IMRT) is the most expensive approach, followed by robot-assisted laparoscopic prostatectomy (RALP).

If the follow-up of focal therapy patients is similar to active surveillance patients, the cost advantage will be quickly lost. On the other hand, if the follow-up is like that of other radically treated patients, focal ther-

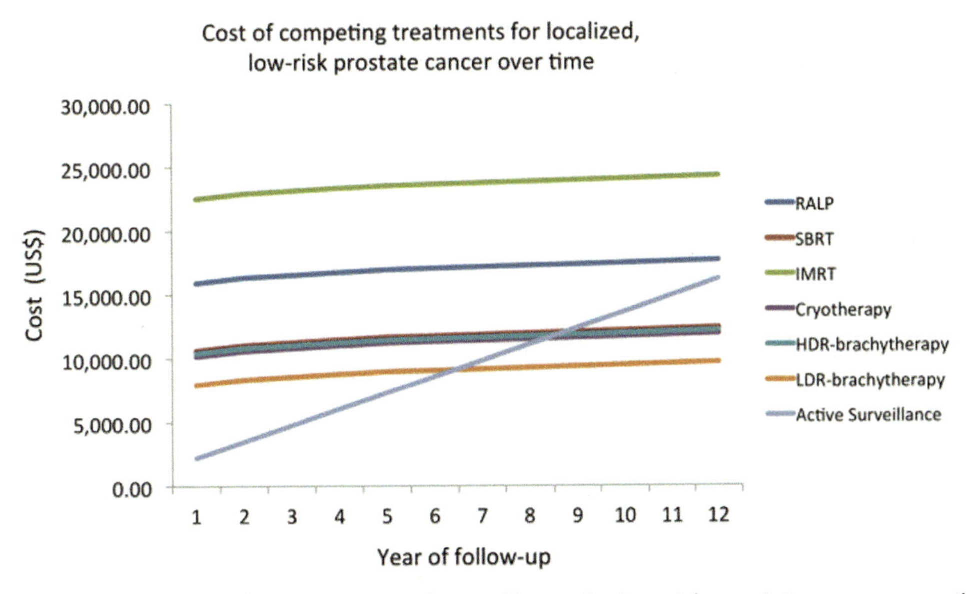

Figure 10–1 Comparative cost of treatments and surveillance for low-risk prostate cancer over time (Laviana et al. 2016). Note: this assumes surveillance is not followed up by any other treatment modality, which is not necessarily the case for all patients.

apy may retain its cost advantage, but only by assuming no failures. The cost of treating a prostate cancer failure is much greater than the cost of the initial radical treatment, particularly if the cancer has metastasized. Because of uncertainties in identifying the sites of disease and accurately targeting those sites rather than treating the whole gland, the failure rate for focal therapy will be higher than for radical therapy. No analyses have yet been reported for focal therapy costs and likely failure rates, which will depend on the risk group profile of the patients selected.

10.2.3 Patient Selection

10.2.3.1 Risk Stratification

An international panel of 15 focal therapy experts reached consensus that the optimum candidates for focal therapy are men with intermediate-risk disease, according to the National Comprehensive Cancer Network (NCCN) classification (Donaldson et al. 2015). The NCCN intermediate-risk group entails either a serum prostate specific antigen (PSA) value between 10 and 20 ng/mL, a Gleason score (GS) of 7 (either 3+4 or 4+3), or a clinical stage of T2b or T2c. Low-risk patients—those with Gleason score ≤6, PSA <10, and stage ≤T2a—are also good candidates, but such patients should be placed on active surveillance if their disease has been well-characterized.

Although high-risk patients treated with ^{103}Pd brachytherapy have a very good 14-year biochemical failure rate of only 8.6%, these patients were also treated with a moderate dose of external-beam radiation therapy (EBRT) because of a high risk of nodal and seminal vesicle involvement (Taira et al. 2013b). Focal therapy is not appropriate in these patients because of the very low likelihood of focal disease.

10.2.3.2 Focality

A plethora of pathology reports on radical prostatectomy specimens have demonstrated that about half of all specimens had multi-focal disease with different foci in the same prostate exhibiting different Gleason patterns. Gleason scores 8 to 10 are high risk and are not considered for focal therapy; GS 6 has been shown to have no metastatic potential (Donin et al. 2013). There may be rare exceptions to the latter conclusion about metastases. In one report, a lethal metastatic clone was tracked to a small, low-grade cancer focus in the original specimen rather than from the higher-grade primary or lymph node metastasis (Haffner et al. 2013). Genomic classifiers are becoming more widely used, and the cell cycle progression score in pathological

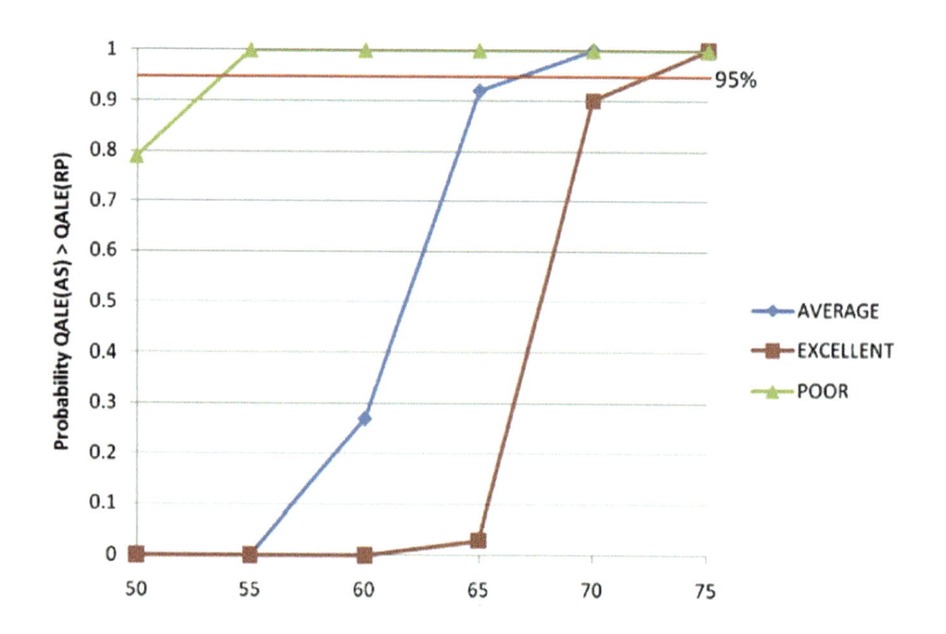

Figure 10–2 Probability that quality-adjusted life expectancy for active surveillance (QALEAS) exceeds that of radical prostatectomy (QALERP) for men with poor, average, or excellent health status (Liu et al. 2012).

GS 6 cancers found some of them displaying molecular patterns of high-risk lethal prostate cancer (Renard-Penna et al. 2015).

If only foci with GS 7 are considered for focal therapy, those foci should be only on the left or right side of the prostate. That way, hemi-ablation will still deliver a reduced dose to the urethra, neurovascular bundles, and rectum.

10.2.3.3 Life Expectancy

Societal guidelines for prostate cancer treatment suggest that patients have a life expectancy of at least 10 years, so the benefits of treatment may outweigh the loss of quality-of-life. A detailed simulation using population data compared quality-adjusted life years for low-risk men on active surveillance versus those receiving radical prostatectomy (Liu et al. 2012). A 65-year-old man of average health status without prostate cancer has 17.4 years of life expectancy (8.6 years for men with poor health or 26.1 years for men in the highest health quartile). For the same man with low-risk prostate cancer, radical surgery adds only 0.3 years of life expectancy (increasing from 16.7 years to 17.0 years), but adds 1.6 years of impotence or incontinence. For this 65-year-old in average health, the quality-adjusted life years achieved by active surveillance are about equal to those achieved by radical prostatectomy at the 95% confidence level (see Figure 10–2). For men in poor health, active surveillance is the preferred option at all ages, while for men with average health status, active surveillance is preferred after the age of 65 and after age 70 for men in excellent health.

A similar analysis comparing focal therapy to whole-gland brachytherapy in intermediate-risk patients cannot be done because there is no data for focal therapy on decrements to life expectancy and improvements in quality of life.

10.2.4 Lesion Localization

10.2.4.1 MRI

With its excellent soft tissue contrast, multi-parametric magnetic resonance imaging (mpMRI) is frequently preferred for target delineation and is becoming valuable in identifying suspicious volumes for biopsy. However, almost all studies of mpMRI prostate targeting have been by done by specialized urologic radiologists at high-volume academic centers, often with higher-field 3T magnets and endorectal coils. Among specialized prostate radiologists, inter-observer agreement was only moderate for highly suspicious lesions (Rosenkrantz et al. 2016). One group of experienced radiologists doubled their detection rate over a 20-month period (Gaziev et al. 2016). Community-based and novice radiologists may face a more extended and difficult learning curve.

Lesions <5 mm diameter are very difficult to detect, but the detection sensitivity increases with increasing GS for tumors >10 mm diameter. It may be beneficial that GS 6 tumors of all sizes rarely look suspicious because men with those tumors should not be treated. However, a study of mpMRI accuracy versus whole-mount prostate pathology missed 25% of cancers with GS 7 or greater and missed 26% of cancers >10 mm diameter (Le et al. 2015).

Getting the gross tumor volume (GTV) correct is necessary for planning the clinical target volume (CTV) and the planning target volume (PTV, see Chapter 1). The gross tumor volume is defined as the demonstrable extent of the tumor (visible, palpable, or demonstrable through imaging); the CTV includes the GTV plus any other tissue presumed to have tumor. The PTV includes the CTV plus a margin around the CTV to account for patient setup and motion uncertainties. However, a recent study of mpMRI with registered whole-mount prostatectomy slides found that the pathologic tumor volume was about three times the contoured MRI volume (Priester et al. 2017). The dimensional mismatch was greatest along the base-apex direction (average 10 mm, 97% frequency) compared to 8 mm for the lateral left-right direction and 7 mm for the anterior-posterior direction. A partial summary of their results is in Table 10–1 and an illustration of the pathology and imaging is in Figure 10–3.

Table 10–1 mpMRI underestimation of prostate cancer volume Based on Priester et al. (Priester et al. 2017)

	Tumor Volume (cm^3)	
	Gleason 3+4	Gleason ≥ 4+3
Number of matched tumors*	61	32
Pathology volume	2.6 ±0.3	3.2 ±0.5
MRI region-of-interest volume	0.7 ±0.1	1.2 ±0.2

* 93 of 141 clinically significant prostate cancer tumors were visible on MRI and 48 of 141 invisible

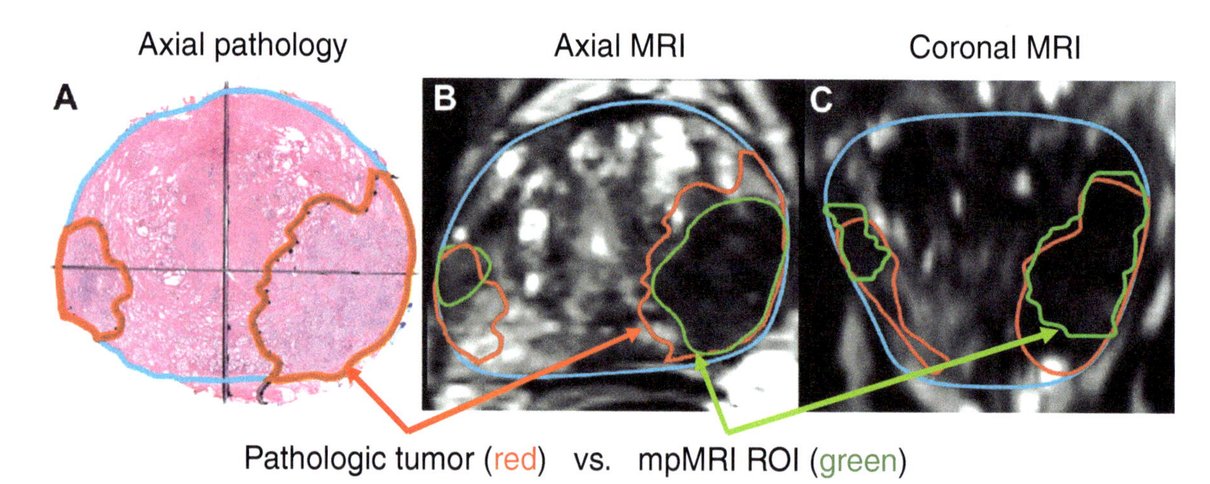

Figure 10–3 Registration of stained prostate tumors (red outline) compared to mpMRI regions of interest (green outline) on (A) whole-mount pathology slice, (B) axial MRI, and (C) coronal MRI. From Priester et al. 2017.

It has also been shown that Magnetic Resonance Spectroscopic Imaging is an effective method to identify dominant intraprostatic lesions. However, if dose planning is done on CT imaging, transferring the data from the MR to the CT can be time-consuming without automated algorithms (Reed et al. 2010).

10.2.4.2 Transperineal Template-guided Mapping Biopsy

Traditional transrectal ultrasound (TRUS)-guided prostate biopsies are noted for missing cancers at the prostate apex and anterior and for misidentifying the locations of cores that are positive for cancer. Because TRUS-guided biopsies are typically performed in a urologist's office without anesthesia, the 6 to 12 cores are obtained quickly and analyzed by pathology at modest cost. However, puncturing the rectum to access the prostate necessarily introduces fecal bacteria along the needle path, and antibiotic prophylaxis is becoming less effective, so what had been a 0.4% hospitalization rate for sepsis 20 years ago is now up to 4% (Borghesi et al. 2017).

Selection of patients and identification of targets for focal brachytherapy is well informed by transperineal, template-guided mapping biopsies (TTMB), and the perineum may be sterilized to keep the infection rate nearly zero. Two groups created 3D computer models of disease characteristics within whole-mount prostatectomy specimens and simulated various biopsy strategies. Using TTMB, they found only one significant cancer missed in 25 patients and an accuracy of 91% for identifying lesions ≥0.2 cm (Crawford et al. 2013; Hu et al. 2012).

A TTMB approach taking an average of 55 cores from 24 predefined prostate anatomic regions had high yield in determining the presence of prostate cancer in men who had at least one prior negative TRUS biopsy (Merrick et al. 2007). Because of its superior ability to identify anterior and apical disease, TTMB has a high yield (73%) when used as the initial biopsy (Bittner et al. 2015). When used to confirm the low-risk status of men whose TRUS biopsy had made them eligible for active surveillance, two studies found 39% of men upgraded to GS 7 or higher by TTMB, which is virtually the same rate of upgrading found in radical prostatectomy pathology of low-risk men (Taira et al. 2013a; Merrick et al. 2017).

The accuracy of TTMB was sufficient to consider it the reference standard in a large, multi-institutional study comparing the diagnostic accuracy of mpMRI and TRUS biopsy (Ahmed et al. 2017). TTMB found prostate cancer in 71% of the men biopsied for the first time for elevated PSA or suspicious digital rectal exam. For clinically significant cancer, mpMRI had a sensitivity of 88% and a specificity of 45%, while systematic TRUS biopsy had a much lower sensitivity of 48% and a high specificity of 99%.

Labeling the cores obtained by TTMB either by anatomic region or by template coordinates and needle offset allows the creation of a patient-specific map of sites positive for cancer.

10.2.5 Incorporation into Brachytherapy

LDR prostate brachytherapy is particularly attractive because of the high conformality of its treatment approach, global agreement about the prescription dose of single, monotherapy implants, and a large base of users experienced in TRUS-guided transperineal needle placement. There is also a large base of HDR prostate brachytherapists, but no wide consensus about dose and fractionation (Mason et al. 2014; Banerjee et al. 2015). HDR plans may sometimes have difficulty conforming to ultra-focal targets, as illustrated in Figure 10–4. The successful plan on the left adequately covers the GTV and PTV with the prescribed dose and spares the rectum. On the right, the tumor was too close to the rectum for the targets to be adequately covered and the rectum adequately spared, and was thus considered a failed plan (though the latter may have benefited from more than just three needles).

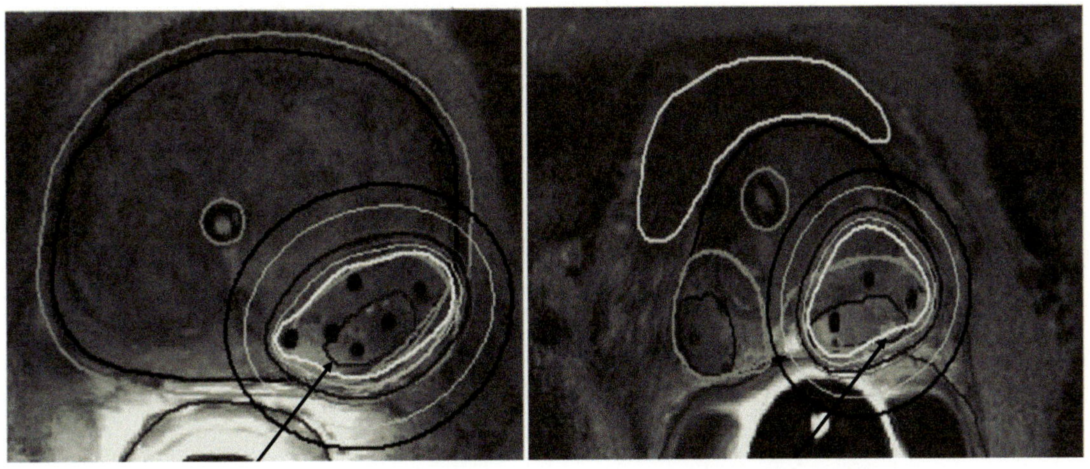

Figure 10–4 Illustration of two HDR prostate brachytherapy plans with similar gross tumor volumes (GTV in black) and planning target volumes (PTV, thick gray line) from Hosni et al. (2017) The plan on the left was successful, while the plan on the right was not able to get adequate coverage.

A consensus report addressed patient selection, target localization, and treatment protocols using ^{125}I seeds (Langley et al. 2012). The suggested protocols were subsequently applied in a retrospective study of ^{125}I treatment plans using whole-gland, hemi-prostate, and focal seed placement approaches (Al-Qaisieh et al. 2015).

10.2.5.1 Volume Strategy and Dominant Lesion Margins

Excellent results are obtained from whole-gland brachytherapy when applying risk-adjusted margins of 5 to 7 mm to the prostate (Taira et al. 2013b). In focal brachytherapy, margins must be even greater, reflecting the lack of well-defined GTV landmarks compared to the relative clarity of the prostate capsule on TRUS. The GTV is the dominant intraprostatic lesion (DIL) volume found on mpMRI, or the location and pathological length of the TTMB core with an additional margin of 3 mm. The DIL should be expanded to a planning target volume (DIL-PTV) by auto-enlargement of the DIL transversely in all directions by 7 mm and superiorly and inferiorly by 10 mm (with Boolean avoidance of the rectum and bladder) to accommodate uncertainties in the target. These include uncertainties in the location of the biopsy core in TTMB or volume averaging in MRI, needle position, ultrasound imaging, and target movement during seed implantation. These structures are illustrated in Figure10–5.

10.2.5.2 Dosimetric Goals

The dosimetric goals for the focal targets are to deliver the same prescription dose as for monotherapy implants of the chosen radionuclide. At least 99% of the primary target should be covered by the prescribed dose, i.e., $V_{100\%} > 99\%$, and the minimum dose covering 90% of the target should be >120% of the prescribed dose, i.e., $D_{90\%} > 120\%$. (See Chapter 1 for more discussion on this terminology.) The urethral $V_{150\%}$ should be less than 10% of the urethral volume and $V_{100\%}$ for the rectum should be <1 cm^3.

10.2.5.3 Seed Placement

Common general rules for seed placement in whole-gland plans—minimize the use of single-seed needles and the placement of seeds back-to-back in needles—are relaxed for focal therapy. Haworth and colleagues (2016) utilized radiobiological optimization to reduce the number of planned ^{125}I seeds by 13% in treating the whole prostate using assumptions on expected tumor cell density and tumor locations. Butler and Mer-

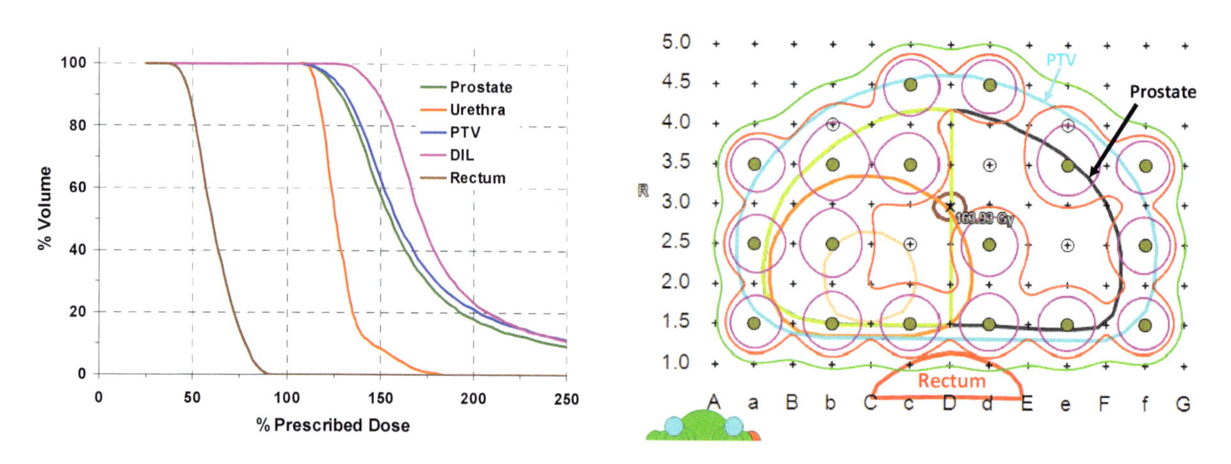

Figure 10–5 Illustration of a whole gland plan—a 30.1 cm^3 prostate enlarged to a 50.6 cm^3 PTV and implanted to a dose of 125 Gy using 86 ^{103}Pd seeds in 22 needles. (Left) Seed distribution at mid-gland. The DIL is contoured in orange comprising the lower left quartile of the prostate contour on the slice shown. (Right) DVHs of the plan. DIL is the dominant intraprostatic lesion.

rick (2016) placed ^{103}Pd seeds manually to meet all dosimetric goals and calculated voxel-based biologically effective dose based on AAPM TG-137 radiobiology parameters for all targets and anatomic structures.

10.2.5.4 Hemi-gland Plan

For patients with unilateral multi-focal disease, hemi-prostate ablation is a feasible approach, even if small foci of GS 6 disease are on the contralateral side. The prostate is bisected sagittally along the urethra to create the hemi-prostate, and that structure may be expanded to a PTV in all directions, except across the sagittal midline. This demarcation preserves the option of treating the other side with radiation in the future if necessary. A 3D view of a hemi-prostate seed placement on a transverse slice and DVHs are shown in Figure 10–6.

The 15.3-cm^3 hemi-prostate had a $V_{100\%}$ = 100%, $D_{90\%}$ = 136% of prescribed dose, and a BED of 153 Gy. The DIL within the hemi-prostate was not explicitly targeted for dose escalation and had $D_{90\%}$ = 138% and a BED of 162 Gy.

10.2.5.5 DIL Plan

The DIL was defined by the length of the mapping biopsy core, 16 mm, and expanded by a radius of 5 mm about the template coordinate sampled, C-2.0. This was further expanded to a DIL-PTV by 7 mm trans-

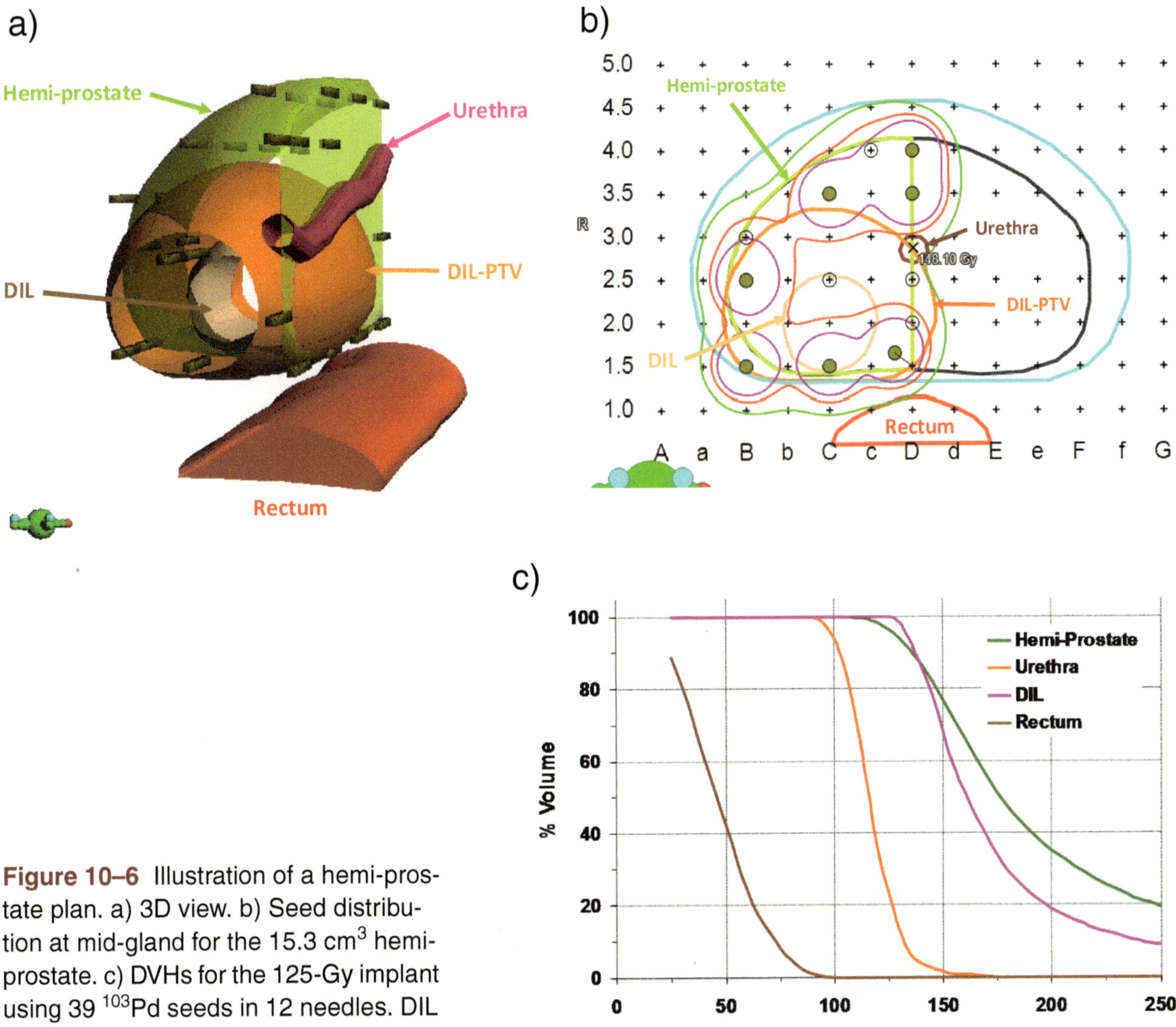

Figure 10–6 Illustration of a hemi-prostate plan. a) 3D view. b) Seed distribution at mid-gland for the 15.3 cm^3 hemi-prostate. c) DVHs for the 125-Gy implant using 39 ^{103}Pd seeds in 12 needles. DIL is the dominant intraprostatic lesion.

versely, but not beyond the posterior border of the prostate, and 10 mm superiorly and inferiorly. The DIL-PTV is sometimes denoted by the expansion parameters as DIL710. This is illustrated in Figure 10–7.

10.2.6 Clinical Dosimetric Outcomes

The dosimetry of 22 men with low- to intermediate-risk disease unilateral on both mpMRI and TTMB who were implanted with the hemi-prostate approach was reported by Laing et al. (2016). The ^{125}I implants were prescribed to 145 Gy, and the mean planned $V_{100\%}$ and $D_{90\%}$ were 98% and 121%, respectively. The post-implant CT imaging was performed within 18 hours of the implant, and the mean day 0 delivered $V_{100\%}$ and $D_{90\%}$ were 93% and 106%, respectively. As expected, there was a significant dose reduction to the contralateral half of the prostate and its OARs.

10.2.7 Conclusion: Treatment Successes, Failures, and Clinical Concerns

Many institutions treating low- and intermediate-risk patients with brachytherapy monotherapy have long-term biochemical failure-free rates exceeding 90% for both groups (Taira et al. 2013b). With such a high level of success for the comparator, any randomized trial trying to prove equivalence (or non-inferiority) of focal therapy to standard LDR brachytherapy would need more than 800 participants. Such a trial is unlikely because the potential gains are so modest and the number of patients needed to treat is so large.

Edema in ultra-focal targets may be more detrimental to dosimetry if the margins are not sufficiently generous. The focal targets will be assaulted by fewer seeds and needles, but given the same curative dose. The amount of edema will probably be similar to that observed in whole-gland implants, but focal targets are more dominated by peripheral dosimetry. The hemi-gland study by Laing et al. (2016) found the hemi-prostate volume increased by a factor of 1.24 on the day 0 imaging.

The small economic advantage of shorter OR time and using fewer seeds and needles in focal therapy could easily be swamped by the cost of re-treatment or salvage if even a slight increase in failure rates occurs. Treatment failures, even when local, will be more difficult and expensive to treat than the initial treatment, and complications will be more likely because of inevitable overlap of treatments.

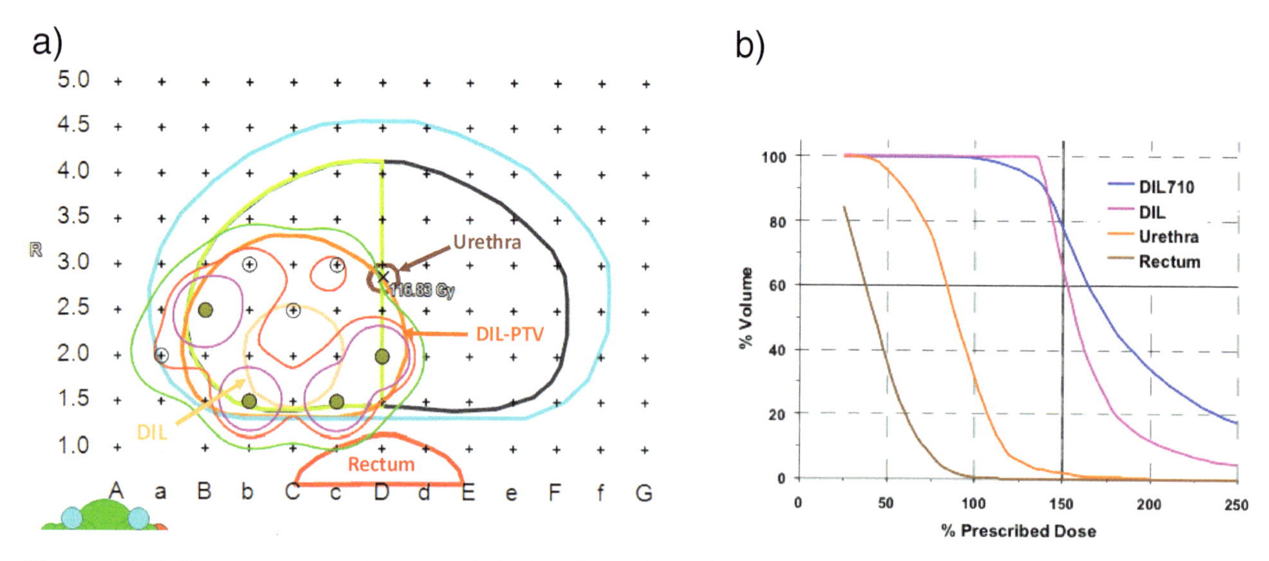

Figure 10–7 The dominant intraprostatic lesion (DIL) plan after expansion to DIL-PTV by adding 7 mm transversely to the DIL (except not beyond the posterior prostate) and 10 mm superiorly and inferiorly to create a DIL710. a) Transverse view of the 1.3 cm^3 DIL and the 11.6 cm^3 DIL-PTV implanted to 125 Gy using 29 ^{103}Pd seeds in 8 needles. b) DVHs for the targets and organs at risk (OAR).

Because half the prostate or less is treated and small foci of disease beyond the index lesion may be left untreated intentionally, definitions of success are elusive. The patient's PSA may nadir, but it will inevitably rise. Follow-up biopsies, even in the treated lobe, are likely to be positive. Quality of life will surely suffer a modest decrease in the short term, followed by a rebound back to pre-implant levels over the first few months. Over the long term, even untreated men suffer a slow decline in sexual and urinary function. Many men shun active surveillance because of fear and uncertainty. For these men, will focal therapy simply be a sham providing psychological solace to satisfy the need to do something over nothing?

There are few, if any, physical challenges that cannot be overcome with careful focal brachytherapy treatment planning. Unanswered, however, are the necessary questions of whether focal therapy over-treats patients with low-risk cancer or under-treats patients with significant multifocal disease, missing the curative window.

10.3 3D Printing

10.3.1 Introduction

Radiotherapy for cancer seeks to deliver therapeutically effective doses to target tissues while sparing healthy tissue unnecessary exposure. This objective is achieved by accurately modeling the relationship between the radiation source and the dose attenuation in the relevant materials (i.e., air, bone, tissue, etc.) and by carefully positioning the source relative to the patient anatomy, both in planning and in delivery.

The American Brachytherapy Society (ABS) states for interstitial brachytherapy for vaginal cancer, "Three-dimensional treatment planning is recommended with CT scan and/or MRI. The treatment plan should be optimized to conform to the clinical target volume and should reduce the dose to critical organs, including the rectum, bladder, urethra, and sigmoid colon" (Beriwal et al. 2012). For vaginal cuff, the ABS states, "A properly fitted brachytherapy applicator should be selected that conforms to the vaginal apex and achieves mucosal contact with optimal tumor and normal tissue dosimetry" (Small et al. 2012). *What does one do when the limited selection of commercially available applicators does not allow meeting ABS applicator conformality recommendations?*

Three-dimensional printing is a method of rapid prototyping that has the potential to improve dose delivery accuracy and consistency, customize patient treatment, reduce the reliance of treatment outcome on physician skill and experience, and expand the use-cases for existing radiotherapy procedures. However, for 3D printing to become clinically relevant, fundamental studies must be performed to characterize the dose attenuation in 3D printing materials as well as their biocompatibility and sterilizability. Once safety and radiation characteristics are established, current planning and delivery work flows can be adapted to utilize the customization options offered by 3D printing.

This section will first introduce the reader to what 3D printing is, followed by a discussion of the tools needed to use it in a clinical setting. Sections 10.3.3 and 10.3.4 cover biocompatibility and sterilization, respectively. Open-source hardware philosophy will be discussed in Section 10.3.5. Finally, Section 10.3.6 will cover some already-implemented clinical applications.

10.3.2 What is 3D Printing?

3D printing, or *additive manufacturing*, is a process used to fabricate solid, three-dimensional objects by laying down successive layers of a building material (most commonly plastic). Three-dimensional printing technology has been around since the 1980s. Its recent surge in popularity and ubiquity is due to the costs associated with the technology falling under the $1,000 mark for hobbyist-level machines. The strength of 3D printing is two-fold: (1) its ability to prototype design ideas quickly and (2) its ability to fabricate three-dimensional objects with complicated internal structure. In contrast, while injection molding—a traditional manufacturing technology used to make plastic objects—is cheaper for mass production, it does not allow for creating objects cheaply with internal structure personalized for each patient.

The material used to build the 3D printed objects is called the *substrate*. The range of materials that can be used as a substrate has grown exponentially in recent years, with anything from concrete to cheese available on specialty printers (Lipson and Kirman 2013). The most common substrates are plastics. In its most simple work flow, a 3D printed object can be fabricated in three steps: (1) design of a virtual object using computer software (computer aided design, CAD); (2) generation of a stereolithography (STL) file for the object for communication of the design to the printer, sometimes called a mesh, since it consists of a 3D wire-frame that outlines all its geometric features; and (3) physical printing of the object on the 3D printer.

Three-dimensional printing is already in service in medicine for a wide range of uses, from fabrication of prostheses to drug delivery devices (Rengier et al. 2010; Schubert et al. 2013; Ventola et al. 2014). Radiation oncology benefits from the strides taken in other medical fields, most critically in the realm of biocompatibility and sterilization.

10.3.3 Biocompatibility

10.3.3.1 Definitions and Regulations

The United States Pharmacopeia (USP) is a non-governmental organization with a self-imposed charge to endorse public health by establishing up-to-the-minute standards to safeguard the quality of medicines and other healthcare technologies. The USP is a large organization that sets standards for quality, purity, strength, and consistency across a wide range of products. This includes evaluating the biocompatibility of materials and establishing standards for the manufacturing industry. The U.S. FDA regulates the sale and manufacturing of these materials based on evaluations by the USP. Vendors of substrates and printers are responsible for establishing the biocompatibility of the materials that they sell.

To establish the biocompatiblity of a material, USP has a set of laboratory tests that place the material in contact with laboratory animals for various lengths of time to assess any adverse effects. Each material is categorized into one of six levels (I to VI) of biocompatibility based on three tests designed around the FDA interpretation of the International Standards Organization (ISO) document ISO-10933 [http://www.fda.gov/RegulatoryInformation/Guidances/default.htm]:

- *Acute Systemic Toxicity Test*: material is introduced into the biological system orally, via inhalation and via dermal exposure.
- *Intra-cutaneous Toxicity Test*: material is administered directly on the tissues it will be in contact with during normal use without protection of the skin or any other body system.
- *Implantation Toxicity Test*: material is implanted into a live animal to test the response of live tissue to the material.

For example, the standard test implantation time required to classify a material as USP Class VI is five days. If, after the five-day period, there is no sign of irritation or toxicity, it meets the implantation requirements of that test. Class I is considered safe for human use, as long as contact is not prolonged. Class VI is considered safe to be in the body permanently.

10.3.3.2 General Application to Brachytherapy Practice

The requirement for the biocompatibility of the printing substrate to be used for brachytherapy depends on what type of tissue the applicator will be in contact with and the length of the contact. Therefore, the type of brachytherapy to be performed determines the biocompatibility class required. For intact skin, only USP Class I is required, while for any permanent implant, Class VI is required. Given the typical environment at the body site where the brachytherapy is to be performed, Table 10–2 summarizes the most suitable biocompatibility level.

Table 10–2 USP class requirements for brachytherapy applications on the skin, in contact with intact mucosal surfaces, any breached surface, and implants

	Skin	Mucosal Surface	Breached Surface	Implants
Biocompatibility				
<1 day	USP Class I	USP Class I	USP Class III	N/A
1–30 days	USP Class I	USP Class III	USP Class V	N/A
>30 days	USP Class I	USP Class V	USP Class VI	USP Class VI
Sterilization				
	Low level disinfection	High level disinfection	Sterilize	Sterilize

Note that the time limit is the aggregate over the course of the treatment. If fabricating an applicator for a breached-skin surface that has a treatment time of 2 minutes per fraction for 6 fractions, the total treatment time is 12 minutes. If setup and removal takes 5 minutes each per fraction, the total time the applicator is in contact with the patient is 72 minutes, (2 min + 5 min + 5 min) × 6 fractions. Therefore, even if the fractionation occurs over 3 days, only USP Class III (<1 day contact) is necessary.

Since the class requirements listed are always the minimum required, any class greater than that listed is also allowed. Many vendors are quickly moving to obtain USP Class VI categorization for the majority of their substrates. Therefore, it is generally a good idea to use Class VI material for all brachytherapy applications.

10.3.4 Sterilization

Cleaning and sterilization is less defined than biocompatibility since the requirements are generally set by each individual hospital or clinic. However, the U.S. Centers for Disease Control has established definitions for three levels of cleaning in its "Guideline for Disinfection and Sterilization in Healthcare Facilities" (Rutala et al. 2008).

Cleaning: removes visible soil (organic and inorganic material) accomplished manually or mechanically using water with detergents or enzymatic products. Thorough cleaning is essential before high-level disinfection and sterilization because inorganic and organic materials interfere with the effectiveness of these processes.

Disinfection: eliminates many or all pathogenic microorganisms, except bacterial spores, on inanimate objects. Common agents in healthcare: liquid chemicals or wet pasteurization.

Low-level disinfectant destroys all vegetative bacteria (except tubercle bacilli), lipid viruses, some nonlipid viruses, and some fungi, but not bacterial spores.

High-level disinfectant kills bacterial spores and all other microorganisms when used in sufficient concentration under suitable conditions.

Sterilization: destroys or eliminates all forms of microbial life by physical or chemical methods. The principal sterilizing agents used in health-care facilities include steam under pressure, dry heat, ethelyne oxide (EtO) gas, hydrogen peroxide gas plasma, and liquid chemicals.

Pros and cons of each of the three most common methods of sterilization (steam, ethylene oxide gas, and hydrogen peroxide gas) are outlined in Table 10–3. EtO has been the standard for decades for plastic-type devices because it does not require high temperatures like that needed for steam. However, three states (CA, NY, and MI) have outlawed its use due to toxicity concerns for the personnel performing the sterilization. Hydrogen peroxide gas (STERRAD® is the most common brand of processing unit) is increasing in popularity, but it is not uncommon to find vendors who have validated their substrates with EtO but not hydrogen peroxide.

Table 10–3 Pros and cons for the three main methods of sterilization in a medical center. Vendors can certify their substrates based on wet-lab tests of the effectiveness of each of these.

	Steam	Ethelyne Oxide	Hydrogen Peroxide
Pros	- Non-toxic - High lumen penetration - Short processing time (4 to 30 min)	- Low temperature (37 to 60 °C) - Adequate lumen penetration	- Non-toxic - Low temperature (37 to 44 °C) - Medium processing time (~50 min) - Adequate lumen penetration
Cons	- High temperature (121 to 132 °C) may cause melting or warping	- Devices should be rinsed before use - Potential health risks to personnel - Restricted from use in CA, NY, & MI - Long processing time (10 to 24 hours)	- Material-dependent lumen penetration

It is the responsibility of the physicist and physician to work with their sterilization department and hospital to determine which method is necessary for each clinical application. If a brachytherapy procedure is occurring in a procedure room in the radiation oncology department or clinic and not in a sterile operating room, the requirements for sterilization are likely less; in this case, cleaning and disinfecting of the applicators may be sufficient.

10.3.5 Clinical Applications

Custom applicators for brachytherapy have been in use for surface and intracavitary brachytherapy for over a century. The next subsections will discuss specific examples of using 3D printing in clinical brachytherapy practice.

10.3.5.1 Dosimetric Properties of Substrates

Before employing any material in a custom applicator in a radiation oncology department, it is critical to evaluate its photon attenuation and properties. Failure to understand a material's properties and incorporate them into the treatment planning system used for planning will lead to erroneous dose distributions in the patient. This, of course, can lead to significant adverse effects for patients. Groups are beginning to evaluate the dosimetric properties of 3D printing substrates. For example, PC-ISO—a polycarbonate compliant with International Standards Organization biocompatibility and sterilization criteria sold by Stratasys (Eden Prairie, MN)—is one material that is almost water-equivalent on CT and for the ^{192}Ir photon spectrum (Cunha et al. 2015).

10.3.5.2 Gynecological Brachytherapy

Tandem and ovoids, tandem and ring, or vaginal cylinder applicators are ubiquitous for HDR brachytherapy of gynecological cancers. Traditionally, dwell positions were loaded to create reproducible applicator-based dose distributions (N.B. the "pear-shaped" distribution to deliver the prescription dose to Point A (ICRU 38) (Pötter et al. 2006). In the era of image-guided, three-dimensional planning, dose is instead prescribed to cover a high-risk critical target volume (HR-CTV, see Chapter 6) at the time of brachytherapy (Haie-Meder et al. 2005) utilizing each of the dwell positions in the applicator as best to tailor the dose the anatomy of the patient.

However, conformality still is restricted, since the relative geometry of the dwell positions within a given applicator are fixed and cannot be modified for each patient. Thus, the coverage provided by the dwell positions within intracavitary applicators may not be ideal. Though adequate dose to the HR-CTV has been shown to correlate with local control, normal tissue toxicities still exist (Dimopoulos et al. 2009).

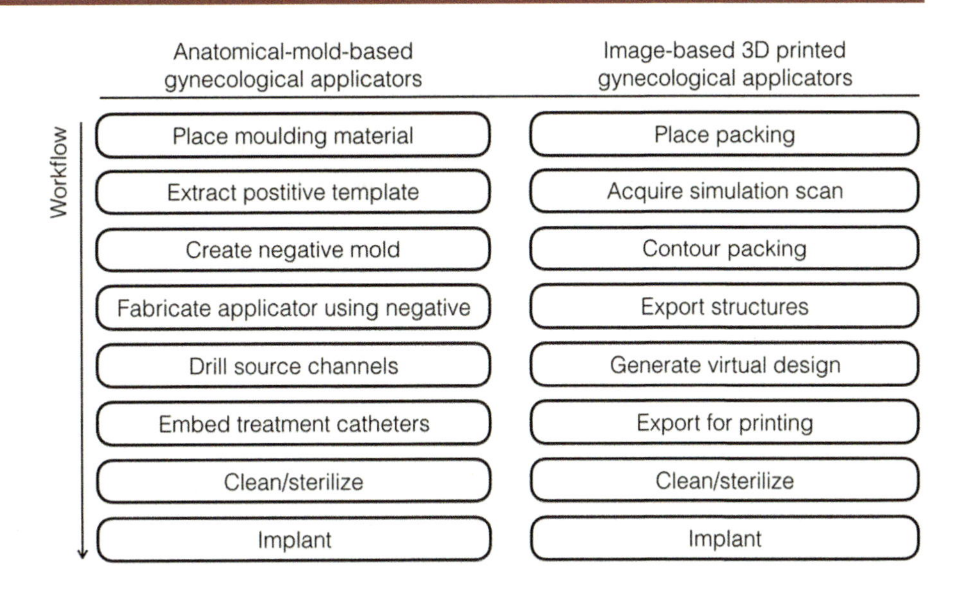

Figure 10–8 An example work flow for generating patient-specific brachytherapy applicators. (Left) Well-established procedure designed using imaging of the patient to be treated. The place packing may not always be necessary. For interstitial head and neck, where the applicator sits on the skin, for example, the design can be constructed.

Poor applicator fit during vaginal intracavitary brachytherapy can lead to air gaps and under-dosing of the target volume (Richardson et al. 2010; Cameron et al. 2008). For interstitial implants, poor applicator fit may lead to poor implant geometry and increased inter-fraction variability when multiple fractions are delivered over a single implant (Pinnaduwage et al. 2013). While commercial applicators are available in a variety of sizes, there remains a substantial and unmet need for customized applicators that allow personalized brachytherapy implants to reduce toxicities without compromising on optimal tumor dose coverage.

10.3.5.3 Simple Applicator Geometries for Better Fit

One of the most straightforward uses of 3D printing is to create vaginal cylinders of custom sizes to better fit the individual patient. This is especially useful for patients with a significantly larger-than-normal or smaller-than-normal vaginal vault. In the latter, standard applicators may not be small enough to fit inside the patient, while in the former the need for large amounts of packing increases the likelihood of applicator movement before or during treatment. Sethi and colleagues (2016) demonstrated the clinical use of several custom-sized cylinders that were better able to fit the patient than what was available commercially. However, an even more patient-specific work flow is possible.

10.3.5.4 Work Flow: Custom Applicators in the Clinic

An example work flow of custom applicator design for gynecological cancers is outlined on the left side of Figure 10–8. This example work flow is based on the work of several groups that acquire imprints of the vaginal cavity to fabricate custom applicators by hand (Albano et al. 2008; Magne et al. 2010; El Khoury et al. 2015; Nilsson et al. 2015). The first step is to obtain a positive imprint of the vaginal cavity. This is done with molding material that is pliable upon insertion but hardens somewhat after placement, e.g., dental putty. (Care must be taken if the introitus is small or the positive template could be difficult to remove.) The positive imprint is then removed and used to cast a negative mold. Once the negative mold is cast, the original positive can be discarded. The negative mold can be used to fabricate multiple positive copies of the vaginal imprint.

A positive applicator is cast from the negative mold using the desired material, then holes are drilled to allow placement of the brachytherapy catheters within or through the applicator. Once the catheters are glued into place, the finished applicator is cleaned and sterilized and is ready for implant. Once implanted, treatment planning follows the standard work flow of imaging, digitization, contouring, and dose planning.

One of the first reported uses of 3D printing in gynecological brachytherapy was to create a custom version of a Vienna applicator that used 3D printing to create a cap mounted on the ring component of tandem and ring, allowing interstitial needles to be inserted through the ring cap. This allowed an interstitial-intracavitary approach for patients with locally advanced tumors (Dimopoulos et al. 2006; Kirisitis et al. 2006; Nomden et al. 2012; Fokdal et al. 2013; Mohamed et al. 2015) without having to replace their already-purchased brachytherapy equipment.

Just as with traditional brachytherapy planning, the most benefit from 3D printing of applicators is achieved when imaging is used to inform the design and construction of the applicators. The work flow for creating *image-based 3D-printed gynecological applicators* derives from the *anatomical-mold-based gynecological applicator* process described above and is outlined on the right side of Figure 10–8.

10.3.5.5 Work Flow: Image-guided Custom Applicator Design

Imaging is desirable to obtain the correct geometry for the applicator. However, imaging the patient in the natural body state does not provide enough information about the elasticity and size of the vaginal cavity. Therefore, prior to acquiring the imaging used for applicator design, packing is placed into the vaginal cavity to facilitate contouring of the shape of the expanded vault. The packing is contoured on a treatment planning system. At this point, cavities can be included to accommodate placement channels for the dose-delivery catheters. The digital structure set (DICOM RS file) is then exported to meshing software to generate the digital design file (STL, or STereoLithography file), and the STL file is sent to the printer for fabrication. Once printed, cleaned, and sterilized, the applicator can be placed in a standard implant work flow.

Applicators designed in this fashion nominally would accept interstitial- or endobronchial-type catheters for source delivery. This makes connection to the afterloader trivial, but connection can be problematic because the catheters have to be secured in place in the applicator. This can be achieved with glue, but Weibe et al. (2015) have shown that this, too, can be incorporated into the applicator design. Figure 10–9 shows an applicator design from their group that uses imaging to generate a personalized applicator for a cervical cancer. The same group has also created custom applicators for penile and anal and perineal cancers.

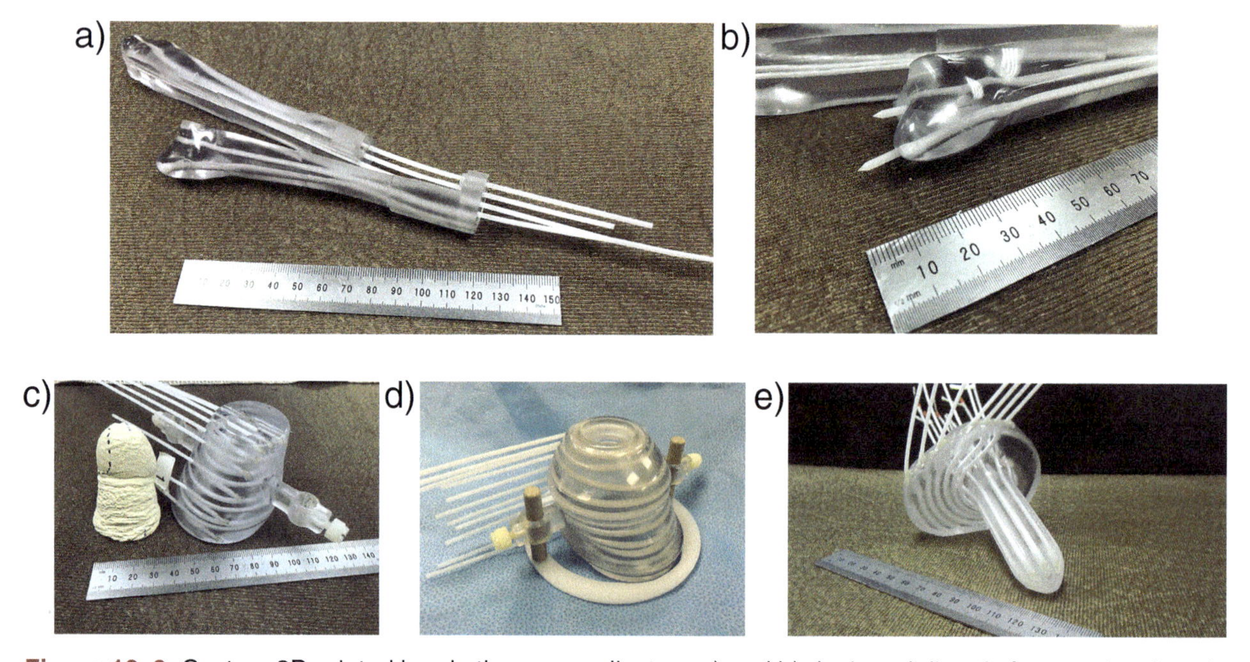

Figure 10–9 Custom 3D printed brachytherapy applicators a) and b) designed directly from patient imaging for cervical cancer, c) and d) designed from molds of the patient for penile cancer, and (e) a custom anal and perineal applicator.

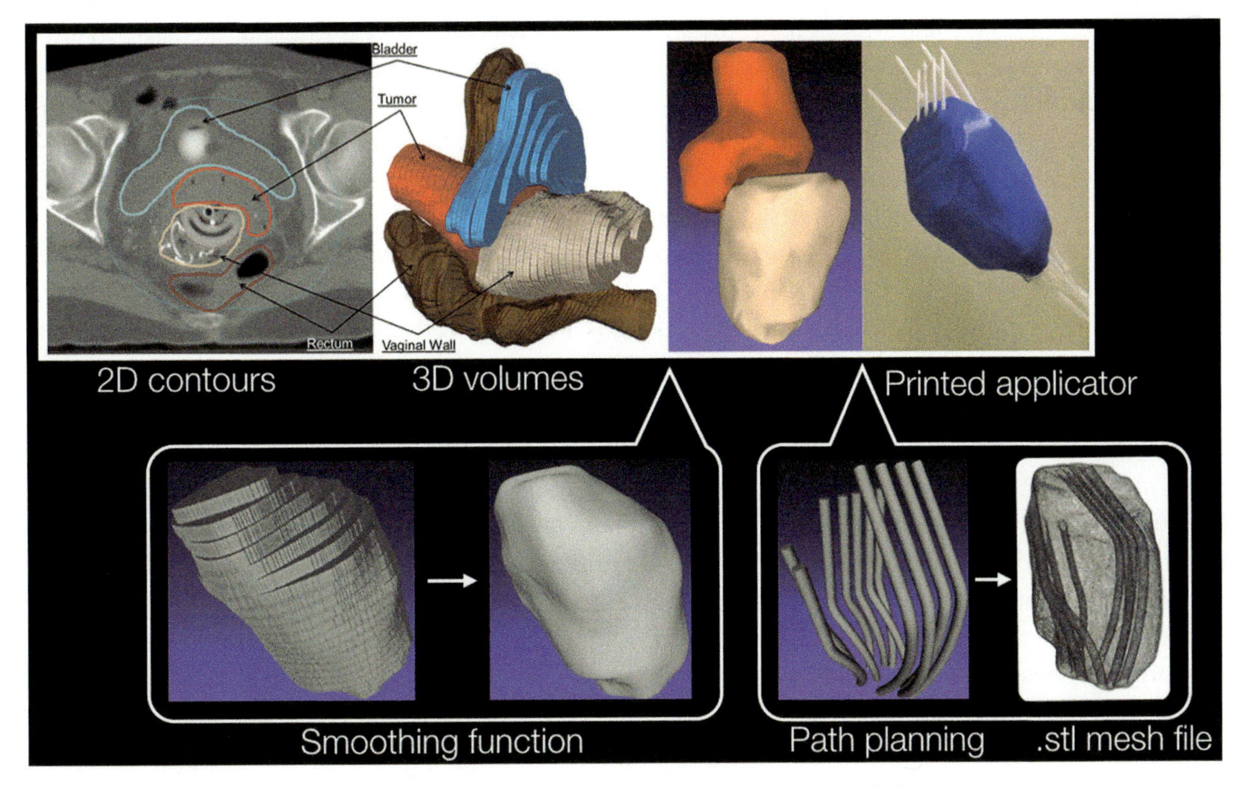

Figure 10–10 Image-guided, dose-optimized custom applicator design work flow.

10.3.5.6 Employing Dose Optimization into Applicator Design

The applicator designs of Weibe et al. are based on a forward-planning type approach. Garg et al. (2013) took the concept to the next level based on the curvilinear needle planning of Xu et al. (2008). They demonstrated a method that uses imaging information (CT, MR, or US) in conjunction with an advanced special optimization algorithm to generate the needle paths within the applicator. Using the contoured CTV and OAR geometries, the needle paths within the applicator are designed at the same time as the dose planning. This allows for applicators tailored not only to the patient's gross anatomy, but also to their individual tumor or normal-structure geometry. This leads to highly conformal dose distributions and can be designed to guide interstitial needles as an intracavitary template.

Figure 10–10 details the work flow for this method. Starting with 2D contours, the 3D volumes from the treatment planning system are fed into a smoothing function to create an anatomically comfortable design. Since the software-designed applicator (pale pink) is customized to the patient, it mates contiguously with the tumor target (red in the figure). Using the *in-silico* shell of the applicator, a custom path-planning algorithm is executed to find the optimal catheter channel geometry (defined as that which provides the most optimal dose distribution). From this, an STL file can be generated to send to the printer to fabricate the custom device (upper right).

10.3.6 Design and Open Source Hardware

The rationale behind the rise in popularity of 3D printing relies heavily on the open-source hardware movement. As the main patents on additive manufacturing lapsed, individuals and groups of people started developing their own 3D printers and sharing their designs openly with others. The Open Source Hardware Association (OSHWA) defines the statement of principles in the following way:

> Open source hardware is hardware whose design is made publicly available so that anyone can study, modify, distribute, make, and sell the design or hardware based on that design. The hardware's source, the design from which it is made, is available in the preferred format for making modifications to it. Ideally, open source hardware uses readily-available components and materials, standard processes, open infrastructure, unrestricted content, and open-source design tools to maximize the ability of individuals to make and use hardware. Open source hardware gives people the freedom to control their technology while sharing knowledge and encouraging commerce through the open exchange of designs. (Website of the Open Source Hardware Association, www.oshwa.org)

In the context of 3D printing for brachytherapy, the field would benefit from an increase in the sharing of designs, experience, and knowledge. Whether it is through the development of applicators for specific needs or the methodology for developing tissue-equivalent materials and phantoms, everyone would benefit from more open sharing of those developments, including the field and, most importantly, the patients. It is important to understand that it remains to the user of the innovation to validate and ensure its appropriate use.

With an increasing interest for open source in medical physics and to increase sharing opportunities of both research and clinical works, a sharing platform (Open Source in Medical Physics, http://www.openmedphys.org) has been developed and is currently gaining momentum in the field (Open Source Hardware Association, www.oshwa.org).

10.4 Needle Tracking

10.4.1 From Surgical Navigation to Brachytherapy Needle Tracking

Tracking technologies have typically been used in the medical field for the navigation of surgical tools inside the patient. Either performed in an imaging suite that does not permit real-time image acquisition (e.g., CT or MRI), or used as a replacement of in-room image guidance, tracking permits one to visualize in real time the location of a surgical instrument inside the patient, relative to the surrounding anatomy. An obvious extension of these techniques is to assist in brachytherapy implantation, where navigation can be used to improve needle positioning inside the tumor and away from organs at risk. Additionally, these technologies can be used for the automatic identification of the needle track, enabling automatic digitization, digitization quality assurance, or treatment verification.

A challenge in brachytherapy needle tracking lies in the presence of two distinct frames of reference: the 3D image frame of reference—which is established by the CT, MR or ultrasound system—and the navigation frame of reference which, depending on the technology and application, may be completely independent from the image frame of reference. In most cases, a registration of the two frames of references will be necessary, and it is crucial to evaluate the performance of a brachytherapy tracking system taking into consideration registration uncertainties. With native accuracy <1 mm for most navigation systems, registration uncertainties may be clinically more important than tracker performance. In this section, four tracking technologies will be discussed: optical, fiber Bragg grating (FBG), electromagnetic tracking (EMT), and active MRI tracking.

10.4.2 Optical Navigation Systems

In interventional radiology, tracking has been used to guide interventions via navigation systems. A mature technology for surgical application is optical tracking, where outside markers connected to the surgical tool are tracked though a series of cameras (Figure 10–11). By registering patient imaging information (e.g., a CT or MRI scan acquired before the intervention) with the frame of reference established by the optical system, it is possible to visualize the position of the surgical tool inside the patient in real time. Such registration can be performed via identification of outside markers that are visible by the optical camera. These markers are either on the surface of the patient, or can be directly associated to the CT scanner in interventional radiology procedures. Reports exist of implementation of optical navigation for HDR brachytherapy of spine lesions (Voros et al. 2016) on a limited number of patients. The use of this technology for the reconstruction of HDR catheters has been limited, though. Brachytherapy catheters can bend inside the patient, and this technology does not typically allow for correct tracking in these situations. Other logistical challenges limit the diffusion of this technology in brachytherapy. Tracking depends on outside optical markers mounted at the base of the needles, which can be bulky. Also, one must have a direct line of sight between the optical camera and the needle mount during insertion.

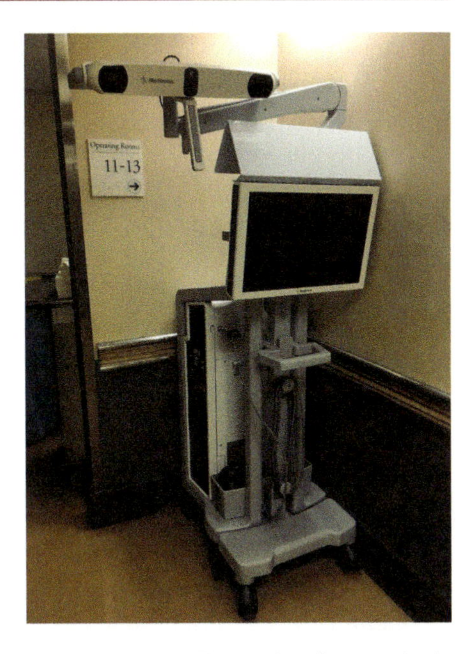

Figure 10–11 Example of an optical navigation system used in spine HDR brachytherapy.

10.4.3 Fiber Bragg Grating

Another possible tracking technology involves the use of FBG sensors (Roesthuis et al, IEEE 2014). Optical fibers with a series of FBG etched on them are inserted into the catheters. Curvatures in FBG etched onto the fibers result in shifts in the wavelength of light passed through the device, allowing the detection of the curvature of the catheters inside the patient (Figure 10–12). This tracking system does not natively provide the location of the

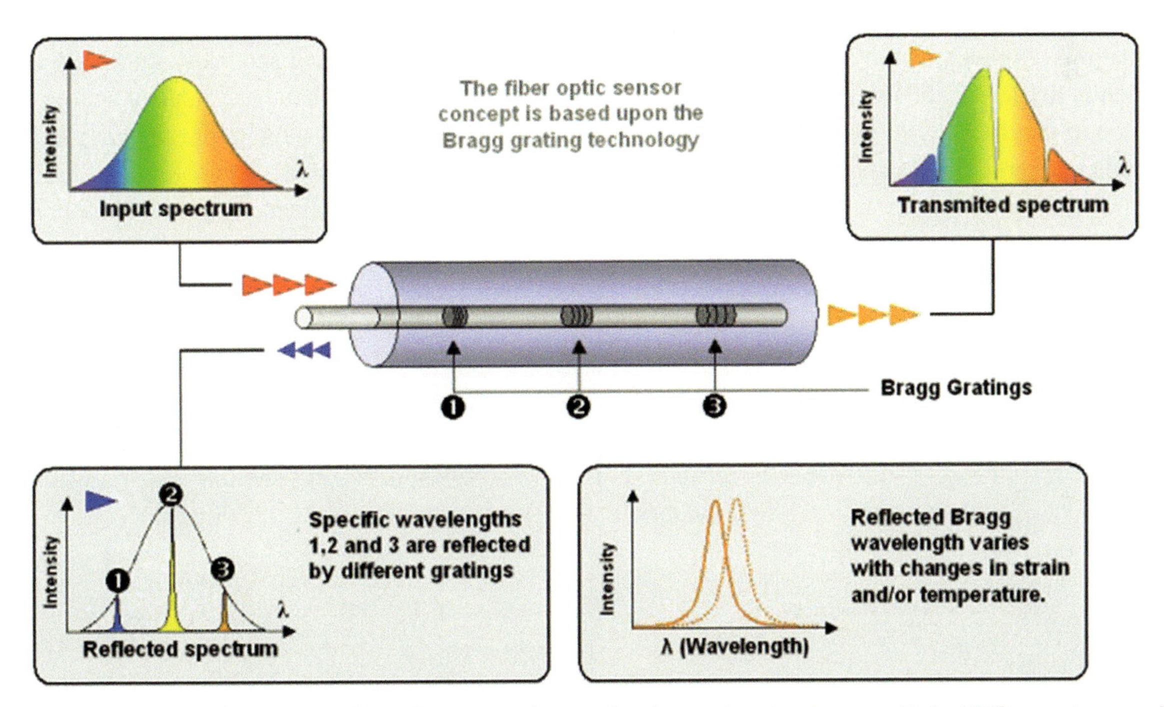

Figure 10–12 Schematic of the Fiber Bragg grating technology, showing how multiple FBG may be used to detect the curvature of a catheter along its length. From: http://www.scaime.com/en/335/article/la-technologie-advoptics.html. Accessed on 3-13-2017.

catheters, only the curvature. To overcome this limitation, a pre-clinical report on FBG tracking application in brachytherapy (de Battisti et al. 2016) describes a calibration work flow permitting one to, in theory, utilize this system in MRI-based brachytherapy implantations. No reports of clinical use of this technology exist at the moment.

10.4.4 Electromagnetic Tracking

As a mature technology, EMT has seen use in many medical procedures (Franz et al. 2014) including EBRT (Willoughby et al. 2006). In the past few years, the application of this technology to brachytherapy has been discussed (Zhou et al. 2013; Mehrtash et al. 2014; Bharat et al. 2014; Damato et al. 2014; Poulin et al. 2015; Bert et al. 2016; Racine et al. 2016). EMT is performed by positioning one or multiple sensors inside an electromagnetic field. Figure 10–13 shows a setup for prostate brachytherapy. Sensors can be placed on the transrectal ultrasound, the template, and inside the needles to map the geometry of the entire brachytherapy system.

The sensors are small coils that are attached to a central control unit. The electromagnetic field, which varied spatially, is provided by a field generator, and it established the reference frame of the tracking system. By measuring the current generated in the coils as they are positioned inside the electromagnetic field, a computerized system provides the spatial coordinates of the coil in the reference system. Some sensor designs allow for the detection values for six independent degrees of freedom.

Electromagnetic tracking does not require line of sight, and miniaturization of the sensors is technically feasible. By inserting the sensors in brachytherapy catheters, the coordinates describing these catheters in the EMT frame of reference can be established. The system's accuracy is <1 mm (Bo et al. 2012). The challenge to brachytherapy application is the registration of the EMT frame of reference to the patient frame of reference, that is, the imaging used for comparison. In theory, the field generator can be mounted precisely compared to the imaging frame of reference, e.g., a field generator mounted at a fixed geometry from a TRUS probe (Figure 10–3). In this case, a-priori registration can be established.

It is technically challenging to fix the two frames of references in clinical settings, though, and it is possible that this approach would introduce meaningful uncertainties to the tracking accuracy. Another common approach is to register the two frames of references based on a set of points that can be easily and precisely identified in the images and by the EMT system. An example, akin to optical navigation systems, is the use of outside markers. The distance between these markers and the tumor site may have the effect of amplifying any rotational errors that may have been introduced during the markers localization. In some applications, registration to points directly inside the implantation site has been described (Damato et al. 2014).

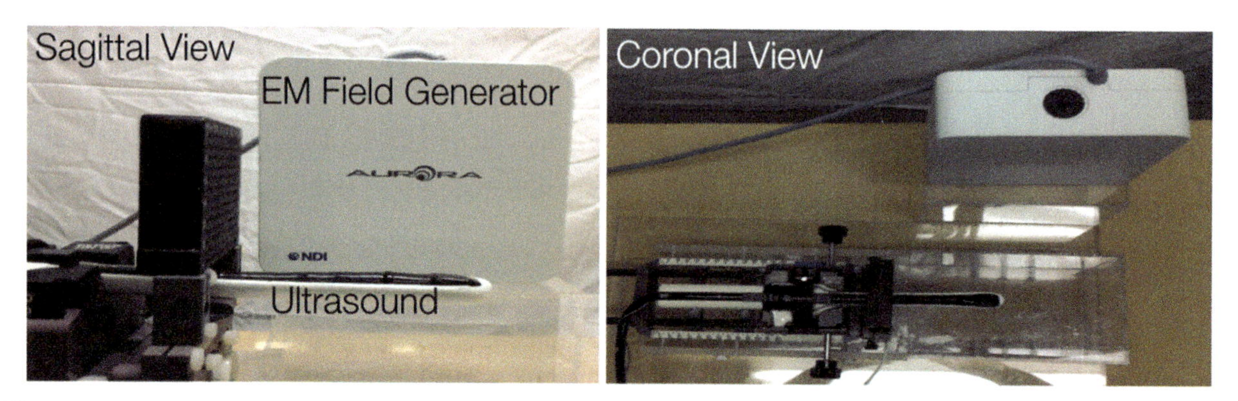

Figure 10–13 An electromagnetic field generator incorporated into a prostate seed implant setup. The usable field generator volume is approximately 30 cm x 30 cm x 30 cm.

For clinical application, it is essential to establish the tracking accuracy, including the expected registration uncertainties. Electromagnetic tracking can also be used to perform *in vivo* measurement of the speed of sound during a brachytherapy procedure to better calibrate the ultrasound image (Samboju 2017). This can be extremely useful when implementing stereotactic robotic brachytherapy (Section 10.6). Electromagnetic tracking has two main limitations. First, distortion to the electromagnetic field due to ferromagnetic materials may introduce artifacts in the tracking data, invalidating the results. It has been shown that good results are possible in clinical settings, but quality assurance of a tracking system in clinical conditions is necessary to evaluate possible data distortions. Second, the field emitter and sensors cannot be manufactured without ferromagnetic materials and, therefore, EMT is fundamentally not compatible with MRI.

10.4.5 Active MRI Tracking

Active MRI tracking works by detecting the position of micro-coils inside the field generated by the MRI scanner. In principle, this technology is similar to EMT, where the micro-coils are the sensors, and the MRI itself is the field generator. This form of tracking has the unique advantage to the ones previously discussed in that it natively shares the same frame-of-reference of the imaging system. This unique feature removes the need for registration, thus eliminating a source of uncertainty. The fabrication of non-ferromagnetic, metallic stylets used to aid in the implantation of brachytherapy plastic needles, equipped with a series of micro-coils, has been described (Wang et al. 2015a).

Preliminary clinical results show that application of this technology to brachytherapy has promise (Wang et al. 2015b). In MRI-guided insertions, active tracking of brachytherapy needles can simplify the insertion process by allowing the combination of superior visibility of the tumor in MRI and an easy identification of the needle location as it is implanted (Damato et al. 2015). Reports on active tracking have thus far come from a single institution, and the commercial unavailability of MRI active trackers and the relative rarity of MRI-guided brachytherapy needle insertions, have limited the diffusion to other centers. As MRI-guided insertions become more common, we can speculate that the use of this or a similar technology will also increase.

10.5 *In vivo* Dosimetry

10.5.1 Context

A recent Vision 20/20 paper in the journal *Medical Physics* defined the scope of *in vivo* dosimetry (IVD) as an approach or method "to detect major errors, to assess clinically relevant differences between planned and delivered dose, to record dose received by individual patients, and to fulfill legal requirements" (Mijnheer et al. 2013). A similar manuscript dedicated to brachytherapy further states that "the initial motivation for performing IVD in brachytherapy was mainly to assess doses to OAR by direct measurements, because precise evaluation of OAR doses was difficult without 3D dose treatment planning" (Tanderup et al. 2013). It is fair to state that IVD is the only approach to measure the true delivered dose to a patient, thus ensuring that the planned dose is indeed equal to the delivered dose. It is also mandatory by law for at least one or all treatment fractions in radiation oncology in many jurisdictions, particularly in Europe.

Close to one third of the incidents reported in the International Atomic Agency *Safety Report No 17* are linked to brachytherapy (IAEA 2000). The International Commission on Radiation Protection *Report 97* identified human errors as one of the main causes of inadequate brachytherapy delivery, including exchanged guide tubes, misadjusted applicators, and reconstruction errors, including misidentification of channel tips (Valentin, ICRP 2005). Therefore, from a quality and safety perspective, the ability to prevent rare but major accidents is an important motivation for IVD.

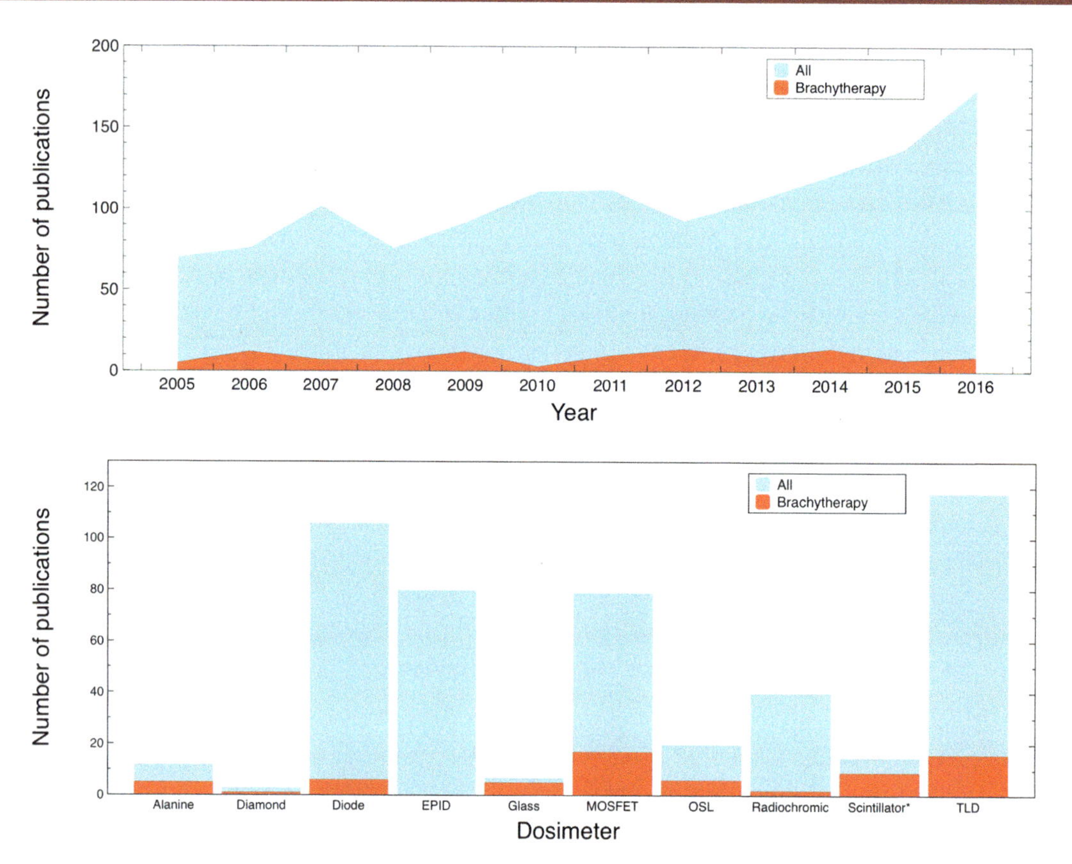

Figure 10–14 Top panel: Number of publications per year using PubMed with the key words "*in vivo*" and "dosimetry" for a given year, as well as adding "brachytherapy" (red curve). Bottom panel: number of publications for various dosimeters used in *in vivo* dosimetry. Here the key words are "*in vivo*," "dosimetry," "<dosimeter type>" and adding "brachytherapy" as a search term when needed.

Figure 10–14 (top panel) shows that the interest for this topic appears to be increasing in the last 10 years. However, brachytherapy application of IVD has remained constant over time and constitutes a small fraction of these publications' reports. This could come as a surprise, since brachytherapy is usually performed with less fractions and higher doses per fraction than EBRT: up to 19 Gy for prostate HDR monotherapy (Krauss et al. 2017). In fact, societal guidance for *in vivo* dosimetry in brachytherapy is extremely limited. In at least three European Society for Radiotherapy and Oncology (ESTRO) booklets, IVD is discussed as an important tool of the quality control chain in radiation therapy (VanDam and Marinello 1994; Huyskens et al. 2001), including brachytherapy (Venselaar and Perez-Calatayud 2004). There is otherwise very little literature providing practical guidelines associated with IVD in brachytherapy. In this context, one should refer to the recent Vision 20/20 papers listed previously, as well as to coverage of this topic in recent books such as *Comprehensive Brachytherapy* (Venselaar et al. 2012) or *Emerging Technologies in Brachytherapy* (Song et al. 2017).

10.5.2 Dosimeters

Over the years, numerous dosimeters and approaches have been proposed to perform IVD in brachytherapy (Tanderup et al. 2013; Kertzscher et al. 2014). A comparison between radiation therapy at large and brachytherapy is provided in the bottom panel of Figure 10–14. One can see that TLD, diodes, EPID, and MOSFET are the most popular dosimeters in EBRT, while MOSFET and TLD are the most often discussed in the literature for brachytherapy. While EPID is specific to EBRT, as such detectors are mounted on every

Table 10–4 Properties of various dosimeters proposed or used for *in vivo* dosimetry in brachytherapy. Adapted from Tanderup et al. (2013).

	TLD	Diode	MOSFET	Alanine	RL	PSD
Size	+	+/–	+/++	–	++	++
Sensitivity	+	++	+	–	++	+/++
Energy Dependence	+	–	–	+	–	++
Angular Dependence	++	–	+	+	++	++
Dynamic Range	++	++	+	–	++	++
Temperature Dependence	++	+	+	++	–	+
Time-resolved Dosimetry	–	++	+	–	++	++
Commercial Availability	++	++	++	++	–	+
Main Advantages	- No cables - Well-studied system	- Commercial systems at reasonable price - Well-studied system	- Small size - Commercial systems at reasonable price	- Limited energy dependence - No cables	- Small size - High sensitivity	- Small size - No angular and energy dependence sensitivity
Main Disadvantages	- Tedious procedures for calibration and readout - No on-line dosimetry	- Angular and energy dependence - Minor temperature dependence	- Limited life of detectors - Energy dependence - Temperature dependence for non-dual bias MOSFET - Possible dose/ LET dependence	- Not sensitive to low doses - Tedious procedures for calibration and readout - No on-line dosimetry - Expensive readout equipment not available in clinics	- Needs frequent recalibration - Stem effect - Minor temperature dependence - Not commercially available	- Stem effect - Minor temperature dependence - Not commercially available for brachytherapy *in vivo* dosimetry

modern linac, the idea of a flat panel imager to track an HDR ^{192}Ir source during treatment has recently been proposed for brachytherapy (Smith et al. 2016). Table 10–4 provides an overview of the key characteristics that make a dosimeter appropriate for brachytherapy, and well as their main advantages and disadvantages (Tanderup et al. 2013). Most dosimeters have one or more good characteristics, such as being small enough to fit in catheters or to be used as rectal dosimeters. However, most dosimeters also have some energy, angular, or temperature dependence. As such, it is crucial to understand factors that can influence the dose measurements and to establish an uncertainty budget associated with a specific dosimeter and clinical process. See (Waldhäusl et al. 2005; Toye et al. 2008).

Depending on the kind of information one is expecting to obtain from performing IVD, the choice of an appropriate dosimeter can be severely reduced. For example, Andersen and colleagues (2009) have demonstrated that measuring the integrated dose over a whole brachytherapy treatment has much less error detection power than performing time-resolved IVD. In the latter, dose is acquired in real time and enables measurement of each dwell position individually, thus more easily catching errors, such as catheter swap that would not be detected from an integrated measurement. However, for time-resolved measurements, two of

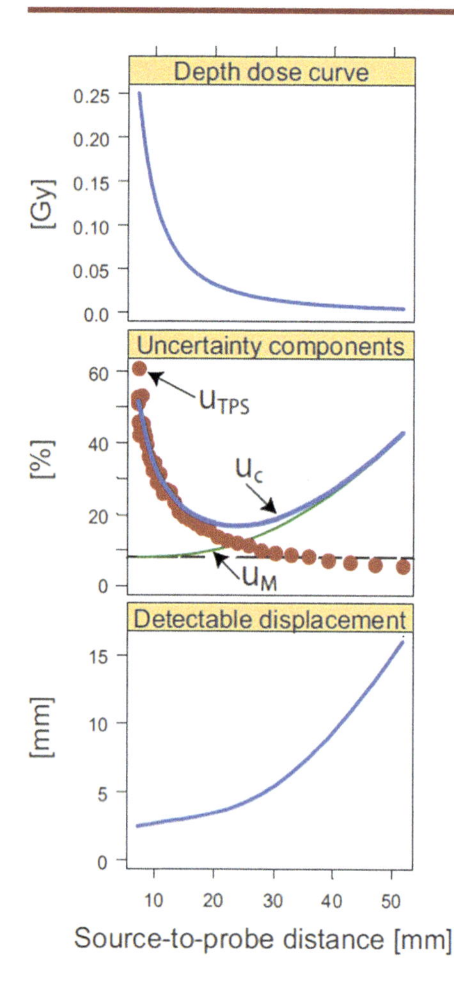

Figure 10–15 a) Calculated depth-dose curve. b) Uncertainty components in the form of a 1-mm positioning uncertainty relative to treatment planning reference value (U_{TPS}), the measurement uncertainty (U_M) (i.e., dosimeter uncertainty budget), and the total combined uncertainty (U_C). c) approximate displacement distances that can be detected. Note: The U depicted here represents an uncertainty; it has no relation to the variable used for brachytherapy source strength (air-kerma strength). From Andersen et al. 2009.

the four dosimeters listed in Table 10–4 are not commercially available at this time. Of the remaining two, one cannot fit inside standard HDR brachytherapy catheters, while also having significant angular and energy dependences.

10.5.3 Uncertainty and IVD

Going back to the bottom panel of Figure 10–14, the third most cited article for brachytherapy is the scintillation dosimeter, for which only a single clinical *in vivo* study has been published (Suchowerska et al. 2011). This brings up the point that most of the publications on IVD in brachytherapy tackle the development of *in vivo* dosimeters and their validation on phantoms rather than their clinical use on patients. Again, a key point mentioned in Table 10–4 is related to the absence of a commercial scintillation dosimeter product for IVD. A recent review by Beaulieu and colleagues reported that, at most, 20 peer-reviewed publications have reported clinical IVD results for brachytherapy patients (Song et al. 2017). These articles covered measurements made with nine different dosimeters and have some key features in common. First, every study tends to either adopt or deduce action levels that are fairly large, with measured difference compared to TPS around 20%. Second, not all observed large dose differences are due to delivery errors, but sometimes simply to the uncertainty in the assumed source-to-dosimeter distances (e.g., due to motion, etc.).

The issue of uncertainty is critical to IVD in brachytherapy. Andersen and colleagues (2009) reframed this discussion around two major components: the positioning uncertainty relative to treatment planning reference value, i.e., the source-to-dosimeter distance (called U_{TPS}) and the dosimeter uncertainty budget, U_M, as displayed in Figure 10–15. In the top panel of Figure 10–15, a depth dose curve is calculated by the treatment-planning system (TPS), which also includes its own uncertainty budget, as described by DeWerd and colleagues (2011) in the TG-138 report.

From Figure 10–15, it can be clearly seen that close to the source, the uncertainty is dominated by the positioning uncertainty. This is due to the strong dose gradient encountered in brachytherapy. The reduction in dose is around 50% after the first 10 mm such that a 1-mm position error (i.e., error in the source-to-dosimeter distance) leads to large dose differences at small distances. At large distances, the measurement uncertainty, U_M, is the major contributor to the total uncertainty, U_C. Obviously, as the dose rate gets lower, the uncertainty associated with the dosimeter response increases. This leads to the result presented in the bottom panel where, for the particular dosimeter used in that study, the minimum displacement detectable in close proximity of the source is 3 mm, increasing to 16 mm at 50 mm distance (Andersen et al. 2009).

Based on these results, it appears obvious that knowing a specific dosimeter's uncertainty budget is mandatory before clinical usage, and this should be performed in a well-controlled geometry. Furthermore, it is also obvious that a single value action threshold is not realistic, but rather depends on the source-to-dosimeter distance and the uncertainty components. As such, better error decision methods need to be adopted (Kertzscher et al. 2011; 2014).

10.5.4 Perspective

The very first question that needs to be answered before invoking IVD: Is IVD really the most appropriate approach for the task to be achieved? An obvious example is validating the source strength, for which a well chamber is the most appropriate tool (not IVD). Similarly, if one does not have easy access to imaging, IVD measurements might be severely constrained by uncertainty on the source-to-dosimeter distances, to the point of making IVD useless.

The following gives a checklist of things that should be explored before starting clinical IVD measurements:

- Get to know your dosimeter of choice. No dosimeter is perfect, so explore its limitations:
    - energy, angular and temperature dependence
    - detection thresholds in terms of dose and dose rate for known distances
- Establish an uncertainty budget in well-controlled conditions.
- Use imaging and, if possible, real-time imaging:
    - establish the dosimeter position relative to organ
    - monitor motion if possible
- Take into account TPS limitations, and use MBDCA, if appropriate (see Chapter 2).
- Use time-resolved IVD, if possible, to gain sensitivity to errors, such as an interchange of transfer tubes.
- Use more than one point of measurement, if possible (triangulation!).

In addition to the elements listed above, new technologies could enable the tracking of the source (Smith et al. 2016) or simplify the measurements of the source-to-dosimeter distance by integrating the dosimeter directly in to the imaging device, as demonstrated recently for ultrasound-based prostate brachytherapy (Carrara et al. 2016).

10.5.5 Conclusion

IVD has an important role to play in brachytherapy: it remains the only measurement that confirms the actual delivered dose either to the target or an organ at risk. At this point in time, commercial dosimeters are available, but they lack integration with better tracking mechanisms and better software that allows intelligent, variable action thresholds. As such, it is possible to state that commercially available IVD systems are lacking. The execution of IVD in a clinical setting still requires a high level of expertise and background preparation in order to be used with confidence.

10.6 Robotics

10.6.1 Introduction

Side effects in brachytherapy result from both excessive radiation to organs at risk and from needle penetration through sensitive structures. In terms of survival rate, brachytherapy is a highly successful method for treating cancer, but it is a high-skill procedure that can produce negative side effects if not performed precisely. Interstitial brachytherapy procedures are highly dependent on physician experience and skill. Tem-

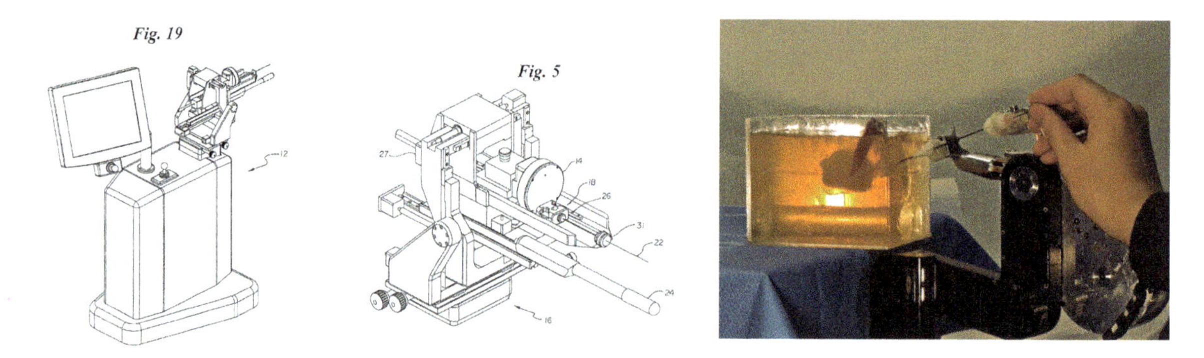

Figure 10–16 Left: a sketch of the controller unit from the 2005 patent for a robotic brachytherapy device. (Elliot 2005). Center: the envisioned needle insertion component (Elliot 2005). Right: A six-degree-of-freedom robot holds a needle in place for insertion by a human (Garg 2013).

plates necessarily restrict implant geometry, disqualifying patients for treatment due to, for example, blocked needle paths by the bones of the pubic arch (pubic arch interference, PAI).

The optimal brachytherapy implantation relies on two factors: seeing where to go and going there. Imaging technology enables the former and robot technology the latter. Robots (Figure 10–16) have the ability to affect almost every part of the brachytherapy implant work flow, thus they can have a significant impact on the ability to perform a high-quality brachytherapy procedure. To this end, the AAPM and GEC-ESTRO convened a task group to present a vision for robots in brachytherapy. The Task Group-192 report (Podder et al. 2014) presents the main goals of robotic brachytherapy as:

1. Improve accuracy of needle placement and seed delivery (i.e., place the needle and seed correctly at the planned location).
2. Improve consistency of the seed implantation procedure (i.e., eliminate inter-clinician variability).
3. Improve avoidance of critical structures (e.g., for prostate implants, urethra, pubic arch, rectum, bladder, and structures of the penis).
4. Improve dose optimization.
5. Reduce the clinician's learning curve.
6. Improve clinician fatigue.
7. Improve staff radiation exposure.
8. Streamline the brachytherapy procedure.

10.6.2 What is a Robot?

10.6.2.1 General Definition

Generally, a robot is a machine that can do the work of a person either automatically or controlled by software. More technically, a robot is a re-programmable, multifunctional manipulator designed to move material, parts, tools, or specialized devices through variable programmed motions for the performance of a variety of tasks. In general, the Robotics Industry Association categorizes a robot into four classes, as summarized in Table 10–5 (www.robotics.org). The classes progress from Class 1 to Class 4; with each increase, the robot acquires more autonomous abilities. Class 1 describes a mechanical device that operates fully under human control. This can include devices that augment human abilities, but that are still fully under human control for all movements. Class 3 robots operate on their own, but within a set of specific operations that are

Table 10–5 The Robotics Industry Association (RIA) classifies robots according to their abilities within an environment

Category	Description
Class 1	Devices that manipulate objects with manual control
Class 2	Automated devices that manipulate objects with predetermined cycles
Class 3	Programmable and servo-controlled robots with continuous point-to-point trajectories
Class 4	Robots of Class 3 that also acquire information from the environment and move intelligently in response

Table 10–6 The Task Group-192 Report classification of devices for robot-assisted brachytherapy are more practical in a clinical environment than those of the RIA. Adapted from Podder et al. 2014.

Category	Description
Level I	A human controls each movement; each machine actuator change is specified by the operator
Level II	A human specifies general moves or position changes and the machine decides specific movements of its actuators
Level III	The operator specifies only the task; the robot manages to complete it independently
Level IV	The machine will create and complete all its tasks without human interaction

defined and static. Class 4 robots acquire and process information about the environment and chose their actions within that environment based on this information.

These definitions, however, can be cumbersome in a clinical environment. AAPM Task Group 192 recommends dividing robotic devices for brachytherapy into four categories, Level I through IV, as summarized in Table 10–6).

A Level I device is a piece of hardware that can be manipulated by the human operator in a fully manual manner. For example, to position a needle-holding device, the human would move the arm into place and lock down its position; then move the needle holder into place and lock down its orientation; then clamp the needle into the gripping actuator. This process is tedious and may not be considered much of an improvement over traditional insertion techniques. However, the process is always fully controlled by the human. This level device is not much more sophisticated than locking jointed arms and Vise-Grip™-style pliers.

A Level II device improves on a Level I device by allowing the human operator to specify a desired end configuration of the device, and the robot engages its motors and actuators to effect that end result. For example, the operator will tell a needle insertion motor to move 2 cm to the right to be in the correct position on the skin surface and then advance the needle 3 cm.

A Level III does not use specific instructions like those of Level II. A Level III robot would most likely be linked to an imaging system. The human operator will tell the robot-controlling software that a seed needs to be placed a specific coordinate within the body. The robot will determine the best way to deliver that seed to the desired location and execute the needed motions by independent activation of its actuators.

It is feasible that robots of Level I through III will enter the clinic within the next five years. Level IV robots, however, operate in a fully autonomous manner. These robots would incorporate a significant level of artificial intelligence in order to assess the anatomical configuration of a given patient, devise a full implant, and execute that implant.

10.6.2.2 Servo vs. Stereotactic

Servo-controlled robots are devices that take human input from controlling devices like a joystick. This is akin to the da Vinci surgical® system by Intuitive Surgical (Sunnyvale, California). This robotic system is fully controlled by the human, and all movements are initiated by the human. Because of this, these devices generally fall under the Class 1 and Level I to II category. In general, since servo-type robots are fully controlled by a human, the device itself does not need to be aware of its position or orientation in the 3D space. The human uses direct line of site or camera-based imaging systems to verify the position and orientation of the actuator performing the action on the patient.

Stereotactic robotic devices, on the other hand, retain information about their position and orientation in space. A CyberKnife® stereotactic radiosurguery robot (Accuray, California, U.S.), for example, is programmed to move the treatment linac to specific positions and orientations in space to deliver radiation to the isocenter from a wide range of directions. Stereotactic devices are generally Class 2 through 4 and Level II through IV.

10.6.3 Integration into the Clinic

10.6.3.1 Safety

Safety is a key concern for integrating robots into the clinic. There is little practical concern for Level I robots since the human operator is in full control of all movements and needle insertions. However, at Level II, III, and IV, additional safety considerations must be taken into account. These include, at the very least, emergency disengagement features, provisions to revert to manual control, and functional reliability. In addition, any device being introduced to the clinic needs to be easily cleaned and sterilized.

10.6.3.2 Fundamental Requirements

The Task Group 192 report lays out the fundamental requirements of a robot used for brachytherapy (Podder et al. 2014):

1. safety for the patient, clinicians, and staff
2. ease of cleaning and decontamination
3. compatibility with sterilization
4. methods for review of planned dose distributions and robot motions before needle placement
5. visual (mandatory) and force (optional) feedback during needle insertion
6. visual confirmation of each needle-tip placement and seed deposition
7. provision for reverting to conventional manual brachytherapy at any time
8. quick and easy disengagement in case of emergency
9. robust and reliable operation
10. ease of operation in the procedure environment

These issues must be incorporated into a work flow that can be safe and effective. Figure 10–17 outlines a detailed basic work flow for robotic brachytherapy.

10.6.4 How Robots Can Improve Brachytherapy

The method of needle insertion is conceptually simple but difficult to do precisely. Beyond simple statements of improved precision, accuracy, reliability, and lowering the barrier for entry, a number of techniques are enabled by robots that can change the way brachytherapy is performed.

10.6.4.1 Non-template-based Needle Patterns

Traditional brachytherapy for prostate consists of templates for needle guidance. Robots can position needles in a wide array of spatial orientations and do not require rigid templates. They can guide needles along non-uniform and non-parallel axes to achieve a continuum of needle positions and angles. The needle insertion process can be designed, for example, to avoid puncture of the penile bulb (Cunha et al. 2009). This same

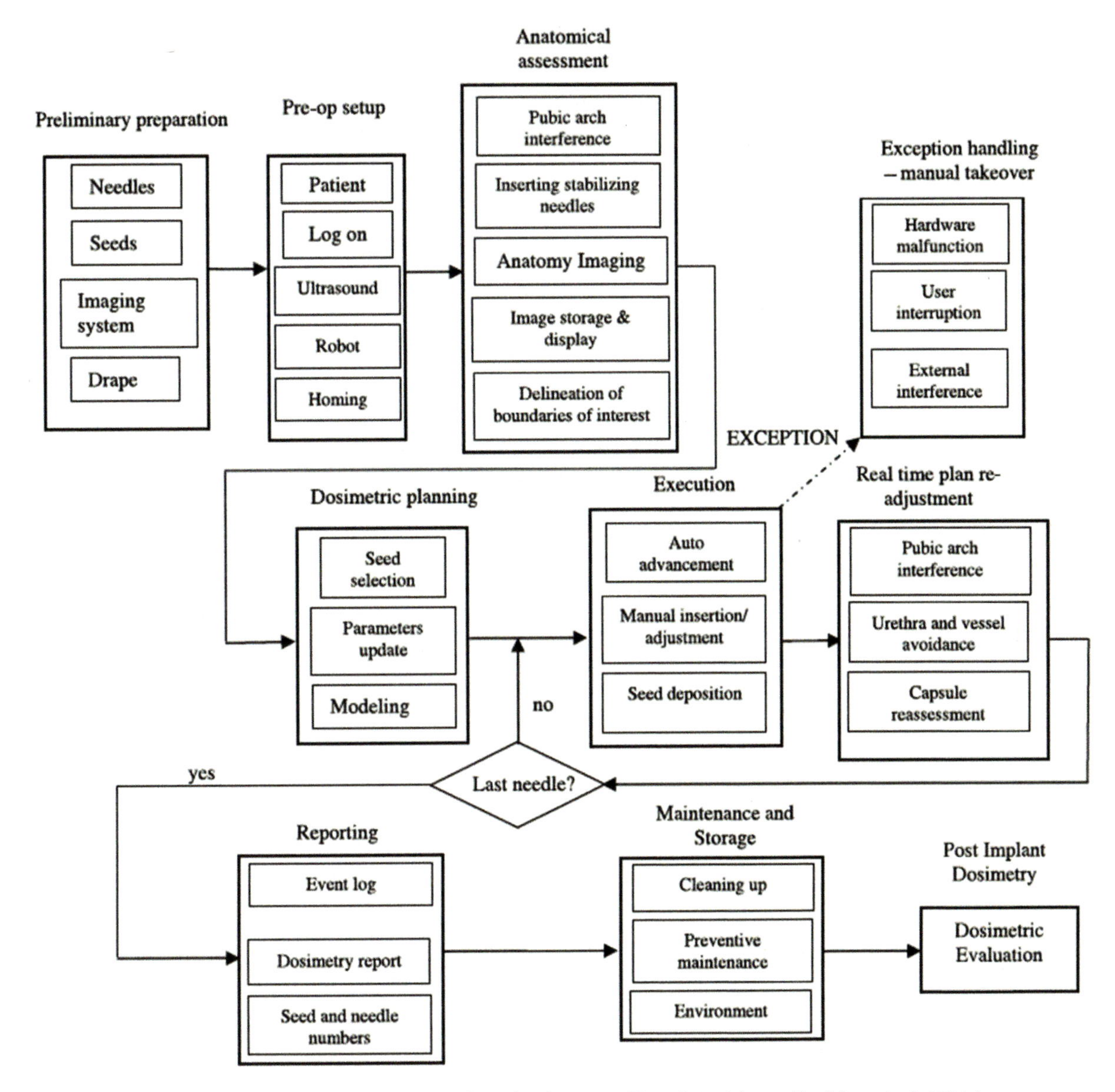

Figure 10–17 Clinical work flow for robotic brachytherapy. Reprinted from Podder et al. 2014.

concept can be used to perform prostate implants on patients with large glands who would traditionally not be good candidates for brachytherapy. In general, a robot would expand the number of possible needle or catheter paths and allow for their placement without the need for the fixed template of needle positions.

10.6.4.2 Integrating into a Stereotactic Imaging System

Stereotactic robotic devices can take full advantage of the plethora of information that is available when using imaging technologies like MRI to allow the physician to identify and incorporate anatomical features into the implant and dosimetry planning by allowing co-registration between the robot's coordinate system and that of the imaging device. This would allow the physician or physicist to explicitly tell the robot to avoid structures seen on the MR imaging (e.g., neurovascular bundles, the seminal vesicles, the penile bulb) or explicitly target other areas for implant (e.g., intra-prostatic lesions). The ability to see the prostate substructure, and perhaps the tumor, allows for more optimal placement of needles or catheters for therapeutic treatment.

Table 10–7 Left: A standard prostate permanent-seed implant (PPI) work flow. Right: A potential adaptive work flow that incorporates in-OR re-optimization of the plan after placing a fraction of the seeds. Stereotactic robots can enable the adaptive procedure on the right by communicating the position of the dropped seeds to an integrated planning system.

Standard PPI Work Flow	Adaptive Work Flow
Image	Image (in procedure room)
Contour	Contour
Plan seed and needle positions	Plan seed and needle positions
Place seeds	Place **some** seeds; record location
Post-implant dose evaluations	Re-image to evaluate contours and seed positions Evaluate current dose distribution Re-optimize Place final seeds Final implant dose evaluation (while in procedure room)

10.6.4.3 Intraoperative Adaptive Brachytherapy

The goal of intraoperative adaptive brachytherapy is to enable real-time planning with dynamic dose calculation to inform an implant procedure and allow on-the-fly changes if the expected goals for the procedure are in failure of being met (Nag et al. 2001). Table 10–7 compares a standard prostate seed implant with a fully adaptive procedure. A stereotactic robot would allow for imaging, planning, and implantation to happen all in one work flow during one procedure. The robot can communicate seed position location upon deposition to the planning system and can enable point-and-click placement of seeds if the resulting dosimetry is not adequate, allowing a physician to identify a region of low dose on the imaging or planning screen and make corrections (Cunha et al. 2010).

10.6.5 Status in the Clinic

Over the last few years, robotic brachytherapy has reached a notable point of maturity. There are a number of devices for which the functionality for clinical integration has been demonstrated. The Task Group 192 report examined 13 robotic brachytherapy devices. Each device has its own strengths and weaknesses, but all had been validated in the context of clinical usefulness and ability. However, only a few of them had been implemented in the clinic. The Nucletron seedSelectron system (see Section 4.4) is a seed insertion mechanism combined with a needle-retraction system that has been available for a decade, but it lacks the functionality for a fully automated procedure (Rivard et al. 2005; Beaulieu et al. 2007). Fichtinger et al. (2008) reported on the use of a parallel-plate needle guidance robot in a Phase I clinical trial of five patients.

It is unfortunate that momentum has slowed on robotic brachytherapy. The main reason for this is that now that feasibility has been demonstrated by researchers, the next step is for vendors to pick up the technology and develop a product that can be marketed. It is likely that the history of reimbursement challenges for brachytherapy has kept vendors from investing heavily in brachytherapy devices. However, the finicky aspect of medical billing can certainly change.

There is some indication that commercial interest is growing. The robot technology developed by Wei et al. (2004) has been licensed to a technology development firm, Eigen (Grass Valley, CA) for development and translation to breast and GYN brachytherapy. The Prosper robot (Baumann et al. 2011) has gone through a recent design iteration and is currently being prepared for its first evaluation in patients. The technology for US-guided and US-tracked needle insertion developed by Krieger et al. (2005) was commercialized with Sentinelle Medical Inc., and Hologic. The Sentinelle product line is part of Philips/Invivo UroNav prostate biopsy technology (Invivo, Gainesville, FL), which is currently on the market.

10.7 Conclusion

Brachytherapy is one of the oldest forms of radiation therapy, and it has a long tradition of effective treatment of cancer. But there is still much improvement that can be made to brachytherapy practice to push forward the state of the art. The five technologies and methods presented here (focal therapy, 3D printing, *in vivo* dosimetry, live needle tracking, and robotics) are just a few of the ways that innovators in the field are pushing the boundaries toward better, more efficient, and more effective management of cancer.

References

Ahmed, H. U., A. El-Shater Bosaily, L. C. Brown, R. Gabe, R. Kaplan, M. K. Parmar, Y. Collaco-Moraes, K. Ward, R. G. Hindley, A. Freeman, A. P. Kirkham, R. Oldroyd, C. Parker, and M. Emberton. (2017). "Diagnostic accuracy of multi-parametric MRI and TRUS biopsy in prostate cancer (PROMIS): a paired validating confirmatory study." *Lancet.* 389(10071):815–22. doi: 10.1016/s0140-6736(16)32401-1.

Al-Qaisieh, B., J. Mason, P. Bownes, A. Henry, L. Dickinson, H. U. Ahmed, M. Emberton, and S. Langley. (2015). "Dosimetry Modeling for Focal Low-Dose-Rate Prostate Brachytherapy." *Int. J. Radiat. Oncol. Biol. Phys.* 92(4):787–93. doi: 10.1016/j.ijrobp.2015.02.043.

Albano, M., I. Dumas, I., and C. Haie-Meder. (2008). "Brachytherapy at the Institut Gustave-Roussy: personalized vaginal mould applicator: technical modification and improvement." *Cancer radiotherapie: journal de la Societe francaise de radiotherapie oncologique*, 12(8):822–26.

Andersen, C. E., S. K. Nielsen, J. C. Lindegaard, and K. Tanderup. (2009). "Time-resolved in-vivo luminescence dosimetry for online error detection in pulsed dose-rate brachytherapy." *Med. Phys.* 36 5033–43.

Bharat, S., C. Kung, E. Dehghan, A. Ravi, N. Venugopal, A. Bonillas, et al. (2014). "Electromagnetic tracking for catheter reconstruction in ultrasound-guided high-dose-rate brachytherapy of the prostate." *Brachytherapy* 13(6):640–50.

Banerjee, R., S. J. Park, E. Anderson, D. J. Demanes, J. Wang, and M. Kamrava. (2015). "From whole gland to hemigland to ultra-focal high-dose-rate prostate brachytherapy: A dosimetric analysis." *Brachytherapy* 14(3):366–72. doi: 10.1016/j.brachy.2014.12.007.

Baumann, M., M. Bolla, V. Daanen, J. L. Descotes, J. Y. Giraud, N. Hungr, et al. (2011). "Prosper: image and robot-guided prostate brachytherapy." IRBM 32(2):63–5.

Beaulieu, L., D.A. Evans, S. Aubin, S. Angyalfi, S. Husain, I. Kay, A. G. Martin, N. Varfalvy, E. Vigneault, and P. Dunscombe. (2007). "Bypassing the learning curve in permanent seed implants using state-of-the-art technology." *Int. J. Radiat. Oncol. Biol. Phys.* 67(1):71–7.

Beriwal, S., D. J. Demanes, B. Erickson, E. Jones, J. F. De Los Santos, R. A. Cormack, C. Yashar, J. J. Rownd, and A. N. Viswanathan. (2012). "American Brachytherapy Society consensus guidelines for interstitial brachytherapy for vaginal cancer." *Brachytherapy* 11:68–75.

Bert, C., M. Kellermeier, and K. Tanderup. (2016). "Electromagnetic tracking for treatment verification in interstitial brachytherapy." *J. Contemp. Brachytherapy* 8(5):448–53.

Bittner, N., G. S. Merrick, A. Bennett, W. M. Butler, H. J. Andreini, W. Taubenslag, and E. Adamovich. (2015). "Diagnostic Performance of Initial Transperineal Template-guided Mapping Biopsy of the Prostate Gland." *Am. J. Clin. Oncol.* 38(3):300–3. doi: 10.1097/COC.0b013e31829a2954.

Bittner, N., G. S. Merrick, K. E. Wallner, J. H. Lief, W. M. Butler, and R. W. Galbreath. (2007). "The impact of acute urinary morbidity on late urinary function after permanent prostate brachytherapy." *Brachytherapy* 6(4):258–66. doi: S1538-4721(07)00247-4 [pii] 10.1016/j.brachy.2007.08.008 [doi].

Borghesi, M., H. Ahmed, R. Nam, E. Schaeffer, R. Schiavina, S. Taneja, W. Weidner, and S. Loeb. (2017). "Complications after systematic, random, and image-guided prostate biopsy." *Eur. Urol.* 71(3):353–65. doi: 10.1016/j.eururo.2016.08.004.

Bø, L. E., H. O. Leira, G. A. Tangen, E. F. Hofstad, T. Amundsen, and T. Langø. (2012). "Accuracy of electromagnetic tracking with a prototype field generator in an interventional OR setting." *Med. Phys.* 39(1)399–406.

Butler, W. M. and G. S. Merrick. (2016). "Focal prostate brachytherapy with ^{103}Pd seeds." *Med. Phys.* 32(3):459–64. doi: 10.1016/j.ejmp.2016.03.012.

Carrara, M., C. Tenconi, G. Rossi, M. Borroni, A. Cerrotta, S. Grisotto, D. Cusumano, B. Pappalardi, D. Cutajar, M. Petasecca, M. Lerch, G. Gambarini, C. Fallai, A. Rosenfeld, and E. Pignoli. (2016). "In-vivo rectal wall measurements during HDR prostate brachytherapy with MOSkin dosimeters integrated on a trans-rectal US probe: Comparison with planned and reconstructed doses." *Radiother. Oncol.* 118:148–53.

Crawford, E. D., K. O. Rove, A. B. Barqawi, P. D. Maroni, P. N. Werahera, C. A. Baer, H. K. Koul, C. A. Rove, M. S. Lucia, and F. G. La Rosa. (2013). "Clinical-pathologic correlation between transperineal mapping biopsies of the prostate and three-dimensional reconstruction of prostatectomy specimens." *Prostate* 73(7):778–87. doi: 10.1002/pros.22622.

Cunha, J. A., I. C. Hsu, J. Pouliot, M. Roach III, K. Shinohara, J. Kurhanewicz, et al. (2010). "Toward adaptive stereotactic robotic brachytherapy for prostate cancer: demonstration of an adaptive work flow incorporating inverse planning and an MR stealth robot." *Minim. Invasive Ther. Allied Technol.* 19(4):189–202.

Cunha, J. A. M., I. C. Hsu, and J. Pouliot. (2009). "Dosimetric equivalence of nonstandard HDR brachytherapy catheter patterns." *Med. Phys.* 36(1):233–39.

Cunha, J. A. M., K. Mellis, R. Sethi, T. Siauw, A. Sudhyadhom, A. Garg, K. Goldberg, I.-C. Hsu, and J. Pouliot. (2015). "Evaluation of PC-ISO for customized, 3D printed, gynecologic ^{192}Ir HDR brachytherapy applicators." *J. Appl. Clin. Med. Phys.*16(1):246–53.

Damato, A. L., A. N. Viswanathan, S. M. Don, J. L. Hansen, and R. A. Cormack. (2014). "A system to use electromagnetic tracking for the quality assurance of brachytherapy catheter digitization." *Med. Phys.* 41(10).

DeWerd, L. A., G. S. Ibbott, A. S. Meigooni, M. G. Mitch, M. J. Rivard, K. E. Stump ,and J. L. M. Venselaar . (2011). "A dosimetric uncertainty analysis for photon-emitting brachytherapy sources: Report of AAPM Task Group No. 138 and GEC-ESTRO." *Med. Phys.* 38:782–801.

de Battisti, B. M., B. Denis de Senneville, M. Maenhout, J. J. Lagendijk, M. van Vulpen, G. Hautvast, et al. (2016). "Fiber Bragg gratings-based sensing for real-time needle tracking during MR-guided brachytherapy." *Med. Phys.* 43(10)"5288–97.

Dimopoulos, J. C., R. Pötter, S. Lang, E. Fidarova, P. Georg, W. Dörr, and C. Kirisits. (2009). "Dose–effect relationship for local control of cervical cancer by magnetic resonance image-guided brachytherapy." *Radiother. Oncol.* 93(2):311–15.

Donaldson, I. A., R. Alonzi, D. Barratt, E. Barret, V. Berge, S. Bott, D. Bottomley, S. Eggener, B. Ehdaie, M. Emberton, R. Hindley, T. Leslie, A. Miners, N. McCartan, C. M. Moore, P. Pinto, T. J. Polascik, L. Simmons, J. van der Meulen, A. Villers, S. Willis, and H. U. Ahmed. (2015). "Focal therapy: patients, interventions, and outcomes—a report from a consensus meeting." *Eur. Urol.* 67(4):771–7. doi: 10.1016/j.eururo.2014.09.018.

Donin, N. M., J. Laze, M. Zhou, Q. Ren, and H. Lepor. (2013). "Gleason 6 prostate tumors diagnosed in the PSA era do not demonstrate the capacity for metastatic spread at the time of radical prostatectomy." *Urology* 82(1):148–52. doi: 10.1016/j.urology.2013.03.054.

Elliott, D.M, et al. "Automated implantation system for radioisotopeseeds," U.S. patent 6869390 B2 (22 March 2005).

Eggener, S. E., P. T. Scardino, P. R. Carroll, M. J. Zelefsky, O. Sartor, H. Hricak, T. M. Wheeler, S. W. Fine, J. Trachtenberg, M. A. Rubin, M. Ohori, K. Kuroiwa, M. Rossignol, and L. Abenhaim. (2007). "Focal therapy for localized prostate cancer: a critical appraisal of rationale and modalities." *J. Urol.* 178(6):2260–67. doi: 10.1016/j.juro.2007.08.072.

El Khoury, C., I. Dumas, A. Tailleur, P. Morice, and C. Haie-Meder. (2015). "Adjuvant brachytherapy for endometrial cancer: Advantages of the vaginal mold technique." *Brachytherapy* 14(1):51–55.

Fichtinger, G., J. P. Fiene, C. W. Kennedy, G. Kronreif, I. Iordachita, D. Y. Song, et al. (2008). "Robotic assistance for ultrasound-guided prostate brachytherapy." *Med. Image Anal.* 12(5):535–45.

Franz, A. M., T. Haidegger, W. Birkfellner, K. Cleary, T. M. Peters, and L. Maier-Hein. (2014). "Electromagnetic tracking in medicine—a review of technology, validation, and applications." *IEEE Trans. Med. Imaging* 33(8):1702–25.

Gaziev, G., K. Wadhwa, T. Barrett, B. C. Koo, F. A. Gallagher, E. Serrao, J. Frey, J. Seidenader, L. Carmona, A. Warren, V. Gnanapragasam, A. Doble, and C. Kastner. (2016). "Defining the learning curve for multiparametric magnetic resonance imaging (MRI) of the prostate using MRI-transrectal ultrasonography (TRUS) fusion-guided transperineal prostate biopsies as a validation tool." *BJU Int.* 117(1):80-6. doi: 10.1111/bju.12892.

Gutman, S., G. S. Merrick, W. M. Butler, K. E. Wallner, Z. Allen, R. W. Galbreath, and E. Adamovich. (2006). "Severity categories of the International Prostate Symptom Score before, and urinary morbidity after, permanent prostate brachytherapy." *BJU Int.* 97(1):62–8.

Haffner, M. C., T. Mosbruger, D. M. Esopi, H. Fedor, C. M. Heaphy, D. A. Walker, N. Adejola, M. Gurel, J. Hicks, A. K. Meeker, M. K. Halushka, J. W. Simons, W. B. Isaacs, A. M. De Marzo, W. G. Nelson, and S. Yegnasubramanian. (2013). "Tracking the clonal origin of lethal prostate cancer." *J. Clin. Invest.* 123(11):4918–22. doi: 10.1172/jci70354.

Haworth, A., C. Mears, J. M. Betts, H. M. Reynolds, G. Tack, K. Leo, S. Williams, and M. A. Ebert. (2016). "A radiobiology-based inverse treatment planning method for optimisation of permanent l-125 prostate implants in focal brachytherapy." *Phys. Med. Biol.* 61(1):430–44. doi: 10.1088/0031-9155/61/1/430.

Hosni, A., M. Carlone, A. Rink, C. Menard, P. Chung, and A. Berlin. (2017). "Dosimetric feasibility of ablative dose escalated focal monotherapy with MRI-guided high-dose-rate (HDR) brachytherapy for prostate cancer." *Radiother. Oncol.* 122(1):103–08. doi: 10.1016/j.radonc.2016.11.011.

Hu, Y., H. U. Ahmed, T. Carter, N. Arumainayagam, E. Lecornet, W. Barzell, A. Freeman, P. Nevoux, D. J. Hawkes, A. Villers, M. Emberton, and D. C. Barratt. (2012). "A biopsy simulation study to assess the accuracy of several transrectal ultrasonography (TRUS)-biopsy strategies compared with template prostate mapping biopsies in patients who have undergone radical prostatectomy." *BJU Int.* 110(6):812–20. doi: 10.1111/j.1464-410X.2012.10933.x.

Haie-Meder, C., R. Pötter, E. Van Limbergen, E. Briot, M. De Brabandere, J. Dimopoulos, et al. (2005). "Recommendations from Gynaecological (GYN) GEC-ESTRO Working Group(I): concepts and terms in 3D image based 3D treatment planning in cervix cancer brachytherapy with emphasis on MRI assessment of GTV and CTV." *Radiother. Oncol.* 74(3):235–45.

Huyskens, D., R. Bogaerts, J. Verstraete, M. Lööf, H. Nyström, C. Fiorino, S. Broggi, N. Jornet, M. Ribas, and D. I. Twaithes. *Practical guidelines for the implementation of in-vivo dosimetry with diodes in external radiotherapy with photon beams (entrance dose).* Brussels: ESTRO, 2001.

International Atomic Energy Agency (IAEA). *Safety Reports Series 17: Lessons learned from accidental exposures in radiotherapy.* Vienna: IAEA, 2000.

Kertzscher, G., C. E. Andersen, K. Tanderup. (2011). "Identifying afterloading PDR and HDR brachytherapy errors using real-time fiber-coupled Al(2)O(3):C dosimetry and a novel statistical error decision criterion." *Radiother. Oncol.* 100:456–62.

Kertzscher, G., A. Rosenfeld, S. Beddar, K. Tanderup, and J. E. Cygler. (2014). "*In-vivo* dosimetry: trends and prospects for brachytherapy." *Br. J. Radiol.* 87:20140206.

Krauss, D. J., H. Ye, A. A. Martinez, B. Mitchell, E. Sebastian, A. Limbacher, and G. S. Gustafson. (2017). "Favorable preliminary outcomes for men with low- and intermediate-risk prostate cancer treated with 19-Gy single-fraction high-dose-rate brachytherapy." *Int. J. Radiat. Oncol. Biol. Phys.* 97:98–106.

Krieger, A., R. C. Susil, C. Ménard, J. A. Coleman, G. Fichtinger, E. Atalar, and L. L. Whitcomb. (2005). "Design of a novel MRI compatible manipulator for image guided prostate interventions." *IEEE Trans. Biomed. Imaging* 52(2):306–13.

Laing, R., A. Franklin, J. Uribe, A. Horton, S. Uribe-Lewis, and S. Langley. (2016). "Hemi-gland focal low dose rate prostate brachytherapy: An analysis of dosimetric outcomes." *Radiother. Oncol.* 121(2):310–15. doi: 10.1016/j.radonc.2016.09.014.

Langley, S., H. U. Ahmed, B. Al-Qaisieh, D. Bostwick, L. Dickinson, F. G. Veiga, P. Grimm, S. Machtens, F. Guedea, and M. Emberton. (2012). "Report of a consensus meeting on focal low dose rate brachytherapy for prostate cancer." *BJU Int.* 109 Suppl 1:7–16. doi: 10.1111/j.1464-410X.2011.10825.x.

Laviana, A. A., A. M. Ilg, D. Veruttipong, H. J. Tan, M. A. Burke, D. R. Niedzwiecki, P. A. Kupelian, C. R. King, M. L. Steinberg, C. R. Kundavaram, M. Kamrava, A. L. Kaplan, A. K. Moriarity, W. Hsu, D. J. Margolis, J. C. Hu, and C. S. Saigal. (2016). "Utilizing time-driven activity-based costing to understand the short- and long-term costs of treating localized, low-risk prostate cancer." *Cancer* 122(3):447–55. doi: 10.1002/cncr.29743.

Le, J. D., N. Tan, E. Shkolyar, D. Y. Lu, L. Kwan, L. S. Marks, J. Huang, D. J. Margolis, S. S. Raman, and R. E. Reiter. (2015). "Multifocality and prostate cancer detection by multiparametric magnetic resonance imaging: correlation with whole-mount histopathology." *Eur. Urol.* 67(3):569–76. doi: 10.1016/j.eururo.2014.08.079.

Liu, D., H. P. Lehmann, K. D. Frick, and H. B. Carter. (2012). "Active surveillance versus surgery for low risk prostate cancer: a clinical decision analysis." *J. Urol.* 187(4):1241–46. doi: 10.1016/j.juro.2011.12.015.

Lipson, H. and M. Kurman. *Fabricated: The New World of 3D Printing.* Indianapolis, IN: John Wiley & Sons, Inc., 2013.

Mijnheer, B., S. Beddar, J. Izewska, and C. Reft. (2013). "*In-vivo* dosimetry in external beam radiotherapy." *Med. Phys.* 40:070903.

Magné, N., C. Chargari, N. SanFilippo, T. Messai, A. Gerbaulet, and C. Haie-Meder. (2010). "Technical aspects and perspectives of the vaginal mold applicator for brachytherapy of gynecologic malignancies." *Brachytherapy* 9(3):274–77.

Mason, J., B. Al-Qaisieh, P. Bownes, D. Thwaites, and A. Henry. (2014). "Dosimetry modeling for focal high-dose-rate prostate brachytherapy." *Brachytherapy* 13(6):611–17. doi: 10.1016/j.brachy.2014.06.007.

Mehrtash, A., A. Damato, G. Pernelle, L. Barber, N. Farhat, A. Viswanathan, et al. (2014). "EM-navigated catheter placement for gynecologic brachytherapy: an accuracy study." In *SPIE Medical Imaging* (pp. 90361F-90361F). International Society for Optics and Photonics.

Merrick, G. S., A. Delatore, W. M. Butler, A. Bennett, R. Fiano, R. Anderson, and E. Adamovich. (2017). "Transperineal template-guided mapping biopsy identifies pathologic differences between very-low-risk and low-risk prostate cancer: Implications for active surveillance." *Am. J. Clin. Oncol.* 40(1):53–59. doi: 10.1097/coc.0000000000000105.

Merrick, G. S., S. Gutman, H. Andreini, W. Taubenslag, D. L. Lindert, R. Curtis, E. Adamovich, R. Anderson, Z. Allen, W. Butler, and K. Wallner. (2007). "Prostate cancer distribution in patients diagnosed by transperineal template-guided saturation biopsy." *Eur. Urol.* 52(3):715–23. doi: S0302-2838(07)00286-2 [pii]. 10.1016/j.eururo.2007.02.041 [doi].

Nag, S., J. P. Ciezki, R. Cormack, S. Doggett, K. DeWyngaert, G. K. Edmundson, et al. (2001). "Intraoperative planning and evaluation of permanent prostate brachytherapy: report of the American Brachytherapy Society." *Int. J. Radiat. Oncol. Biol. Phys.* 51(5):1422–30.

Nilsson, S., Z. Moutrie, R. Cheuk, P. Chan, C. Lancaster, T. Markwell, et al. (2015). "A unique approach to high-dose-rate vaginal mold brachytherapy of gynecologic malignancies." *Brachytherapy* 14(2):267–72.

Perera, M., N. Krishnananthan, U. Lindner, and N. Lawrentschuk. (2016). "An update on focal therapy for prostate cancer." *Nat. Rev. Urol.* 13(11):641–53. doi: 10.1038/nrurol.2016.177.

Pinnaduwage, D. S., J. A. Cunha, V. Weinberg, D. Krishnamurthy, M. Nash, I. C. Hsu, and J. Pouliot. (2013). "A dosimetric evaluation of using a single treatment plan for multiple treatment fractions within a given applicator insertion in gynecologic brachytherapy." *Brachytherapy* 12(5):487–94.

Podder, T. K., L. Beaulieu, B. Caldwell, R. A. Cormack, J. B. Crass, A. P. Dicker, et al. (2014). "AAPM and GEC-ESTRO guidelines for image-guided robotic brachytherapy: Report of Task Group 192." *Med. Phys.* 41(10):101501.

Poulin, E., E. Racine, D. Binnekamp, and L. Beaulieu. (2015). "Fast, automatic, and accurate catheter reconstruction in HDR brachytherapy using an electromagnetic 3D tracking system." *Med. Phys.* 42(3):1227–32.

Pötter, R., C. Haie-Meder, E. Van Limbergen, I. Barillot, M. De Brabandere, J. Dimopoulos, et al. (2006). "Recommendations from gynaecological (GYN) GEC ESTRO working group (II): Concepts and terms in 3D image-based treatment planning in cervix cancer brachytherapy—3D dose volume parameters and aspects of 3D image-based anatomy, radiation physics, radiobiology." *Radiother. Oncol.* 78(1)67–77.

Priester, A., S. Natarajan, P. Khoshnoodi, D. J. Margolis, S. S. Raman, R. E. Reiter, J. Huang, W. Grundfest, and L. S. Marks. (2017). "Magnetic Resonance Imaging Underestimation of Prostate Cancer Geometry: Use of Patient Specific Molds to Correlate Images with Whole Mount Pathology." *J. Urol.* 197(2):320–26. doi: 10.1016/j.juro.2016.07.084.

Racine, E., G. Hautvast, D. Binnekamp, and L. Beaulieu. (2016). "Real-time electromagnetic seed drop detection for permanent implants brachytherapy: Technology overview and performance assessment." *Med. Phys.* 43(12):6217.

Reed, G., J. A. Cunha, S. Noworolski, et al. (2011). "Interactive, multi-modality image registrations for combined MRI/MRSI-planned HDR prostate brachytherapy." *J. Contemp. Brachytherapy* 3(1):1–6. doi:10.5114/jcb.2011.21040.

Rengier, F., A. Mehndiratta, H. von Tengg-Kobligk, C. M. Zechmann, R. Unterhinninghofen, H. U. Kauczor, and F. L. Giesel. (2010). "3D printing based on imaging data: review of medical applications." *Int. J. Comput. Assist. Radiol. Surg.* 5(4):335–41.

Rivard, M. J., D. A. R. Evans, and I. Kay. (2005). "A technical evaluation of the Nucletron FIRST system: Conformance of a remote afterloading brachytherapy seed implantation system to manufacturer specifications and AAPM Task Group report recommendations." *J. Appl. Clin. Med. Phys.* 6(1):22–50.

Rutala, W. A., D. J. Weber, and HICPAC. (2008). "Guideline for Disinfection and Sterilization in Healthcare Facilities, 2008." www.cdc.gov/hicpac/pubs.html.

Renard-Penna, R., G. Cancel-Tassin, E. Comperat, J. Varinot, P. Leon, M. Roupret, P. Mozer, C. Vaessen, O. Lucidarme, M. O. Bitker, and O. Cussenot. (2015). "Multiparametric magnetic resonance imaging predicts postoperative pathology but misses aggressive prostate cancers as assessed by cell cycle progression score." *J. Urol.* 194(6):1617–23. doi: 10.1016/j.juro.2015.06.107.

Rosenkrantz, A. B., L. A. Ginocchio, D. Cornfeld, A. T. Froemming, R. T. Gupta, B. Turkbey, A. C. Westphalen, J. S. Babb, and D. J. Margolis. (2016). "Interobserver Reproducibility of the PI-RADS Version 2 Lexicon: A Multicenter Study of Six Experienced Prostate Radiologists." *Radiology* 280(3):793–804. doi: 10.1148/radiol.2016152542.

Samboju, V., M. Adams, V. Salgaonkar, C. J. Diederich, and J. A. M. Cunha. (2017). "Improved accuracy of ultrasound-guided therapies using electromagnetic tracking: in-vivo speed of sound measurements." SPIE BiOSInternational Society for Optics and Photonics. Vol. 10066:100660P-7.

Schubert, C., M. C. van Langeveld, and L. A. Donoso. (2014). "Innovations in 3D printing: a 3D overview from optics to organs." *Br. J. Opthamol.* 98(2):159–61.

Sethi, R., A. Cunha, K. Mellis, T. Siauw, C. Diederich, J. Pouliot, and I. C. Hsu. (2016). "Clinical applications of custom-made vaginal cylinders constructed using three-dimensional printing technology." *J. Contemp. Brachytherapy* 8(3):210–16.

Small Jr., W., S. Beriwal, D. J. Demanes, K. E. Dusenbery, P. Eifel, B. Erickson, E. Jones, J. J. Rownd, J. F. De Los Santos, A. N.Viswanathan, and D. Gaffney. (2012) "American Brachytherapy Society consensus guidelines for adjuvantvaginal cuff brachytherapy after hysterectomy." *Brachytherapy* 11:58–67.

Smith, R. L., A. Haworth, V. Panettieri, J. L. Millar, and R. D. Franich. (2016). "A method for verification of treatment delivery in HDR prostate brachytherapy using a flat panel detector for both imaging and source tracking." *Med. Phys.* 43:2435–42.

Song, W. Y., K. Tanderup, and B. Pieters. *Emerging Technologies in Brachytherapy.* Boca Raton, FL: CRC Press, 2017.

Suchowerska, N., M. Jackson, J. Lambert, Y. B. Yin, G. Hruby, and D. R. McKenzie. (2011). "Clinical trials of a urethral dose measurement system in brachytherapy using scintillation detectors." *Int. J. Radiat. Oncol. Biol. Phys.* 79:609–15.

Tanderup, K., S. Beddar, C. E. Andersen, G. Kertzscher, and J. E. Cygler. (2013). "*In-vivo* dosimetry in brachytherapy." *Med. Phys.* 40:070902.

Toye, W., R. Das, T. Kron, R. Franich, P. Johnston, and G. Duchesne. (2008). "An *in-vivo* investigative protocol for HDR prostate brachytherapy using urethral and rectal thermoluminescence dosimetry." *Radiother. Oncol.* 91:243–48.

Taira, A. V., G. S. Merrick, A. Bennett, H. Andreini, W. Taubenslag, R. W. Galbreath, W. M. Butler, N. Bittner, and E. Adamovich. (2013a). "Transperineal template-guided mapping biopsy as a staging procedure to select patients best suited for active surveillance." *Am. J. Clin. Oncol.* 36(2):116–20. doi: 10.1097/COC.0b013e31823fe639 [doi].

Taira, A. V., G. S. Merrick, W. M. Butler, R. W. Galbreath, R. Fiano, K. E. Wallner, and E. Adamovich. (2013b). "Time to failure after definitive therapy for prostate cancer: implications for importance of aggressive local treatment." *J. Contemp. Brachytherapy* 5(4):215–21. doi: 10.5114/jcb.2013.39210.

Taira, A. V., G. S. Merrick, R. W. Galbreath, W. M. Butler, K. E. Wallner, B. S. Kurko, R. Anderson, and J. H. Lief. (2009). "Erectile function durability following permanent prostate brachytherapy." *Int. J. Radiat. Oncol. Biol. Phys.* 75(3):639–48. doi: S0360-3016(08)03828-5 [pii]10.1016/j.ijrobp.2008.11.058 [doi].

Valerio, M., H. U. Ahmed, M. Emberton, N. Lawrentschuk, M. Lazzeri, R. Montironi, P. L. Nguyen, J. Trachtenberg, and T. J. Polascik. (2014). "The role of focal therapy in the management of localised prostate cancer: a systematic review." *Eur. Urol.* 66(4):732–51. doi: 10.1016/j.eururo.2013.05.048.

Valentin, J., International Commission on Radiation Protection (ICRP). (2005). "Prevention of high-dose-rate brachytherapy accidents." ICRP Publication 97. *Ann ICRP*35 1–51.

VanDam, J. and G. Marinello. *Methods for in-vivo dosimetry in external radiotherapy.* Brussels: ESTRO, 1994.

Venselaar, J. and J. Perez-Calatayud. *A practical guide to quality control of brachytherapy equipment: ESTRO Booklet No. 8.* Brussels: ESTRO, 2004.

Venselaar, J., D. Baltas, A. S. Meigooni, and P. J. Hoskin, Eds. *Comprehensive Brachytherapy: Physical and Clinical Aspects.* Boca Raton, Florida: Taylor & Francis, 2012.

Ventola, C. L. (2014). "Medical applications for 3D printing: Current and projected uses." *P & T: a peer-reviewed journal for formulary management* 39(10):704–11.

Voros, L., G. Cohen, M. Zaider, and Y. Yamada. (2016). "To navigate, or not to navigate: HDR BT in recurrent spine lesions.'" *Med. Phys.* 43(6):3367.

Waldhäusl, C., A. Wambersie, R. Pötter, and D. Georg. (2005). "*In-vivo* dosimetry for gynaecological brachytherapy: physical and clinical considerations." *Radiother. Oncol.* 77:310–17.

Wang, W., C. L. Dumoulin, A. N. Viswanathan, Z. T. Tse, A. Mehrtash, W. Loew, I. Norton, J. Tokuda, R. T. Seethamraju, T. Kapur, and A. L. Damato. (2015a). "Real-time active MR-tracking of metallic stylets in MR-guided radiation therapy." *Magn. Reson. Med.* 73(5):1803–11.

Wang, W., A. N. Viswanathan, A. L. Damato, Y. Chen, Z. Tse, L. Pan, et al. (2015b). "Evaluation of an active magnetic resonance tracking system for interstitial brachytherapy." *Med. Phys.* 42(12):7114–21.

Wei, Z., G. Wan, L. Gardi, G. Mills, D. Downey, and A. Fenster. (2004). "Robot-assisted 3D-TRUS guided prostate brachytherapy: System integration and validation." *Med. Phys.* 31(3):539–48.

Wiebe, E., H. Easton, G. Thomas, L. Barbera, L. D'Alimonte, and A. Ravi. (2015). "Customized vaginal vault brachytherapy with computed tomography imaging-derived applicator prototyping." *Brachytherapy* 14(3):380–84.

Willoughby, T. R., P. A. Kupelian, J. Pouliot, K. Shinohara, M. Aubin, M. Roach, et al. (2006). "Target localization and real-time tracking using the Calypso 4D localization system in patients with localized prostate cancer." *Int. J. Radiat. Oncol. Biol. Phys.* 65(2):528–34.

Xu, J., V. Duindam, R. Alterovitz, J. Pouliot, J. A. M. Cunha, I.-C. Hsu, and K. Goldberg."Planning fireworks trajectories for steerable medical needles to reduce patient trauma." In IEEE/RSJ International Conference on Intelligent Robots and Systems, 2009. (IROS). 2009, pp. 4517–22.

Zhou, J., E. Sebastian, V. Mangona, and D. Yan. (2013). "Real-time catheter tracking for high-dose-rate prostate brachytherapy using an electromagnetic 3D-guidance device: A preliminary performance study." *Med. Phys.* 40(2):021716.

Example Problems

(Answers are found at the end of the book.)

1. What are the two key uncertainty components in IVD?

   a. patient positioning and volume definition

   b. source-to-dosimeter positioning uncertainty and measurement uncertainty

   c. TPS uncertainty and organ deformation

   d. source strength determination and energy response

2. True or false (and why): As with EBRT, dose output measurement, measuring dose close to a brachytherapy source to better than 1% is easy?

3. Which of the following sterilization procedures is considered in all 50 states to be non-toxic to the personnel performing the sterilization: (1) steam, (2) ethelyne oxide, (3) hydrogen peroxide?

   a. (1) only

   b. (2) only

   c. (3) only

   d. (1) and (2)

   e. (1) and (3)

4. True or false: Any FDA-approved biocompatible 3D printing material is suitable for brachytherapy.

5. A robot is designed to hold a needle and position it in space with a programmed orientation. This robot does not insert the needle; this is reserved for the physician operator. What category does this robot best fit into?

   a. Level I

   b. Level II

   c. Level III

   d. Level IV

6. Regarding candidates for focal therapy, which of the following examinations comes closest to radical prostatectomy in accurately determining the location and aggressiveness of prostate cancer?

   a. genetic testing for prostate cancer using a commercial panel.

   b. multi-parametric magnetic resonance imaging of the prostate at 3 tesla.

   c. systematic transrectal ultrasound-guided needle biopsy.

   d. transperineal ultrasound-guided mapping needle biopsy.

7. Focal brachytherapy should meet which of the following criteria?

   a. The dosimetric goals for $D_{90\%}$ and $V_{100\%}$ for the focal volume should be the same as for whole prostate brachytherapy.

   b. The focal target has margins added to account for imaging and source placement uncertainty.

   c. The target volume has the same prescription dose as for whole prostate monotherapy.

   d. All of the above are true.

8. The optimal candidate for focal prostate therapy is which of the following?
   a. high-risk patients
   b. intermediate-risk patients
   c. low-risk patients
   d. patients from all three risk groups are good candidates.
9. True or false: Electromagnetic tracking systems can be used to perform *in vivo* calibration of a transrectal ultrasound probe?
10. Identify which of the main brachytherapy catheter reconstruction errors below cannot be mitigated by incorporating electromagnetic tracking into a brachytherapy implant and treatment planning work flow.
    a. catheter swap error
    b. catheter length error
    c. catheter mix error
    d. catheter shift error

PROBLEM ANSWERS

Chapter 1: General Planning

1. T Reference: See Section 1.2.3.1
2. d Reference: See Section 1.2.2.1
3. b Reference: See Section 1.3.3
4. b Reference: See Section 1.8.1.3
5. c Reference: Holloway et al. 2013
6. b Reference: Kirisits et al. 2014
7. b Reference: Nath et al. 2009
8. c Reference: ICRU 2016
9. a Reference: See Section 1.2.3.2
10. b Reference: Shah et al. 2013

Chapter 2: MBDCA

1. d References: Section 2.6.1 and Beaulieu et al. 2012
   Choices a), b), and c) all relate to radiological physics interactions. Choice d) is another topic of sophisticated investigation for brachytherapy physics. However, it is not covered in the TG-186 report.
2. c References: Section 2.6.1 and Beaulieu et al. 2012
   It is possible for a physicist to receive sources and applicators that differ from what is in the TPS or applicator library. Therefore, it is necessary that s/he evaluate the specific equipment in that medical center.
3. F References: Section 2.6.2 and Beaulieu et al. 2012
   The accuracy of an independent check of an MBDCA calculation using a TG-43 hand calculation will depend on the magnitude of the difference between the clinical circumstances and the TG-43 conditions of a liquid water environment, no intersource attenuation, and equilibrium scatter conditions. Part of the lure of MBDCAs for brachytherapy is their accuracy improvements over TG-43 dose calculations. As there currently are no software tools beyond physicist-driven TG-43 hand calculations for independently checking MBDCA dose calculations, the physicist should evaluate anatomic-specific circumstances and establish expected results towards commissioning independent check methods for brachytherapy treatment plans using MBDCAs.
4. b References: Section 2.6.3 and Beaulieu et al. 2012
   The current TPSs using MBDCAs use either dose in medium to water *or* dose in medium to medium. Therefore, this reason is inapplicable.
5. T Reference: Afsharpour et al. 2010 and Landry et al. 2010
   For low-energy sources, the dose to the PTV can be changed as much as 30% to 35% depending on the amount of fatty or muscular tissue is found in the breast.

6. d Reference: Suit et al. 1960
Boundary conditions, tissue variability, and presence of air may influence the dose in the 1% to 4% region. However, the use of shielded ovoids typically provides a dose reduction in the 20% to 30% range in the shadow of the ovoid.
7. T References: Section 2.2.1 and Beaulieu et al. 2012
Mass energy absorption coefficient relative to water given the maximum differences for various tissues in the low-energy photon brachytherapy and x-ray imaging range.
8. b References: Section 2.3 and P. Papagiannis et al. 2014
For current state of the art brachytherapy treatment planning dosimetry algorithms, see *Br. J. Radiol.* 87:20140163.
9. F Reference: E. E. Lewis and W. F. Miller. *Computational Methods of Neutron Transport.* New York: John Wiley and Sons, Inc., 1984
10. T Reference: E. E. Lewis and W. F. Miller. *Computational Methods of Neutron Transport.* New York: John Wiley and Sons, Inc., 1984

Chapter 3: eBT

1. b Reference: Thomadsen et al. 2009 (AAPM TG-152 Report)
2. c Reference: Hiatt et al. 2008
3. d References: Rivard et al. 2006 and Hiatt et al. 2015
4. b Reference: DeWerd et al. 2015
5. b Reference: INTRABEAM user's manual

   "The determination of treatment time (and subsequently the dose delivered) is based on a TG-61 reference dose rate, and then corrected by measured depth dose curves and applicator transfer functions. Outside of an annual check of these values, no addition dose optimization is currently employed. The treatment planning computer is able to visualize anticipated dose distributions based on the measured source output, depth dose values, and isotropy values."
6. a Reference: Thomadsen et al. 2009

   "Having electrically-generated radiation avoids special licensing required for radionuclide-based sources containing nuclear byproduct materials and there are less regulator concerns (Thomadsen et al. 2009). Further, the useful combination of high-dose-rate (HDR) with low-energy photons allows the choice for a variety of treatment rooms without requiring a massive treatment room bunker as is required for high-energy HDR brachytherapy sources such as ^{192}Ir."
7. a Reference: INTRABEAM User's Manual

   "The XRS4 source generates a spherical dose distribution at the tip of a thin probe (diameter = 3.2 mm). To produce the low-energy bremsstrahlung photon spectrum, electrons are accelerated and steered down through the probe and strike a gold target."
8. a Reference: Rusch et al. 2007
9. b Reference: Rusch et al. 2007
10. d References: Ma et al. 2001, Fulkerson et al. 2014, and Candela et al. 2015
11. c Reference: Dickler et al. 2009

Chapter 4: LDR Prostate

1. c Reference: Pfeiffer et al. 2008
2. d Reference: Nath et al. 2009
3. b Reference: Crook et al. 2011
4. d Reference: Davis et al. 2012
5. b Reference: Rivard et al. 2005
6. a References: El-Bared et al. 2016, Major et al. 2014, and Hinnen et al. 2010
7. d Reference: Nath et al. 2009
8. a References: Blasko 2006, Bowes and Crook 2011, and Yoshida et al. 2013
9. d References: Todor et al. 2003, Westendorp et al. 2007, and Racine et al. 2016
10. e Reference: Nath et al. 2009
11. c Reference: Nath et al. 2009
12. c Reference: Nath et al. 2009

Chapter 5: HDR Prostate

1. T Reference: Brenner and Hall 1999
2. T Reference: Brenner and Hall 1999
3. c Reference: Ares et al. 2009
4. F References: Rivard et al. 2004 and Perez-Calatayud et al. 2012
5. b Reference: Hoskin et al. 2013
6. c Reference: Pfeiffer et al. 2008
7. F Reference: Holly et al. 2011
8. b Reference: Turkbey et al. 2016
9. T Reference: Batchelar et al. 2014
10. a, b, d, and f References: Yamada et al. 2012 and Hoskin et al. 2013
11. T Reference: Morton 2015

Chapter 6: GYN

1. c Reference: Small et al. 2012
2. d Reference: Small et al. 2012
3. b Reference: Harkenrider et al. 2016
4. b References: Small et al. 2012 and http://pbadupws.nrc.gov/docs/ML1426/ML14260A235.pdf
5. c Reference: Viswanathan et al. 2012
6. b Reference: ICRU Report 89
7. c Reference: Viswanathan et al. 2012
8. e Reference: Hellebust et al. 2010
9. a Reference: ICRU Report 89
10. a Reference: Grover et al. 2016

Chapter 7: Skin

1. a Reference: Ouhib et al. 2015
2. T Reference: Evans et al. 1997
3. a Reference: AAPM TG-253 report (under review)
4. b Reference: Beaulieu et al. 2012
5. d Reference: Ouhib et al. 2015
6. c Reference: Ouhib et al. 2015
7. a Reference: Ouhib et al. 2015
8. d Reference: Rivard et al. 2004
9. a Reference: AAPM TG-253 report (under review)
10. d Reference: Ouhib et al. 2015

Chapter 8: Breast

1. d Reference: Hepel et al. (2012)
2. c Reference: Gurdalli et al. (2011)
3. h Reference: Tang et al. (2014).
4. a F
   b F
   c T
   d F
   e T
   References: Shah et al. 2013, Vicini et al. 2016, and RTOG 0413/NSABP B-39 Protocol 2016
5. b Reference: RTOG 0413/NSABP B-39 Protocol 2016
6. F Reference: Cheng et al. 2005
7. c Reference: Section 8.3.4 and Hepel et al. 2014
8. T Reference: Section 8.4.5 and Section 8.7 (Trifiletti et al. 2015)
9. $\Delta t \approx 8.1 \times 10^{-4}$ °C. Second answer: No
10. 1.91 Gy

Chapter 9: IMBT

1. b Reference: Section 9.1.1
   Typically, IMBT is achieved with a shielded applicator, either static or dynamic.
2. b Reference: Section 9.2.2.a
3. a & c References: Section 9.3.1 and Webster et al. 2013
   DMBT has been defined in two ways: Direction-modulated brachytherapy and Dynamic-modulated brachytherapy. Direction-modulated brachytherapy is achieved through classic shielding design. Dynamic-modulated brachytherapy uses motion to translate or rotate the applicator/shielding, or both, or can use a sliding window to paint dose.
4. c References: Section 9.3.2.3 and Price et al. 2009
5. a References: Section 9.3.2.1 and Adams et al. 2014
6. c References: Section 9.3.2.2 and Liu et al. 2015

7. b & c References: Tables 9–1 and 9–12 of Section 9.4.2 and Heredia 2013
8. d References: Table 9–2 of Section 9.4.2 and Heredia 2013
9. b References: Section 9.4.2, Lin 2006, Lin et al. 2008, and Heredia 2013
10. a References: Section 9.4.3, Aima et al. 2015, and CivaTech Oncology, Inc. (http://www.civatechoncology.com/civasheet.htm)

Chapter 10: Early Clinical

1. b Reference: Andersen et al. 2009
2. F Dose gradient close to a brachytherapy source is many %/mm such that a fraction of a millimeter error in distance easily yields +5% difference in dose.
3. e Ethylene oxide is considered a toxic chemical in CA, NY, and MI,
4. F The class of biocompatible material needed for the brachytherapy application depends on the type and location of the brachytherapy. See Table 10–5.
5. b Reference: Podder et al. 2014
6. d Reference: Ahmed et al. 2017
7. d Reference: Butler and Merrick 2016
8. b Reference: Donaldson et al. 2015
9. T Reference: Samboju et al. 2017
10. b Reference: Damato et al. 2014